Laboratory Fish in Biomedical Research

AF766783

Laboratory Fish in Biomedical Research

Biology, Husbandry and Research Applications for Zebrafish, Medaka, Killifish, Cavefish, Stickleback, Goldfish and Danionella Translucida

Edited by

Livia D'Angelo
Dept. Veterinary Medicine and Animal Production
University of Naples Federico II
Italy

Paolo de Girolamo
Dept. Veterinary Medicine and Animal Production
University of Naples Federico II
Italy

ELSEVIER

ACADEMIC PRESS
An imprint of Elsevier

Academic Press is an imprint of Elsevier
125 London Wall, London EC2Y 5AS, United Kingdom
525 B Street, Suite 1650, San Diego, CA 92101, United States
50 Hampshire Street, 5th Floor, Cambridge, MA 02139, United States
The Boulevard, Langford Lane, Kidlington, Oxford OX5 1GB, United Kingdom

Copyright © 2022 Elsevier Inc. All rights reserved.

No part of this publication may be reproduced or transmitted in any form or by any means, electronic or mechanical, including photocopying, recording, or any information storage and retrieval system, without permission in writing from the publisher. Details on how to seek permission, further information about the Publisher's permissions policies and our arrangements with organizations such as the Copyright Clearance Center and the Copyright Licensing Agency, can be found at our website: www.elsevier.com/permissions.

This book and the individual contributions contained in it are protected under copyright by the Publisher (other than as may be noted herein).

Notices
Knowledge and best practice in this field are constantly changing. As new research and experience broaden our understanding, changes in research methods, professional practices, or medical treatment may become necessary.

Practitioners and researchers must always rely on their own experience and knowledge in evaluating and using any information, methods, compounds, or experiments described herein. In using such information or methods they should be mindful of their own safety and the safety of others, including parties for whom they have a professional responsibility.

To the fullest extent of the law, neither the Publisher nor the authors, contributors, or editors, assume any liability for any injury and/or damage to persons or property as a matter of products liability, negligence or otherwise, or from any use or operation of any methods, products, instructions, or ideas contained in the material herein.

Library of Congress Cataloging-in-Publication Data
A catalog record for this book is available from the Library of Congress

British Library Cataloguing-in-Publication Data
A catalogue record for this book is available from the British Library

ISBN: 978-0-12-821099-4

For information on all Academic Press publications visit our website at https://www.elsevier.com/books-and-journals

Publisher: Andre Gerhard Wolff
Acquisitions Editor: Linda Versteeg-buschman
Editorial Project Manager: Pat Gonzalez
Production Project Manager: Kiruthika Govindaraju
Cover Designer: Mark Rogers

Working together to grow libraries in developing countries
www.elsevier.com • www.bookaid.org

Typeset by TNQ Technologies

Contents

Section 2
Other small freshwater fish

Section 3
Databases

Contributors

Claire Allen, Bateson Centre, University of Sheffield, Sheffield, United Kingdom

Francesco Argenton, Dipartimento di Biologia, Università degli Studi di Padova, Padua, Italy

Gwendoline Astre, Department of Genetics, The Silberman Institute, The Hebrew University of Jerusalem, Jerusalem, Israel

Diana P. Baumann, Reptile & Aquatics, Stowers Institute for Medical Research, Kansas City, MO, United States

Filipek-Górniok Beata, Department of Organismal Biology, Uppsala University, Uppsala, Sweden

John Bischof, Department of Mechanical Engineering, University of Minnesota Twin Cities, Minneapolis, MN, United States; Department of Biomedical Engineering, University of Minnesota Twin Cities, Minneapolis, MN, United States; NSF Center for Advanced Technologies for the Preservation of Biological Systems (ATP-Bio)

Ellen Blaker, Centre for Environment, Fisheries and Aquaculture Science (Cefas), The Nothe, Barrack Road, Weymouth, Dorset, United Kingdom

Ayelén M. Blanco, Department of Bioloxía funcional e ciencias da saúde, University of Vigo, Vigo, Pontevedra, Spain; Department of Veterinary Biomedical Science, Western College of Veterinary Medicine, University of Saskatchewan, Saskatoon, SK, Canada

Radim Blažek, Institute of Vertebrate Biology, Czech Academy of Sciences, Brno, Czech Republic; Department of Botany and Zoology, Faculty of Science, Masaryk University, Brno, Czech Republic

Marco Brocca, Aquatic Solutions Tecniplast Spa., Buguggiate (VA), Italy

Livia D'Angelo, Department of Veterinary Medicine and Animal Production, University of Naples Federico II, Naples, Italy

Paolo de Girolamo, Department of Veterinary Medicine and Animal Production, University of Naples Federico II, Naples, Italy

Filippo Del Bene, Sorbonne University; INSERM, CNRS, Institut de la Vision, Paris, France; Institut Curie, PSL Research University; INSERM, CNRS, Paris, France

Karine Duroure, Sorbonne University; INSERM, CNRS, Institut de la Vision, Paris, France; Institut Curie, PSL Research University; INSERM, CNRS, Paris, France

Iva Dyková, Department of Botany and Zoology, Faculty of Science, Masaryk University, Brno, Czech Republic

Tim Ellis, Centre for Environment, Fisheries and Aquaculture Science (Cefas), The Nothe, Barrack Road, Weymouth, Dorset, United Kingdom

Chika Fujimori, Laboratory of Physiology, Atmosphere and Ocean Research Institute, The University of Tokyo, Kashiwa, Chiba, Japan

Robert Gerlai, University of Toronto Mississauga, Department of Psychology, Toronto, ON, Canada

Itamar Harel, Department of Genetics, The Silberman Institute, The Hebrew University of Jerusalem, Jerusalem, Israel

Andrew Ingalls, Reptile & Aquatics, Stowers Institute for Medical Research, Kansas City, MO, United States

Tokiro Ishikawa, Department of Biophysics, Graduate School of Science, Kyoto University, Sakyo-ku, Kyoto, Japan

Shinji Kanda, Laboratory of Physiology, Atmosphere and Ocean Research Institute, The University of Tokyo, Kashiwa, Chiba, Japan

Ioanna Katsiadaki, Centre for Environment, Fisheries and Aquaculture Science (Cefas), The Nothe, Barrack Road, Weymouth, Dorset, United Kingdom

Daichi Kayo, Department of Aquatic Bioscience, Graduate School of Agricultural and Life Sciences, The University of Tokyo, Tokyo, Japan

Kanav Khosla, Department of Mechanical Engineering, University of Minnesota Twin Cities, Minneapolis, MN, United States; NSF Center for Advanced Technologies for the Preservation of Biological Systems (ATP-Bio)

Masato Kinoshita, Division of Applied Biosciences, Graduate School of Agriculture, Kyoto University, Sakyo-ku, Kyoto, Japan

Almut Köhler, Safety and Environment, Karlsruhe Institute of Technology (KIT), Karlsruhe, Germany

Bräutigam Lars, Comparative Medicine, Karolinska Institutet, Stockholm, Sweden

Jean-Philippe Mocho, DanioVet, Joint Production System Ltd, United Kingdom

Eitan Moses, Department of Genetics, The Silberman Institute, The Hebrew University of Jerusalem, Jerusalem, Israel

Yu Murakami, Department of Fisheries, Graduate School of Agriculture, Kindai University, Nara-shi, Nara, Japan

Mikoto Nakajo, Department of Biological Sciences, Graduate School of Science, The University of Tokyo, Tokyo, Japan; Department of Physiology, Division of Life Sciences, Faculty of Medicine, Osaka Medical and Pharmaceutical University, Takatsuki, Osaka, Japan

Kiyoshi Naruse, Laboratory of Bioresources, National Institute for Basic Biology, Okazaki, Aichi, Japan

Bruce Newell, Aquatics, Aviary and Wildlife Supervisor, Deakin University, Geelong, VIC, Australia

Yoshitaka Oka, Department of Biological Sciences, Graduate School of Science, The University of Tokyo, Tokyo, Japan

Jana Oltova, Department of Cell Differentiation, Institute of Molecular Genetics of the Czech Academy of Sciences, Prague, Czech Republic; CZ-OPENSCREEN, Institute of Molecular Genetics of the Czech Academy of Sciences, Prague, Czech Republic

Nuno Pereira, Chronic Diseases Research Center (CEDOC), Nova Medical School, Lisboa, Portugal; Faculty of Veterinary Medicine, Lusófona University, Lisboa, Portugal; Instituto Gulbenkian de Ciência, Oeiras, Portugal; ISPA - Instituto Universitário, Lisboa, Portugal; Oceanário de Lisboa, Lisboa, Portugal

Matej Polačik, Institute of Vertebrate Biology, Czech Academy of Sciences, Brno, Czech Republic

Gokul Rajan, Sorbonne University; INSERM, CNRS, Institut de la Vision, Paris, France; Institut Curie, PSL Research University; INSERM, CNRS, Paris, France

Martin Reichard, Institute of Vertebrate Biology, Czech Academy of Sciences, Brno, Czech Republic; Department of Ecology and Vertebrate Zoology, Faculty of Biology and Environmental Protection, University of Łodz, Łodz, Poland; Department of Botany and Zoology, Faculty of Science, Masaryk University, Brno, Czech Republic

Paul Georg Schroeder, Department of Biomedical Services, Oxford University, United Kingdom

Marion Sebire, Centre for Environment, Fisheries and Aquaculture Science (Cefas), The Nothe, Barrack Road, Weymouth, Dorset, United Kingdom

Natascia Tiso, Dipartimento di Biologia, Università degli Studi di Padova, Padua, Italy

Benjamin Tsang, University of Toronto Mississauga, Department of Psychology, Toronto, ON, Canada

Chie Umatani, Department of Biological Sciences, Graduate School of Science, The University of Tokyo, Tokyo, Japan

Suraj Unniappan, Department of Veterinary Biomedical Science, Western College of Veterinary Medicine, University of Saskatchewan, Saskatoon, SK, Canada

Ana M. Valentim, i3S - Instituto de Investigação e Inovação em Saúde, Universidade do Porto, Portugal; Laboratory Animal Science, IBMC – Instituto de Biologia Molecular e Celular, Universidade do Porto, Porto, Portugal

Fredericus J. Van Eeden, Bateson Centre, University of Sheffield, Sheffield, United Kingdom

Zoltán M. Varga, Zebrafish International Resource Center, University of Oregon, Eugene, OR, United States

Jakub Žák, Institute of Vertebrate Biology, Czech Academy of Sciences, Brno, Czech Republic; Department of Zoology, Faculty of Science, Charles University, Prague, Czech Republic

Fish as model systems

Paolo de Girolamo, Livia D'Angelo
Department of Veterinary Medicine and Animal Production, University of Naples Federico II, Naples, Italy

Fishes exist in a multitude of forms and have many unique physiological, behavioral, and ecological specializations, thanks to the enormous evolutionary radiation of fishes comprising at least 27,000 species (Nelson, 2006). Thus they are useful indicators of environmental quality and ecological integrity, and their individual adaptations and physiological specializations make them suitable for use as physiological and biomedical models. Although fish diverged from humans more than 400 million years ago, there are enough commonalities to justify conducting research that is relevant to humans in these animals (Schartl, 2014; D'Angelo et al., 2016). Small fish have several advantages for biomedical research:

1. embryos develop externally in many species, allowing easy visualization of embryological processes;
2. generation times are relatively fast allowing genetic experiments to be performed within a short time frame;
3. development of genetic modification tools has sped up the rate at which genes related to human diseases can be tested in fish through the generation of mutant models;
4. large embryo clutch sizes enable many individuals to be tested simultaneously and it is considerably easier to reach statistically significant experimental numbers;
5. these animals are ideal models for large-scale analyses, i.e., any soluble chemical compound can be tested in a multiwell plate approach by submersion of animals.

In addition, fish can be bred and maintained in large numbers easily and at low cost.

Therefore the use of small fish in biomedical research has grown quickly, aided also by the acquisition and application of the principle of relative replacement, suggesting wherever possible the use of animals with a simpler central nervous system (Russell & Burch, 1959).

In terms of scientific rigor, the fish species chosen for experimental study must be appropriate to the scientific objectives. Therefore choice of species requires knowledge of their life history, behavior, physiology, and genetics along with husbandry requirements for laboratory studies (Sneddon et al., 2017), to ensure reproducibility and repeatability of results. In addition, husbandry and experimental conditions require standardized parameters to control undesirable effects of factors that may affect results and the challenge of replicability from study to study. Standardization aims also to reduce the number of subjects and increase the comparability of results within and between laboratories (Lieggi et al., 2020). The parameters that require standardization depend on the experimental objections, the sensitivity of the readouts, and the possible impact of standardization on animal welfare. However, a strictly standardized study may fail to detect the effects of experimental manipulations, either because the environmental and developmental conditions and/or the test environment may not promote the expression of a phenotype, or because the test environment provides conditions hardly ever seen in replication but under which an experimental outcome becomes visible (Lieggi et al., 2020). The consistency of research reproducibility and repeatability relies on further relevant aspects, such as variations in husbandry and environmental conditions, including the environmental conditions before and during experiments. Variations in any biotic and abiotic environmental parameters (Fig. 1) can potentially induce physiological responses in the animal and may

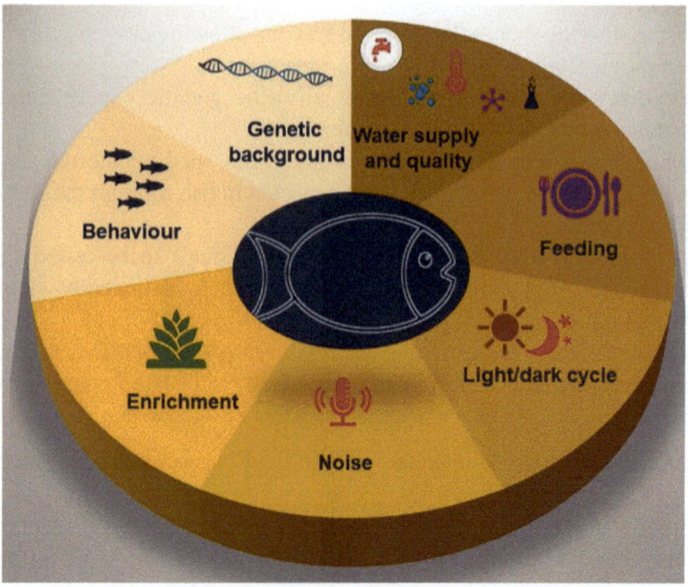

FIGURE 1 Representation of biotic and abiotic factors that can affect physiological responses in fish and may consequently affect the experimental results.

consequently affect the experimental results. The control and standardization of environmental parameters are thus crucial to ensure the welfare of animals, other than the quality and reproducibility of the scientific outcome. Depending on the species considered, knowledge of the effects of such variations may be more or less incomplete, and therefore not completely predictable. The natural history of each organism must be used as a template for developing appropriate husbandry practices and ensuring animal welfare. Welfare assessment implies deep knowledge of the species-specific behavioral responses to the different environmental situations, other than the traditional definitions of production in terms of fecundity and growth.

Multiple environmental factors affect fish health and welfare, including housing systems, water supply and quality, feeding, circadian rhythms, noise, and the environment within and surrounding the tank.

Housing tanks vary in size, shape, and materials (glass, fiberglass, polycarbonate, etc.) according to the fish species, its ecology, and research needs. Tank size has influence on fish activity levels and behavior during different developmental stages.

Water supply and quality depends on the local water supply and on the water system in place. Most housing systems are recirculating, in which pumps feed water into the tanks and, through an overflow system, remove an equal amount of water. These systems are fully integrated with filter systems, germicidal irradiation and light and temperature control units, and rate of water flow. The water flow is an important variable particularly for those species living in slow-flowing water bodies, such as zebrafish, where higher water flow increases aggression between conspecifics (Suriyampola et al., 2017). Water quality parameters must be set according to the different life stages and physiological status of each species. Among water parameters to consider and control are:

- oxygen: fish, as with all aerobic organisms, require oxygen for breathing. The concentration of O_2 dissolved in water affects fish activity and metabolism and alters the swimming behavior with related effects on many aspects of fish life;
- nitrogenous compounds: derive as waste products from the amino acid catabolism, and the most abundant nitrogen products of excretion are ammonia (sum of NH_3 and NH_4^+) and urea;
- pH: the great majority of freshwater aquatic organisms live at pH 6.5–8.5, which corresponds to the same range found in most freshwater lakes, streams, and ponds. pH variations affect the metabolism and homeostasis of cells and the whole organism;
- salinity: fish species are subdivided in euryhaline, able to tolerate large variations in salt concentration, and stenohaline, more sensitive to salinity changes. The high tolerability is due to their regulatory systems of ions and

water (gills, digestive system, kidney) able to undergo structural and functional reorganizations in response to altered salinity;
- temperature: fish are ectothermic or poikilothermic organisms whose body temperature corresponds to water temperature, and thus variations in ambient temperature strongly affect fish biology and influence growth rate, food consumption, feed conversion, physiology, and behavior along with other body functions in addition to modifying the pH and environmental parameters.

Feeding includes the amount and quality of food sufficient to ensure the intake of calories and nutrients necessary to meet the metabolic needs of the animal without producing excessive waste in the aquatic system. Based on feeding habits, fish species are grouped as herbivorous, carnivorous, detritivorous, and omnivorous, and acquisition of food is a process based on different sensory systems, including vision, chemoreception, acoustics, lateral line, and electroreception all of which may contribute to aspects of the feeding behavior in fish.

Light/dark cycle is relevant for the entire life cycle, from embryonic development to sexual maturation into adulthood and regulation of the circadian rhythm and activity of species. Fish must be maintained under an appropriate photoperiod where natural light does not allow a suitable light/dark cycle and controlled lighting with an intensity adapted to the reared species must be provided to satisfy biological requirements (Toni et al., 2019).

Noise is an important parameter affecting the welfare of animals. Teleost fish can be separated into two nontaxonomic groups based on their sensitivity to sound: hearing specialists and hearing generalists (Popper, 2003). The hearing specialists, such as the goldfish, have small bony connections (Weberian ossicles) or other structures that bridge the swim bladder with the inner ear, enabling these species to detect higher-frequency sounds. Hearing generalists, which are the majority of fish species, lack these specialized connections and only perceive frequencies below 500–1000 Hz (Popper, 2003). Intense noise (over 140 dB) in fish may induce temporary hearing loss, damage in the inner ear sensory epithelium, and endocrine stress responses (Amoser & Ladich, 2003).

Enrichment of environment, either structural and social, is an important variable to consider for ensuring reproducible and valid data (Volgin et al., 2018). Structural enrichment (such as plants, gravel, images depicting gravel substrates, etc.) exposes fish to sensory stimuli (visual, motor, cognitive, and somatosensory) and represents an alternative and/or complementary approach to reduce stress and promote animal welfare, with a positive effect also on reproduction and survivorship of larvae (Volgin et al., 2018). Social enrichment depends on the ethology of housed fish species. Zebrafish, for instance, are highly social animals and display typical shoaling behavior (swim in aggregated groups). Interestingly, shoaling behavior can be observed among fish that are separated by glass, remaining in visual but not in physical contact (Kistler et al., 2011).

In addition to these abiotic factors, in the aquatic environment it is equally important to consider normal species behavior, group dynamics, stocking density, and importantly the genetic background, all aspects that can influence research results. Behavioral patterns such as locomotion, hunting, prey avoidance, exploration, reproduction, and territory protection are tightly dependent on the biology and ecology of the fish species and require a profound knowledge to interpret experimental data.

The genetic background of fish is of high importance within certain model species. There are genetic strains that may differ phenotypically (Crim & Lawrence, 2021) and the most appropriate strain for the experimental question must be considered. When choosing a fish species, the genetic background is critically important to experimental outcomes, interpretation, reproducibility, and replicability. This is still an unresolved issue with regards of some fish species, including zebrafish (van den Bos et al., 2019). Standardization of genotypes using inbred strains or selected lines, consisting of genetically identical "clones" that are in principle precisely replicated across studies, is commonly applied to a wide range of vertebrate model organisms, including mammals and some fish species, i.e., Medaka and platyfish, which have been inbred for more than 100 generations. However, in the case of zebrafish, the common laboratory fish, inbred strains are not in place yet, although zebrafish wild-type lines are often reported and treated as inbred strains, implying a uniformity of genetic background that does not exist. Different zebrafish lines display a wide array of differences, including various behavioral traits, swimming ability, growth, and gene expression to strain (van den Bos et al., 2019).

Altogether, recognition of all these abiotic and biotic factors is essential to ensure fish welfare and a high-quality study design. Empirical evidence for nociception has demonstrated the evolutionary conservation of nociception and pain from invertebrates to vertebrates, and the biology of the nociceptive system is strikingly similar between fish and mammals (Sneddon, 2019). Evidence of nociceptors in fishes comes from molecular biology, neurobiology, and anatomy, as well as from the whole animal behavioral responses. When experiencing a painful stimulus, fish do not show appropriate fear or antipredator responses. Furthermore, changes in behavior are prevented by drugs that provide effective analgesia, providing evidence that these changes are driven by nociception and pain mechanisms (Sneddon, 2019).

In conclusion, the wonderful diversity of fish and their amenability to conduct research, together with the acquisition and application of the principle of relative replacement (Nelson, 2006), have led to an explosion of fish as model systems in science. However, rigorous approaches respectful of species-specific welfare needs, environmental and husbandry conditions, as well as genetic background represent the pillar to design experiments with fish as well as ensure reproducibility and repeatability of research data.

References

Amoser, S., Ladich, F., 2003. Diversity in noise-induced temporary hearing loss in otophysine fishes. J. Acoust. Soc. Am. 113 (4 Pt 1), 2170–2179.

Crim, M.J., Lawrence, C., 2021. A fish is not a mouse: understanding differences in background genetics is critical for reproducibility. Lab. Anim. 50, 19–25.

D'Angelo, L., Lossi, L., Merighi, A., de Girolamo, P., 2016. Anatomical features for the adequate choice of experimental animal models in biomedicine: I. Fishes. Ann. Anat. 205, 75–84.

Kistler, C., Hegglin, D., Wurber, H., et al., 2011. Preference for structured environment in zebrafish (*Danio rerio*) and checker barbs (*Puntius oligolepis*). Appl. Anim. Behav. Sci. 135, 318–327.

Lieggi, C., Kalueff, A.V., Lawrence, C., Collymore, C., 2020. The influence of behavioral, social, and environmental factors on reproducibility and replicability in aquatic animal models. ILAR J. 60 (2), 270–288.

Nelson, J.S., 2006. Fishes of the World. Wiley, Hoboken, New Jersey.

Popper, A.N., 2003. Effects of anthropogenic sounds on fishes. Fisheries 28 (10), 24–31.

Russell, W.M.S., Burch, R.L., 1959 (as reprinted 1992). The Principles of Humane Experimental Technique. Universities Federation for Animal Welfare, Wheathampstead (UK).

Schartl, M., 2014. Beyond the zebrafish: diverse fish species for modeling human disease. Dis. Model. Mech. 7, 181–192.

Sneddon, L.U., 2019. Evolution of nociception and pain: evidence from fish models. Phil. Trans. R. Soc. B 374, 20190290.

Sneddon, L.U., Halsey, L.G., Bury, N.R., 2017. Considering aspects of the 3Rs principles within experimental animal biology. J. Exp. Biol. 220, 3007–3016.

Suriyampola, P.S., Sykes, D.J., Khemka, A., Shelton, D.S., Bhat, A., Martins, E.P., 2017. Water flow impacts group behavior in zebrafish (*Danio rerio*). Behav. Ecol. 28 (1), 94–100.

Toni, M., Manciocco, A., Angiulli, E., Alleva, E., Cioni, C., Malavasi, S., 2019. Assessing fish welfare in research and aquaculture, with a focus on European directives. Animal 13 (1), 161–170.

van den Bos, R., Althuizen, J., Tschigg, K., Bomert, M., Zethof, J., Filk, G., Gorissen, M., 2019. Early life exposure to cortisol in zebrafish (*Danio rerio*): similarities and differences in behaviour and physiology between larvae of the AB and TL strains. Behav. Pharmacol. 30, 260–271.

Volgin, A.D., Yakovlev, O.V., Demin, K.A., et al., 2018. Understanding the role of environmental enrichment in zebrafish neurobehavioral models. Zebrafish 15 (5), 425–432.

Section 1

Zebrafish (*Danio rerio*)

Chapter 1

Biology and research applications

Bräutigam Lars[1], Filipek-Górniok Beata[2]

[1]*Comparative Medicine, Karolinska Institutet, Stockholm, Sweden;* [2]*Department of Organismal Biology, Uppsala University, Uppsala, Sweden*

The biology of the zebrafish

The zebrafish was first described by Francis Hamilton in 1882 who found this "beautiful fish with several blue and silver stripes on each side" near the Ganges river in India (Hamilton, 1822). After George Streisinger first used the zebrafish in the 1980s to study vertebrate development, it became one of the most important laboratory animals at an unprecedented pace. Detailed knowledge of the biology, ecology, and natural behavior of zebrafish is essential for optimizing zebrafish husbandry in research laboratories and for improving animal welfare, which in turn is an essential part of generating reliable scientific data.

Geographic distribution

The zebrafish is native to much of the Indian subcontinent, stretching from Pakistan in the west across India, Nepal, and Bangladesh to Myanmar in the east. The zebrafish can be found in a diverse range of habitats, from fast-flowing streams on the foothills of the Himalayan mountain range to the natural wetlands of the Indo-Gangentic plains as well as rice fields, ponds, and drainage ditches (Spence et al., 2006). Those habitats are all located in the monsoon region with an extreme seasonal climate variation, from cold and dry winter months to the hot and extremely wet period of the monsoon that brings heavy rains that submerge much of the floodplains of the Indian subcontinent. The zebrafish is clearly well adapted to drastic environmental changes and field expeditions have found zebrafish in water habitats with an astonishingly broad range of water conditions and floor substrates (Spence et al., 2006; Arunachalam et al., 2013). Most of these habitats were associated with aquatic vegetation; however, zebrafish were also found in barren waters. Aquatic vegetation likely provides larval and adult zebrafish shelter as well as a reliable source of food (Spence et al., 2007).

Laboratory Fish in Biomedical Research. https://doi.org/10.1016/B978-0-12-821099-4.00013-4
Copyright © 2022 Elsevier Inc. All rights reserved.

In summary, the zebrafish occupies a wide range of natural habitats with highly variable environmental conditions, which makes the fish well suited to being bred in captivity. Although zebrafish thrive in diverse environments, standard conditions for husbandry and care are required to generate reliable and repeatable scientific data (Alestrom et al., 2020).

Zebrafish appearance and life span

The adult zebrafish is between 2 and 4 cm long and is characterized by a fusiform, laterally compressed shape (Fig. 1.1). Whereas the male is torpedo shaped (Fig. 1.1B), the female has a characteristically whitish belly (Fig. 1.1A). Both sexes have dark-blue stripes on each side interspaced with either gold (male) or silver (female) stripes.

The zebrafish possesses a pair of pectoral and pelvic fins and a single dorsal, anal, and caudal fin. The large caudal fin is often partly amputated for genotyping purposes and has, as have many other tissues of the zebrafish, a remarkable potential for regeneration. Wild zebrafish are thought to be primarily an annual species, most likely due to the extreme seasonal variation that occurs in their geographic distribution (Spence et al., 2007). In captivity, however, zebrafish can live for 3–4 years (Gerhard et al., 2002) even though most colony fish are sacrificed at 2 years of age for health management and breeding reasons (Kent et al., 2020).

Zebrafish phylogeny and genetics

The zebrafish, *Danio rerio*, is one of many members of the genus *Danio*, the taxonomy of which has only recently been resolved with the profound impact of molecular taxonomy (McCluskey and Postlethwait, 2015). The genus *Danio* itself is part of the cypriniforms, which belong to the teleostei infraclass (Betancur et al., 2017). Importantly, the ancestor of the teleostei infraclass underwent an additional whole-genome duplication. After a partial rediploidization, a substantial number of zebrafish genes remain duplicated and those paralogs often show functional diversification or neofunctionalization (Force et al., 1999), which needs to be kept in mind when dissecting gene function and conservation. Besides *Danio rerio*, many other *Danio* species

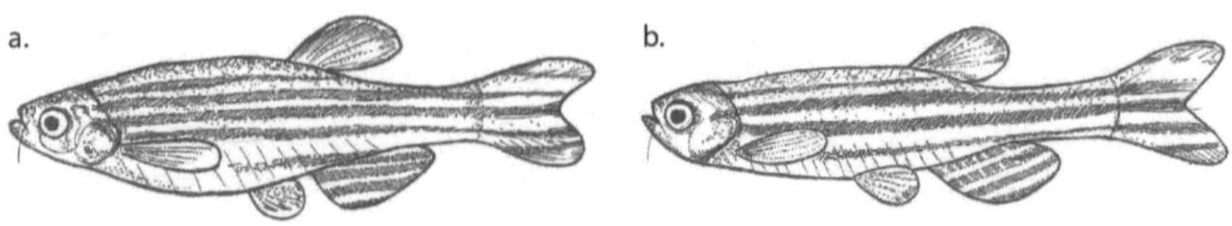

FIGURE 1.1 Representative drawing of the adult zebrafish: (A) female; (B) male.

have been described (McCluskey and Postlethwait, 2015). All those strains populate the same habitats and share most of the intrinsic traits that make *Danio rerio* such a powerful model organism. Among them is *Danionella translucida*, an almost completely transparent member of the *Danio* species, which will be described in detail in this book.

Zebrafish embryonic development

The zebrafish egg is relatively large (~0.7 mm), completely transparent, and fertilized ex utero. The sperm enters the egg through the micropyle, which corresponds to the future animal pole (Amanze and Iyengar, 1990). The vegetal pole is located on the opposite site and the animal–vegetal axis determines the anterior–posterior axis that forms during gastrulation. After the sperm has entered the egg, the first cell appears at the animal pole within 30 min. A meroblastic cleavage period starts, i.e., the blastula phase, during which the embryonic cells divide synchronically every 20 min. This incredible speed of division is only possible since the cells skip the G1 and G2 phase and solely run through M- and S-phase (Kimmel and Law, 1985). Three hours into embryonic development the first milestone is reached: the *maternal-to-zygotic transition*. At that stage four dramatic changes take place: (1) an increasing number of zygotic genes are transcribed, (2) the cell divisions become asynchronous and start to include an interphase (G1/G2), (3) the cells become increasingly restricted in their fates, and (4) the first cellular rearrangements take place, which mark the onset of gastrulation and the establishment of the three germ layers. During this gastrulation phase, the cells start to migrate toward the vegetal pole around the yolk in a process called epiboly. When the cells cover approximately 50% of the egg, the *shield* stage is reached and a thick ring called the *germ ring* is formed. The cells located in the germ ring subsequently begin to involute and move underneath the surface cell layer. During that process, the endoderm, mesoderm, and ectoderm are specified. Besides the macroscopic cellular rearrangements, a complex pattern of signaling pathways is established, among others from the organizer during the shield stage, which determines the cellular fate of most of the embryonic cells (Shih and Fraser, 1996). After the gastrulation phase is completed, somitogenesis starts during which one segment appears every 20 min. The number of segments is often used for precise staging of the early embryo. The precursors of almost all organs are formed during somitogenesis and after only 24 h postfertilization (hfp) the zebrafish embryo has a functional axonal network. Soon thereafter the heart starts to beat (at ca. 30 hpf) and the zebrafish hatches (at ca. 48 hpf). See also Fig. 1.2.

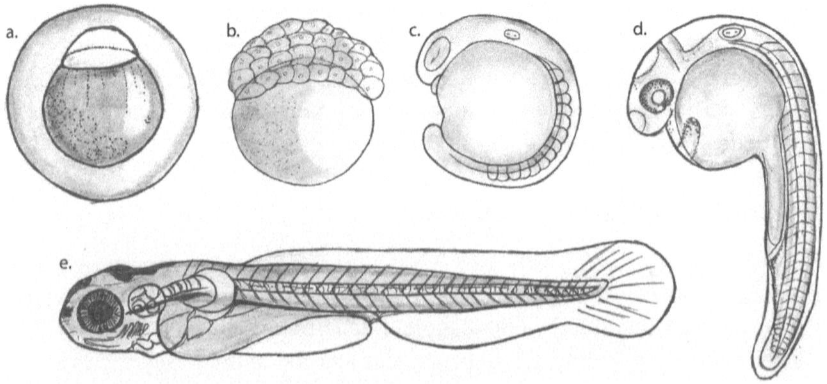

FIGURE 1.2 (A) One-cell stage (0.5 h postfertilization [hpf]), (B) 64-cell stage (2 hpf), (C) 14-somite stage (16 hpf), (D) prim-6 stage (25 hpf), (E) 72 hpf.

The history of the zebrafish as laboratory model

Behind each new idea and discovery stands a person with a great mind driven by its hunger for knowledge, determination, and an urge to seek new, better solutions.

Very few researchers have contributed to the emergence of a new vertebrate model system as has George Streisinger, establishing this amazing new tool of research for thousands of scientists worldwide. Zebrafish had already been used since 1950 to study embryology and toxicology (van Raamsdonk et al., 1980; Weis, 1968; Weis, 1968; Laale, 1977; Hisaoka and Battle, 1958); however, it is George Streisinger who is recognized as the founder of zebrafish model-based research. His primary efforts in the field date back to the late 1960s when his work on recessive mutations pushed him to seek and establish a new functional vertebrate model. He and his group considered several species of fish with zebrafish ultimately being selected due to the same reasons for which they are still valued today. Zebrafish are easy and fast to breed in laboratory conditions year-round. The transparent eggs are big, easy to manipulate, and externally fertilized, same as the embryos which are translucent and large, enabling manipulation as well as phenotypical and behavioral observations. One of Streisinger's biggest achievements was the development of a method for the production of homozygous diploids of zebrafish (Streisinger et al., 1981).

Another researcher whose name has gained well-deserved respect and recognition in establishing zebrafish as a model system is Charles Kimmel, whose primary neurobiology focus eventually resulted in the development of fluorescent techniques allowing for neuronal lineage tracing and the discovery of the zebrafish brain's segmental structure (Grunwald and Eisen, 2002). Few publications have been so influential for the zebrafish field as the paper from

Kimmel and colleagues on the stages of embryonic development of zebrafish published in 1995 (Kimmel et al., 1995). It described the early embryonic development in unprecedented detail and provided the conceptional and husbandry framework for all laboratories worldwide, enabling standardization of conditions and comparability of data.

A name not to be forgotten among those who have made a great impact on the zebrafish research field is Christiane Nüsslein-Volhard, Nobel Prize winner and leader of the ambitious ENU (*N*-ethyl-*N*-nitrosourea) point mutation mutagenesis project, which had been inspired by a similar study from her group conducted in *Drosophila*. Her project involved 65 scientists and students conducting embryonic screens and resulted in the discovery of more than 4000 novel mutants.

"(...) when I realized that zebrafish was a system that might allow us to exploit the genetic approach in a vertebrate, I saw a fascinating challenge, particularly as genetics had worked so successfully in Drosophila to dissect early patterning processes," said Christiane Nüsslein-Volhard for the Spotlight series of *Development* articles. Nearly 1200 mutants displayed phenotypes affecting development or organogenesis, which, with immense effort, were characterized and classified. The final results were summarized into a *Development* special issue containing 37 papers gathered on 481 pages published in 1996, becoming one of the most recognized sources of early zebrafish development references (Nusslein-Volhard, 2012).

Resources for the zebrafish model

Growing interest in using zebrafish as a research animal model led to the first small zebrafish meeting hosted at Streisinger's home campus in Oregon in 1990, followed by the foundation of the Zebrafish International Resource Center (ZIRC). Later, in 1994, the online-based Zebrafish International Network (ZFIN) was founded, setting the goal to give access to data resources, including genetic, genomic, and developmental information, to the research community (https://zfin.org/). In practice, the ZFIN website provides a comprehensive linkage between genomic and expressional data, publications, references, protocols, training availabilities, job opportunities, and researchers' laboratory descriptions. The availability of manually cured and detailed genomic DNA sequence is crucial to most, if not all, studies involving zebrafish. The first efforts to sequence and annotate the zebrafish genome were initiated in 2001 at the Wellcome Trust Sanger Institute (Howe et al., 2013). The latest assembly coordinated by the Genome Reference Consortium and an international collaboration between ZFIN and several other big research institutes was released in 2018. This assembly included nearly 26,000 coding and an additional 6600 noncoding genes (http://genomeref.blogspot.com/2020/05/zfin-and-genome-reference-consortium.html).

The zebrafish as a model for development and pathology

The zebrafish as model in developmental biology

The zebrafish is a member of the phylum Chordata, more commonly known as vertebrates, and has proven to be an excellent model system to unveil answers to essential questions asked by developmental biologists. The seemingly chaotic processes of cell division, differentiation, cell–cell communication, migration, and morphogenesis have been in the scope of their interest. The zebrafish has been used as a tool to help understand the fundamental aspects of functional organism formation, as high levels of body organization and genetic conservation between different vertebrate species, therefore allowing for broader conclusions, based on findings in this model. The zebrafish has been utilized in studies on early development ranging from cell cleavage, determination of body axis, trunk segmentation, different aspects of nervous system development, organogenesis, and cardiovascular and lymphatic system formation, while investigations of Kupffer's vesicle and its ciliation, equivalent to mammalian embryonic node, has allowed for better understanding of organ laterality (Gokey et al., 2016). The administration of dyes and the development of transgenesis techniques, such as the Tol2 or CRISPR/Cas9 system, allowed for the generation of thousands of transgenic lines. Prominent examples are strains such as Tg(cmlc2:EGFP) (Huang et al., 2003), Tg(elavl3:EGFP) (Kim et al., 1996), and Tg(fli1a:EGFP) (Lawson and Weinstein, 2002) in which the precursor cells of the heart, the neurons, and the blood vessels are labeled with green fluorescent protein (GFP) providing unprecedented insights into the key organ systems of vertebrates. Those techniques also enabled whole-brain functional imaging with single-neuron resolution, giving insights into the neuronal activity of a living organism (Ahrens et al., 2013). Furthermore, the growing field of single-cell RNA transcriptomics combined with CRISPR/Cas9 technology led to the development of new tools for large-scale cell lineage tracing (Ahrens et al., 2013). Those advanced techniques, complemented with laser-ablation methods using the KillerRed protein, which allows for cell-specific optogenetic ablations dissect the role of specialized cells such as pericytes, immune cells, and renin cells (Buckley et al., 2017).

The zebrafish as model in pathology

Due to the large mutagenesis screens published in the 1990s (Driever et al., 1996), the sequencing of the zebrafish genome (Howe et al., 2013) and the outstanding technical advances thereafter, the field of zebrafish research has also grown to include a countless number of disease models. Those models can generally be divided into three different categories: (1) genetic models in which a gene has been mutated to recapitulate a human disease-causing genetic variant, (2) chemical models in which a pathological condition is mimicked by exposure to chemical molecules, and (3) transplantation models in which cancer cell lines or patient-derived xenografts are transplanted into the zebrafish. Besides powerful disease models, many transgenic lines are available that allow for the screening of phenotypic and physiological changes upon exposure to small molecules. The Tg(cmlc2:EGFP) line, which expresses GFP in the heart, can be used to study the cardiotoxicity similarly as the numerous reporter strains, which are used for ecotoxicology screens (Bambino and Chu, 2017). Those disease and screening models can be analyzed either with imaging-based methodology or with behavior-based analytical tools.

The zebrafish as a model for cancer research

Cancer is one of the major health condition that kills millions of people every year. Although the number of cancer cases is continuing to rise worldwide, it is not a "modern" disease. The ancient Egyptians suffered from cancer and the medical name came from Hippocrates around 400 BCE because of its similarities to crabs: "karkinos" in Greek. Since ancient times, humans have tried to find a cure for cancer, but it was in the mid to late 1940s when Sidney Farber started his famous experiments with antifolates to treat childhood acute lymphoblastic leukemia (ALL) that chemotherapy became a weapon in the fight against cancer. Soon, animal models were needed not only to test thousands of chemicals for their antitumorigenic activity, but also to understand the underlying cause of cancer. In the 1980s, a new laboratory animal, a small tropical fish, started to appear in the scientific literature. After the seminal mutagenesis screens from Christiane Nüsslein-Volhard and Wolfgang Driever in the 1990s (Driever et al., 1996), and the sequencing of the zebrafish genome (Howe et al., 2013), it became clear that genetic programs regulating both tissue and organ development as well as oncogenes and tumor suppressors were remarkably conserved in zebrafish (Howe et al., 2013). The extensive conservation of organs and tissues together with the technical advances for creating transgenic zebrafish paved the way for the zebrafish to enter the arena of cancer modeling.

Transgenic cancer models

In 2003, Langenau and colleagues reported the first zebrafish cancer model in a seminal publication in which they described that the overexpression of c-myc under the control of a lymphoid-specific promotor led to the development of ALL in zebrafish—similar to human patients. After that groundbreaking work, numerous stable transgenic cancer models rapidly appeared. Again, the similarity between zebrafish cancer models and human patients was striking. Transgenic zebrafish expressing the $BRAF^{V600E}$ mutation develop pigmented moles, just like human patients. Combining the expression of this oncogene with the knockout of the tumor suppressor p53 led to the development of malignant melanomas in almost 100% of the stable double transgenic zebrafish. Emerging techniques, like CRISPR/Cas9 for specific gene knockdown (Hwang et al., 2013) or the Tol2 system for rapid creation of the transgenic zebrafish (Kawakami, 2007), opened new ways for combinatorial transgenic cancer modeling in zebrafish. Multiplexing CRISPR/Cas9 guide mRNAs were shown to efficiently recapitulate complex genetic interactions in human cancers and placing the Cas9 sequence under a tissue specific promotor allowed for the creation of tissue-specific lesions or mutations. Furthermore, combining CRISPR/Cas9 with the Tol2 system for efficient generation of transgenic zebrafish enabled researchers to study different oncogenic mutation profiles of malignant melanoma or T-ALL (Burns et al., 2018) among others.

Engraftment cancer models

Zebrafish embryos are particularly suitable for engrafting cancer cells. Husbandry costs for zebrafish are much lower than rodents and a single female can lay more than 200 eggs per week. Due to the optical transparency of embryos, fluorescent transplants can easily be monitored and the interaction with host cells can be studied in hundreds of reporter strains that are readily available through zebrafish resource centers. Zebrafish embryos develop very fast and although most components of the innate immune response are already differentiated in 2-day-old embryos, the adaptive immune system is not functional until 2 weeks after fertilization so no immunosuppressants are needed for embryonic engraftment. The small size of zebrafish embryos allows for in vivo high-throughput drug discovery screens or precision medicine applications.

The first engraftment of human and murine cancer cells into zebrafish embryos was reported in 2006 and 2007. Remarkably, those xenotransplants were found to induce tumor neovascularization by secretion of fibroblast growth factor and vascular endothelial growth factor, exemplifying the high conservation of human signaling pathways in zebrafish (Nicoli et al., 2007). The successful engraftment of many other cancer types was published during the following years and culminated in 2017, when the group of M. Ferreira described the first patient-derived xenotransplantation (PDx) in zebrafish embryos together with a phenotypic drug-response screen (Fior et al., 2017). With that outstanding work, the zebrafish entered the field of precision cancer medicine. Today, after more than a decade of optimizing cell culturing and transplantation techniques, zebrafish PDx models show an accurate prediction of patient response and have begun to revolutionize precision cancer medicine (Xiao et al., 2020) (Fig. 1.3). Embryonic engraftment models have also significantly added to our understanding of the tumor microenvironment, especially hypoxia, which is an important contributor to cancer metastasis, cell quiescence, and therapy resistance. Zebrafish embryos can, in contrast to mice and rats, be exposed to severe hypoxic conditions and, due to their optical transparency, provide a unique platform to study tumor microenvironment in disease progression and drug resistance.

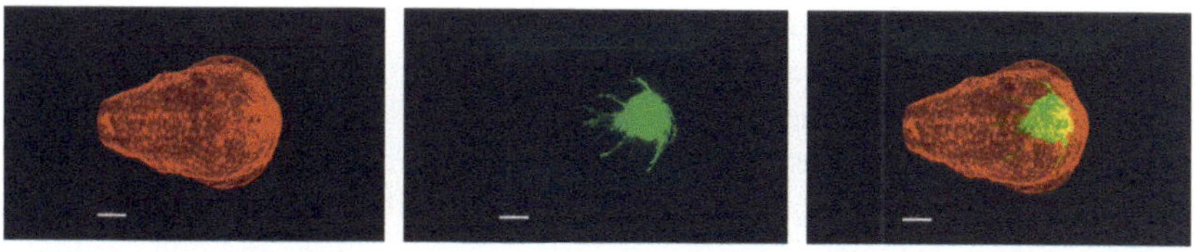

FIGURE 1.3 Human primary glioblastoma cells transplanted intracranially into zebrafish embryos. The tumor microtubules are clearly visible. *Pictures reproduced from Pudelko, L., et al., 2018. An orthotopic glioblastoma animal model suitable for high-throughput screenings. Neuro Oncol. 20, 1475–1484. https://doi.org/10.1093/neuonc/noy071.*

There are, however, inherent limitations to using the zebrafish embryo as a model for malignancies. The early zebrafish embryo is relatively small allowing for only around 150 cancer cells to be transplanted, which may not recapitulate the intrinsic heterogeneity and drug response observed in human tumors. Furthermore, transplantation experiments must be carried out at temperatures below 37°C, which may lower the proliferation rate compared to malignancies in humans. Lastly, the embryo is quickly developing with cascades of developmental-specific signaling pathways being activated as opposed to the adult human patient.

Some of those issues can be circumvented by performing xenografts into adult zebrafish. Recently, the Langenau group generated an adult zebrafish transplantation model that allowed for long-term engraftment of human cancer cells. In this model, two genes essential for the adaptive immune system are homozygously inactivated so that chemical immunosuppressants or irradiation are not required (Yan et al., 2019). Adult zebrafish can be transplanted with significantly more tumor cells and, importantly, they can be kept at 37°C when slowly acclimatized.

Since the 1980s, when George Streisinger first used the zebrafish animal model to understand the genetics of vertebrate development, the zebrafish has, despite inherent limitations, matured into an indispensable platform to understand and dissect cancer genetics, the cancer microenvironment, and to design, develop, and optimize novel anticancer therapies. The remarkable conservation of tissue and organ architecture, signaling pathways, and tumor pathology will pave the way for zebrafish in the field of precision cancer medicine.

The zebrafish embryo in drug discovery and high-throughput screens

Two fundamentally different strategies are employed in drug-development pipelines: *target-based* and *phenotypic-based* screens. In a *target-based* screen, druggable target molecules (e.g., signaling molecules, enzymes, hormones) involved in pathological processes are identified and characterized in the laboratory. Subsequently, chemical libraries are screened in vitro to find substances modulating those targets. In a *phenotypic-based* screen, chemical libraries are tested in a disease model (either in cellulo or in vivo) and molecules that alter the disease phenotype, i.e., killing cancer cells, are selected for further characterization. In both strategies, "hits" which are identified, undergo intense medical chemistry, in vitro in vivo testing, and clinical trials before they reach the market. The zebrafish, with its intrinsic traits favorable for translational and clinical research, genetic accessibility, conservation of disease-causing genes, and availability of transgenic lines and disease models, has become an important link in the bench-to-bedside process of drug discovery. It is employed in (1) target identification and validation, (2) lead optimization and (3) toxicology/side-effect characterization (Cully, 2019).

Robotics and tools for automated screening

All drug screens involving zebrafish embryos require the ability to produce a large number of developmentally staged embryos. For that, chambers for mass production of zebrafish eggs (Adatto et al., 2011) as well as robotic systems for sorting fertilized and unfertilized eggs have been published and commercialized (https://bionomous.ch). The optical clarity and the small size of zebrafish embryos make them ideal for image-based automatic screening to follow the onset and progression of a disease or the efficacy of a drug treatment in vivo and in real time. Several robotic solutions for automated imaging are available. Zebrafish embryos, either alive or fixed, can be arrayed in 96-well plates by means of agarose molds that keep them fixed in a dorsal or ventral position (Wittbrodt et al., 2014). These plates can be handled in common automated imaging systems designed for cell-based screens. Another solution for automated imaging, the VAST system, has been published and commercialized by UnionBiometrica. This system aspirates embryos automatically into a glass capillary, which can be rotated along the x-axis to precisely position the animal for imaging or filming. Besides image-based methods for screening, the zebrafish embryo has behavioral traits that make them highly amenable for large-scale analysis. Embryos can be positioned in multiwell plates and the swimming pattern of up to 96 embryos can be recorded and analyzed simultaneously with commercially available systems. Among others, this approach has been used to find and characterize novel anticonvulsive compounds (Dinday and Baraban, 2015; Baxendale et al., 2012). The swimming behavior of zebrafish embryos can not only be altered genetically or chemically, but a highly stereotypical escape response can be triggered with light or sound. This reaction to stimuli as well as short- and long-term memory and habituation is used in neuropharmacological screens (Basnet et al., 2019) as well as in drug screens for Parkinson's disease and other movement disorders (Vaz et al., 2018). Lately, the current technology for image- and behavioral-based screening has been complemented with microfluidic devices. This lab-on-a-chip technology offers fascinating possibilities for precise orientation of embryos, exposure to small molecules, exposure to precisely controlled compound gradients and compound combinations, as well as embryo manipulation and imaging (Khalili and Rezai, 2019). The increased importance of zebrafish in contemporary drug screening pipelines propels the development not only of novel instruments but also of analytical software tools for quantitative assessment of morphological features (Teixido et al., 2019; Schutera et al., 2016).

Relevance and limitation of the zebrafish embryo for chemical screening

The zebrafish has become an attractive animal model from bridge cell culture studies to traditional rodent animal models in drug discovery screens. About 80% of human disease-causing genes are conserved (Howe et al., 2013) and much of the underlying genetics, pathology, and physiology is surprisingly similar in zebrafish embryos (MacRae and Peterson, 2015). Furthermore, important structures like the blood–brain barrier have a similar architecture in zebrafish (Quinonez-Silvero et al., 2020). However, despite the clear advantages of this small vertebrate for preclinical drug discovery, it has intrinsic limitations and lacks characterization in important aspects. The zebrafish embryo develops very fast and the regulatory circuits are mainly primed on growth, development, and differentiation in contrast to those in adult patients. Also, the adaptive immune response is lacking in embryonic zebrafish. The different regulatory pathways as well as the lack of the adaptive immune system provides an environment that clearly differs from human patients despite the overall conservation of physiological processes. In cancer drug discovery, the zebrafish provides a powerful model, but the small size of the embryo limits the number of cancer cells that can be transplanted, which in turn might not reflect the complex tumor heterogeneity. Additionally, zebrafish embryos cannot survive at temperatures above 34°C, which might affect the growth and division speed of malignant cells. Furthermore, although most chemical substances are taken up from the embryo, the update routes and pharmacokinetics (uptake speed, metabolism, and excretion) are not as well characterized as for rodents (Cassar et al., 2020).

The zebrafish as a model for toxicology

The field of toxicology is of ever-increasing importance due to the expanding repertoire of chemicals, including plastics and nanomaterials, being produced by today's society. The zebrafish model offers a powerful in vivo platform to address central questions of environmental and developmental toxicology such as acute toxicity, latent and transgenerational effects, biomonitoring, and insights into the mechanism of action of toxicants. The intrinsic traits of the zebrafish, such as their optical clarity and ex utero development, provide a unique possibility to study the teratogenic effects of toxins, chemicals, and drugs during embryonic development, which is of extraordinary importance for medications administered to pregnant females or babies (Felix et al., 2019). After initial studies more than 60 years ago (Battle and Hisaoka, 1952), pharmaceutical companies joined efforts in 2009 to harmonize zebrafish toxicity assays and the classification of teratogenic effects (Gustafson et al., 2012). Technological and methodological advances improved those screening efforts and today robotic imaging, computational approaches for data analysis, and the availability of transgenic reporter strains make the zebrafish indispensable for state-of-the-art toxicological analysis (Bambino and Chu, 2017). The most common readouts for toxicity profiling in zebrafish are (1) gross morphological changes, (2) changes in embryonic behavior, or (3) changes in reporter gene expression. It is the generation of reporter strains that has propelled the zebrafish into the center of toxicological screening. In those strains, specific biosensors regulate the expression of fluorescent proteins upon exposure to chemicals, pollutants, and toxins that can be monitored and evaluated with high-throughput imaging systems. Zebrafish reporter lines allowing exposure redouts for aromatic hydrocarbons (Xu et al., 2015), heavy metals (Liu et al., 2016), endocrine disrupting substances (Gorelick et al., 2016), bisphenol A (Cano-Nicolau et al., 2016), and others are well established and readily available from international stock centers such as ZIRC and the European Zebrafish Resource Center (www.zebrafish.org; www.ezrc.kit.edu). Besides reporter lines for specific substance groups, more general cellular stress reporters are available that provide amenable first-line models for pollution monitoring. Those reporters react to different chemicals with distinct patterns of GFP expression, which indicate the cell types and organs being affected (Lee et al., 2014).

However, there are certain limits in using the zebrafish as a model for toxicology. The most prominent gene family for xenobiotic metabolism and biotransformation contains the cytochrome p450 genes. The zebrafish genome encodes for 87 *cyp* genes; however, besides the overall strong conservation of those genes, the distribution pattern and substrate specificity might differ and/or is not yet characterized (Saad et al., 2016). Humans take up many chemicals through the gastrointestinal system, whereas zebrafish embryos are mostly exposed by immersion, which foremost recapitulates dermal exposure (embryonic zebrafish do not swallow water for breathing). The different exposure route can substantially affect absorption, tissue distribution, and accumulation of pollutants and chemicals. Moreover, chemicals that are difficult to solubilize in water would need to be injected into the fish—a clear limit for high-throughput screens. Another aspect to consider is that drug-discovery toxicology requires absorption, distribution, metabolism, and excretion profiling of drug candidates. However, sampling zebrafish embryo plasma and urine is difficult to impossible although technical advances are being made (Villacrez et al., 2018; Grech et al., 2019).

The zebrafish as model for neurodegenerative diseases

Neurodegenerative diseases (NDs) are a wide range of disorders involving progressive structural and functional degeneration of the central and peripheral nervous system leading to movement malfunctions (ataxias) or affected mental function (dementias). ND causes can vary between metal poisonings, mitochondrial disfunctions, or genetic mutations.

The zebrafish system has several advantages, which make this animal model relevant for studies that give insights into ND pathologies. Based on comparative neuroanatomy, the human and zebrafish nervous systems exhibit similar organization of the forebrain, midbrain, hindbrain, as well as the sensory and motor functions of the peripheral nervous system (Lieschke and Currie, 2007). Like other higher vertebrates, the zebrafish has a functional blood–brain barrier from as early as 2.5–3 days postfertilization, which makes this fish a relevant model for pharmaceutical studies conducted in higher vertebrates (Quinonez-Silvero et al., 2020; Jeong et al., 2008). Furthermore, all primary neurotransmitter systems critical to sustaining brain function, affecting behavior and memory and learning ability known to play role in both Parkinson's and Alzheimer's disease (AD), have been characterized in the zebrafish model (Horzmann and Freeman, 2016). Despite the many similarities, the organizational brain regions, including the amygdala, substantia nigra, and hippocampus, are missing in the zebrafish with their functions believed to be carried out by other brain regions, including the lateral pallium (Kalueff et al., 2014). Furthermore, the absence of the corticospinal and rubrospinal tracts in the zebrafish central nervous system limits the relevance of studies of upper motor neuron disorders (Babin et al., 2014). Another important aspect to consider when using the zebrafish as a model for neurodegeneration is that the zebrafish possesses a remarkable capacity to regenerate neurons and brain structures, which is much greater compared to nonhuman primates and humans (Kizil et al., 2012).

Alzheimer's disease

Among neurodegenerative diseases, AD and Parkinson's disease have the highest prevalence in society (Ballard et al., 2011). As societies get older and the number of people affected by these diseases increases, so does the demand for therapeutic solutions. The zebrafish is a powerful model for preclinical drug-discovery projects, closing the gap between cell culture studies and experimentation in murine models (Santana et al., 2012). The majority of the zebrafish gene orthologs involved in familial Alzheimer's, including *psen1*, *psen2*, *appa*, *appbpsenen*, *ncstn*, *aph1b*, *bace1*, *bace2*, *mapta*, *maptb*, *apoea*, and *apoeb*, have been identified, which allows for the genetic modeling of the disease in zebrafish. Furthermore, several transgenic and chemical models mimicking AD pathologies have already been established (Koehler and Williams, 2018).

Amyotrophic lateral sclerosis

The zebrafish has also served as an animal model for genetic studies on the etiology of amyotrophic lateral sclerosis (ALS), a fatal upper and lower motor neuron degenerative disease. The disease is typically diagnosed in individuals after 40 years of age and leads to death within 2–5 years after the diagnosis (Sathasivam, 2010). Several genetic mutations have been associated with human ALS. To address the heritability of the disease, several zebrafish morpholino knockdown as well as mRNA injection-based studies have been conducted. Among others, a *zc9orf72* loss-of-function model, which has been associated with axonal degeneration in motor neurons, exemplifies the usefulness of this system for future therapeutic screens (Ciura et al., 2013). Similarly, the zebrafish *vapb* knockdown has been shown to lead to axonal defects and motor impairment (Kabashi et al., 2013). More recently, CRISPR/Cas9 technology combined with single-stranded oligodeoxynucleotide donor template injections, with the latter enabling site-directed single nucleotide editing, resulted in the generation of two missense mutations, $tardbp^{A379T}$ and fus^{R536H}, which have been associated with ALS (Armstrong et al., 2016).

Lessons from the zebrafish model on muscle dystrophies

Muscular dystrophies (MDs) have a prevalence of approximately 20–25 cases per 100,000. Patients with MDs exhibit symptoms such as weakening of muscles and joints, which can have profound effects. Phenotypes can vary from severe congenital muscular dystrophy to mild limb-girdle muscular dystrophy with adult onset (Theadom et al., 2014). Efforts to create mammalian models, including pigs, dogs, and rodents, allowing for the dissection of the pathogeneses behind MDs, have been conducted for decades (Larcher et al., 2014). However, considering the large number of genes that have been associated with MDs (Bansal et al., 2003), the zebrafish model offers a unique opportunity for large-scale genetic studies to unravel the cellular and molecular details of MD phenotypic variability. The first report on the usage of the zebrafish as an MD model dates back to 2003, when Bassett and colleagues identified the zebrafish *sapje* gene as an ortholog of the *Duchenne muscular dystrophy* (DMD) gene (Bassett et al., 2003). More models for DMD followed, including *sapje-like* and *dmdpc2* mutants, exemplifying the importance of the zebrafish for both understanding MDs and finding potential cures (Waugh et al., 2014; Kawahara et al., 2014; Giacomotto et al., 2013). Importantly, several rodent models were not able to recapitulate the severity of human disease, while the zebrafish model served this purpose much better. One example of this is a study on Ullrich congenital muscular dystrophy where zebrafish was used for the preliminary novel therapeutic screens (Telfer et al., 2010). Zebrafish have also been widely used for both the identification and validation of gene candidates for orphan MDs (Schindler et al., 2016).

References

Ahrens, M.B., Orger, M.B., Robson, D.N., Li, J.M., Keller, P.J., 2013. Whole-brain functional imaging at cellular resolution using light-sheet microscopy. Nat. Methods 10, 413–420. https://doi.org/10.1038/nmeth.2434.

Alestrom, P., et al., 2020. Zebrafish: housing and husbandry recommendations. Lab. Anim 54, 213–224. https://doi.org/10.1177/0023677219869037.

Amanze, D., Iyengar, A., 1990. The micropyle: a sperm guidance system in teleost fertilization. Development 109, 495–500.

Armstrong, G.A., et al., 2016. Homology directed knockin of point mutations in the zebrafish *tardbp* and *fus* genes in ALS using the CRISPR/Cas9 system. PLoS One 11, e0150188. https://doi.org/10.1371/journal.pone.0150188.

Arunachalam, M., Raja, M., Vijayakumar, C., Malaiammal, P., Mayden, R.L., 2013. Natural history of zebrafish (*Danio rerio*) in India. Zebrafish 10, 1–14. https://doi.org/10.1089/zeb.2012.0803.

Babin, P.J., Goizet, C., Raldua, D., 2014. Zebrafish models of human motor neuron diseases: advantages and limitations. Prog. Neurobiol. 118, 36–58. https://doi.org/10.1016/j.pneurobio.2014.03.001.

Ballard, C., et al., 2011. Alzheimer's disease. Lancet 377, 1019–1031. https://doi.org/10.1016/S0140-6736(10)61349-9.

Bambino, K., Chu, J., 2017. Zebrafish in toxicology and environmental health. Curr. Top. Dev. Biol. 124, 331–367. https://doi.org/10.1016/bs.ctdb.2016.10.007.

Bansal, D., et al., 2003. Defective membrane repair in dysferlin-deficient muscular dystrophy. Nature 423, 168–172. https://doi.org/10.1038/nature01573.

Basnet, R.M., Zizioli, D., Taweedet, S., Finazzi, D., Memo, M., 2019. Zebrafish larvae as a behavioral model in neuropharmacology. Biomedicines 7. https://doi.org/10.3390/biomedicines7010023.

Bassett, D.I., et al., 2003. Dystrophin is required for the formation of stable muscle attachments in the zebrafish embryo. Development 130, 5851–5860. https://doi.org/10.1242/dev.00799.

Battle, H.I., Hisaoka, K.K., 1952. Effects of ethyl carbamate (urethan) on the early development of the teleost *Brachydanio rerio*. Cancer Res. 12, 334–340.

Baxendale, S., et al., 2012. Identification of compounds with anti-convulsant properties in a zebrafish model of epileptic seizures. Dis. Model Mech. 5, 773–784. https://doi.org/10.1242/dmm.010090.

Betancur, R.R., et al., 2017. Phylogenetic classification of bony fishes. BMC Evol. Biol. 17, 162. https://doi.org/10.1186/s12862-017-0958-3.

Buckley, C., et al., 2017. Precise spatio-temporal control of rapid optogenetic cell ablation with mem-KillerRed in Zebrafish. Sci. Rep. 7, 5096. https://doi.org/10.1038/s41598-017-05028-2.

Burns, M.A., et al., 2018. Hedgehog pathway mutations drive oncogenic transformation in high-risk T-cell acute lymphoblastic leukemia. Leukemia 32, 2126–2137. https://doi.org/10.1038/s41375-018-0097-x.

Cano-Nicolau, J., et al., 2016. Estrogenic effects of several BPA analogs in the developing zebrafish brain. Front. Neurosci. 10, 112. https://doi.org/10.3389/fnins.2016.00112.

Cassar, S., et al., 2020. Use of zebrafish in drug discovery toxicology. Chem. Res. Toxicol. 33, 95–118. https://doi.org/10.1021/acs.chemrestox.9b00335.

Ciura, S., et al., 2013. Loss of function of C9orf72 causes motor deficits in a zebrafish model of amyotrophic lateral sclerosis. Ann. Neurol. 74, 180–187. https://doi.org/10.1002/ana.23946.

Cully, M., 2019. Zebrafish earn their drug discovery stripes. Nat. Rev. Drug Discov. 18, 811–813. https://doi.org/10.1038/d41573-019-00165-x.

Adatto, I., Lawrence, C., Thompson, M., Zon, L.I., 2011. A new system for the rapid collection of large numbers of developmentally staged zebrafish embryos. PLoS One 6, e21715. https://doi.org/10.1371/journal.pone.0021715.

Dinday, M.T., Baraban, S.C., 2015. Large-scale phenotype-based antiepileptic drug screening in a zebrafish model of dravet syndrome. eNeuro 2. https://doi.org/10.1523/ENEURO.0068-15.2015.

Driever, W., et al., 1996. A genetic screen for mutations affecting embryogenesis in zebrafish. Development 123, 37–46.

Felix, L., Coimbra, A.M., Valentim, A.M., Antunes, L., 2019. Review on the use of zebrafish embryos to study the effects of anesthetics during early development. Crit. Rev. Toxicol. 49, 357–370. https://doi.org/10.1080/10408444.2019.1617236.

Fior, R., et al., 2017. Single-cell functional and chemosensitive profiling of combinatorial colorectal therapy in zebrafish xenografts. Proc. Natl. Acad. Sci. U. S. A. 114, E8234–E8243. https://doi.org/10.1073/pnas.1618389114.

Force, A., et al., 1999. Preservation of duplicate genes by complementary, degenerative mutations. Genetics 151, 1531–1545.

Gerhard, G.S., et al., 2002. Life spans and senescent phenotypes in two strains of Zebrafish (*Danio rerio*). Exp. Gerontol. 37, 1055–1068. https://doi.org/10.1016/s0531-5565(02)00088-8.

Giacomotto, J., et al., 2013. Chemical genetics unveils a key role of mitochondrial dynamics, cytochrome c release and IP3R activity in muscular dystrophy. Hum. Mol. Genet. 22, 4562–4578. https://doi.org/10.1093/hmg/ddt302.

Gokey, J.J., Ji, Y., Tay, H.G., Litts, B., Amack, J.D., 2016. Kupffer's vesicle size threshold for robust left-right patterning of the zebrafish embryo. Dev. Dyn. 245, 22–33. https://doi.org/10.1002/dvdy.24355.

Gorelick, D.A., Pinto, C.L., Hao, R., Bondesson, M., 2016. Use of reporter genes to analyze estrogen response: the transgenic zebrafish model. Methods Mol. Biol. 1366, 315–325. https://doi.org/10.1007/978-1-4939-3127-9_24.

Grech, A., et al., 2019. Generic physiologically-based toxicokinetic modelling for fish: integration of environmental factors and species variability. Sci. Total Environ. 651, 516–531. https://doi.org/10.1016/j.scitotenv.2018.09.163.

Grunwald, D.J., Eisen, J.S., 2002. Headwaters of the zebrafish – emergence of a new model vertebrate. Nat. Rev. Genet. 3, 717–724. https://doi.org/10.1038/nrg892.

Gustafson, A.L., et al., 2012. Inter-laboratory assessment of a harmonized zebrafish developmental toxicology assay - progress report on phase I. Reprod. Toxicol. 33, 155–164. https://doi.org/10.1016/j.reprotox.2011.12.004.

Hamilton, F., 1822. An Account of the Fishes Found in the River Ganges and its Branches. Printed for A. Constable and company; etc..

Hisaoka, K.K., Battle, H.I., 1958. The normal developmental stages of the zebrafish, *Brachydanio rerio* (Hamilton-Buchanan). J. Morphol. 102, 311–327. https://doi.org/10.1002/jmor.1051020205.

Horzmann, K.A., Freeman, J.L., 2016. Zebrafish get connected: investigating neurotransmission targets and alterations in chemical toxicity. Toxics 4. https://doi.org/10.3390/toxics4030019.

Howe, K., et al., 2013. The zebrafish reference genome sequence and its relationship to the human genome. Nature 496, 498–503. https://doi.org/10.1038/nature12111.

Huang, C.J., Tu, C.T., Hsiao, C.D., Hsieh, F.J., Tsai, H.J., 2003. Germ-line transmission of a myocardium-specific GFP transgene reveals critical regulatory elements in the cardiac myosin light chain 2 promoter of zebrafish. Dev. Dyn. 228, 30–40. https://doi.org/10.1002/dvdy.10356.

Hwang, W.Y., et al., 2013. Efficient genome editing in zebrafish using a CRISPR-Cas system. Nat. Biotechnol. 31, 227–229. https://doi.org/10.1038/nbt.2501.

Jeong, J.Y., et al., 2008. Functional and developmental analysis of the blood-brain barrier in zebrafish. Brain Res. Bull. 75, 619–628. https://doi.org/10.1016/j.brainresbull.2007.10.043.

Kabashi, E., et al., 2013. Investigating the contribution of VAPB/ALS8 loss of function in amyotrophic lateral sclerosis. Hum. Mol. Genet. 22, 2350–2360. https://doi.org/10.1093/hmg/ddt080.

Kalueff, A.V., Echevarria, D.J., Stewart, A.M., 2014. Gaining translational momentum: more zebrafish models for neuroscience research. Prog. Neuro-Psychopharmacol. Biol. Psychiatry 55, 1–6. https://doi.org/10.1016/j.pnpbp.2014.01.022.

Kawahara, G., et al., 2014. Dystrophic muscle improvement in zebrafish via increased heme oxygenase signaling. Hum. Mol. Genet. 23, 1869–1878. https://doi.org/10.1093/hmg/ddt579.

Kawakami, K., 2007. Tol2: a versatile gene transfer vector in vertebrates. Genome Biol. 8 (Suppl. 1), S7. https://doi.org/10.1186/gb-2007-8-s1-s7.

Kent, M.L., Sanders, J.L., Spagnoli, S., Al-Samarrie, C.E., Murray, K.N., 2020. Review of diseases and health management in zebrafish *Danio rerio* (Hamilton 1822) in research facilities. J. Fish Dis. https://doi.org/10.1111/jfd.13165.

Khalili, A., Rezai, P., 2019. Microfluidic devices for embryonic and larval zebrafish studies. Brief. Funct. Genomics 18, 419–432. https://doi.org/10.1093/bfgp/elz006.

Kim, C.H., et al., 1996. Zebrafish elav/HuC homologue as a very early neuronal marker. Neurosci. Lett. 216, 109–112. https://doi.org/10.1016/0304-3940(96)13021-4.

Kimmel, C.B., Law, R.D., 1985. Cell lineage of zebrafish blastomeres. I. Cleavage pattern and cytoplasmic bridges between cells. Dev. Biol. 108, 78–85. https://doi.org/10.1016/0012-1606(85)90010-7.

Kimmel, C.B., Ballard, W.W., Kimmel, S.R., Ullmann, B., Schilling, T.F., 1995. Stages of embryonic development of the zebrafish. Dev. Dyn. 203, 253–310. https://doi.org/10.1002/aja.1002030302.

Kizil, C., Kaslin, J., Kroehne, V., Brand, M., 2012. Adult neurogenesis and brain regeneration in zebrafish. Dev. Neurobiol. 72, 429–461. https://doi.org/10.1002/dneu.20918.

Koehler, D., Williams, F.E., 2018. Utilizing zebrafish and okadaic acid to study Alzheimer's disease. Neural Regen. Res. 13, 1538–1541. https://doi.org/10.4103/1673-5374.237111.

Laale, H.W., 1977. The biology and use of zebrafish, *Brachydanio rerio* in fisheries research. J. Fish Biol. 10, 121–173. https://doi.org/10.1111/j.1095-8649.1977.tb04049.x.

Larcher, T., et al., 2014. Characterization of dystrophin deficient rats: a new model for Duchenne muscular dystrophy. PLoS One 9, e110371. https://doi.org/10.1371/journal.pone.0110371.

Lawson, N.D., Weinstein, B.M., 2002. In vivo imaging of embryonic vascular development using transgenic zebrafish. Dev. Biol. 248, 307–318. https://doi.org/10.1006/dbio.2002.0711.

Lee, H.C., et al., 2014. Zebrafish transgenic line huORFZ is an effective living bioindicator for detecting environmental toxicants. PLoS One 9, e90160. https://doi.org/10.1371/journal.pone.0090160.

Lieschke, G.J., Currie, P.D., 2007. Animal models of human disease: zebrafish swim into view. Nat. Rev. Genet. 8, 353–367. https://doi.org/10.1038/nrg2091.

Liu, L., Yan, Y., Wang, J., Wu, W., Xu, L., 2016. Generation of *mt:egfp* transgenic zebrafish biosensor for the detection of aquatic zinc and cadmium. Environ. Toxicol. Chem. 35, 2066–2073. https://doi.org/10.1002/etc.3362.

MacRae, C.A., Peterson, R.T., 2015. Zebrafish as tools for drug discovery. Nat. Rev. Drug Discov. 14, 721–731. https://doi.org/10.1038/nrd4627.

McCluskey, B.M., Postlethwait, J.H., 2015. Phylogeny of zebrafish, a "model species," within Danio, a "model genus". Mol. Biol. Evol. 32, 635–652. https://doi.org/10.1093/molbev/msu325.

Nicoli, S., Ribatti, D., Cotelli, F., Presta, M., 2007. Mammalian tumor xenografts induce neovascularization in zebrafish embryos. Cancer Res. 67, 2927–2931. https://doi.org/10.1158/0008-5472.CAN-06-4268.

Nusslein-Volhard, C., 2012. The zebrafish issue of development. Development 139, 4099–4103. https://doi.org/10.1242/dev.085217.

Pudelko, L., et al., 2018. An orthotopic glioblastoma animal model suitable for high-throughput screenings. Neuro Oncol. 20, 1475–1484. https://doi.org/10.1093/neuonc/noy071.

Quinonez-Silvero, C., Hubner, K., Herzog, W., 2020. Development of the brain vasculature and the blood-brain barrier in zebrafish. Dev. Biol. 457, 181–190. https://doi.org/10.1016/j.ydbio.2019.03.005.

Saad, M., et al., 2016. Xenobiotic metabolism in the zebrafish: a review of the spatiotemporal distribution, modulation and activity of Cytochrome P450 families 1 to 3. J. Toxicol. Sci. 41, 1–11. https://doi.org/10.2131/jts.41.1.

Santana, S., Rico, E.P., Burgos, J.S., 2012. Can zebrafish be used as animal model to study Alzheimer's disease? Am. J. Neurodegener. Dis. 1, 32–48.

Sathasivam, S., 2010. Motor neurone disease: clinical features, diagnosis, diagnostic pitfalls and prognostic markers. Singapore Med. J. 51, 367–372 quiz 373.

Schindler, R.F.R., et al., 2016. POPDC1S201F causes muscular dystrophy and arrhythmia by affecting protein trafficking. J. Clin. Invest. 126, 239–253. https://doi.org/10.1172/JCI79562.

Schutera, M., et al., 2016. Automated phenotype pattern recognition of zebrafish for high-throughput screening. Bioengineered 7, 261–265. https://doi.org/10.1080/21655979.2016.1197710.

Shih, J., Fraser, S.E., 1996. Characterizing the zebrafish organizer: microsurgical analysis at the early-shield stage. Development 122, 1313–1322.

Spence, R., et al., 2006. The distribution and habitat preferences of the zebrafish in Bangladesh, 69, 1435–1448. https://doi.org/10.1111/j.1095-8649.2006.01206.x.

Spence, R., Fatema, M.K., Ellis, S., Ahmed, Z.F., Smith, C., 2007. Diet, growth and recruitment of wild zebrafish in Bangladesh. J. Fish Biol. 71, 304–309. https://doi.org/10.1111/j.1095-8649.2007.01492.x.

Streisinger, G., Walker, C., Dower, N., Knauber, D., Singer, F., 1981. Production of clones of homozygous diploid zebra fish (*Brachydanio rerio*). Nature 291, 293–296. https://doi.org/10.1038/291293a0.

Teixido, E., et al., 2019. Automated morphological feature assessment for zebrafish embryo developmental toxicity screens. Toxicol. Sci. 167, 438–449. https://doi.org/10.1093/toxsci/kfy250.

Telfer, W.R., Busta, A.S., Bonnemann, C.G., Feldman, E.L., Dowling, J.J., 2010. Zebrafish models of collagen VI-related myopathies. Hum. Mol. Genet. 19, 2433–2444. https://doi.org/10.1093/hmg/ddq126.

Theadom, A., et al., 2014. Prevalence of muscular dystrophies: a systematic literature review. Neuroepidemiology 43, 259–268. https://doi.org/10.1159/000369343.

van Raamsdonk, W., Tekronnie, G., Pool, C.W., van de Laarse, W., 1980. An immune histochemical and enzymic characterization of the muscle fibres in myotomal muscle of the teleost

Brachydanio rerio, Hamilton-Buchanan. Acta Histochem. 67, 200–216. https://doi.org/10.1016/S0065-1281(80)80024-9.

Vaz, R.L., Outeiro, T.F., Ferreira, J.J., 2018. Zebrafish as an animal model for drug discovery in Parkinson's disease and other movement disorders: a systematic review. Front. Neurol. 9, 347. https://doi.org/10.3389/fneur.2018.00347.

Villacrez, M., et al., 2018. Evaluation of drug exposure and metabolism in locust and zebrafish brains using mass spectrometry imaging. ACS Chem. Neurosci. 9, 1994–2000. https://doi.org/10.1021/acschemneuro.7b00459.

Waugh, T.A., et al., 2014. Fluoxetine prevents dystrophic changes in a zebrafish model of Duchenne muscular dystrophy. Hum. Mol. Genet. 23, 4651–4662. https://doi.org/10.1093/hmg/ddu185.

Weis, J.S., 1968. Analysis of the development of nervous system of the zebrafish, *Brachydanio rerio*. I. The normal morphology and development of the spinal cord and ganglia of the zebrafish. J. Embryol. Exp. Morphol. 19, 109–119.

Weis, J.S., 1968. Analysis of the development of the nervous system of the zebrafish, *Brachydanio rerio*. II. The effect of nerve growth factor and its antiserum on the nervous system of the zebrafish. J. Embryol. Exp. Morphol. 19, 121–135.

Wittbrodt, J.N., Liebel, U., Gehrig, J., 2014. Generation of orientation tools for automated zebrafish screening assays using desktop 3D printing. BMC Biotechnol. 14, 36. https://doi.org/10.1186/1472-6750-14-36.

Xiao, J., Glasgow, E., Agarwal, S., 2020. Zebrafish xenografts for drug discovery and personalized medicine. Trends Canc. 6, 569–579. https://doi.org/10.1016/j.trecan.2020.03.012.

Xu, H., et al., 2015. Generation of *Tg(cyp1a:gfp)* transgenic zebrafish for development of a convenient and sensitive in vivo assay for aryl hydrocarbon receptor activity. Mar. Biotechnol. 17, 831–840. https://doi.org/10.1007/s10126-015-9669-1.

Yan, C., et al., 2019. Visualizing engrafted human cancer and therapy responses in immunodeficient zebrafish. Cell 177, 1903–1914.e14. https://doi.org/10.1016/j.cell.2019.04.004.

Chapter 2

Housing and maintenance of zebrafish, new technologies in laboratory aquatic systems and considerations for facility design

Bruce Newell[1], **Marco Brocca**[2]
[1]*Aquatics, Aviary and Wildlife Supervisor, Deakin University, Geelong, VIC, Australia;* [2]*Aquatic Solutions Tecniplast Spa., Buguggiate (VA), Italy*

While this section mainly references the care, housing, and maintenance of zebrafish, the context of this chapter is applicable to the care of other small teleost species utilized in biomedical research. The scope of this chapter is not extensive, and references are included for further reading. The target audience for this section is technicians/facility managers and entry-level researchers.

Introduction to the model

Zebrafish (*Danio rerio*) are endemic to a range that covers much of India, Bangladesh, and into Nepal. They are found in the tributaries of the Ganges and Brahmaputra rivers in the north through to paddies, irrigation, and flood plain areas in the south. The range extends from as far west as Pakistan through to Myanmar (Spence et al., 2008; Engeszer et al., 2007). There is a documented temperature range of 8–35°C within the species' natural range (Spence et al., 2006; McClure et al., 2006). Under laboratory conditions the zebrafish has demonstrated an ability to survive temperature ranges from 6.7 through to 41.7°C (Cortemeglia and Beitinger, 2005). Zebrafish are a shoaling species with groups in the wild ranging from a few animals to over 2000 individuals. Water flow appears to have an influence on the number of animals in any given group. The higher the flow is, the larger the group, with several leadership and fission–fusion changes (Suriyampola et al., 2016; Shelton et al., 2020).

Laboratory Fish in Biomedical Research. https://doi.org/10.1016/B978-0-12-821099-4.00018-3
Copyright © 2022 Elsevier Inc. All rights reserved.

The small size and fecundity enable large numbers of embryos to be produced within a tight footprint, achieving a high level of economy relative to mammal models. The species is a nonseasonal, crepuscular breeder with a strong preference, like many cyprinids, for shallow water areas and pebble substrates (Sessa et al., 2008; Schroeder et al., 2014). Hence, the species can be easily manipulated for production of timed embryos.

Defining your needs from an animal model and the impact on facility design

In setting up an aquatic facility, the design will be impacted by the research needs and these should be considered during the planning stages. Will the research involve behavioral work, or is the focus on genetics, toxicology, or development? Are resources to be shared with other research groups? What is the start-up and ongoing budget availability? Large-scale behavioral works are likely to require considerably increased floor space than a centralized high-density housing system targeting high levels of embryo production.

Heavily centralized recirculating aquaculture systems (RAS) may not be the best choice where water quality influences in a study, may affect neighboring animals. In those specific cases, flow-through systems or multiple smaller independent RAS may be more suitable.

System choice: Investing in top-quality husbandry equipment should be a very high priority, but this is not always a possibility for every research group. There are distinct benefits to utilizing the services and latest offerings available from technology leaders, as this provides the best chance of implementing stable, reliable, and repeatable husbandry results. However, aquatic research does not exclude those working on a tight budget.

A small-scale research program can potentially be operated from one or more independently filtered "hobbyist-style" aquaria used to house adult zebrafish for embryo production or to investigate behavioral questions. This approach is impractical for large-scale support but may be an entry path into teleost research for some groups, or it may be an appropriate solution for some simple projects. Simple support systems will generally leave researchers looking for better, more scalable solutions as the limitations are quickly realized. Variation in water quality in basic installations tends to be greater than where larger volumes of water are utilized. Stocking densities that are able to be held in smaller "hobbyist" tanks are likely to prove to be a limiting factor and the labor inputs required to maintain multiple small tanks become impractical as the scope of the project increases.

The logical step forward is to move to RAS that allow for centralized filtration and control of water parameters (Fig. 2.1).

Small to medium-scale RAS can be developed using resources available to hand, either as a "self-build" or by utilizing the services of one of the numerous small to medium enterprises operating in the aquaculture industry.

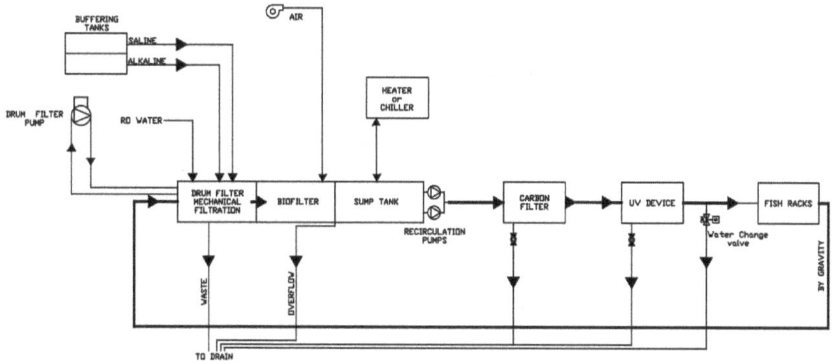

FIGURE 2.1 A typical advanced recirculating aquaculture system design layout showing the key elements and water flow patterns. *RO*, reverse osmosis; *UV*, ultraviolet. *Drawing: Courtesy of Tecniplast SpA.*

These providers may be able to assist in the engineering, design, and creation of systems of various levels of complexity, refinement and automation. If pursuing this path, it would be wise to do a thorough and realistic costing of the build for comparison against options from the specialist technology leaders. A list of some specialist providers is included in the references of this chapter (Fig. 2.2).

The technology leaders have vast experience in the research field and have invested time and resources in the production and refinement of dedicated, research-specific RAS. Solutions range from small benchtop racks, through intermediately sized scalable systems that can be expanded as budget and research allows, to large-scale turnkey solutions. These vendors will be able to provide detailed technical drawings and dedicated room layouts to assist with facility design and construction and can often provide solutions for other aquatic research models such as medaka, killifish, platyfish, cavefish, stickleback, goldfish, and *Danionella translucida*.

As the system design is formulated, it is very important to consult with other facilities to seek opinions and experience. Since there is a high degree of cross-over in the husbandry of teleosts, much can be learnt from those working with a variety of species. Gaining insights into the advancement of installation phases can prevent expensive design flaws that may lead to bottlenecks and potentially expensive reworks. Be aware of facility-level biocontainment requirements in your locale, as failing to meet compliance can be a costly mistake and is easily avoided by careful forward planning. When designing a facility, it may be wise to include a containment engineer within the design team as well as working with designers/architects who have broad experience in this type of work, particularly for larger projects.

FIGURE 2.2 (1) Two "self-build" recirculating aquaculture systems (RAS) at the Federal University of Jatai. Brazil. Photo Prof. Monica Rodrigues Ferreira Machado. (2) A commercially produced "Educational" standalone RAS. Photo Aquaneering Inc. (3) A large-scale, highly automated rack system with mass spawning devices attached. Photo Bruce Newell, Deakin University. (4) A small zebrafish set-up utilizing "hobbyist" equipment used for approximately 12 months before expansion into a commercially produced RAS system. Photo Dr. Suzita Mohd Noor, Universiti Malaya. (5) Automation control panel monitoring a centralized RAS. Photo Tecniplast SpA.

The most common facility layout for zebrafish housing racks tends to be within a single large room rather than in a series of smaller spaces (Zynda, 2020). This assists in efficient management of heat and humidity by the heating ventilation air conditioning (HVAC) system. The volume of water within a housing system can create a large thermal mass that helps with stability of room ambient conditions.

A factor that is always a driving force in facility design is the effective use of space, with an emphasis on grouping similar equipment, research, and operational spaces together. This can provide a better idea of personnel and equipment flow, which is important to consider in facility biosecurity design.

At the time of writing (2020), the SARS-Cov-2 pandemic has brought attendant requirements for social distancing and reduced occupancy limits, highlighting problematic constraints within many existing facilities. Workflow patterns in some units have proven to be difficult to manage where equipment or resources are heavily concentrated in small spaces. It is hard to predict the situation long term, but those preparing for a facility built in the current public health and social climate should evaluate where the bottlenecks and workflow problems are likely to occur.

Aisle widths in multirack set-ups have proven to be difficult, not allowing for suitable physical distance when two or more people are simultaneously working. Moreover, workstations for scientific procedures have been traditionally clustered by use either within, or very close to, the main animal housing area and are not able to provide a necessary degree of separation in many cases.

Consider how the organization of workstations may provide temporal spacing. Fin clipping, for instance, is not time limited in the same way microinjection is. Interspersing workstations that are likely to be used at different times can result in greater physical separation without requiring more space. Ensuring that activities that are not as time critical are moved into the "off-peak" time slots can have large positive impacts on space utilization.

Personnel movements through building entries and exits have shown to be "pinch points" that impact effective flows of staff in a socially distanced manner.

While it is unlikely that a facility already laid out with rack systems will substantially change its configuration, when thinking about a new layout, if the space is available, decentralizing animal holding spaces to a degree may be advantageous in light of recent public health restrictions. There are distinct advantages from a building maintenance, operations workflow, HVAC and cost perspective to having all of the animal housing in one space, but it remains to be seen if this continues to outweigh public health considerations in a post-pandemic world.

Smaller racks in suites that can be set up for one or two operators may incur a considerable cost in duplications and floor space, but could possibly end up being seen as advantageous in small facilities due to the increased level of flexibility this may provide. Filtration systems can still be centralized and provide the same degree of life support and quality control.

Facility design is a complex and multifaceted subject that relies on many different disciplines for success. Professional facility design companies exist, and the specialist technology leaders can assist in facility layouts, including provision of 3D drawings of completed systems.

Further reading can be found in Zynda (2020).

Primary housing

The choice of tank designs will be determined by the manufacturer if you are using a specialist supplier to deliver a turnkey solution. For those who are "self-building," directing funds to obtain tanks from a specialist manufacturer is highly recommended. The primary housing units (tanks) in which zebrafish are maintained on a RAS are principally constructed of molded thermoplastics such as polycarbonate or polysulfone; however, glass, fiberglass, polyethylene, or acrylic may also be utilized for more specific purposes. Primary housing units of a mix of configurations and sizes can be found in most facilities. The specific mix will depend upon the needs of the research and the groupings in which the animals may be required to be held. Tanks must be easily cleanable, and have integrated methods of solids removal designed in, as rapid and efficient removal of solids is of high importance. Most primary housing units utilize an arrangement that allows for solids to be collected from the bottom of the tank (Cockington, 2020). There are a large range of styles and tank formats available with new variations being developed and released regularly. Spawning solutions have been developed ranging from static housing tanks of less than 1 L designed for small group spawning through to large-scale mass spawning devices capable of holding hundreds of fish. Detailed information on spawning solutions can be found in breeding and larviculture (Chapter 3 of this publication).

Tank hygiene

Attention to the efficient cleaning and washing of primary housing units has seen a huge increase over the years. The move from static glass tanks into banks of injection-molded, lightweight plastic tanks has allowed for easier and safer handling. The aim for cleaning is to remove biofilms, algae, and other debris efficiently and effectively from the tank surfaces. Manual scrubbing with sponges and tap water followed by a rinse step with reverse osmosis water is often adopted by facilities that do not have the benefit of automated washing systems. Manual options can be ineffective, time consuming, and a potential source of ergonomic issues for personnel. There is also potential for abrading the plastic surfaces of the tanks. Scratches in the surfaces increase the areas that biofilms can lodge making the cleaning process more difficult. Laboratory glassware washers and domestic/small commercial dishwashers have been utilized with varying degrees of success. The usage of chemical products for fish tank washing was often considered risky from a residual contamination aspect, and this has resulted in some washers being run on a chemical-free hot water cycle, often after a presoaking routine.

Recently, it has been demonstrated that a dedicated mix of chemicals, including liquid hydrogen peroxide (H_2O_2), can be safely used, especially when combined with efficient rinsing cycles and appropriate pressurized washing. This approach can provide a certain degree of disinfection by combining these chemical and temperature profiles accurately. With reduced physical damage to surfaces, improved removal of algae and knockdown of potential pathogenic organisms, the time between required wash cycles may be extended. There are aquatic-specific tank-washing solutions emerging that provide high-level sanitation in an automated format (Tecniplast SpA, 2020).

For further information on tank washing see Sanders (2020).

Aquatic concepts

Moving toward the more technical aspects of aquatic husbandry, it is important to address the basic concepts. These allow for successful maintenance of an animal that has vastly different respiratory mechanisms and home environment requirements to most other vertebrate laboratory research models.

Mechanical filtration

Effective mechanical filtration is a major factor that must be considered when designing laboratory aquatic model holding systems, both within the primary housing unit and in any subsequent filtration systems. An arrangement that allows the accumulation of debris will likely suffer detrimental effects and not perform as expected, with potential negative consequences on water quality and on the health of the housed aquatic species. Mechanical filtration is a generalized reference to the physical trapping of particulate matter, which allows for the removal of solids such as fecal casts, uneaten foods, and other organic and inorganic debris. Effective mechanical filtration, usually down to below 50 μm, helps maintain water clarity, reduces the build-up of dissolved nitrogenous compounds, and assists in the suppression of heterotrophic bacterial populations, opportunistic organisms, and potential pathogenic species.

Methods of mechanical filtration can range from basic gravel filter beds and foam sponge elements of the type utilized in "hobbyist" equipment, through filter cloths and screens of various mesh sizes formed into cartridge or bag filters, parabolic screens, or automated drum filters. Other methods of mechanical filtration include depth-based media such as sand filters and bead filters and techniques including foam fractionation and vortex swirl separators. It is the job of the design engineer to select the most appropriate combination of technologies considering scale, available space, and scope of the system, together with the budget for both capital equipment and for ongoing maintenance.

Mechanical filters will have some degree of cross-over to other filtration styles as biological action is likely to occur on the surfaces of equipment as the microorganisms that act upon the nitrogenous and organic wastes have a chance to colonize these areas. Chemical filtration elements can also be coopted into housings designed for mechanical filtration.

Maintenance of mechanical filtration elements is of critical importance, with the level of input required and the cost varying by equipment choice. The key to efficient mechanical filtration lies not only in the trapping capabilities, but also the effective and timely removal of the captured materials. A device featuring automatic removal of trapped materials is likely a more effective unit than one that captures and holds materials within the water column exposed to biological action able to break that waste down. Flushing will assist in the removal of accumulated colloidal solids, refractory organics, and nitrogenous wastes, thus increasing RAS overall performance. These options will come at the cost of water and buffering compound consumption, as well as energy expenditure, particularly if these are not well engineered. Automated backwashing may not be a suitable choice in an environment where water conservation is a key criterion. On any aquatic research system there will be a water budget that must be considered. The water budget needs to account for not only the monetary cost of water usage, but also the environmental costs. If an automated backflushing system uses less than the allotted water budget for the system, the impact is reduced and may not be significant.

The selection of mechanical filtration elements for a given task is something that should be discussed and evaluated with the assistance of experienced system designers and facility operators. A combination of different approaches may be incorporated into the design of a functioning zebrafish RAS.

Chemical filtration

Chemical filtration is the use of a compound that through adsorption, catalysis, oxidation, or dissolution will chemically change the makeup of water. The most commonly used forms of chemical filtration in zebrafish housing systems are granular activated carbon (GAC) and aragonite or other calcareous materials.

GAC is used to adsorb chemicals from the water column; it is widely adopted for the neutralization of chlorine from water supplies used in aquatic culture. GAC will adsorb volatile organic compounds, tannins, and active or inactivated medications from the water column. Activated carbon will also bind some pheromones from the water column (Jonker and Van Mourik, 2014).

When utilizing GAC, it is important to choose a dedicated carbon for aquaculture purposes that will not have an effect on the pH of the recirculating water.

When the activated carbon reaches its saturation point, the material is no longer capable of efficiently binding compounds from the water column. This is most frequently judged on a time interval basis, and often this may not be based on quantitative information. In some facilities the use of empirical measurement of the adsorptive capacity of activated carbons is being undertaken in an attempt to extend usage cycles on carbon filters (ASTM international, 2013; Baumann, 2020). In RAS, activated carbon is seen by some facilities as a somewhat optional element, especially when water sources such as reverse osmosis are utilized for water supply. GAC and chemical media should be installed downstream of mechanical filtration steps to prevent rapid clogging.

There are various adsorptive and ion exchange resins that may prove useful in specific situations. For example, Cuprisorb (Seachem Laboratories Inc., 2020) will preferentially take up copper and other heavy metals from a water column. This type of material can be very useful for some applications where an environmental contaminant may be present within a system that is not easily controlled by other methods.

Water disinfection and biological control

Broadly broken into two mainstays of the aquaculture industry, the main options here are ozone and ultraviolet C (UVC) radiation, with the latter being by far the more prevalent in small-scale, indoor aquaculture. Ozone gains a brief mention because, while it is not a common option in research zebrafish RAS, it may provide benefits in certain applications. One of the main reasons UVC radiation is so widely and commonly used in RAS is because nothing but energy is applied to the water stream. There are no potentially dangerous chemicals involved. The energy addition will add heat to the recirculating water and there are potential occupational health issues with exposure to unshielded UV tubes.

For clarity, a UVC radiation device is often referred to with the misnomer "ultraviolet sterilizer" (UVS); a more appropriate way of thinking of these devices is "ultraviolet disinfection." The effect is not sterilization of the water, but rather a suppression of the load of microorganisms able to survive post-treatment. UVC radiation (200–280 nm) disrupts the DNA of exposed microorganisms, fungi, algal cells, etc.

The key figure that has to be examined is the calculated "irradiation dosage" usually expressed as $\mu W/cm^2/s$ that organisms are exposed to during their path through the UVC device. This is a complex matter and is impacted not only by the wattage and design of the reaction chamber, but also by the flow rates and clarity of water moving through the device. In RAS for biomedical research purposes, independently from the volume of water treated, a usual disinfection value ranges between 100,000 and 150,000 $\mu W/cm^2/s$ at the end of the operational life of the bulb.

Effective mechanical filtration is a critical part of this picture, since bacteria and other microbials can be shielded by suspended solids. This leads to reduced system performance and has the potential to negatively impact the health of the housed animals.

UVC devices must be maintained regularly following the manufacturer's specifications. Particular attention needs to be paid to cleaning of the quartz glass sleeves and the timely replacement of the irradiation sources. In most UVC units likely to be found in a small to medium RAS, the radiation source takes the form of low-pressure fluorescent tubes. UVC wavelength LEDs do exist but have not yet found wide application in aquatics.

Ozone (O_3), when used at the correct concentrations and with appropriate controls in place, should present few problems to human health; however, in the event of failures of design or control, there is the potential for exposure to an aggressively oxidative gas, particularly in indoor applications. This ionized form of oxygen may produce irritation to respiratory and mucous membranes (McDonnell et al., 1983).

Ozone disinfection works predominantly via the oxidation of organics on contact. Being highly reactive, the ozone molecule is usually produced close to the point of use and does not tend to remain as a residual component in the downstream water if control and dosage systems are appropriately implemented. Ozone use will encourage microflocculation of particulates within the water column and it is often integrated into foam fractionation as the flocculation improves organics removal (Timmons and Ebeling, 2007). The two main options for generation of O_3 are high-voltage corona discharge or UV radiation of 140–190 nm to excite oxygen molecules into the O_3 form. Corona discharge is the more energy efficient process of the two.

Automated control in an ozone system is essential; this is usually achieved by monitoring and responding to variations of the redox potential of the water. A freshwater system should be maintained at around 300 mV redox potential but the needs of the species being cultured should be investigated. If considering ozone for a research system, it is strongly suggested to obtain advice from a systems design engineer familiar with its application. The use of ozone and UVC equipment is not mutually exclusive.

An emerging field of photocatalytic reactors that may develop useful techniques for water disinfection is also covered briefly in the "New and emerging technologies" section.

Biological filtration

Biological filtration is the reliance on a cohort of chemoautotrophic bacteria, and archaea that oxidize nitrogenous wastes from cultured specimens and associated organic inputs to our colonies, in an oxic environment. For convenience the author will refer to these cohorts as "biofilters" or "biological filtration"; to delve into the dynamics of the specific organisms involved in these oxidizing processes is outside the scope of this chapter. If this subject is of interest, further reading can be found (Sauder et al., 2011; Hovanec et al., 1998; Silke Ehrich et al., 1995).

Effective biological filtration is a crucial factor in the captive maintenance of aquatic species.

While it is possible to dilute, capture, or oxidize ammonia from a culture environment by nonbiological means (Zhou and Boyd, 2014; Ben-Asher and Lahav, 2016; Gendel and Lahav, 2013), by far the most typical approach is reliance on the biofilter to oxidize total ammonia nitrogen (TAN) into nitrite (NO_2) and then further into nitrate (NO_3).

Heterotrophic bacteria are also involved in the breakdown processes of wastes, and these bacterial populations can compete with the chemoautotrophs and archaea for oxygen supply within a culture medium (Timmons and Ebeling, 2007; Hammer, 2020).

The advantage of encouraging autotrophic breakdown of wastes over heterotrophic breakdown is that autotrophic processes tend to create lower quantities of bacterial biomass in comparison to heterotrophic processes and result in lower carbon dioxide (CO_2) production (Timmons and Ebeling, 2007).

The nitrification process is a large consumer of both oxygen and carbonates from a water column. For each gram of ammonia-N oxidized in a system, 4.57 g of oxygen as well as 7.05 g of a carbonate source ($CaCO_3$) will be consumed (Timmons and Ebeling, 2007).

Establishing biofiltration

It is very important to understand how to best manipulate microorganisms in the biofilter, as the health of the animal colonies relies upon it. If a new or restarted system is to house animals of the same health status of an existing colony, then cross-seeding of filtration material from the existing colony can be used to quickly introduce the biofilter organisms and establish effective nitrification, but there are distinct disadvantages with this.

Cross-seeding will transfer a host of organisms between the systems and can compromise the health status of the new system. If this is not desired, a different method of "starting" the biofilter will need to be undertaken.

One method is the "cold start" (DeLong and Losordo, 2012). This involves introducing a small number of animals to an "unseeded" environment. In this scenario the animals that are introduced to the new environment will carry microorganisms with them that will be suited to forming a functional biofilter, with the waste those animals produce, providing the base ammonia compounds for biofilter development.

This approach also has disadvantages, both from a time taken and an ethical point of view. The establishment of suitable numbers of microbes can be slow, and the accumulation of potentially toxic nitrogenous compounds within the water column may result in undue stress, potential disease, or even death of fish exposed to this process.

The more effective and certainly more ethically appropriate method is the use of specifically grown starter cultures, which are then fed using a source of ammonia other than excreted wastes. One example of a biofilter starter culture is "one and only" (DrTim's Aquatics, LLC, 2020) but numerous others are available.

Ammonium chloride (NH_4Cl) is used to feed biofilter populations in the absence of fish.

Levels of ammonia in a culture that exceed 5 mg/L will impede the development of biofilter populations and must be avoided. It is possible to calculate the ammonia production of any given population based upon food inputs, so from this it is possible to fully mature a filter to deal with any load of stock without exposing any animals to risk.

If the provider of the biological starter culture presents specific instructions for use, then follow that set of guidelines. It is important to ensure that water conditions are within a suitable range for the microorganisms to develop and multiply.

A carbon source is needed, and the most common approach is to ensure that a level of carbonate hardness (KH) in the range of 150–250 mg/L is available (Timmons and Ebeling, 2007; DeLong and Losordo, 2012). Sodium bicarbonate ($NaHCO_3$) is the most frequently used additive for increasing carbonate levels in solution, but the addition of potassium hydrogen carbonate ($KHCO_3$) or sodium carbonate (Na_2CO_3) may also be used. Biological filtration in fresh water will establish most rapidly in a pH range of 7.0–8.0 and a temperature between 25 and 30°C (Timmons and Ebeling, 2007).

Denitrification

Denitrification processes, while utilized in some forms of aquaculture, appear to still be an uncommon approach in the zebrafish research husbandry field. As the management and conservation of water resources continues to gain importance, these processes may become more widely incorporated into system designs (Lee et al., 2000).

Husbandry

Water parameter targets and dealing with anomalies

The water parameters for zebrafish husbandry tend to fall within a range rather than adherence to a standardized set point. This may be as a response to local factors such as available water sources, budgetary pressures, or a result of protocol definition by research needs. The increased inclusion of water parameters and husbandry details within the methods of research publications, highlights the widely perceived need for a more standardized approach. Once water parameter target values are established, sudden and large changes should be avoided. Well-designed RAS will include dedicated water monitoring systems able to constantly oversee the main water parameters and deliver dedicated alarm notifications when values are outside preset optimal safe ranges. In most advanced set-ups, these systems will automatically control dosing of buffering solutions to normalize parameters.

The following section looks at the individual water parameters, how they interact, and basic troubleshooting for each parameter.

When problems arise:

1. DON'T PANIC.
2. Take a moment to think before reacting.
3. Cross-check to make sure what you *think* you are seeing is *actually* the case. A rapid response to an apparent problem may make things worse if the cause has been misinterpreted.
4. Withhold feeding during problem evaluation and resolution.

pH

pH is measured on a logarithmic scale that expresses the inverse log of the presence of hydrogen ions (H^+) within a solution. The scale runs from 1 (acid) to 14 (alkaline). The higher the concentration of H^+, the lower the pH reading. A pH reading of 7.0 indicates a neutral solution that is neither acidic nor basic.

pH has effects not only on the animals, but also on the practical efficiencies of biofilters.

Data from wild habitats generally place the zebrafish in slightly alkaline waters (McClure et al., 2006; Spence et al., 2008; Harper and Lawrence, 2011) but seasonal variations in pH occur during monsoonal conditions (Engeszer et al., 2007). An optimal pH for the zebrafish has not yet been determined. While the zebrafish may have a tolerance for a wide pH range, stability within a range of 7–8 is generally considered an appropriate compromise for this species in a research situation (Harper and Lawrence, 2011; Masser et al., 1992; Aleström et al., 2019). Very high (11) or low (4) pH exposures will cause excessive mucous production at the gill epithelium and the lower pH ranges will interfere with the ability of hemoglobin to bind with oxygen (Hammer, 2020; Zahangir et al., 2015).

In an aquatic environment, there are constant pressures from oxidation of ammonia, respiration and decay of organic compounds, driving the pH of the water toward acidic values.

If it becomes necessary to alter pH values, then the process should be done gradually, as alterations of pH levels come at an energetic cost to the animals. pH variations greater than 0.2–0.5 units within a 24-h period should be avoided. Testing of pH should occur at least once per day.

Troubleshooting low pH values

1. Water aeration, surface agitation, and surface films: Poor gas diffusion can lead to a build-up of CO_2, which can drive pH down. Degassing will help to restore proper pH values.
2. KH should be tested, as high inputs of organics into a system may deplete KH reserves. If automated buffer dosing is implemented, then ensure that these are operational. Test dosing pumps and ensure dosing hoses are not blocked or that dosing pump intake hoses have not suffered a puncture, which may cause loss of prime. Very low pH levels will negatively affect biofilters. It is imperative to test ammonia levels prior to raising pH.
3. Check water influx points for potential low pH/low KH water that may contribute to the pH drop.
4. Overfeeding: If a pH drop is associated with an overfeeding event, actively remove as much uneaten food and organics as possible from the system.

Compounds to raise pH include sodium bicarbonate ($NaHCO_3$) and sodium carbonate (Na_2CO_3).

Troubleshooting high pH values

1. Check dosing systems: Dosing system failures are the most common cause of an elevated pH in a zebrafish RAS. A likely cause will be a failure of the KH doser that has allowed excess buffer compound to be added to the system. Check the content of the buffer reservoir to establish if it has been significantly emptied. Test the KH of the water. As KH dosing systems are run by monitoring pH values, then something that affects the reading delivered by a pH probe may cause an automated dosing system to continually deliver buffer.

Compounds to lower pH are monosodium phosphate (NaH_2PO_4), dilute hydrochloric acid (HCl), and dilute phosphoric acid (H_3PO_4). Addition of acidic compounds will also lower KH.

Temperature

Zebrafish are poikilothermic animals and as such temperature affects metabolic rates, growth, appetite, physical activity, waste production, oxygen demand, reproductive processes, sex determination, and more.

Temperature will also affect environmental parameters in the water, with increased temperature directly influencing oxygen saturation (Timmons and Ebeling, pp. 397–410). Higher temperatures decrease oxygen-carrying capacity while causing an increase in the animal's biological demand for oxygen due to an elevation of metabolic rate. The equilibrium of ammonia between ionized and unionized is also impacted by temperature (refer to the "Nitrogenous wastes" section).

The zebrafish is one of the most eurythermal fish on record, but the temperature range for zebrafish culture is most frequently limited to between 23 and 29°C (Matthews et al., 2002; Westerfield, 2007; Harper and Lawrence, 2011). Most facilities the author is aware of target close to 28°C.

Early developmental works, including embryo staging, were conducted at 28.5°C (Kimmel et al., 1995), and staging points seen as standard in zebrafish are based upon this temperature point. 28°C is seen as close to the optimum for growth rates in the species as well (Harper and Lawrence, 2011, p. 110; Schaefer and Ryan, 2006; Hammer, 2020).

Temperature control is a critical factor in the care and stability of a research colony. When designing housing systems it is better to include multiple heating elements within the life support, this will provide some redundancy in the event of failure. In facilities where ambient temperatures are likely to exceed the target set point at times throughout the year, then chiller units may be a requirement to maintain the target temperature. Testing of temperature should occur at least once per day.

Temperature troubleshooting

Low temperature: Unless you are dealing with a species with a very narrow temperature tolerance, a failure of heating elements in a culture situation is likely to be less harmful than a failure of control systems that allows water to overshoot the target temperature. Lower temperatures result in a higher dissolved oxygen content, combined with a decreased demand for oxygen from both the fish and the biofilter. If lowering temperature, keep in mind that the appetite of the fish will decrease in line with the temperature drop and therefore will require adjustments to feed inputs. As a guide, a drop of 5°C will result in a reduction of metabolic rate of around 50% (Dr. Robert Jones, personal communication, 2020).

In the event of a complete failure of heating systems, one approach to maintain temperature is to fill sealed containers from the building hot water supply, these are immersed in the culture water to help maintain temperature. This is practical only in the smallest of systems; if a failure of this kind should occur in a large RAS installation, then adjusting the operation of HVAC systems may provide a short-term fix.

Overheating is much more problematic. As temperature increases, dissolved oxygen decreases rapidly while metabolic oxygen demand is rising. If thermostat failure is suspected, isolate the heating elements until a replacement or repair can be affected. If the high temperature is due to environmental conditions such as a failure of HVAC systems during extreme weather, then a similar approach can be used to the previous paragraph, but with the substitution of ice or very cold water inside the sealed containers. Increasing airflow across the surface of the water will increase evaporation, this decreases the temperature of the remaining water. Heavy water aeration or surface agitation will additionally aid evaporation.

Total hardness

The language traditionally used around hardness can be potentially confusing: total hardness is a measure of the combined carbonate hardness (KH) and general hardness of water.

Carbonate hardness (KH) is sometimes referred to as alkalinity, or temporary hardness. If a sample of water is boiled, the KH components will be predominantly driven off, hence the "temporary hardness" reference. General hardness (GH) is occasionally referred to as permanent hardness as boiling does not change this significantly.

Carbonate Hardness (KH)

KH is closely interlinked to the pH of water, as it is a measure of bases available that are able to neutralize acids. KH can be thought of as the buffering capacity of water and its resistance to changes in pH.

KH is most often expressed in mg/L (ppm) of calcium carbonate ($CaCO_3$), but it may also be expressed in dKH (German degrees hardness) or meq/L (milliequivalents per litre) depending upon the manufacturer of the testing equipment (1^0 dKH = 18 mg/L) (1 meq/L = 50 mg/L) (Timmons and Ebeling, 2007, p. 56). The most important components of KH in freshwater are carbonates, bicarbonates, and hydroxides (Hammer, 2020). With the constant creation of acids there is a corresponding depletion of KH. The KH level will determine if the pH change is significant or not under any given loading and feeding regime. With ammonia oxidation using carbonates at a 7:1 ratio it is very important to ensure a KH reserve is maintained, particularly in an aquatic system where stocking densities and feeding rates are high. A suggested target range of 50–75 mg/L is appropriate, and a higher level may not be detrimental if the pH is stable (Hammer, 2020, p. 327).

Testing of the KH value should be undertaken at least weekly. Most dedicated zebrafish RAS builders include some form of automated dosing systems that monitor pH and will deliver sodium bicarbonate solution into the water in a measured manner to help replenish depleted carbonates. It is important that these systems are cleaned and calibrated in accordance with manufacturers' recommendations both in manner of servicing and frequency.

Aragonite, crushed coral skeleton or crushed oyster shell, are all potential sources for increasing both GH and KH within an aquatic system. Often a small amount of calcareous material will be included somewhere in the water flow within the system. There is a release of carbonates in this reaction contributing to the temporary hardness (KH), though typically the amount of carbonates released is not enough to meet all the needs of the system. Calcareous materials will dissolve slowly if the pH is below 7.8. Sodium bicarbonate additions are by far the most common method of maintaining the KH within a zebrafish system. Other compounds that may be used for raising KH include sodium carbonate (Na_2CO_3) and potassium hydrogen carbonate ($KHCO_3$).

General hardness

GH is a measure of divalent ions within the water, these being predominantly calcium and magnesium, but selenium, manganese, and iron also play a role (Hammer, 2020; Harper and Lawrence, 2011). GH of naturally occurring water bodies will be variable due to the local geography and geology.

A reliance on groundwater or municipal water supplies in an area with a high limestone geology, may result in naturally high hardness levels in incoming water. If using reverse osmosis or deionized water as a basis for system water, there will be minimal to zero hardness ions in the incoming water, and these ions will need to be added to the water separately. Research the specific needs of the subject species.

The ions involved in GH support important biological functions within fish and need to be supplied either from the culture water or from food sources, with some diet manufacturers adding specific mineral blends. Calcium is required for bone and scale development, blood clotting, brain function and muscular and neurological processes (Wurts, 2002; Hammer, 2020). Maintenance of GH levels can be achieved in many ways. When using calcareous materials, solubility can be accelerated using a reaction chamber that allows fluidization, or enhanced dissolution of the aragonite by the addition of small amounts of carbon dioxide. Use of these "calcium reactors" is not a common approach in zebrafish husbandry but may be worth evaluating in large-scale systems or with hard water species.

Most automated dosing systems will add calcium and magnesium as components of artificial sea salts that are dosed to control conductivity. It is possible that these additions will not provide sufficient quantities of calcium and magnesium. The ratio of sodium to calcium in these artificial sea salt blends is also potentially skewed beyond what may be found naturally in freshwater environments. Testing of GH should be performed weekly. As the optimum hardness level is not well defined for the zebrafish, a target range of 75−200 mg/L is generally accepted for zebrafish culture (Hammer, 2020; Harper and Lawrence, 2011; CHEN et al., 2003; Lawrence, 2007; Wurts, 2002).

Specific targeting of calcium, magnesium, and other mineral profiles can be undertaken in culture environments by blending mineral "conditioners" to match the natural environment of a subject species. This type of species/biotope-specific mineral blend is commercially available from numerous manufacturers. Some compounds used for this type of conditioning blend include calcium chloride ($CaCl_2$), sodium chloride (NaCl), and magnesium sulfate ($MgSO_4$).

Conductivity/salinity

Conductivity is a measure of the ability of water (or any material) to conduct electricity, this is directly related to the total dissolved ions within the water and this includes the ions involved with hardness. Conductivity is measured in a unit called micro-Siemens per centimeter (μS/cm).

Ultrapure water will have a conductivity reading of close to zero μS/cm, whereas seawater will be approaching 50,000 μS/cm or above. The target range for conductivity will vary with species. Zebrafish will tolerate very low conductivities of 0.2 μS/cm but this is not conducive to egg production and also has impacts on other metabolic processes of the animal (Uliano et al., 2010), (Boisen et al., 2003). Zebrafish are found in ranges from 10 to 2000 μS/cm and in laboratory conditions are generally held within a range of 200−3000 μS/cm (Harper and Lawrence, 2011). Maintaining fish at lower conductivities has an energetic cost for the animal, as freshwater fish live in a hypotonic solution and constantly need to excrete water as dilute urine to enable them to maintain osmotic balance. Increased salinities are sometimes employed to reduce the osmoregulatory demands during times of stress such as transport and disease outbreaks (Harper and Lawrence, 2011). FELASA guidelines suggest a range between 150 and 1700 μS/cm.

The unit part per thousand (ppt) is often used when measuring dissolved salts: 1 ppt = 1,000 mg/L.

Larvae are regularly raised in a brackish salt solution of 3–5 ppt in a polyculture with *Brachionus plicatilis* rotifers (Best et al., 2010).

Conductivity troubleshooting: Generally, addressing conductivity issues is straightforward. In the case of elevated levels, check that the water change regime is being followed and there are no problems with dosing systems overrunning or syphoning content from buffer dosing vats. Check that buffer dosing vats contain the correct solutions. For low conductivities, again check buffer vats for content and dosing systems for operation and ensure that there is not an uncontrolled inflow of low conductivity water.

Dissolved gases

Dissolved oxygen

Dissolved oxygen is one of the most limiting factors within an aquatic culture environment. While atmospheric oxygen is present at around 21% (210,000 mg/L), it has relatively low solubility in water. Dissolved oxygen is most often expressed as a percentage of saturation rather than as mg/L and saturation is interrelated to temperature, atmospheric pressure and altitude.

One hundred percent saturation level of dissolved oxygen in water under typical zebrafish culture conditions equates to around 7.7 mg/L.

The gill membranes of a fish are well adapted to obtain oxygen at these levels, but this does come at a higher energetic cost compared to a terrestrial animal.

Oxygen consumption by both the respiratory process of our culture animals and the biofilter will drive this level downward constantly. RAS engineers will incorporate methods of ensuring that dissolved oxygen levels remain close to saturation during normal operation within a system design. For those looking to create "build your own" solutions, the inclusion of heavily cascading overflows within the water path is beneficial, as are the use of degassing towers or areas of a technical sump that include heavy aeration. When using "home aquaria," or nonrecirculating systems, it is critical to ensure that there is direct agitation of the water surface to ensure adequate gas exchange can occur. Aeration or direction of powered filtration outlets across the surface of water will prove helpful. Be aware that some diets can result in the accumulation of lipid films that blanket the surface of a tank and can impinge on efficient gas exchange. This in turn can result in a low pH value as carbon dioxide accumulates in the water (Table 2.1).

TABLE 2.1 Chart of dissolved oxygen in mg/L at 100% saturation in fresh water and brackish water.

Saturation of dissolved oxygen in mg/L in fresh water and at 7.2 ppt salinity					
Temperature (°C)	0 ppt salinity	7.2 ppt salinity	Temperature (°C)	0 ppt salinity	7.2 ppt salinity
1	14.24	13.54	16	9.89	9.44
2	13.84	13.18	17	9.67	9.26
3	13.45	12.84	18	9.47	9.07
4	13.09	12.51	19	9.28	8.88
5	12.75	12.18	20	9.11	8.7
6	12.44	11.86	21	8.93	8.54
7	12.13	11.58	22	8.75	8.38
8	11.85	11.29	23	8.60	8.22
9	11.56	11.02	24	8.44	8.07
10	11.29	10.77	25	8.27	7.92
11	11.05	10.53	26	8.12	7.78
12	10.80	10.29	27	7.98	7.64
13	10.56	10.07	28	7.84	7.51
14	10.33	9.86	29	7.69	7.38
15	10.10	9.64	30	7.56	7.25

Adapted from information in Timmons, M.B., Ebeling, J.M., 2007. *Recirculating Aquaculture.* Cayuga Aqua Ventures, Ithaca New York, p. 52. (Table by Bruce Newell, Deakin University).

Carbon dioxide

Carbon dioxide (CO_2) forms carbonic acid (H_2CO_3) when dissolved in water. CO_2 accumulations may become problematic in tanks where there is an inadequate disruption of surface tension to allow for adequate gas exchange. Carbon dioxide is produced in aquatic environments not only by the respiration of the fish, but also from the breakdown of organic materials and the acid–base reactions that occur within an aquaculture system (Summerfelt et al., 2000). Carbon dioxide has a relatively high solubility in water, and accumulation can cause physiological problems for teleosts, including interference with the ability of blood to transport oxygen. CO_2 has an anaesthetic effect on fish as levels approach 15–20 mg/L (Timmons and Ebeling, 2007; Hammer, 2020). It is best to maintain levels as low as possible, and certainly below 10 mg/L. Heavily aerating water to disrupt surface tension will degas CO_2 from the water.

CO_2 levels can become elevated in a transport container resulting in a lowering of pH in the transport water. A lower pH reduces toxicity of ammonia by ionization. If ammonia has accumulated during transport, this can rapidly become more toxic when the transport container is opened and the CO_2 disperses. Some shippers may include "biofilter starter culture" in the transport container to help control ammonia accumulation in this phase.

Gas supersaturation

There are situations where atmospheric gases will be present in the water at levels higher than would naturally occur. Supersaturation events can rapidly cause lethal damage to your animals as the oversaturated gases come out of solution within the fish's tissues forming emboli that block capillaries and damage tissue. This is commonly known as Gas Bubble Syndrome (Fig. 2.3).

Total gas pressure (TGP) monitoring devices are available and included in the most advanced and professionally built RAS to constantly monitor gas saturations. TGP devices can be programmed to automatically shut down systems as a response to elevated total dissolved gas pressure levels.

The critical action when a supersaturation event is suspected, is to immediately stop water circulation by shutting down system pumps. Examine water levels within the sumps to ensure that air is not being drawn into the intake of pumps. Provide agitation to the housing tanks to assist in off-gassing the excess dissolved gases. Carefully examine the plumbing on the intake side of pumps with particular attention to alignment of fittings and presence and condition of O-rings and security of barrel unions. Signs of gas supersaturation may include small bubbles forming on tank surfaces, microbubbles visible in the water flow from the tank inlets, and the appearance of emboli within tissues of deceased animals.

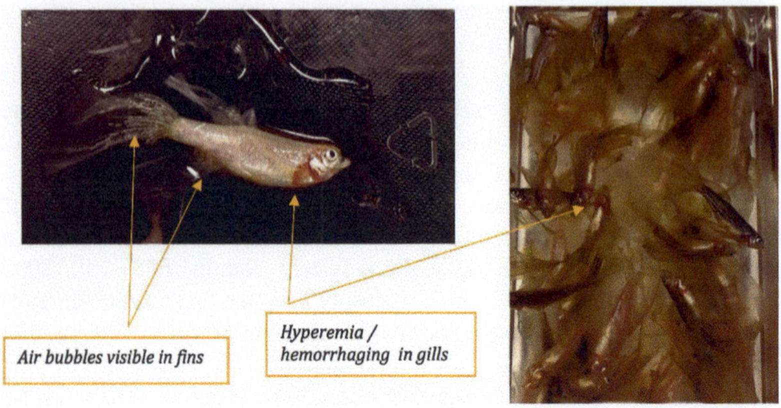

FIGURE 2.3 Fish exhibiting gross clinical signs of a gas supersaturation event. Note the visible emboli in the fin tissue and the inflammation around the opercula. *Photo: Gillian Lawrence. University of Queensland.*

Nitrogenous wastes

As seen in biological filtration, nitrogenous compounds are normally found in RAS in three forms: ammonia TAN (NH_3/NH_4), nitrites (NO_2), and nitrates (NO_3).

It is suggested to test nitrogenous waste levels at least once per week and more frequently if problems are experienced, biofiltration is being established, or when large increases in stocking densities occur.

Ammonia is the principal source of waste from teleost metabolism and is primarily excreted across the gill epithelium by passive diffusion (Schreier et al., 2010; Shih et al., 2008; Wilkie, 2002). Ammonia is also formed from the breakdown of organic materials such as animal tissue, foods, and fecal material.

Ammonia is found in solution both as ionized ammonium (NH_4) and unionized ammonia (NH_3), in an equilibrium influenced predominantly by temperature and pH. The combined summary of both the ionized and unionized forms is referred to as TAN (Total Ammonia Nitrogen) or ammonia-N.

Most test kits will provide results in TAN and this figure must be read in conjunction with conversion charts such as Table 2.2.

Ionized ammonium is relatively harmless to fish life below certain concentrations (Eshchara et al., 2006). Unionized ammonia is far more toxic to fish and must be removed from the culture media. Levels exceeding 0.05 mg/L are generally considered problematic for aquatic animals.

To use Table 2.2 to calculate the unionized ammonia content in a sample of water with TAN 1.2 mg/L, pH 7.2, temperature 26°C, multiply the TAN value by the conversion factor 0.0096. This results in a NH_3 concentration of 0.0015 mg/L. The same example at a pH of 8.4 results in a NH_3 concentration of 0.159 mg/L, which is far more problematic. TAN should be maintained as low as possible. Unionized ammonia should not be allowed to exceed 0.05 mg/L and overall TAN should not be allowed to exceed 1 mg/L (Timmons and Ebeling, 2007). If TAN readings exceed 0.25 mg/L on a regular basis, investigate possible causes, and cross-check the testing equipment you are using.

Nitrite

Nitrite accumulation is more problematic in freshwater than in marine environments as chloride and hydrocarbonate ions help to mitigate some of the effects of nitrite toxicity. Nitrite should be maintained as close to 0 mg/L as practicable and if levels approaching 0.5 mg/L are detected, then the underlying cause should be identified. Nitrite crosses the gill epithelium and binds with hemoglobin forming methemoglobin impinging on the ability of the blood to carry oxygen to bodily tissues. Methemoglobin turns blood into a brownish color as the nitrite preferentially binds to the hemoglobin precluding oxygen uptake. Nitrite toxicity can cause respiratory distress and possibly death.

TABLE 2.2 Ammonia equilibrium factors at various temperatures and pH.

pH	Temperature Celsius								
	16	18	20	22	24	26	28	30	32
7.00	0.0029	0.0034	0.0039	0.0046	0.0052	0.0060	0.0069	0.0080	0.0093
7.20	0.0046	0.0054	0.0062	0.0072	0.0083	0.0096	0.0110	0.0126	0.0150
7.40	0.0073	0.0085	0.0098	0.0114	0.0131	0.0150	0.0173	0.0198	0.0236
7.60	0.0016	0.0134	0.0155	0.0179	0.0206	0.0236	0.0271	0.0310	0.0369
7.80	0.0182	0.0211	0.0244	0.0281	0.0322	0.0370	0.0423	0.0482	0.0572
8.00	0.2860	0.0330	0.0381	0.0438	0.0502	0.0574	0.0654	0.0743	0.0877
8.20	0.0445	0.0514	0.0590	0.0676	0.0772	0.0880	0.0998	0.1129	0.1322
8.40	0.0688	0.0790	0.0904	0.1031	0.1171	0.1326	0.1495	0.1678	0.1948
8.60	0.1048	0.1197	0.1361	0.1541	0.1737	0.1950	0.2178	0.2422	0.2768
8.80	0.1566	0.1773	0.1998	0.2241	0.2500	0.2774	0.3062	0.3362	0.3776
9.00	0.2273	0.2546	0.2836	0.3140	0.3456	0.3783	0.4116	0.4453	0.4902

From information in Emerson, K., Russo, R., Lund, R., Thurston, R., 1975. Aqueous ammonia equilibrium calculations: effects of pH. J. Fish. Res. Board Can. 32 (12), 2379–2383. (Table by Bruce Newell, Deakin University).

Nitrate

Zebrafish have been shown to be moderately tolerant of nitrate at levels up to 200 mg/L in the larval rearing period (Learmonth and Carvalho, 2015). Nitrate has a similar hemoglobin-binding quality to nitrite (Camargo et al., 2005), but nitrate does not cross the gill epithelium as easily as nitrite, hence exhibiting lower toxicity than nitrite (Hammer, 2020, p. 330). Nitrate levels should also be maintained as low as practicable and in the absence of denitrification strategies, dilution by water exchange is the most common approach. Where water resources are particularly scarce, then denitrification strategies should be pursued. The recent FELASA guidelines for zebrafish culture suggest maintaining nitrate levels at less than 25 mg/L (Aleström et al., 2019).

Nitrogenous waste troubleshooting

"The solution to pollution is dilution": Water changing can assist greatly in reducing overall concentrations of an unwanted compound. Ammonia and nitrite levels can become elevated due to many different causes such as overfeeding, breakdown of the biofilter, and massive pH changes. Numerous commercially available compounds will bind ammonia and nitrite and reduce acute toxicity. As always, research the contents of the products you keep for these contingencies, and ensure they will not adversely affect research if you need to use them. In an emergency situation these can prove very valuable. Products that may be of value in your emergency kit include Prime (Seachem Industries Inc., 2020) and Ammo Lock (Mars or Affiliates, 2020). Zeolite in any of its commercially available forms, or other compounds capable of controlling ammonia levels, may be used to mitigate the effects of ammonia toxicity while remedial actions are undertaken.

Important actions for reducing the severity of ammonia toxicity are pH reduction close to neutral, as this will improve the ammonia/ammonium equilibrium in your favor. Cease feeding for the short term while investigations and remedies are put into place and accelerate the normal water change program. While dilution is an effective method to reduce concentrations, be aware of the pH of the incoming water. It may be prudent to allow the temperature to drop in the system by a couple of degrees if it is expected that the resolution will take some time. This will lower the metabolic rate of the animals reducing ammonia output, as well as having marginal improvements on the equilibrium reaction. As nitrite toxicity is mediated by the sodium concentration in the culture water, increasing conductivity levels may have a slightly beneficial effect.

Chlorine

It is not unheard of to have free chlorine breakthrough into tanks either by engineering failure, human error, unanticipated surge dosing of chlorine/chloramine by municipal water suppliers, or failure of prefiltration systems for incoming water. The most effective treatment for a chlorine breakthrough is the use of sodium thiosulfate ($Na_2S_2O_3$) solution. Sodium thiosulfate will rapidly bind up chlorine when dosed at a rate of 7 mg/L per mg/L of free chlorine present. If the source problem is a breakthrough of chloramine, then there will be a secondary ammonia problem to deal with. Previously mentioned ammonia blockers may be helpful in this situation. When utilizing a municipal water source, research the disinfection agent utilized in your locality. Strong aeration will disperse chlorine, but it is mostly ineffective against chloramine.

Lighting

Zebrafish are crepuscular spawning animals that are easily manipulated by regular lighting patterns.

The most common practice is to utilize a 14:10 light:dark cycle and it is important when designing facilities that the main housing space is protected from unwanted light sources (Adatto et al., 2016; Villamizar et al., 2014; Di Rosa et al., 2015). Pay particular attention in the design phase to eliminate unwanted light sources such as exit signs and backlit light switches and displays. Ensure that the building engineers understand that you need to create a "darkroom" in the animal housing zone. This may require that you receive exemptions from certain building regulations around exit sign illumination.

With LED lighting becoming far more common in installations, there is an opportunity to manipulate light intensities easily. The authors facility uses a system with gradual dawn and dusk features as well as the ability to very gradually ramp intensities automatically when the room is not occupied for long periods. When choosing LED lighting pay attention to the color temperature of the unit (° kelvin). To achieve good color rendition and contrast, recommendations are for a light with a color temperature from 5000 to 6500 °K. Targets of between 54 and 334 lux at the tank face are suggested (Matthews et al., 2002; Aleström et al., 2019). It is a good idea to put a data logger into the tank holding spaces to ensure the night/day cycle is not interrupted as this can have highly detrimental effects on fish physiology and breeding performance.

Feeding

Feeding and nutrition is an area undergoing considerable research and development (Fowler et al., 2020).

In reality, it is not at all clear that feeding practices that are typically in use in zebrafish research settings today achieve the stated goal of producing and maintaining "representatively normal, healthy" research subjects. One primary reason for this is that the nutritional requirements of zebrafish are still largely unknown (Watts et al., 2012, 2016).

The original zebrafish husbandry works undertaken by Streisinger relied upon commercial flake foods. These flake foods are still utilized in some zebrafish facilities and may be more common with other species. Flake foods will quickly leach water-soluble vitamins into the water column in comparison to pelletized, crumbled, or encapsulated food stocks. Due to a high surface/volume ratio, flake foods will tend to oxidize more rapidly. For these reasons flake foods have to a large degree been replaced in the zebrafish field. Feeds utilized for laboratory fish will vary depending upon species requirements and availability in any particular locale.

At the time of writing there are a number of suppliers providing zebrafish-specific diets; however, none have achieved a universal uptake within the research community. A nutritionally defined, purified, open-source reference diet has yet to be fully developed for the zebrafish in the same way that has been undertaken for other research models used in biomedical settings, e.g., rodent models, but work on this is continuing. Nutritional studies clearly demonstrate that variations in dietary ingredients and corresponding nutrient composition among commercially available diets can affect growth, reproduction, disease, and consequently response to experimental manipulation (Fowler et al., 2019; Smith et al., 2013; Siccardi et al., 2009; Gonzales, 2012; Gonzales and Law, 2013; Markovich et al., 2007) (Table 2.3).

For some methods of delivering dry and live foods to zebrafish in controlled manners, see the "New and emerging technologies" section.

Live foods

Live foods are an area constantly undergoing review. Traditionally there has been a high reliance on paramecium, *Artemia*, and rotifers in the culture of zebrafish and other species. This has been predominantly due to a combination of the balanced nutritional content, digestibility, and small gape size of the zebrafish at the onset of exogenous feeding (Kaushik et al., 2011).

Some microparticulate dry foods have been demonstrated to provide reasonable growth and survival levels, and some facilities use dry feeds exclusively.

There are some potential benefits from an environmental enrichment point of view in feeding live foods and this features as a suggestion in the FELASA guidelines (Aleström et al., 2019).

Organisms used for live feeding can serve as concentration vectors for pathogens such as *Mycobacteria marinum* (Peterson et al., 2013; Chang et al., 2019). There are a growing number of facilities moving away from the use of live foods, particularly of *Artemia*, as a response to the biosecurity issues presented by the use of a wild collected organism in a research environment. Rotifer cysts that have been produced in cultured conditions and undergone surface disinfection with ethanol are available (Florida AquaFarms Inc., 2016). Decapsulating *Artemia* can reduce the risk to a degree by exposing the cysts to

TABLE 2.3 **A range of feeds commonly in use in the zebrafish field can be seen here: while some of these feeds are manufactured specifically for research using zebrafish, others are adapted to use from other finfish aquaculture applications. Commercial aquaculture for finfish tends to be targeted at maximum food conversion to body mass and it is unclear that this is the best strategy for a laboratory animal.**

Manufacturer	Product name	Size range (μm)	Format	Production location
DanioLab	ZRD-1	80–100	Granule	Los Angeles, USA
	ZRD-2	100–200	Granule	Los Angeles, USA
	ZRD-3	200–300	Granule	Los Angeles, USA
	ZRD-4	300–500	Granule	Los Angeles, USA
INVE	NRD 3/5	300–500	Granule	Nonthaburi, Thailand
	NRD 5/8	500–800	Granule	Nonthaburi, Thailand
	NRD G8	800	Granule	Nonthaburi, Thailand
	NRD G12	1200	Granule	Nonthaburi, Thailand
INVE (O-Range)	START-S	100–200	Granule	Nonthaburi, Thailand
	START-L	200–300	Granule	Nonthaburi, Thailand
	WEAN-S	200–400	Granule	Nonthaburi, Thailand
	WEAN-L	300–500	Granule	Nonthaburi, Thailand
	GROW-L	500–800	Granule	Nonthaburi, Thailand
Marubeni Nisshin Feed Company	Otohime A1	75–150	Granule	Japan
	Otohime A2	150–250	Granule	Japan
	Otohime B1	250–360	Granule	Japan
	Otohime S1	600–1400	Granule	Japan
Skretting	Gemma Micro 75	50–100	Granule	Fontaine Les Vervins, France
	Gemma Micro 150	100–200	Granule	Fontaine Les Vervins, France
	Gemma Micro 300	200–500	Granule	Fontaine Les Vervins, France
	Gemma Micro 500	400–700	Granule	Fontaine Les Vervins, France
Sparos	Zebrafeed	<100	Granule	Olhao, Portugal
	Zebrafeed	100–200	Granule	Olhao, Portugal
	Zebrafeed	200–400	Granule	Olhao, Portugal
	Zebrafeed	400–600	Granule	Olhao, Portugal

TABLE 2.3 A range of feeds commonly in use in the zebrafish field can be seen here: while some of these feeds are manufactured specifically for research using zebrafish, others are adapted to use from other finfish aquaculture applications. Commercial aquaculture for finfish tends to be targeted at maximum food conversion to body mass and it is unclear that this is the best strategy for a laboratory animal.—cont'd

Manufacturer	Product name	Size range (μm)	Format	Production location
Zebrafish Management	ZM-100	80–200	Granule	Hampshire, United Kingdom
	ZM-200	200–300	Granule	Hampshire, United Kingdom
	ZM-300	300	Granule	Hampshire, United Kingdom
	ZM-400	400	Granule	Hampshire, United Kingdom
	ZM-500	500	Granule	Hampshire, United Kingdom
Zeigler	Zebrafish Feed	1 mm	Granule	Gardners, Pennsylvania, USA
	Aquatox	Variable	Flake	Gardners, Pennsylvania, USA
	AP100 1	<50	Granule	Gardners, Pennsylvania, USA
	AP100 2	<100	Granule	Gardners, Pennsylvania, USA
	AP100 3	100–150	Granule	Gardners, Pennsylvania, USA
	AP100 4	150–250	Granule	Gardners, Pennsylvania, USA
	AP100 5	250–450	Granule	Gardners, Pennsylvania, USA

Produced with kind assistance from Carrie Barton. (Table by Bruce Newell, Deakin University).

sodium hypochlorite and sodium hydroxide; however, this is an exothermic reaction that carries inherent occupational health and safety issues. The chorion of *Artemia* is relatively indigestible and can cause gut impaction particularly in juvenile animals (Fig. 2.4).

Feeding frequency

As ad libitum feeding of fish is problematic in most cases beyond larval stages, the most commonly adopted approach is to feed to satiation or to a ration, either singly or multiple times per day. Recent work suggests that feeding routines affect behaviors and anxiety responses in fish (Dametto et al., 2018).

Adult zebrafish have been shown to perform well on a single larger feed per day (Lawrence et al., 2012); however, many facilities feed smaller feeds at multiple points during the day. In some instances, it is possible to automate this process with the use of robotics. This can be particularly helpful in raising juvenile animals

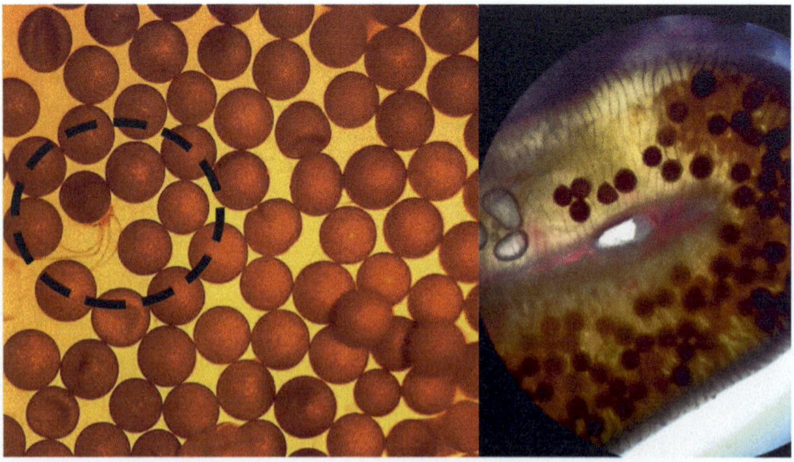

FIGURE 2.4 (L) A photograph of a *Hydra* sp. organism attached to the outer chorion of a hydrated *Artemia* cyst (*circled*). (R) Unhatched *Artemia* cysts in the gut of an adult zebrafish. *(L) Photo Carrie Barton, Sinnhuber Institute. (R) Photo Marco Brocca. Tecniplast.*

as multiple small feeds can be delivered over the course of a day (Erik Sanders, personal communication, November 2020; Tecniplast SpA, 2017).

A previous section mentioned withholding foods as a response to system failures and upsets in biological filtration where a reduction in ammonia production would be advantageous. This is an approach that should only be considered for very short timeframes and only in an emergency situation. Fasting zebrafish for between 7 and 21 days is an area that has undergone work in the past (Jia et al., 2019; Ahi et al., 2019; Meyer et al., 2013).

There appears to be variations in responses to fasting between genetic lines in the zebrafish, but overall, in an emergency situation, fasting animals for up to 48–72 h is unlikely to cause long-term problems to animals. It is best to achieve a speedy resolution to the problem and resume normal feed as quickly as possible. A consideration is that the resilience of very young animals to fasting will be appreciably lower than it would be for older animals.

Maintenance programs

> *Preventative maintenance safeguards against equipment failure and can avoid potential loss of valuable research animals.*
>
> Pandolfo and Aliucci (2020).

There is a need to maintain mechanical and biological systems to ensure reliable and optimized operation. This is of course particularly critical when dealing with systems that are responsible for the life support of our animals. The specifics of any maintenance program will vary depending on design and implementation of the venture.

It is important that there are thorough and accurate records of both routine maintenance and repair works. The age of existing equipment and critical spare part stock holdings are to be given proper consideration as well. These records can be helpful to create a more robust maintenance program. Record keeping can take many forms and does not need to be a centralized record set. Some aspects such as environmental monitoring of the housing rooms, water parameter test results, feed records, etc. may suit being recorded in a log book within the facility, whereas records of service visits from technicians, changing of UV tubes and pumps, etc. may be more appropriately maintained in an electronic log where events with a lower frequency can be easily discerned or searched for. The approach and the tools used to achieve the outcome are open to individual preference of those designing the maintenance program.

Whatever approach is undertaken, precise records will enable detailed running costs and efficiencies to be evaluated, predictions to be made, and give foresight of problems that may develop.

A preventive rather than a reactive maintenance program is best.

Timely replacement of equipment may prevent an avoidable failure that could threaten the life of research animals; however, replacement of equipment that still has a long service life is wasteful and counterproductive. It is important to carry a range of spare parts on hand for items that will cause serious problems in the event of a failure, but it is unrealistic for most facilities to maintain a complete inventory of every part that could ever fail.

Holding spare parts on a shelf for a long period of time can be detrimental as equipment may perish while in storage and provide a false sense of security. Poor storage conditions could cause bearings or seals in a pump to seize or pit and the unit may be nonfunctional when needed. Equally, rubber O-rings and seals stored for years may not be in a usable condition if they have been stored in a poor environment.

Ideally, systems should have a level of functional redundancy built in, particularly around the key items of water parameter management. Having two pumps capable of supplying most or all the water needed, stands a far better chance of avoiding loss of animals in the event of a pump failure. This approach allows for the regular changeover of the pumps to avoid excessive wear on one unit in a "lead/lag" arrangement. When a pump needs to be serviced in this arrangement, there should be no impact upon the performance of the system. Similarly, if heating is supplied from two separate heating elements, the chance of both of those failing at the same time is much lower than that of a single unit.

Evaluating the duty cycle of equipment is important. Most manufacturers of quality equipment will provide documentation of the life expectancy of the equipment they produce. Plan to replace equipment before the predicted failure is reached. This may not be a straightforward process when equipment does not operate continually. A dosing pump may only operate for a total of 3 or 4 min a day; the manufacturer may state the diaphragm has a life expectancy of 3000 h. At 3 min per day, this amounts to a predicted working life of over 160 years. This would influence the frequency of service intervals and whether you would hold spare parts.

Some degree of risk evaluation and management must be taken into account. When replacement parts are readily and quickly available, the impetus to carry stock is lower. If parts are scarce, hard to come by, or may involve considerable transit times, then the impetus to have spares to hand increases.

It is important to factor in repair costs and spare part inventories in the original and ongoing equipment budgets.

One essential part of a maintenance program that is often overlooked is the experience of the operators within the space. An observant, well-trained, and interested operator who has been taught to seek out changes in operation of equipment will potentially spot an emerging problem long before it becomes a critical failure. Noticing small changes such as a shift in audible tone or operating temperature of a pump can make a huge difference. Encourage staff to observe, investigate, and report these changes.

Continuity planning

Anything that can go wrong, will go wrong.

Continuity planning or "disaster planning" warrants extensive consideration and preparation. Every facility and program are different; two facilities equipped in exactly the same manner are unlikely to face exactly the same challenges, so each facility continuity plan will be unique. There are obvious recurrent themes seen between different facilities; a series of prompts and scenarios that highlight the major points to consider is presented.

Human

The first goal must always be the health and welfare of the people involved, with animal welfare a very close second. Legislative guidelines exist that set out minimum requirements for human occupational health, welfare, and safety, as well as animal welfare requirements and it is important to be familiar with the relevant local requirements.

Healthy, well-cared for animals that are managed in a repeatable and well-documented manner will help the third goal, which is research. Avoidance of nonprotocol-induced variations is a key factor in the gathering of robust data. Ensuring rapid, effective responses to problems will reduce these variables in research.

The more complex a system is, the greater the number of components that may individually break, fail, or not perform to expectations. Proactively seek out these potentials for failure and plan effective responses before a failure occurs. This applies not only to mechanical systems but also to human systems. Develop a proactive relationship with the local service engineer of your system manufacturer. A collaborative approach assists in knowledge sharing and helps strengthen understanding of system dynamics.

Ask "What might go wrong, and how can those eventualities be best prepared for?"

From a human point of view, consider: What motivates the people involved in the research; where do they live; are they traveling long distances; if so, how? During environmental challenges such as extreme weather events or natural disasters, who among the team is going to be able to help out, and are they capable of filling gaps that may be left in the workgroup?

Environmental conditions and natural threats are widely variable depending on location. Floods, storms, blizzards, dust storms, bushfires, hurricanes, and earthquakes are all realities for people living in various parts of the world. Consider the local exposures; with changes occurring in long-term climate patterns, adverse weather events are expected to become more frequent and severe. Search the historical record for the locale. Consider local history to determine probabilities of different disasters. At the time of writing, the impact of public health emergencies is being highlighted.

Evaluate the skills that each team member possesses and identify overlaps or potential gaps.

Researching this part of your continuity plan highlights the need for effective and useful standard operating procedure (SOP) documents that will enable tasks to be completed by operators unfamiliar with a task should the need arise. Detailed, functional SOP documents are invaluable in an emergency situation. Pictorial guides and even video guides can be very useful for a novice. Technologies such as quick response codes that link to web or cloud-hosted videos can be very useful, particularly for complex tasks. These should not form a replacement for well-written documents as the technology has multiple points of potential failure, but can serve as a useful adjunct.

Ensure that the people who you may come to rely upon in emergency situations are aware of what you do. Having technical and facility support people who understand the priorities required in keeping a research colony alive and healthy is very important. Also consider the services you may not think of straight away. Having local emergency services on side when things go severely wrong may make a difference.

Mechanical and infrastructure

From a mechanistic point of view, you have a great reliance on factors that are totally or partially outside your control. Some of these factors may include:

Continuity of electrical supply, communications channels, reticulated water supply. You may suffer supply chain problems for items that you normally take for granted, or failure of mechanical components within your housing and supply systems, shutdowns of HVAC systems etc.

Electrical supply continuity is a major issue: The ability to withstand loss of power without losing animals will vary based on system design, stocking densities and other factors. It is highly recommended to ensure that a power supply such as a generator, which is independent of the local grid, can be connected to ensure life support system continuity. If your facility has its own backup generator, ensure it is serviced regularly and fueled. If you share a backup system with other users, then make sure all are aware of what you do and why continuity is critical for you. Load shedding is often required, so make sure that your critical equipment is not sidelined by developing a predetermined hierarchy of importance. If you are running a very small-scale facility or are in a situation where you do not have facility support, consider purchasing a small generator capable of running the essentials and test the generator regularly.

Load-shedding priorities: Consider reducing HVAC support first, as this is generally a high-energy consumption item. The next high-energy consideration would be water heating; water has a high thermal inertia and a gradually decreasing water temperature will slow down the metabolism of the animals with an attendant drop in ammonia production and oxygen demand. The highest priority for power continuity is biofilter mechanisms and water circulation pumps to maintain dissolved oxygen levels and minimize loss of biological function.

Reticulated water supply: There are likely to be many points of failure to consider. If you are reliant on reverse osmosis water, how will you cope with a failure of your equipment? If you are on well or artesian water supplies, do you have a plan if you experience quality issues with this water? If you use municipal water, do you have systems in place to provide a buffer for a period? Are you familiar with the supply systems, and who do you contact when things go wrong? Document and make it a regular timed event to revise your plan and contacts. Make direct contact with the people who control your water supply and let them know your reliance on the water. This contact may result in some degree of forewarning of upcoming changes to your incoming water supply.

Spare parts

Equipment failure: Identify what is most likely to fail in your system. Develop an understanding of how your systems work. Have a set of spares available on hand for the critical items. A fractured quartz sleeve on a UV unit can be a simple thing to replace if you have a spare. At the start of a long weekend or protracted holiday break when no spares are available, the system may be compromised for a significant period of time. Likewise, plan for how a failure of a main circulation pump can be addressed. If you are designing a system for a "self-build," it is critically important to include the ability to bypass sections of the filtration unit in a breakdown situation.

Spare parts to keep in stock should include: Spare UV tubes and quartz sleeves, main circulation pump if there are no inbuilt redundancies, an emergency repair kit that contains spare barrel unions, O-rings, spare valve bodies, and a quantity of replacement hoses and fittings for pressure applications. Spare heating elements and fuses, a spare air pump rated to your system, a spare dosing pump and connection hoses can all be very valuable. If the system is reliant on an integrated electronic control system, then consider having a spare power supply to hand.

It is important to have an independent method of checking pH, conductivity, and temperature. If these are digital instruments, then they must be regularly maintained and calibrated, so they are ready and reliable when needed. A well-thought-out "emergency kit" can be invaluable.

Talk to the system manufacturer, to other facility operators, and to your staff and facility support people. Ask both simple and complex questions and pay attention to the responses.

Continuity planning should consider the value of a well-organized cryopreservation program. It is possible to do this in house if you have the resources, and protocols can be found through ZIRC (Matthews et al., 2018). ZIRC also provides line preservation services (ZIRC, 2020). There is at least one commercial provider of cryopreservation services working in the zebrafish field who can service facilities on site in some countries (Cryogenetics AS Hamar, Norway, 2020).

New and emerging technologies

There are emerging technologies and products that are currently making impacts within the aquatic field, or look to hold promise as they are developed. As space is limited for this section, pointers to investigate have been provided, if these are of interest. In most cases, web addresses can be found in the references section.

As the uptake and reliability of 3D printers improves, methods of rapidly and easily creating equipment are advancing. Additive printing can reduce long lead times and high developmental costs. Prototyping using these types of printing devices can enable fine tuning of designs to be undertaken with minimal fuss and sharing of designs by computer code. Some uses for this technique are feed dispensers such as those described in Sketchfab (2020) and Tangara et al. (2019) and isolation caging for zebrafish during dosing for chemical screening (Monstad-Rios et al., 2018). The scope of uses for this technology is very broad. A factor to consider is the material used by the 3D printing process. Some polymers used for this work may present toxicity issues, so it is important to test the compatibility of your polymers prior to use (Walpitagama et al., 2019).

The type of three-axis movement involved with these printers is also being utilized to create equipment such as the automated zebrafish embryo sorter described in Breitwieser et al. (2018).

Of course, there are a number of feed delivery solutions available off the shelf. Examples are units by Danio Lab (2020), Gigafloaquatics (2020), Pentair AES (2020).

In line with the 3R's in research is the "Zebrafish Embryo Genotyper" from wFluidX (2020). This device is based on research work at the University of Utah (Lambert et al., 2018). The device allows for the nonlethal harvesting of cells from a zebrafish embryo to enable polymerase chain reaction (PCR) genotyping to be performed before the onset of exogenous feeding.

The use of PCR techniques also has a positive influence for the 3Rs when it comes to health and environmental monitoring. Traditionally, sentinel animals and environmental samples are used for health monitoring, and some sentinel programs involve large numbers of animals to achieve statistical significance. Tecniplast SpA (2020) has recently released a standardized method of environmental screening for pathogens using the outfall water from a RAS. This sampling/testing arrangement relies on a set of graduated screens (patent pending) that are installed on a system for a minimum period of 5 weeks before the screens are packed in a standard 50 mL centrifuge tube and dispatched for PCR screening. Bacteriology as well as parasitology screening can be achieved using this process by adding a defined quantity of liquid Cary—Blair transport media to the centrifuge tube. Freezing of the sample may not be required if delivery times are less than a week in moderate temperatures. This will result in considerably lower freight costs in some circumstances, and may assist with reduction of animal numbers required to reliably monitor the microbial status of RAS.

For those using *Artemia* as part of their feeding regime, Inve Technologies (2020) provide a magnetically coated *Artemia* cyst that allows for separation of unhatched cysts and chorion debris from a culture using magnetic attraction. Separation of cyst debris reduces the chance of gut impaction in larval fish that are consuming *Artemia*.

Shallow profile tank systems for breeding and housing are becoming popular. The zebrafish is a shoaling animal and can benefit from a more lateral environment during breeding and shoaling, particularly where an enriched bottom structure is included. A new, horizontally oriented tank that encourages more natural spawning interactions in shoals of zebrafish than usually found in more vertical housing in laboratory set-ups was recently released. This unit is adaptable for use across multiple manufacturers' platforms (Tecniplast SpA, 2020).

A water-testing solution that is making inroads into the zebrafish field is Spin Touch (LaMotte Company, 2020). Originally developed for the pool industry, this device is capable of producing results for GH, KH, pH, TAN, nitrite, nitrate, and phosphate using an integrated disk that allows all tests to be run in parallel in approximately 2 min. This can provide considerable time saving for facilities that are operating multiple systems. Disks are available for both fresh and saltwater applications.

While not necessarily new, imaging systems for zebrafish movement tracking for a wide range of behavioral and screening purposes. There are a number of suppliers, including Noldus Information Technology BV (2020) and Viewpoint Life Sciences Inc. (2020) among others.

An area of research that is developing currently is photocatalytic processes for water treatment.

These processes rely on compounds such as TiO_2, ZnO, WO_3, $BiVO_4$, and silicon doped with various nanoparticles. These processes involve creating hydroxyl radicals that then react with organic substances, or by directly stripping electrons from compounds resulting in degradative oxidation. An electrophotocatalytic process currently in developmental research by Exciton Clean LLC (2020) relies upon titanium dioxide nanotube materials as an electrode within a UVC chamber (Kropp et al., 2009). This technology provides a combined benefit from the UVC exposure chamber providing disinfection as well as the capability to oxidize other compounds within the water column.

An overview of some uses of photocatalytic processes can be found in Peleyejuaband and Arotiba (2018). At the time of writing, these technologies show some promise in control not only of microorganisms but also of some nuisance organisms such as bryozoans. There is a potential for electrophotocatalytic processes at some point to obviate reliance on microbial populations for ammonia oxidation, but time will tell how these develop and if they will translate well into the zebrafish field or wider aquaculture industry.

At the time of writing there is at least one company trialing a gel-based diet delivery system that appears to offer a more ad libitum feeding profile in comparison to bolus feeding using pelleted or granulated foods. The gel matrix can be cut to size and manipulated for drug and compound delivery and dietary manipulations if required (https://www.clearh2o.com/, 2020).

The following are specialist suppliers of husbandry equipment: Aquaneering Incorporated (2020), Aqua Schwarz GmbH (2020), Aquatic_Enterprises_ Inc. (2020), Danio Lab (2020), Iwaki Aquatic (2020), and Tecniplast SpA (2020).

Web addresses can be found in the references section.

Acknowledgments

I would like to thank the following people for their direct and indirect support in compiling this section: foremost to Kathy Newell for her love, patience and ongoing support, Dr. Rod Collins, Lyn Crammond, David Taylor, and Deakin University for the time to be able to complete this work, and the following people who have contributed data, references, reviews, or insight and encouragement for this work: Diana Baumann, Andrew Ingalls, Carrie Barton, Gareth Barber, Dr. Suzita Mohd Noor, Gillian Lawrence, Bobbi Bauer, Rory Francis, Dr. Rob Jones, Christian Lawrence, Lauren Pandolfo, Erik Sanders, and the work of so many others.

References

Adatto, I., Krug, L., Zon, L.I., 2016. The red light district and its effects on zebrafish reproduction. Zebrafish 13 (3).

Ahi, E., Brunel, M., Tsakoumis, E.e. a., 2019. Transcriptional study of appetite regulating genes in the brain of zebrafish (*Danio rerio*) with impaired leptin signalling. Sci. Rep. 9, 20166.

Aleström, P., et al., 2019. Zebrafish: Housing and Husbandry Recommendations. Laboratory Animals Limited, Eye, Suffolk England.

Aqua Schwarz GmbH, 2020. Contact and Company Data. Aqua Schwarz. Available at: http://www.aquaschwarz.com/flycms/en/html/2/-/Contact.html. (Accessed 23 August 2020).

Aquaneering incorporated, 2020. Contact Information: Aquaneering Inc. Available at: http://www.aquaneering.com/contact_information.php. (Accessed 23 August 2020).

Aquatic_Enterprises_inc., 2020. Aquarius Fish Systems. Available at: http://aquaticenterprises.com/research/aquarius. (Accessed 20 October 2020).

ASTM international, 2013. Standard Guide for Estimating Carbon Saturation by Temperature Rise upon Immersion. ATSM International, West Conshohocken Philadelphia.

Baumann, D., November 9, 2020. Evaluation of GAC Adsorption Capacity at Stowers Institute for Medical Research.

Ben-Asher, R., Lahav, O., 2016. Electrooxidation for simultaneous ammonia control and disinfection in seawater recirculating aquaculture systems. Aquacult. Eng. 72–73, 77–87.

Best, J., et al., 2010. A novel method for rearing first-feeding larval zebrafish: polyculture with type L saltwater rotifers (*Brachionus plicatilis*). Zebrafish 7 (3).

Boisen, A., Amstrup, J., Novak, I., Grosell, M., 2003. Sodium and chloride transport in soft water and hard water acclimated zebrafish (*Danio rerio*). Biochim. Biophys. Acta Biomembr. 1618 (2), 207–218.

Breitwieser, H., et al., 2018. Fully automated pipetting sorting system for different morphological phenotypes of zebrafish embryos. SLAS Technol. 23 (2), 128–133.

Camargo, J.A., Alonso, A., Salamanca, A., 2005. Nitrate toxicity to aquatic animals: a review with new data for freshwater invertebrates. Chemosphere 58 (9), 1255–1267.

Chang, C.T., Benedict, S., Whipps, C.M., 2019. Transmission of *Mycobacterium chelonae* and *Mycobacterium marinum* in laboratory zebrafish through live feeds. J. Fish Dis. 42, 1425–1431.

Chen, Y.-Y., Lu, F.-I., Hwang, P.-P., 2003. Comparisons of calcium regulation in fish larvae. J. Exp. Zool. 295A (2), 127–135.

Cockington, J., 2020. Aquatic housing. In: Cartner, S., et al. (Eds.), The Zebrafish in Biomedical Research; Biology, Husbandry, Diseases and Research Applications. Academic Press (Elsevier), London, pp. 279–298.

Cortemeglia, C., Beitinger, T.L., 2005. Temperature tolerances of wild-type and red Transgenic Zebra Danios. Trans. Am. Fish. Soc. 134 (6), 1431–1437.

Cryogenetics AS Hamar, Norway, 2020. Cryogenetics. Available at: https://www.cryogenetics.com/. (Accessed 18 October 2020).

Dametto, F.S., Fior, D., Idalencio, R., et al., 2018. Feeding regimen modulates zebrafish behavior. PeerJ 6 (pe.5343).

Danio Lab, 2020. Danio Lab. Available at: https://daniolab.com/. (Accessed 23 August 2020).

DeLong, D.P., Losordo, T.M., 2012. How to Start a Biofilter. Southern Regional Aquaculture Centre, Stoneville Mississippi.

Di Rosa, V., et al., 2015. The light wavelength affects the ontogeny of clock gene expression and activity rhythms in zebrafish larvae. PLoS One 10 (7).

DrTim's Aquatics, LLC, 2020. Contact Us: Dr Tim's Aquatics, LLC. Available at: https://store.drtimsaquatics.com/crm.asp?action=contactus. (Accessed 23 August 2020).

Emerson, K., Russo, R., Lund, R., Thurston, R., 1975. Aqueous ammonia equilibrium calculations: effects of pH. J. Fish. Res. Board Can. 32 (12), 2379–2383.

Engeszer, R.E., Patteron, L.B., Rao, A.A., Parichy, D.M., 2007. Zebrafish in the wild: a review of natural history and new notes from the field. Zebrafish 4 (1).

Eshchara, M., et al., 2006. Intensive fish culture at high ammonium and low pH. Aquaculture 255 (1 - 4), 301–313.

Exciton Clean LLC, 2020. Exciton Clean. Available at: https://excitonclean.com/. (Accessed 24 October 2020).

Florida AquaFarms inc., 2016. Resting Rotifer Cultures. Available at: http://floridaaquafarms.com/shop/resting-rotifer-cultures/. (Accessed 29 October 2020).

Fowler, L.A., et al., 2019. Influence of commercial and laboratory dietson growth, body composition, and reproductionin the zebrafish *Danio rerio*. Zebrafish 16 (6), 508–521.

Fowler, L., Williams, M., D'Abramo, L., Watts, S., 2020. Zebrafish nutrition - moving forward. In: Farmer, S.C. (Ed.), The Zebrafish in Biomedical Research; Biology, Husbandry, Diseases and Research Applications. Academic Press (Elsevier), London, pp. 379–402.

Gendel, Y., Lahav, O., 2013. A novel approach for ammonia removal from fresh-water recirculated aquaculture systems, comprising ion exchange and electrochemical regeneration. Aquacult. Eng. 52, 27–38.

Gigafloaquatics, 2020. Gigaflo Aquatics. Available at: https://www.gigafloaquatics.com/. (Accessed 24 October 2020).

Gonzales, J.M.J., 2012. Preliminary evaluation on the effects of feeds on the growth and early reproductive performance of zebrafish (*Danio rerio*). J. Am. Assoc. Lab. Anim. Sci. 51 (4), 412–417.

Gonzales, J.M.J., Law, S.H.W., 2013. Feed and feeding regime affect growth rateand gonadosomatic index of adult zebrafish (*Danio rerio*). Zebrafish 10 (4), 532–540.

Hammer, H., 2020. Water quality for zebrafish culture. In: Farmer, S.C. (Ed.), The Zebrafish in Biomedical Research: Biology Husbandry Diseases and Research Applications. Academic Press, London England, pp. 321–333.

Harper, C., Lawrence, C., 2011. The Laoratory Zebrafish. CRC Press, Boca Raton, Florida.

Hovanec, T.A., Taylor, L.T., Blakis, A., Delong, E.F., 1998. Nitrospira-like bacteria associated with nitrite oxidation in freshwater aquaria. Appl. Environ. Microbiol. 64 (1), 258–264.

https://www.clearh2o.com/, 2020. Come Discover ClearH2O's New Gel Diet. Available at: https://www.clearh2o.com/. (Accessed 30 October 2020).

Inve Technologies, 2020. Inve Artemia. Available at: https://www.inveaquaculture.com/product/speciality-artemia/. (Accessed 30 October 2020).

Iwaki Aquatic, 2020. Home; Iwaki Aquatics. Available at: https://www.iwakiaquatic.com/. (Accessed 23 August 2020).

Jia, J., Qin, J., Yuan, X., et al., 2019. Microarray and metabolome analysis of hepatic response to fasting and subsequent refeeding in zebrafish (*Danio rerio*). BMC Genomics 20, 919.

Jonker, M.T., Van Mourik, L., 2014. Exceptionally strong sorption of infochemicals to activated carbon reduces their bioavailability to fish. Environ. Toxicol. Chem. 33 (3), 493–499.

Kaushik, S., Georga, I., Koumoundouros, G., 2011. Growth and body composition of zebrafish (*Danio rerio*) larvae fed a compound feed from first feeding onward: toward implications on nutrient requirements. Zebrafish 8 (2), 87–95.

Kimmel, C.B., et al., 1995. Stages of embryonic development of the zebrafish. Dev. Dyn. 203, 253–310.

Kropp, R., et al., 2009. A device that converts aqueous ammonia into nitrogen gas. Aquacult. Eng. 41 (1), 28–34.

Lambert, C.J., et al., 2018. An automated system for rapid cellular extraction from live zebrafish embryos and larvae: development and application to genotyping. PLoS One 13 (3).

LaMotte Company, 2020. LaMotte. Available at: https://lamotte.com/. (Accessed 18 October 2020).

Lawrence, C., 2007. The husbandry of zebrafish (*Danio rerio*): a review. Aquaculture 269 (1–4), 1–20.

Lawrence, C., Best, J., James, A., Maloney, K., 2012. The effects of feeding frequency on growth and reproduction in zebrafish (*Danio rerio*). Aquaculture 368–369, 103–108.

Learmonth, C., Carvalho, A.P., 2015. Rearing, acute and chronic toxicity of nitrate to early life stages of zebrafish—setting nitrate safety levels for zebrafish. Zebrafish 12 (4).

Lee, P., Lea, R., Dohmann, E., Prebilsky, W., 2000. Denitrification in aquaculture systems: an example of a fuzzy logic control problem. Aquacult. Eng. 23 (1–3), 37–59.

Markovich, M.L., Rizzuto, N.V., Brown, P.B., 2007. Diet affects spawning in zebrafish. Zebrafish 4 (1), 69–74.

Mars or Affiliates, 2020. Ammo-Lock. Available at: https://apifishcare.com/product/ammo-lock. (Accessed 24 October 2020).

Masser, M.P., Rakocy, J., Losordo, T.M., 1992. Recirculating Aquaculture Tank Production Systems Management of Recirculating Systems, s.L. Southern Region Aquaculture Centre.

Matthews, J.L., et al., 2018. Changes to extender, cryoprotective medium, and in vitro fertilization improve zebrafish sperm cryopreservation. Zebrafish 15 (3), 279–290.

Matthews, M., Trevarrow, B., Matthews, J., 2002. A virtual tour of the guide for zebrafish users. Lab. Anim. 31 (3), 34–40.

McClure, M., McIntyre, P., McCune, A., 2006. Notes on the natural diet and habitat of eight danionin fishes, including the zebrafish *Danio rerio*. J. Fish Biol. 69 (2), 553–570.

McDonnell, W., et al., 1983. Pulmonary effects of ozone exposure during exercise: dose-response characteristics. J. Appl. Physiol. 54 (5), 1345–1352.

Meyer, B.M., Froehlich, J.M., Galt, N.J., Biga, P.R., 2013. Inbred strains of zebrafish exhibit variation in growth performance and myostatin expression following fasting. Comp. Biochem. Physiol. Mol. Integr. Physiol. 164 (1), 1–18.

Monstad-Rios, A.T., Watson, C.J., Kwon, R.Y., 2018. ScreenCube: a 3D printed system for rapid and cost-effective chemical screening in adult zebrafish. Zebrafish 15 (1).

Noldus Information Technology BV, 2020. Zebrafish Video Tracking. Available at: https://www.noldus.com/applications/zebrafish-video-tracking. (Accessed 18 October 2020).

Pandolfo, L., Aliucci, J., 2020. Zebrafish aquatic systems: preventative maintenance and troubleshooting. In: Farmer, S.C. (Ed.), The Zebrafish in Biomedical Research; Biology, Husbandry, Diseases and Research Applications. Academic Press (Elsevier), London, pp. 309–320.

Peleyejuaband, M.G., Arotiba, O.A., 2018. Recent trend in visible-light photoelectrocatalytic systems for degradation of organic contaminants in water/wastewater. Environ. Sci. Water Res. Technol. 4 (10), 1389–1411.

Pentair AES, 2020. Handheld Granular Feed Dispenser. Available at: https://pentairaes.com/handheld-granular-feed-dispenser.html. (Accessed 24 October 2020).

Peterson, T.S., et al., 2013. *Paramecium caudatum* enhances transmission and infectivity of *Mycobacterium marinum* and *M. chelonae* in zebrafish *Danio rerio*. Dis. Aquat. Org. 106, 229–239.

Sanders, E., 2020. Cleaning and disinfection of life systems. In: Farmer, S. (Ed.), The Zebrafish in Biomedical Research Biology, Husbandry, Diseases and Research Applications. Academic Press (Elsevier), London, pp. 299–308.

Sauder, L.A., et al., 2011. Aquarium nitrification revisited: thaumarchaeota are the dominant ammonia oxidizers in freshwater aquarium biofilters. PLoS One 6. Ontario Canada.

Schaefer, R., Ryan, A., 2006. Stages of embryonic development of the zebrafish. Fish Biol. 69 (3), 722–734.

Schreier, H.J., Mirzoyan, N., Saito, K., 2010. Microbial diversity of biological filters in recirculating aquaculture systems. Curr. Opin. Biotechnol. 21 (3), 318–325.

Schroeder, P., Jones, S., Young, I.S., Sneddon, L.U., 2014. What do zebrafish want? Impact of social grouping, dominance and gender on preference for enrichment. Lab. Anim 48 (4), 328.

Seachem Industries inc., 2020. Prime. Available at: https://www.seachem.com/prime.php. (Accessed 24 October 2020).

Seachem Laboratories inc., 2020. Cuprisorb. Available at: https://www.seachem.com/cuprisorb.php. (Accessed 23 October 2020).

Sessa, A.K., et al., 2008. The effect of a depth gradient on the mating behavior, oviposition site preference, and embryo production in the zebrafish, *Danio rerio*. Zebrafish 5 (4), 335–339.

Shelton, D.S., et al., 2020. Collective behavior in wild zebrafish. Zebrafish 17 (4), 243–252.

Shih, T.-H., Horng, J.-L., Hwang, P.-P., Lin, L.-Y., 2008. Ammonia excretion by the skin of zebrafish (*Danio rerio*) larvae. Am. J. Physiol. Cell Physiol. 295 (6), C 1625–C 1632.

Siccardi, A.J.I., et al., 2009. Growth and survival of zebrafish (*Danio rerio*) fed different commercial and laboratory diets. Zebrafish 6 (3), 275–280.

Silke Ehrich, S., et al., 1995. A new obligately chemolithoautotrophic, nitrite-oxidizing bacterium, *Nitrospira moscoviensis* sp. nov. and its phylogenetic relationship. Arch. Microbiol. 164, 16–23.

Sketchfab, 2020. Sketchfab Fish Feeder Version 2. Available at: https://sketchfab.com/3d-models/fish-feeder-version-2-0f08d672a35e4109ac1414a993825ed2. (Accessed 24 October 2020).

Smith, D.J., et al., 2013. Dietary protein source influence on body size and body composition in growing zebrafish. Zebrafish 10 (3), 439–446.

Spence, R., et al., 2006. The distribution and habitat preferences of the zebrafish in Bangladesh. J. Fish Biol. 69 (5), 1435–1448.

Spence, R., Gerlach, G., Lawrence, C., Smith, C., 2008. The behaviour and ecology of the zebrafish, *Danio rerio*. Biol. Rev. 83, 13–34.

Summerfelt, S., Vinci, B., Piedrahita, R., 2000. Oxygenation and carbon dioxide control in water reuse systems. Aquacult. Eng. 22, 87–108.

Suriyampola, P.S., et al., 2016. Zebrafish social behavior in the wild. Zebrafish 13 (1), 1–8.

Tangara, A., et al., 2019. An Open-Source and Low-Cost Feeding System for Zebrafish Facilities. University of France (IBENS), Paris France.

Tecniplast spa., 2017. Tritone - Automatic Feeding. Available at: https://www.tecniplast.it/us/product/tritone-automatic-feeding-system.html. (Accessed 5 November 2020).

Tecniplast SpA, 2020. Aquatic Solutions. Tecniplast SpA. Available at: https://www.tecniplast.it/en/company-profile.html. (Accessed 23 August 2020).

Timmons, M.B., Ebeling, J.M., 2007. Recirculating Aquaculture. Cayuga Aqua Ventures, Ithaca New York.

Uliano, E., et al., 2010. Effects of acute changes in salinity and temperature on routine metabolism and nitrogen excretion in gambusia (*Gambusia affinis*) and zebrafish (*Danio rerio*). Comp. Biochem. Physiol. Mol. Integr. Physiol. 157 (3), 283–290.

Viewpoint Life Sciences inc., 2020. Viewpoint Behaviour Technology. Available at: http://www.viewpoint.fr/en/home. (Accessed 18 October 2020).

Villamizar, N., Vera, L.M., Foulkes, N.S., Sanchez-Vazquez, F.J., 2014. Effect of lighting conditions on zebrafish growth and development. Zebrafish 11 (2).

Walpitagama, M., et al., 2019. Additives migrating from 3D-printed plastic induce developmental toxicity and neuro-behavioural alterations in early life zebrafish (*Danio rerio*). Aquat. Toxicol. 213.

Watts, S.A., Lawrence, C., Powell, M., D'Abramo, L.R., 2016. The vital relationship between nutrition and health in zebrafish. Zebrafish 13 (1), S72–S76.

Watts, S.A., Powell, M., D'Abramo, L.R., 2012. Fundamental approaches to the study of zebrafish nutrition. ILAR J. 53 (2), 144–160.

Westerfield, M., 2007. The Zebrafish Book: A Guide for the Laboratory Use of Zebrafish. University of Oregon, Eugene Oregon U.S.A.

wFluidX, 2020. wFluidX Know Your Fish. Available at: https://www.wfluidx.com/. (Accessed 24 October 2020).

Wilkie, M.P., 2002. Ammonia excretion and urea handling by fish gills: present understanding and future research challenges. J. Exp. Zool. 293 (3), 284–301.

Wurts, W.A., 2002. Alkalinity and Hardness in Production Ponds. Kentucky State University CEP at UK Research and Education Center, Princeton, Kentucky.

Zahangir, M., Haque, F., Mostakim, G., Islam, S.M., 2015. Secondary Stress Responses of Zebrafish to Different pH: Evaluation in a Seasonal Manner, Mymensigh 2202, Bangladesh: Aquaculture Reports. Elsevier Press.

Zhou, L., Boyd, C.E., 2014. Total ammonia nitrogen removal from aqueous solutions by the natural zeolite, mordenite: a laboratory test and experimental study. Aquaculture 432, 252–257.

ZIRC, 2020. Zirc Importation Procedures and Cost Estimates. Available at: https://zebrafish.org/documents/strainsCosts.php. (Accessed 25 October 2020).

Zynda, J.R., 2020. Aquatics facility design considerations: incorporating aquatics into an animal facility. In: Farmer, S.C. (Ed.), The Zebrafish in Biomedical Research; Biology, Husbandry, Diseases and Research Applications. Academic Press (Elsevier), London, pp. 265–277.

Further reading

Howe, K., Clark, M.D., Torroja, C.F., et al., 2013. The zebrafish reference genome sequence and its relationship to the human genome. Nature 496, 498–503.

Chapter 3

Breeding and larviculture of zebrafish (*Danio rerio*)

Benjamin Tsang, Robert Gerlai
University of Toronto Mississauga, Department of Psychology, Toronto, ON, Canada

Introduction

Zebrafish (*Danio rerio*), or as it has been known in the aquarium hobby, zebra danio, has been bred in captivity for several decades. It has become one of the most common aquarium fish species as a hobby, but has also been gaining recognition and popularity in scientific research. Although the earliest publications of zebrafish appeared almost a century ago (Goodrich and Nichols, 1931; Creaser, 1934), until about four decades ago the zebrafish was not very well known to science. This, however, changed when a developmental biologist, George Straisinger, started to employ the species (Streisinger et al., 1981). By now zebrafish have been utilized in practically every subfield of biology, from cancer research (Hason and Bartůněk, 2019) to behavioral neuroscience (Levin and Cerutti, 2009). One reason for the popularity of this species is that it is easy to keep and breed in captivity. In fact, in the aquarium hobby it is considered a beginner fish. Indeed, it can be found in most aquarium pet stores, and is rather cheap to buy. It tolerates a wide range of water conditions and temperatures, and accepts all sorts of dried and live fish foods. It is small, highly social, so a large number of zebrafish can be housed in small tanks, and is extremely prolific. Despite the ease of its maintenance and prolific nature, numerous publications have discussed the methods of its breeding and husbandry. In this chapter, we distill what is known about these aspects of zebrafish and add our own personal perspective. Our goal is to provide detailed information on how to keep and breed this species. We also discuss the reasons why optimization of these procedures and keeping their parameters standard may be of value to the scientist.

Laboratory Fish in Biomedical Research. https://doi.org/10.1016/B978-0-12-821099-4.00015-8
Copyright © 2022 Elsevier Inc. All rights reserved.

Should we mimic nature, or should we ignore it in laboratory breeding and husbandry of zebrafish: theoretical considerations from the perspective of experimental biology

The first topic we briefly discuss is the natural habitat of the zebrafish. Why do we start a chapter on breeding and larviculture with this topic? Numerous facilities and laboratories have bred and kept zebrafish under conditions that are distinctly different from those in the natural habitat of this species. Proponents of such conditions argue that "it does not matter how these fish live in nature" because "from an aquaculture standpoint" these conditions make good sense. One example is salinity, a point to which we will return later. Here, we just note that some zebrafish facilities employ NaCl to raise salinity to 2000 and sometimes 3000 μS. Their argument is that this is useful as it limits bacterial growth and thus keeps the water cleaner and the fish healthier. We believe that using such artificial conditions is a mistake. If we force zebrafish to breed, grow, and live under unnatural conditions, we may be testing the limits of their physiology and their ability to maintain homeostasis without stress. Zebrafish have so far been complying with our methods, and thus some may argue that there is nothing wrong with the way we breed and keep them, even though the conditions do not resemble those in nature. This is a strong argument, except that up to recent times, zebrafish were only used to uncover robust biological effects, e.g., major genes affecting organogenesis. Variations in salinity, for example, are unlikely to be able to appreciably modify major gene effects in studies where mutation-induced alterations of fundamental developmental processes were examined (Grunwald and Streisinger, 1992). However, nowadays we are probing genes and biological mechanisms underlying such processes as learning and memory (Norton and Bally-Cuif, 2010), effects of trace amounts of toxicants in the environment (Dai et al., 2013), mild embryonic alcohol exposure (Facciol et al., 2019b), and so on. In such studies, subtle changes in biological processes resulting from experimental manipulations may be significantly altered by unnatural environmental conditions.

Let us first briefly examine how, in principle, unnatural conditions may be deleterious. Although zebrafish have been found in a variety of habitats living under a wide range of conditions (briefly reviewed later), pushing water conditions outside of this already broad range may make the fish more vulnerable to fluctuations in the laboratory environment. Taking salinity as a hypothetical example, if the upper tolerable level of salt concentration is, say, 3100 μS, keeping the fish at 3000 μS means that occasionally, due to, e.g., crowding, excess food, lack of appropriate water changes coupled with evaporation-induced water loss, actual salinity levels may reach or exceed the upper threshold of tolerability. At that moment, homeostatic processes cannot compensate for the high salt concentration and the fish suffer physiological stress. As this stress manifests in certain but not all fish tanks, and perhaps as this stress affects the weaker, smaller, and perhaps already less healthy fish more, the population as a whole will experience increased variability. This variability will manifest as increased error variance in research results, which then leads to reduced statistical power to find effects of experimental manipulations significant. Furthermore, elevated error variance may also lead to reduced reproducibility of findings, because higher sampling error may seep into the result. In sum, for these reasons it is a good idea to keep and breed zebrafish under environmental conditions that are in the mid-range of the values one can observe in the natural habitat of this species.

Zebrafish in nature

What are these values? A few studies have already explored the natural habitat of zebrafish and started to address this question (Engeszer et al., 2007; Sundin et al., 2019). Thus, we know that zebrafish live across a wide range of habitats whose abiotic features vary in terms of geographic location as well as seasonality. Zebrafish are found in India, Pakistan, Nepal, Bangladesh, and Myanmar in freshwaters near sea level all the way up to 1500 m elevation (Spence et al., 2006, 2007a, 2007b; Engeszer et al., 2007; Whiteley et al., 2011; Arunachalam et al., 2013; Suriyampola et al., 2016 Daniels, 2002). The temperature of the water in these habitats ranges between 12 and 39°C, pH between 5.9 and 9.2, and conductivity between 10 and 271 μS (for review, see Engeszer et al., 2007; Parichy and Postlethwait, 2020). Organic waste load, e.g., levels of ammonia, nitrite, or nitrate, likely also vary between being completely absent to high levels depending on the habitat.

Based upon these observations, in a recent paper we recommended the following optimal environmental parameters for zebrafish maintenance: temperature 27−29°C, pH 6.5−7.5, conductivity 100−200 μS, ammonia, nitrite, and nitrate levels 0 ppm (Tsang et al., 2020). However, we emphasize that the parameters of zebrafish maintenance have not been systematically analyzed using controlled parametric empirical studies, and thus our recommendations, just like those of others, are based upon experience and our best guesses. We also note that most zebrafish facilities focus on the absolute value of these parameters and ignore the question of the speed with which they change. This is a mistake. For example, the salinity may be out of range, but bringing it back to the desired level must be achieved slowly and gradually. The speed of change is a particularly acute problem for pH, low pH that is, which is often corrected by dumping sodium bicarbonate (baking soda) into the water, inducing an abrupt jump in its value. Adjustment of out-of-range parameters must be achieved gradually. For most parameters this means several hours, or perhaps even better days. We also note that the desired salinity in almost all zebrafish facilities we know of and almost all zebrafish publications we reviewed are achieved by adding NaCl, a salt that is practically absent in natural waters of the zebrafish. Calcium carbonate ($CaCO_3$) is the salt that is found almost ubiquitously in freshwater habitats of the zebrafish. It has an advantage over NaCl also in that it is able to buffer the water against low and fluctuating pH. Thus we have recommended the use of aragonite, a crystal form of $CaCO_3$ that is relatively easy to obtain (from pet stores, for example) and dissolves well in water (when placed in filters or system water reservoirs in mesh bags) (Tsang et al., 2020). However, we note again that systematic parametric analysis of environmental parameters, including salt concentration/composition and pH, has not been carried out.

Light is also an important factor in zebrafish maintenance. In their natural habitat, zebrafish live in shallow water. These natural waters usually have a lot of aquatic vegetation, and thus the zebrafish can hide from open and well-illuminated waters. Nevertheless, the fish are often exposed to strong and direct sunlight. The question of light versus dark preference has been controversial in the literature because investigators often confuse background shade (black vs. white) with level of illumination (dark vs. well lit) among other factors (Facciol et al., 2019a). According to our own studies, zebrafish appear to prefer tanks with a dark background/bottom but with strong illumination (e.g., Facciol et al., 2019a). In our hands, zebrafish thrive in well-illuminated tanks. The light cycle in the natural habitat of zebrafish is around 14 h light and 10 h dark, which most zebrafish facilities match.

Assuming we have established these ideal environmental conditions, the next most important question is how to get the parental fish up to good breeding condition.

Preparing zebrafish for breeding

Aquarists, and by now zebrafish scientists too, know that perhaps the most important aspect of breeding fish is properly preparing them for spawning. Keeping the parental zebrafish under optimal conditions, feeding them well and thus bringing them up to excellent spawning condition is a must. We have briefly dealt with maintenance conditions, including water chemistry parameters and temperature; here, we further examine some of these questions more specifically according to their relevance to breeding, and we will also discuss how to feed the parental fish.

Zebrafish have been found to breed throughout the year in nature; nevertheless, as with other tropical freshwater species, breeding activity of zebrafish is expected to be highest during the rainy season (Spence et al., 2007b). Rain brings organic nutrients, which leads to elevation of microorganisms as well as macroorganisms in the water that serve as food for both adult and juvenile zebrafish. The rainy season also brings two important changes in water parameters: salinity decreases and temperature increases. Mimicking such changes in the laboratory also enhances breeding success. Thus a couple of days before the expected spawning, we reduce salinity from 200 to 100 μS and increase the temperature from 27 up to 30°C. We emphasize that these changes are gradual and occur over the period of 2 days preceding spawning, as abrupt alterations of water chemistry or temperature have been found stressful for the zebrafish (Abozaid et al., 2020).

An appropriate amount and quality of food are perhaps even more important in keeping zebrafish healthy and preparing them for spawning. The mouth structure of the zebrafish suggests that this species forages near or from the surface of the water. In nature, zebrafish consume a variety of foods, including nonaquatic organisms that fall into the water as well as aquatic organisms, for example, insects and crustaceans. Some of the aquatic organisms zebrafish consume forage on algae, and thus zebrafish eat this foodstuff indirectly too. In captivity, zebrafish accept a variety of foods, including live as well as dried foods. Most zebrafish facilities employ dried foods, in the form of pellets or flakes. Live foods are also available, but their use risks the spread of bacterial or other infections. There are dried foods specifically formulated for zebrafish available from commercial vendors (e.g., Zeigler, Aquaneering). Systematic comparison of the appropriateness of such foods is often lacking, but the question of food quality and consistent quality control for commercially available dried fish food has started to be discussed (Certal, 2016). In our facility, we feed our adult zebrafish with a 50%:50% mixture of Omega One (or Tetramin) flakes and Jehmco Spirulina, but we also had success with Zeigler tropical fish micropellets. In addition, and especially for zebrafish that we expect to breed, we supplement this dried food diet by offering our parental fish, once a day, freshly hatched nauplii of brine shrimp (*Artemia salina*). Last, again borrowing from what we know in nature, we note that the frequency of feeding is almost as important as the quality of food offered. In nature, the food items zebrafish capture and eat are distributed both spatially and temporally, i.e., these fish tend to consume food throughout the day continuously. We have seen zebrafish facilities where the fish are fed once a day. This latter method is rather suboptimal as fish cannot consume enough food in one eating that they would need for optimal growth and optimal breeding conditions. Although, unlike with rodents, continuous ad libitum feeding is not possible with tropical fish, increasing the frequency of feeding (and decreasing the amount of food per feeding occasion) is highly recommended. We found that feeding three or four times a day is a reasonable compromise. This may seem like a tremendous amount of work in large facilities; however, technical solutions have already been worked out to achieve it. For example, automated robotic feeding systems have been developed. If these expensive systems are not feasible for financial or technical reasons, a cheap and simple alternative is to employ battery-powered tropical fish feeders available from aquarium pet stores. Some of these feeders can dispense dried food (crushed flakes and small pellets) four to six times per day—a reasonably high frequency of feeding for zebrafish.

Another question we consider for how to prepare the parental zebrafish for spawning is their sex-specific housing. In nature, zebrafish form mixed sex groups and thus separating the sexes before spawning is undoubtedly an unnatural manipulation. Nevertheless, it ensures that spawning can be timed precisely, and also facilitates spawning if done properly. In a mixed sex holding tank, zebrafish will spawn without the knowledge and control of the experimenter. The deposited eggs will be quickly eaten by the fish in the tank, and thus females with ripe eggs will likely be less willing/able to spawn when subsequently placed in their breeding tank. Thus separating the sexes before the scheduled spawning may be a good idea. However, separating the sexes for prolonged periods of time carries a risk, as females can become "egg bound," i.e., can develop serious health issues as a result of the unreleased aging eggs inside. We have found a 3–4-day long isolation of the sexes to be optimal.

The last question we consider in this segment is the age of the parents. Under ideal conditions, zebrafish may reach sexual maturity by their age of 3 months. However, the fish start spawning reliably only after 4 months of age. Zebrafish, similarly to several other fish species, continue to grow throughout their life and remain sexually active until they die. However, the ability to spawn usually does decline with age. The best egg yield and fertilization ratio can be achieved with fish that are between 5 and 10 months of age.

We already mentioned that spawning may be achieved by moving the females and males into a breeding tank. However, this is only one of the possible breeding methods, a question we discuss next.

The breeding tank

There are two fundamentally different ways zebrafish may be bred: (1) in a small breeding or spawning tank in which only a few males and females are placed, and (2) in large breeding cylinders. The latter is employed as a continuous breeding method, providing a large number of fertilized eggs/embryos in a consistent basis and thus does not require separating the sexes. The former is employed if the goal is to breed specific individuals, e.g., mutants, or if the precise timing of fertilization must be controlled, or when continuous supply of a large number of fertilized eggs is not required. We also note that breeding cylinders, due to their large volume and the number of breeding zebrafish inside, will allow preservation of genetic variance, if desired, whereas breeding only a few zebrafish, as employed in small breeding tanks, will likely lead to random genetic drift and unintended inbreeding.

We first describe the large-scale breeding cylinder method. The concept is simple. Place a large number of adult female and male zebrafish into a large cylindrical tank. Place a mesh underneath the adults that would allow eggs to fall down but would not allow the adults to follow and eat the eggs, and collect the eggs in the conically shaped bottom of the tank. As zebrafish spawn in nature at dawn, all one needs to do with the breeding cylinder method is to harvest the eggs every morning. The volume of these large-scale cylinders varies between 15 and 100 L, and depending on this, they can accommodate 20–250 breeders. There are, of course, many other considerations one has to be aware of when designing or purchasing such large-scale breeding systems. For example, filtration of water can be achieved via external canister filters, or by connecting the breeding cylinder to the central filtration system of the zebrafish facility. Temperature, light, pH, and salinity must all be controlled and maintained at desired values. Feeding must also be provided as discussed earlier. It is also important to consider what happens with the eggs after they have been released and fertilized. Properly designed mass-breeding cylinders should provide appropriate oxygenation to the eggs collecting in the bottom. In nature, the eggs would be scattered around and would thus be able to breathe well. In the collection cone of the cylinder, however, they accumulate in large numbers and are packed on each other, and thus may experience hypoxia, unless appropriate oxygenation via, e.g., an air compressor-driven air stone is provided.

The alternative method to breeding cylinders is the more often-employed small spawning tank (Figs. 3.1 and 3.2). These tanks are usually only 2–3 L in volume and would have a total of two to six adult breeders at a time. The modern zebrafish spawning tank usually contains a center divider that allows the experimenter to separate males and females. The spawning tank is set up the night before the expected spawning with females on one and males on the other side of the divider. The central divider is removed either after dusk (after lights turn off) or before dawn (before lights turn on the next morning). The spawning tank is also usually equipped with a mesh bottom that allows the eggs to fall through thereby protecting them from being eaten by the parents. Newer designs have a sloping bottom mesh, creating a shallow "beach," which mimics the natural habitat and utilizes natural behavior of the zebrafish. Zebrafish have been observed to spawn in the shallowest areas of their natural waters, likely because eggs deposited in this area are less accessible to fish that would eat them.

Although numerous spawning tanks specifically developed for the zebrafish are available from commercial vendors, a regular pet store variety glass tank may also be sufficient. In fact, in the past, these tanks were more often employed. The bottom of the tank was usually covered with marbles and thus the eggs falling down to the glass bottom in between the tightly spaced marbles were protected from hungry parents. These tanks, as well as the newer professional zebrafish spawning tanks, can be equipped with artificial (plastic) plants. These plants mimic the natural environment of the zebrafish and also provide hiding places for subordinate males or females that are not yet ready to spawn.

The sex ratio in large breeding cylinders is usually kept at 50:50. In small breeding tanks, this ratio is debated. Some argue that having more females than males minimizes the potentially negative, stress-inducing, effect of being

FIGURE 3.1 Panel (A): A typical zebrafish holding tank used in high-density zebrafish rack systems. This particular tank is 1.5 L in volume and can be employed to hold up to five zebrafish. A single adult zebrafish (Tubingen longfin strain) is shown in the tank for relative size comparison. A tank like this may be appropriate for isolating the sexes for a few days before spawning to increase spawning success. Panel (B): A typical small-scale breeding tank (this particular example was manufactured by Aquaneering Inc., San Diego, CA). The volume of such breeding tanks is usually 2–3 L. The tank can hold two to six zebrafish. Note the center divider in the middle of the tank,

constantly chased by males for the female. Others argue that having a larger number of males increases the ratio of fertilized eggs. Although no systematic studies on the optimization of spawning techniques and the factors that influence spawning have been published, decades of experience and anecdotal evidence allow us to make recommendations. Based upon this and our own experience, a 2–3 female:1–2 male ratio should work well, but we suspect this value may change according to the age and, perhaps more importantly, the genotype (strain or population origin) of the zebrafish.

Raising zebrafish juveniles is not as easy as often stated

Although adult zebrafish are easy to maintain, and making the adults spawn is also not that difficult, rearing juveniles is another matter. Unlike in the case of other favorites of biomedical research like the house mouse and the rat, the initial difficulty with larviculture may surprise those new to this species, given the general consensus from the literature that paints this fish as "easy to keep and easy to breed." The novice may encounter issues of early mortality and/or morbidity, and/or a large number of unfertilized eggs. The underlying issues may be plenty, but most often include inappropriate water parameters (pH, hardness, organic waste, temperature) and suboptimal food or feeding regime. Next, we discuss several such issues along with methods that we believe would make rearing conditions optimal for the zebrafish juveniles. We start the discussion with how to hatch the eggs collected from the spawning tanks.

Hatching the eggs and caring for the larvae

After collecting the eggs, experimenters may employ different methods for hatching them. Perhaps the most often-employed method is to collect the eggs into large Petri dishes and keep the eggs in them until they hatch. At 28–29°C this happens at around 3 days postfertilization (dpf). To ensure temperature is kept constant at this optimal level, the Petri dishes may be placed into an incubator or into a thermostat-controlled heated water bath. It is also important to remember that the photoperiod of 14 h light 10 h dark must also be maintained for the eggs and hatching larvae. Without this photoperiodicity,

which is kept lowered to separate the sexes until the morning of desired spawning. A perforated bottom layer is also visible. The perforated bottom is employed so that the fertilized eggs can fall through its openings to protect them from being eaten by the parents. Panel (C): The breeding tank components set apart. Clockwise, starting from top left: breeding tank insert with the perforated bottom, breeding tank outer shell, tank cover with feeding and airline holes, center divider, plastic plants. The plastic plants are usually employed to provide hiding places, especially for females, to reduce stress. Panel (D): An alternative breeding tank set-up in which three breeding tank inserts are lowered into a larger (10 L) breeding tank outer shell. This set-up allows simultaneous breeding of three breeding groups of zebrafish. The eggs from the three spawnings fall into the bottom of the tank and thus will mix.

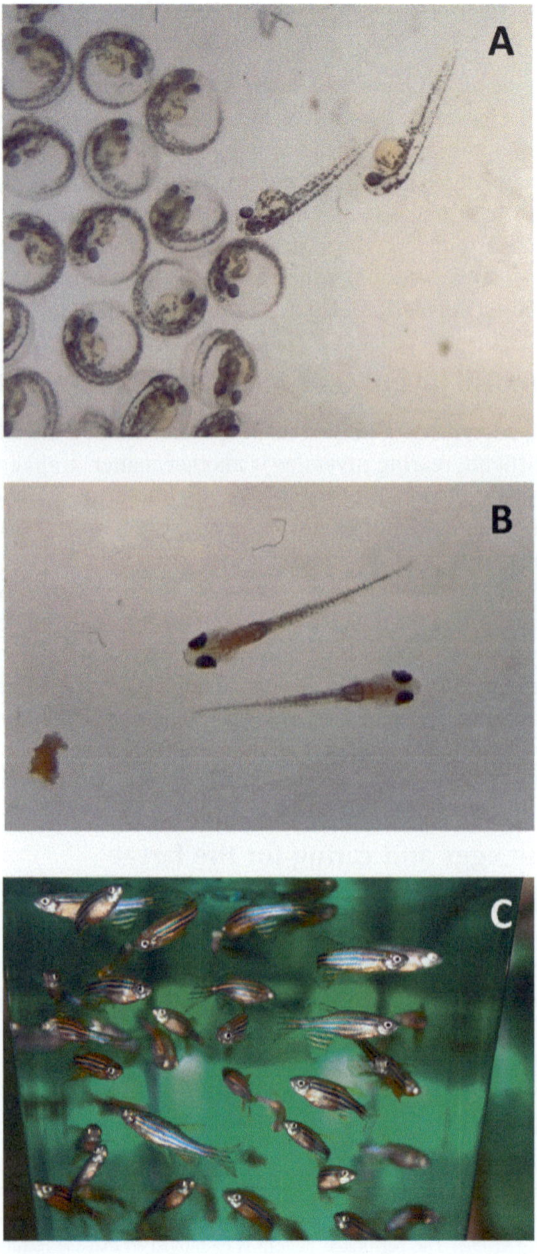

FIGURE 3.2 Examples of zebrafish at different stages of their development. Panel (A): Zebrafish embryos at 3 days postfertilization (dpf). At this age, the chorion of most eggs is intact, i.e., the embryos are still inside the egg, but a few have already hatched. Note the visible yolk sac in the hatched embryo, which will provide nutrients for the embryo until it reaches free swimming stage at around 5 dpf. Panel (B): 14 dpf old zebrafish measuring about 5.5 mm standard length photographed from above. At this age, juvenile zebrafish are free swimming, capture food, respond to all sorts of stimuli, but do not yet show strong shoaling responses. Panel (C): Young adult, 3 months old, zebrafish of a genetically heterogeneous wild-type population. At this age, zebrafish are becoming sexually mature and, if kept under optimal conditions, may start spawning. Note that these particular fish carry a mutation that manifests as enlarged fins.

increased lethality and numerous development abnormalities have been observed (e.g., Villamizar et al., 2014). It is important to realize that in the Petri dish, there is no water filtration or oxygenation employed. Thus to allow proper oxygen exchange and to enhance water quality, it is advisable to use larger Petri dishes (e.g., 10–15 cm diameter) and to fill the dish only to a low depth (about 1 cm), i.e., to increase the surface-to-volume ratio. It is also important to replace the water once a day with fresh water of identical chemistry and temperature.

Opinions are divided as to the requirement of sterile conditions (e.g., bleaching the eggs and using sterile media), and whether antifungal or antibacterial agents (e.g., methylene blue, antibiotics) should be employed for the embryo medium used in the Petri dish. According to our experience, such measures are usually not necessary for regular healthy zebrafish of standard laboratory or wild-type strains and populations. Although for certain studies they may be required (e.g., Stagaman et al., 2020), the potentially long-lasting consequences of sterile environment and also the consequences of application of disinfecting agents during development are not fully understood.

After hatching, i.e., around 3 dpf, the embryo enters the larval stage, called "wiggler" in the aquarium hobby. At this stage, although the larval zebrafish is already outside the chorion (eggshell) and responds to stimuli and is mobile, it cannot swim and fully relies on its yolk sac for nutrition. This stage lasts until about 5 dpf, by which time the larva has become free swimming; a fully formed zebrafish measuring about 3 mm in length. The zebrafish literature calls these little fish "larvae" up to about their age of 2–3 weeks post-fertilization (Fig. 3.1), and thus we will include discussion on maintenance of zebrafish of this age in this chapter under the term larviculture. However, we note that this terminology (larva and larviculture) that has become accepted in zebrafish research is misleading. Zebrafish do not undergo metamorphosis after hatching. Although numerous developmental changes occur after 5 dpf, and the fry (the colloquial term for baby fish) also undergo rapid growth, these changes do not require substantial reorganization of structures/functions or the body plan of the organism, as it happens in metamorphosis, for example, in insects. For these reasons, in this chapter we call the stage from 5 to 21 dpf early or young juvenile stage.

Caring for young juveniles

At 5 dpf, the young juvenile zebrafish is fully mobile. Its organs are formed, and among other systems, it has a well-developed central nervous system that allows the fish to respond to stimuli. It hunts for food, avoids danger, and, overall, exhibits a fairly complex behavioral repertoire (Kalueff et al., 2013). Because at this stage the juvenile fish consumes food and excretes waste, starting filtration of its holding water is important. For this reason, moving the fry as soon as they reach free swimming stage out of the Petri dish and into a filtered nursery tank is a good idea. There are commercially available standalone zebrafish nursery racks that are designed to provide an appropriate level of filtration/oxygenation without the danger of damaging or filtering out the tiny fry. Regular zebrafish tanks retrofitted to serve the needs of the fry are also available, and these converted nursery tanks may be placed on, and integrated to utilize the filtration system of, standard zebrafish racks (shelf systems) zebrafish facilities often employ. One usual problem with the commercially available nursery tanks and nursery racks is setting the water flow (filtration) rate. Too much water flow can physically damage or stress the young juveniles and can also remove dispensed food too fast, i.e., before the fry can capture and eat it. We have successfully used an alternative method that solves these issues. We often raise our young juveniles in 40-L glass tanks (available from most pet stores). These tanks are equipped with sponge filters, a sponge cylinder into which an air stone is lowered. Air pressed through the stone serves two functions: (1) it provides oxygenation to the water, and (2) as the air bubbles move upward, they move water from the internal lumen of the sponge cylinder out to the top opening of the cylinder, thereby sucking water from the outside through the wall of the sponge. The large surface area of the sponge on which beneficial bacteria can live provides excellent biological filtration in addition to mechanical filtration, but the force with which water is moved into the sponge is distributed throughout the sponge's surface and thus it is weak enough so the small fry can easily escape from it. Furthermore, the relatively large volume of this tank allows for increased stability of water parameters, including temperature, salinity, and pH. One potential downside of the large volume of this tank is that food may be diluted/distributed (but not filtered out), and thus the small fry may not be able to find it as efficiently as in smaller nursery tanks.

Perhaps the most important factor in larviculture of zebrafish is the type of food the small juveniles are given. There are a variety of dried as well as live foods available. In nature, zebrafish fry eat a variety of microorganisms, and thus live food is what these fish prefer in the laboratory too. We start our discussion with this food type.

The most often-employed live food in tropical fish aquaculture is brine shrimp (*A. salina*). The nauplii of brine shrimp are readily consumed by fry of a variety of tropical fish species. However, the 5 dpf old zebrafish juvenile is too small to eat the nauplii and thus this food should not be offered before the zebrafish juveniles are about 14–15 dpf old. The first live food the zebrafish fry may be able to eat include paramecium and rotifers. As our laboratory employs the latter, we will briefly describe the pros and cons of cultivating rotifers. Rotifers are tiny, often microscopic, invertebrates that numerous zebrafish facilities across the globe have adopted for larviculture in part due to their rapid reproduction and our ability to gut load them with additional nutrients. There is a variety of both freshwater and saltwater rotifers commercially available. In our facility, we employ the SS-type *Brachionus rotundiformis*, obtained from Reed Mariculture. This species is a saltwater rotifer that has a body length of approximately 100 μm. Details of saltwater rotifer culture set-up and maintenance have been given elsewhere (Lawrence et al., 2016). Here, we only emphasize some of the pros and cons of employing these rotifers. Rotifer cultures require daily harvesting to maintain both water chemistry and rotifer population in check. Rotifer cultures are susceptible to crashes if the water chemistry fluctuates as a result of rapid overpopulation. Thus it is often the cause of complete colony losses requiring the repurchasing of starter colonies. For any zebrafish facility considering implementing rotifer cultures for raising young juveniles, we recommend to first evaluate whether a rotifer culture is needed by considering the manual labor required and the operational expenses of rotifer feed and maintenance of the salt medium. Second, we suggest scaling the rotifer culture appropriate to the needs of the facility to reduce unnecessary waste of rotifers and their associated costs. For smaller facilities with one to four racks (about a 1000–5000 zebrafish capacity), a DIY bottle method using emptied 2L soda bottles can be sufficient. For our facility, we find the use of two 18-L buckets of SS-type rotifers more than sufficient for maintaining a 15-standalone six-shelf/rack operation (about 40,000 adult zebrafish and their offspring), with one bucket often kept as a backup colony.

The advantage of these live food items is that they are readily consumed by young zebrafish and also that rotifers stay alive for hours in the water and provide a steady source of nutrition. For these reasons, they can robustly speed development and growth of the fry. The disadvantage of these food items is that they, and also their culture medium, may potentially carry infectious agents, e.g., bacteria that can cause disease. To avoid such infections, some zebrafish facilities employ only dried food, which can be sterilized and kept in the freezer to practically eliminate the possibility of introducing diseases. There are a variety of dried foods available for the zebrafish sometimes labeled "artificial plankton" or "artificial rotifers." Irrespective of these names, the key feature of these food items is their size. They tend to be not larger than 100 μm in diameter and thus even 5 dpf old juvenile zebrafish can consume them. Another advantage of dried foods is that they may be dispensed using automated feeders. However, a major disadvantage of these dried foods is that they are not as readily consumed by the zebrafish juvenile as the aforementioned live foods, and thus the juvenile fish fed on them may not grow as fast. Furthermore, in case of overfeeding, the excess amount of leftover food may impair water quality, unless promptly removed manually. Despite these complications, however, we have successfully employed dried foods in our larviculture.

Similarly to other environmental factors, optimization of feeding strategies and types of food employed in larviculture of zebrafish have not been systematically explored by empirical studies. Nevertheless, based upon decades of experience from zebrafish laboratories and on our own, we recommend the following. Start feeding the zebrafish juveniles with rotifers. Ideally, offer this food at least two to three times a day. Continue with this food and feeding regimen up to 15 dpf. Start offering a small amount of dried food from 10 dpf by administering the food once a day. Time this so that you offer the dried food before administering live food, i.e., when the juveniles are most hungry. Observe whether the juveniles accept the dried food. If they do, slowly increase the amount of dried food offered over days, and slowly decrease the amount of live food offered. After the fish reach 15 dpf, start offering artemia nauplii instead of rotifers, but continue offering dried food too. At this age (15 dpf), you may switch to the type of dried food the adult zebrafish receive in the laboratory, but make sure this dried food is ground/milled to a small enough size so that the juveniles can consume it.

The last question we consider in this segment is housing density. Usually, this factor is not discussed in zebrafish larviculture as the number of fish per volume of tank ratio is less limiting for the tiny juvenile zebrafish. For adults, it is a different matter, however, as recent studies have uncovered alarming evidence that the current industry standard of keeping adult zebrafish densely packed into tiny tanks is not an optimal husbandry practice (Shams et al., 2017), as it may induce both physiological and psychological stress. Because the small size and the little overall organic waste output the small juveniles produce, a much larger number of them may be housed in small tanks than would be possible with adults. Nevertheless, crowding-induced stress has already been noted in juvenile ("larval") zebrafish screening (Stephen Ekker, personal communication). Our recommendations for housing density for juvenile zebrafish are as follows. In the aforementioned 40-L juvenile housing tank, 500–1000 5 dpf old zebrafish juveniles may be housed. This number, however, must be reduced by moving some fish out to additional tanks as they grow. For example, by age 14 dpf, the number per 40-L volume should be down to about 200, and by the age of 1 month, this number should not exceed 100. As a comparison, consider that the same (40 L) tank could house up to about 25 adult zebrafish.

Standardization of breeding and larviculture practices: the question of replicability and reproducibility

The last topic of this chapter considers the question whether we should standardize our breeding and larviculture practices. Up until now, we implicitly assumed that standardization is required. This is because, as we argued, the zebrafish performs best somewhere in the middle in between its homeostatic tolerance limits. Thus the environmental parameters and procedures that maintain them should be the same for all laboratories. When bred, raised, and kept under these specifically controlled optimal conditions, we argued, zebrafish should yield reliable results when employed in all our experiments.

The question of standardization, however, is not this trivial. Its pros and cons have been discussed and debated in more general terms both for the zebrafish (Gerlai, 2019) and for other laboratory organisms of biomedical research (Kaflkafi et al., 2018). We will not rehash all the points here, and will only discuss what we believe are the central questions. These questions concern how best to achieve replicability and reproducibility (Gerlai, 2019). For example, if breeding and larviculture procedures are standardized and thus employed in a highly similar, or identical, manner across laboratories, some argue that results obtained with zebrafish from these laboratories should be more comparable. Thus a repeated experiment, for example, is expected to reproduce the results of the original one, which we call reproducibility. Similarly, if the breeding and maintenance conditions and procedures across facilities are standardized, experiments asking similar questions with the use of different techniques should give compatible/comparable results, which we call replicability. However, others point out that local conditions always vary across laboratories, and thus standardization is not really possible. Proponents of laboratory-specific methods also argue that laboratory-specific idiosyncratic conditions are appropriate from the standpoint of experimental biology because both the treatment and control groups experience the same conditions. This argument, however, implicitly assumes that there is no interaction between the employed treatment and facility-specific conditions, i.e., it assumes that these effects are additive. The validity of this assumption, however, is almost never checked, and thus laboratory-specific conditions may influence the difference between the treatment and control groups. Should, thus, we standardize? Perhaps not, according to yet another argument. This argument concerns the robustness of our results and perhaps is best illustrated by discussing the difference between replicability and reproducibility.

Assume everything is standardized. All breeding, rearing methods, and environmental parameters are kept identical within and across laboratories. If achieved, this would reduce environmentally induced error variation both within the laboratory and across laboratories. This is a good thing from the perspective of gaining statistical power as the reduced error variation will enhance our ability to find significant effects of our experimental manipulation, the effect of treatment we employ compared to control. But do we really want this? If the goal is reproducibility of even the least important result, then yes. However, if the goal is to find significance of experimental treatment effects only when these effects are robust and truly meaningful, perhaps it is not the best idea to eliminate environmental error variation. If the treatment effect is so "finicky" that it can be reproduced only if everything is done precisely the same manner, perhaps it is not worth discovering at all, goes the argument by those who are against standardization. The concept of replicability cuts to the heart of the matter. The concept of replicability, unlike that of reproducibility, does assume variations in procedures, and despite such variations expects similar, i.e., compatible, results.

Thus should we or should we not standardize our breeding and larviculture parameters and procedures for zebrafish? Given the lack of systematic analyses that could answer this question, we can only provide our personal perspective. We believe that identification of the optimal conditions and the most appropriate procedures for zebrafish breeding and larviculture is crucial. Identification of the optimal values of environmental parameters and the methods of their establishment should be based upon the features of the natural habitat of the zebrafish. Nevertheless, optimization of such values and the methods of their laboratory control must be achieved/developed by systematic parametric empirical analyses. Once identified, an optimal range of values for the environmental factors/parameters should be established as standards across all zebrafish laboratories. This should be our baseline. This baseline condition set should be employed in all our pioneering studies, making them comparable within and across laboratories. That is, we should strive for reproducibility. However, once the pioneering studies have been performed, we should allow, and in fact we should explicitly explore, deviations from the baseline and how these deviations may alter our results. Only this way we will be able to tell whether our findings are robust across variable conditions.

References

Abozaid, A., Tsang, B., Gerlai, R., 2020. The effects of small but abrupt change in temperature on the behavior of larval zebrafish. Physiol. Behav. 227, 113169. https://doi.org/10.1016/j.physbeh.2020.113169.

Arunachalam, M., Raja, M., Vijayakumar, C., Malaiammal, P., Mayden, R., 2013. Natural history of zebrafish (*Danio rerio*) in India. Zebrafish 10 (1), 1–14. https://doi.org/10.1089/zeb.2012.0803.

Certal, A., 2016. Different feeds and feeding regimens have an impact on zebrafish larval rearing and breeding performance. Int. J. Mar. Biol. Res. 1 (1), 1–8. https://doi.org/10.15226/24754706/1/1/00101.

Creaser, C.W., 1934. The technic of handling the zebra fish (*Brachydanio rerio*) for the production of eggs which are favorable for embryological research and are available at any specified time throughout the year. Copeia 1934, 159–161.

Dai, Y., Jia, Y., Chen, N., Bian, W., Li, Q., Ma, Y., et al., 2013. Zebrafish as a model system to study toxicology. Environ. Toxicol. Chem. 33 (1), 11–17. https://doi.org/10.1002/etc.2406.

Daniels, R.J.R., 2002. Freshwater Fishes of Peninsula India. Universities Press, Hyderabad.

Engeszer, R., Patterson, L., Rao, A., Parichy, D., 2007. Zebrafish in the wild: a review of natural history and new notes from the field. Zebrafish 4 (1), 21–40. https://doi.org/10.1089/zeb.2006.9997.

Facciol, A., Iqbal, M., Eada, A., Tran, S., Gerlai, R., 2019a. The light-dark task in zebrafish confuses two distinct factors: interaction between background shade and illumination level preference. Pharmacol. Biochem. Behav. 179, 9–21. https://doi.org/10.1016/j.pbb.2019.01.006.

Facciol, A., Tsang, B., Gerlai, R., 2019b. Alcohol exposure during embryonic development: an opportunity to conduct systematic developmental time course analyses in zebrafish. Neurosci. Biobehav. Rev. 98, 185–193. https://doi.org/10.1016/j.neubiorev.2019.01.012.

Gerlai, R., 2019. Reproducibility and replicability in zebrafish behavioral neuroscience research. Pharmacol. Biochem. Behav. 178, 30–38.

Goodrich, H.B., Nichols, R., 1931. The development and the regeneration of the color pattern in *Brachydanio rerio*. J. Morphol. 52, 513–523.

Grunwald, D.J., Streisinger, G., 1992. Induction of recessive lethal and specific locus mutations in the zebrafish with ethyl nitrosourea. Genet. Res. 59, 103–116.

Hason, Bartůněk, 2019. Zebrafish models of cancer—new insights on modeling human cancer in a non-mammalian vertebrate. Genes 10 (11), 935. https://doi.org/10.3390/genes10110935.

Kaflkafi, N., Agassi, J., Benjamini, Y., Chesler, E., Crabbe, J., Crusio, W., Eilam, D., Gerlai, R., et al., 2018. Reproducibility and replicability of mouse phenotyping in pre-clinical studies. Neurosci. Biobehav. Rev. 87, 218–232.

Kalueff, A., Gebhardt, M., Stewart, A., Cachat, J., Brimmer, M., Chawla, J., et al., 2013. Towards a comprehensive catalog of zebrafish behavior 1.0 and beyond. Zebrafish 10 (1), 70–86. https://doi.org/10.1089/zeb.2012.0861.

Lawrence, C., Best, J., Cockington, J., Henry, E., Hurley, S., James, A., et al., 2016. The complete and updated "rotifer polyculture method" for rearing first feeding zebrafish. J. Vis. Exp. 107 https://doi.org/10.3791/53629.

Levin, E.D., Cerutti, D.T., 2009. Behavioral neuroscience of zebrafish. In: Buccafusco, J.J. (Ed.), Methods of Behavior Analysis in Neuroscience, second ed. Taylor & Francis, CRC Press, Boca Raton, United States (Chapter 15).

Norton, W., Bally-Cuif, L., 2010. Adult zebrafish as a model organism for behavioural genetics. BMC Neurosci. 11 (1) https://doi.org/10.1186/1471-2202-11-90.

Parichy, D.M., Postlethwait, J.H., 2020. The biotic and abiotic environment of zebrafish. In: Gerlai, R.T. (Ed.), Behavioral and Neural Genetics of Zebrafish. Elsevier, Academic Press, San Diego, United States, pp. 17–32.

Shams, S., Seguin, D., Facciol, A., Chatterjee, D., Gerlai, R., 2017. Chronic social isolation affects anxiety-related behaviors, cortisol, and monoamines in adult zebrafish. Behav. Neurosci. 131, 492–504.

Spence, R., Fatema, M., Ellis, S., Ahmed, Z., Smith, C., 2007a. Diet, growth and recruitment of wild zebrafish in Bangladesh. J. Fish. Biol. 71 (1), 304–309. https://doi.org/10.1111/j.1095-8649.2007.01492.x.

Spence, R., Gerlach, G., Lawrence, C., Smith, C., 2007b. The behaviour and ecology of the zebrafish, *Danio rerio*. Biol. Rev. 83 (1), 13–34. https://doi.org/10.1111/j.1469-185x.2007.00030.x.

Spence, R., Fatema, M., Reichard, M., Huq, K., Wahab, M., Ahmed, Z., Smith, C., 2006. The distribution and habitat preferences of the zebrafish in Bangladesh. J. Fish. Biol. 69 (5), 1435–1448. https://doi.org/10.1111/j.1095-8649.2006.01206.x.

Stagaman, K., Sharpton, T.J., Guillemin, K., 2020. Zebrafish microbiome studies make waves. Lab. Anim. 49, 201–207.

Streisinger, G., Walker, C., Dower, N., Knauber, D., Singer, F., 1981. Production of clones of homozygous diploid zebra fish (*Brachydanio rerio*). Nature 291 (5813), 293–296. https://doi.org/10.1038/291293a0.

Sundin, J., Morgan, R., Finnøen, M., Dey, A., Sarkar, K., Jutfelt, F., 2019. On the observation of wild zebrafish (*Danio rerio*) in India. Zebrafish 16 (6), 546–553. https://doi.org/10.1089/zeb.2019.1778.

Suriyampola, P., Shelton, D., Shukla, R., Roy, T., Bhat, A., Martins, E., 2016. Zebrafish social behavior in the wild. Zebrafish 13 (1), 1–8. https://doi.org/10.1089/zeb.2015.1159.

Tsang, B., Ansari, R., Gerlai, R.T., 2020. Maintenance and breeding of zebrafish, with some ethological and ecological considerations in mind. In: Gerlai, R.T. (Ed.), Behavioral and Neural Genetics of Zebrafish. Elsevier, Academic Press, San Diego, United States, pp. 17–32.

Villamizar, N., Vera, L., Foulkes, N., Sánchez-Vázquez, F., 2014. Effect of lighting conditions on zebrafish growth and development. Zebrafish 11 (2), 173–181. https://doi.org/10.1089/zeb.2013.0926.

Whiteley, A., Bhat, A., Martins, E., Mayden, R., Arunachalam, M., Uusi-Heikkilä, S., et al., 2011. Population genomics of wild and laboratory zebrafish (*Danio rerio*). Mol. Ecol. 20 (20), 4259–4276. https://doi.org/10.1111/j.1365-294x.2011.05272.x.

Chapter 4

Health monitoring, disease, and clinical pathology

Jean-Philippe Mocho[1], **Nuno Pereira**[2,3,4,5,6]

[1]*DanioVet, Joint Production System Ltd, United Kingdom;* [2]*Chronic Diseases Research Center (CEDOC), Nova Medical School, Lisboa, Portugal;* [3]*Faculty of Veterinary Medicine, Lusófona University, Lisboa, Portugal;* [4]*Instituto Gulbenkian de Ciência, Oeiras, Portugal;* [5]*ISPA - Instituto Universitário, Lisboa, Portugal;* [6]*Oceanário de Lisboa, Lisboa, Portugal*

Introduction

In 2017, about 500,000 zebrafish (*Danio rerio*) were "used for the first time (naïve animals) for research and testing" under the European Directive 2010/63/EU (European Commission, 2020). This makes the species the third most used in the European Union, behind mice and rats. One can argue that it is an underestimation since wild-type fish only used to produce eggs, and all larvae terminated before they are capable of independent feeding, are not represented in these statistics. Still, it reveals the importance of aquatic animals in modern research. This European Directive sets requirements for the daily care of animals: "Animals shall be checked at least daily by a competent person. These checks shall ensure all sick or injured animals are identified and appropriate action is taken"; and more specifically about their health monitoring: "Establishments shall have a strategy in place to ensure that a health status of the animals is maintained that safeguards animal welfare and meets scientific requirements. This strategy shall include regular health monitoring, a microbiological surveillance program and plans for dealing with health breakdowns and shall define health parameters and procedures for the introduction of new animals."

Beyond legislation compliance, the justification for such a program relies on three pillars: preserving animal welfare, scientific data, and staff health. Animal welfare can be scrutinized through the Five Freedoms spectrum (Farm Animal Welfare Council, 2012). This includes the need to protect animals from unnecessary suffering, and to aim for their freedom from pain, injury, and disease by prevention of contamination, rapid detection of disease, and treatment—albeit remediation to infection often means euthanasia for laboratory zebrafish. Scientific data can also be compromised by fish disease because the animal physiology is disturbed and the experiment's

Laboratory Fish in Biomedical Research. https://doi.org/10.1016/B978-0-12-821099-4.00014-6

Copyright © 2022 Elsevier Inc. All rights reserved.

reproducibility and translatability are compromised (Lieggi et al., 2020; Midttun et al., 2020). Finally, staff health can be affected by transmission of pathogens from fish to humans, i.e., zoonotic disease. For example, Mason et al. (2016) described a *Mycobacterium marinum* outbreak in a zebrafish laboratory, significantly affecting fish and personnel.

The European Directive summarizes the aims of a health monitoring program and sets questions we will address in this chapter. First, we will focus on good husbandry practices to monitor population health, i.e., daily care. Then, we will detail relevant zebrafish pathogens and disease, and sample types and diagnostic techniques. This will allow us to introduce the concepts of epidemiological unit (EU) (i.e., the fish population to health monitor) and specific pathogen free (SPF), to calculate the number of fish to sample, and to propose screening patterns with number and frequency of samples, including environmental samples. Nevertheless, the microbiological surveillance program must be accompanied with a biosecurity program to preserve the EU health status. This involves the introduction of fish in a bio-contained quarantine, and triaging imports to avoid excluded pathogens. Biosecurity barriers also prevent contamination from any introduced goods and personnel, and from other epidemiological units in the facility, which is particularly relevant in multispecies set-ups.

Controlling and monitoring population health

Husbandry, environmental conditions, and infection pressure

Husbandry is a key process to master for the control of fish health. Good practices for zebrafish have been described (Harper and Lawrence, 2011; Alestrom et al., 2019; Lee et al., 2019). A sound husbandry warrants stable water conditions (e.g., filtration, nitrogen, temperature, and pH) and feed supply, whereas variations in these key parameters distract fish physiology and immunity from optimal defense against diseases. Clinical signs of noninfectious disease may then appear (e.g., buoyancy issue following abrupt change of diet), or commensal pathogens may trigger opportunistic infections. Pathogens may be present in the environment as biofilm (Chang et al., 2019b), in live feed diets (Chang et al., 2019a), or already contaminating fish at a low prevalence (Mocho, 2016). The more pathogens present in the system, the higher the infection pressure, the more an outbreak is likely to affect fish. Therefore husbandry processes should aim at controlling two aspects potentially interfering with fish health: stable environmental conditions and the reduction of pathogen load in the system by thorough cleaning and disinfection routines and prompt removal of biofilms and cadavers.

Monitoring performance

Feeding regimens have an obvious direct impact on growth rates. However, nutritional requirements of zebrafish remain unknown, and distribution protocols (i.e., quantity and frequency), as well as tank fish density according to developmental stages, are not standardized. Water parameters vary too (Alestrom et al., 2019). In consequence, there is an opportunity for laboratories to define their growth curves, at least for their most used wild-type lines. This can be done by setting time points at which fish wet weights are measured without sedation or anesthesia. For example, a container is weighed, 10 fish are added, and the container is weighed again. The difference divided by 10 gives an average wet weight per fish. Similarly, body condition scoring is a noninvasive method to assess relative weight to fat cover of zebrafish (Clark et al., 2018). Unexpected changes in growth curves between clutches, or departure from body condition score expected range, allows detection of deviation from expected husbandry output (Lee et al., 2019), thus potentially triggering further investigation regarding husbandry practices and potential pathogen contamination (Sanders et al., 2020). Fish performance can also be monitored by recording trends in egg production and embryo survivability, although this might be affected by other parameters not directly related to fish health, like spawning contexts (Wafer et al., 2016).

Daily recording of morbidity and mortality

Researchers, animal caretakers, technicians, and veterinarians take part routinely in monitoring and controlling laboratory fish health. Detection of animals with clinical signs should be recorded. A plan of action depending on the welfare impact of clinical signs and their severity should be in place, allowing action to be taken promptly upon detection. Morbidity recording helps monitoring the percentage of a population that is affected by a specific clinical sign. For example, a line may be affected with deterioration of operculum cover. This may be due to inbreeding and an increased prevalence of genetic defect within the line. An outbreeding scheme could be adopted in consequence. Monitoring more severe signs of disease (e.g., ulcers, chronic buoyancy disorder) and mortality is key to detect a surge in infections (e.g., *Mycobacterium* spp., contamination of the EU by a pathogen) or noninfectious disease, for example, due to uncontrolled husbandry parameters. First, a baseline mortality rate needs to be determined for specific husbandry conditions, period of recording, and population defined by an age group (e.g., 0–5 days postfertilization [dpf], from 5 dpf to juvenile tank splitting, adults, fish over 18 months of age). Variation from this baseline can then trigger further veterinary investigation to identify the cause of increased mortality (Mocho, 2016).

Pathogens and diseases to monitor

Zebrafish can be contaminated by many microbes (we use the terms microbes and microorganisms to regroup bacteria, fungi, parasites, and viruses). The pathogens relevant to monitor in zebrafish laboratories depend on the fish sources of the facility, the import and quarantining process, and the biosecurity measures in place. For example, fish imported from pet shops or the wild are usually of an unknown microbiological status, whereas fish from research resource centers are rederived and intensively screened (Kent et al., 2011). The risk of contamination in the former case is increased compared to the latter. In consequence, it is advisable to screen twice a year for pathogens like *Edwardsiella ictaluri*, *Flavobacterium columnare*, *Ichthyophthirius multifiliis*, *Piscinoodinium pillulare*, and *Pleistophora hyphessobryconis* zebrafish colonies deemed at low risk of contamination, and more frequently in the case of unknown microbiological status of fish sources.

Similarly, pathogenic viruses are not commonly detected in *Danio rerio* (Cartner et al., 2020). However, multispecies facilities may want to screen their zebrafish for potential viral contamination from other fish colonies. For example, nervous necrosis virus and infectious spleen and kidney necrosis virus have triggered high mortality events in *Danio rerio* laboratories (Binesh, 2013; Bermudez et al., 2018). Another virus has been extensively detected in zebrafish colonies: picornavirus (Altan et al., 2019), though the impact on fish health and science has not been determined yet. Another prevalent microbe, *Myxidium streisingeri*, can be classified in the same category for which contamination has unknown consequences. Researchers would decide whether such pathogens are a risk for their animal models, and whether it is therefore necessary to include them in the health monitoring program.

Further information on fish pathogens are extensively available in the literature (Noga, 2010; Smith, 2019) with proposition of eradication programs (Erlacher-Reid, 2018; Cartner et al., 2020). Once pathogens are detected in the colonies, options are to treat and rederive fish and to eradicate the microbe from the system and EU, or to accept its presence with mitigation measures to prevent further impact on fish welfare, science, and staff health. Further screening may thereafter help monitor pathogen prevalence and control mitigation measure efficiency. Presence of microorganism deemed of lesser relevance may be acceptable with no further need to pursue specific screening against the microbe.

Here, we will focus on the most relevant laboratory zebrafish pathogens. These microorganisms are the most frequently detected pathogens in submissions to the Zebrafish International Resource Center diagnostic laboratory (Zebrafish International Resource Center, 2020a). They impact fish health and scientific output of experiments, and some present a zoonotic risk. We recommend that, whatever the context, *Danio rerio* colonies are monitored quarterly for these pathogens.

Mycobacterium spp.

Zebrafish as well as other fish species are commonly infected with mycobacteria. Up to eight *Mycobacterium* species, all potentially zoonotic, are described in zebrafish mycobacteriosis: *Mycobacterium abscessus*, *Mycobacterium chelonae*, *Mycobacterium fortuitum*, *Mycobacterium gordonae*, *Mycobacterium haemophilum*, *Mycobacterium marinum*, *Mycobacterium peregrinum*, and *Mycobacterium saopaulense* (Cartner et al., 2020). These species and their strains can have different epidemiological, clinical, and prognostic traits. Mycobacteria in aquatic settings are ubiquitous in water and biofilms. Hence, it is expected to detect mycobacteria in fish and environmental samples. These environmental bacteria would not always be the strains inducing fish disease (Chang et al., 2019b). It is therefore necessary to monitor species prevalence. For example, *M. marinum* and *M. haemophilum*, less commonly found in zebrafish colonies, are more problematic due to their potentially higher morbidity and mortality (Mason et al., 2016; Mocho, 2016).

In zebrafish, mycobacteriosis is frequently detected with a subclinical presentation. However, it can also induce acute to chronic clinical infections. Common clinical signs and lesions can be loss of body condition, cutaneous ulcers, aerocystitis, and coelomitis. Diagnosis can start at gross necropsy with detection of yellow to whitish granulomas in gills and in other organs. Acid-fast stains such as Ziehl–Neelsen modified according to Fite can reveal acid-fast positive bacilli, mostly mycobacteria, at cytology (e.g., tissue imprint) or histology, either in the granulomas or nongranuloma associated (e.g., diffuse presentation is not uncommon for *M. haemophilum*). Use of a Ziehl–Neelsen stain modified according to Fite is more effective than the classic Ziehl–Neelsen stain for the observation of acid-fast bacilli in fish (Romano et al., 2020). Not all granulomas are associated with acid-fast bacilli. Diagnostic confirmation and speciation can be done swiftly by polymerase chain reaction (PCR). Bacterial culture is a slower approach due to the mycobacterial fastidious growth. *Mycobacterium* species from biofilms, water, water surface, or tank sludge can also be identified by PCR (Crim et al., 2017; Mocho et al., 2017) (Fig. 4.1).

No antibiotic treatment is advisable due to the risk of increasing antibiotic resistance in these relevant zoonotic pathogens. In case of high mortality, depopulation and disinfection of the system may be necessary (Rácz et al., 2019). It is thus of utmost importance to prevent outbreak of clinical disease, even more so considering the zoonotic risk. Transmission can be through ingestion. The presence of infected material or heavily contaminated biofilms should be minimized by husbandry measures such as prompt removal of dead or sick animals, and optimized cleaning and disinfection of the system (e.g., ultraviolet filter). Live feed culture should be controlled too (e.g., paramecia, brine shrimp, and rotifers). Mycobacteriosis affects predominantly immunocompromised fish and older animals, which may also increase pathogen shedding in the system. Older (over 18 months) and immunocompromised fish should be removed from recirculation and isolated in bio-contained epidemiological units (Kent et al., 2009). Import of fish protocols should be designed to mitigate entrance of contaminated animals (e.g., exclusion of

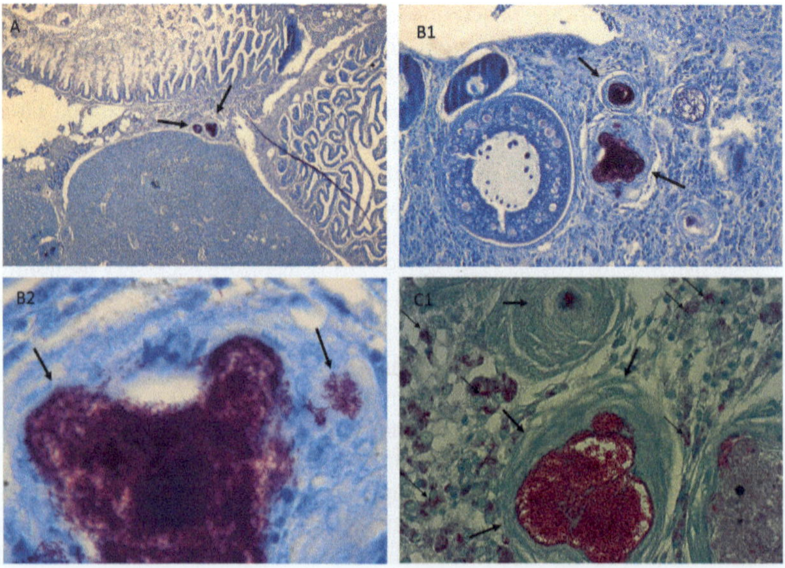

FIGURE 4.1 **Histology images of acid-fast bacilli stained red by Ziehl–Neelsen (ZN), visible inside typical zebrafish mycobacteriosis granulomas or with an intracellular location.** (A) Two clearly visible granulomas (→) located in the mesentery near the liver (ZN, 10×). (B1) Two granuloma (→) located in the ovarian stroma. Granuloma without acid-fast bacteria are frequently observed in ovaries and not necessarily linked with egg-associated inflammation (ZN, 20×). (B2) Same granuloma of B1 with visible acid-fast bacilli (→) (ZN, 40×). (C) Two granulomas (*large arrows*) in different stages of evolution and both with visible acid-fast bacteria. Intracellular bacilli in the surrounding macrophages (*thin arrows*) (ZN, 40×). *Images from zebrafish of Instituto Gulbenkian de Ciência, Oeiras and Chronic Diseases Research Center (CEDOC), Nova Medical School, Lisboa.*

M. marinum and *M. haemophilum*, introduction of surface disinfected eggs only) (Borges et al., 2016; Cartner et al., 2020; Mason et al., 2016).

Pseudocapillaria tomentosa

The intestinal nematode *Pseudocapillaria tomentosa* can infest a broad number of fish species and can reach up to 12 mm in length. Infection occurs by ingestion of parasite eggs, which have distinctive bipolar plugs and can be seen inside female parasite, free in the fish intestinal lumen, or at the tank bottom. Infected fish may display chronic clinical emaciation. Morbidity and mortality can be severe. Detected lesions are chronic enteritis and coelomitis (Gaulke et al., 2019; Murray and Peterson, 2015). The parasite is associated in zebrafish with an increased incidence of tumors (intestinal carcinomas). Even a subclinical infection may therefore impact on fish welfare and experimental outcomes (Gaulke et al., 2019).

Diagnosis can be done by environmental screening of the sump sludge or quarantine tank. Eggs are easily identifiable under a compound microscope using fecal flotation techniques (Mocho et al., 2017; Murray and Peterson, 2015).

Worms and eggs can also be easily detected in the intestine wet mounts at necropsies and in intestinal histology slides. Furthermore, they can be identified by PCR in fish and environmental samples (Crim et al., 2017; Mocho et al., 2017). Observation of adult nematodes in the sludge does not allow diagnosis since there are many other free-living metazoans, often nonpathogenic, that thrive in aquatic systems (Mocho, 2016). Several treatments against *P. tomentosa* are described in the literature for the fish (Kent et al., 2019a; Cartner et al., 2020) and at the scale of the whole water system (Kent et al., 2019b).

Pseudoloma neurophilia

Pseudoloma neurophilia is a pyriform parasite infecting several fish species. Affected zebrafish frequently present a subclinical form of disease. Clinical presentation varies from loss of body condition, emaciation, spinal deformities, and, more rarely, diminished fecundity and discrete mortality (Murray et al., 2011; Ramsay et al., 2009; Sanders et al., 2020). Due to the presence of parasites in the central nervous system and associated tissue inflammation, the parasite infection, even subclinical, can cause disruption of normal behavior and impact scientific data. Whenever possible, *P. neurophilia*-free animals should be used in behavior and neuroscience studies (Kent et al., 2011; Midttun et al., 2020).

At histological examination, *P. neurophilia* is detected as clusters/aggregates or free spores, mainly in the hindbrain, spinal cord, and nerve roots, and also in gonads (intra- and extrafollicle), muscles, or other organs (Murray et al., 2011; Ramsay et al., 2009; Sanders et al., 2020). Histology, contrary to PCR, allows evaluation of the disease extension in the fish (Murray et al., 2011; Sanders et al., 2020). Spores will stain with hematoxylin and eosin stain but are better highlighted by other stains as Ziehl−Neelsen, Gram, and in particular Luna (Murray et al., 2011; Sanders et al., 2020). Using more than one of these stains in histology for zebrafish health screening increases the probability of diagnosing light infections, due to the better visualization of sparse located spores, and by the fact that more tissue is covered in the various stains/slides per fish. With practice, spores can also be detected with the compound microscope in wet mounts of brain and spinal cord (Murray et al., 2011). Spores can be highlighted by phase-contrast microscopy to decrease the risk of false-negative results. Environmental PCR can be performed in the water, namely in spawning tanks, but may lead to a significant number of false negatives (Crim et al., 2017). Fish PCR is therefore preferred (Kent et al., 2011; Murray et al., 2011) (Fig. 4.2).

There is no reported treatment against *P. neurophilia*. Control measures include depopulation or isolation of affected fish, and epidemiological investigation of horizontal and (pseudo)vertical transmissions. The former is by ingestion of infected tissues. Hence, the importance of promptly removing

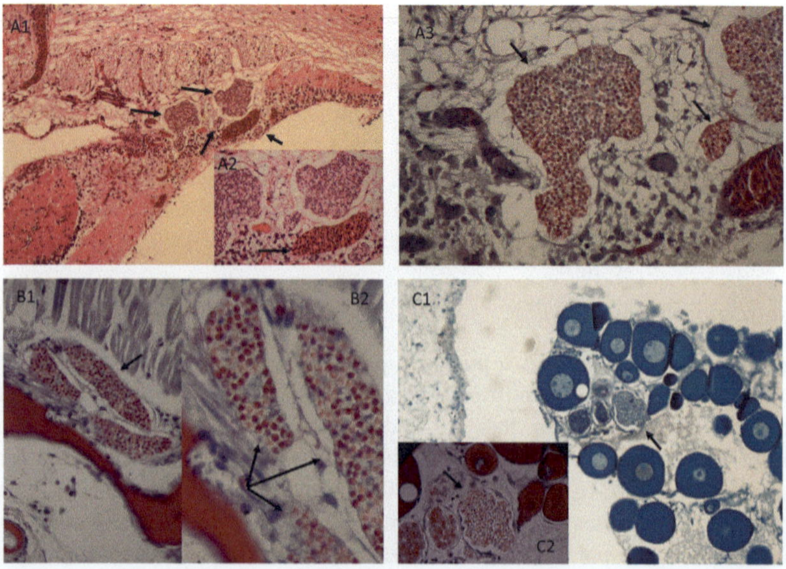

FIGURE 4.2 Histology images of *Pseudoloma neurophilia* aggregates of spores in different locations. (A1) Four aggregates (→) in the spinal cord (hematoxylin and eosin [HE], 10×). (A2) Magnification of these aggregates around a blood vessel (→) (HE, 20×). (A3) Spores highlighted in red by Luna staining (→) (Luna, 20×). (B1) Aggregates in the spinal nerve roots (→) (Luna, 20x). (B2) Same in a higher magnification (Luna, 40×). (C1) One aggregate in the ovarian stroma (→) (Ziehl–Neelsen, 10×). (C2) Same aggregate stained with Luna (→) (Luna, 20×). *Images from zebrafish of Instituto Gulbenkian de Ciência, Oeiras and Chronic Diseases Research Center (CEDOC), Nova Medical School, Lisboa.*

cadavers to prevent cannibalism (Murray et al., 2011). Spawning released extraovum spores can lead to pseudovertical transmission (Murray et al., 2011). The parasite is also transmissible by true vertical transmission (intra-ovum), implying that egg surface disinfection cannot inactivate spores inside eggs (Ramsay et al., 2009). Still, it is recommended to only introduce, in quarantine or in main holding systems, surface-disinfected eggs (Murray et al., 2011). Progression of the disease can be accelerated in the presence of stress, and optimized husbandry helps limiting disease prevalence and morbidity (Ramsay et al., 2009). Immunosuppressed fish can have a disseminated presentation and old animals act as a reservoir (Murray et al., 2011; Ramsay et al., 2009). Therefore husbandry and biosecurity measures should be implemented to decrease transmission risk and reduce parasite presence in the facility, for example, by isolating more sensitive populations (Murray et al., 2011).

Noninfectious diseases

Diseases with noninfectious causes are frequently encountered in zebrafish facilities, often associated with husbandry problems. Comprehensive descriptions of these diseases are available in the literature. Only essential aspects are described here, mostly husbandry related (Harper and Lawrence, 2011; Cartner et al., 2020). So-called "environmental diseases" are directly connected with poor water quality and parameters imbalance (e.g., ammonia, nitrite, chlorine, chloramine and heavy metal toxicities, and supersaturation). Diagnosis in fish is done by clinical observation and histology of affected tissues, and corroborated by water analysis.

Nephrocalcinosis, another common environmental disease, is elusive in early stages. It is diagnosed histologically by the observation of calcium precipitates in the renal tubules and ducts. It is mainly associated with high concentration of dissolved carbon dioxide. It usually has a subclinical presentation in zebrafish, but, in other fish species, advanced stages show lesions responsible for clinical renal insufficiency. Nephrocalcinosis lesions can also be secondary to other diseases. If prevalence of these renal lesions increases, an increase of dissolved calcium dioxide should be suspected, and measures should be taken (e.g., control eventual stock density, increase water aeration, and exchanges) (Harper and Lawrence, 2011; Cartner et al., 2020).

Gill inflammatory lesions, branchitis, associated with environmental disease can be common in zebrafish facilities and present an acute or chronic evolution. Gill epithelial hypertrophy and hyperplasia with eventual clubbing and fusion of gill lamellae, diagnosed mainly by histology, can be caused by chemical pollution (e.g., nitrogen, chlorine) or bacteria (e.g., gliding bacteria) similar to bacterial gill disease of salmonids (Smith, 2019; Cartner et al., 2020).

Other diseases may have unknown causes. The most common idiopathic disease is egg-associated inflammation and fibroplasia. It is associated with egg retention and the diagnosis is done by histological observation of degenerated egg material (e.g., follicular atresia) with associated inflammation. The lesions can extend to periovarian tissues, as the coelomic wall, with transdermal migration and the formation of ulcers. Granulomatous lesions are often present and acid-fast staining should be used to rule out mycobacteriosis. The subclinical disease may reach a clinical presentation when fishes display a distended coelomic cavity, compromised fecundity, and cutaneous ulcers. To decrease incidence of the disease, egg retention should be minimized by maintaining both females and males in tanks and by spawning females regularly (Harper and Lawrence, 2011; Cartner et al., 2020) (Fig. 4.3).

Finally, neoplasia are frequently encountered in zebrafish facilities, with a subclinical or clinical presentation. The majority have an internal location. Definitive diagnosis is done by histology. The most common lesions are seminomas, soft tissue sarcomas as peripheral nerve sheath tumors, and intestinal tumors which can be associated with pathogen infections (e.g., *P. tomentosa*) (Spitsbergen et al., 2012; Cartner et al., 2020).

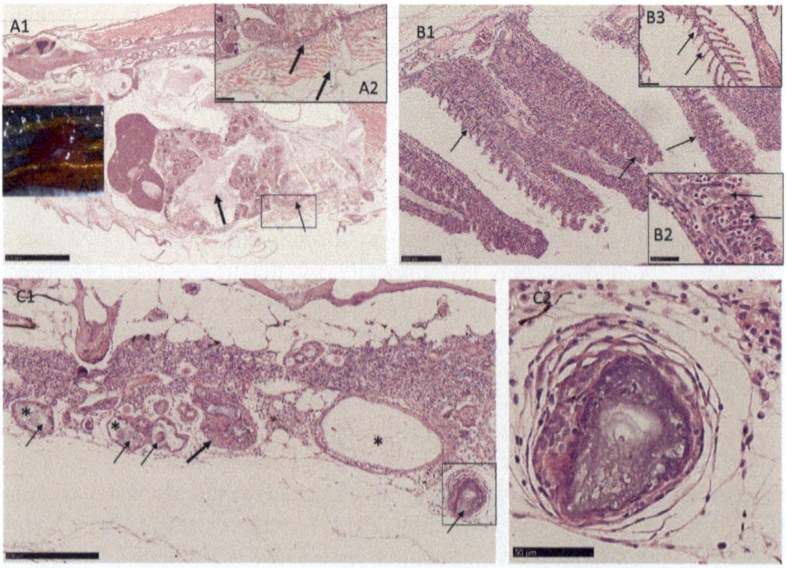

FIGURE 4.3 Histology images of noninfectious diseases lesions in zebrafish. (A1) Egg-associated inflammation and fibroplasia (EAIF). Abundant eosinophilic material depots are seen in the ovary, surrounded by oocytes (*large arrow*), associated with inflammatory cell infiltration of the visceral cavity extending to periovarian tissues (*thin arrow*) (hematoxylin and eosin [HE], bar = 2,5 mm). (A2) Inflammatory adhesions to the coelomic wall. Muscle and skin infiltration by inflammatory cells (→) (HE, bar = 10 μm). This transdermal migration can in a later phase produce skin ulcers like that shown in (A3) (HE, bar = 250 μm). (B1) Gill inflammatory lesions, branchitis, with extensive diffuse hyperplasia of the epithelial cells with clubbing and fusing of the space between the secondary lamellae (→) (HE, bar = 100 μm). (B2) Detail of the loss of the interlamellar spaces through fusion via epithelial cell hyperplasia (→) (HE, bar = 25 μm). (B3) Normal gills with preserved interlamellar spaces are depicted for comparison (→) (HE, bar = 50 μm). (C1) A zebrafish kidney with nephrocalcinosis within the collecting tubules. The tubular lumen shows basophilic deposits of calcium salts precipitate (*thin arrows*), which can cause tubular blockage and tubular dilation (*). (C2) Detail of a renal tubule with the lumen filled with the calcium basophilic deposits (HE, bar = 50 μm). *Images from zebrafish of Instituto Gulbenkian de Ciência, Oeiras and Chronic Diseases Research Center (CEDOC), Nova Medical School, Lisboa.*

Fish and environmental samples

Colony fish

When a microbial contamination hits the EU, it usually enters in a tank, fish, or live feed culture, etc. It does not affect all fish at once. If fish are sampled when the contamination happens, it is likely that the contamination will not be detected. When sampling fish from the colonies, we rely on the pathogens to spread to the sampled fish. It takes time for the pathogens to spread to other fish and tanks; this is why the health screening should be repeated regularly. Statistically, as detailed later, the chances of detecting a pathogen depend on the pathogen prevalence and on the number of sampled fish. However, the selection of colony fish to sample can be directed toward detection of pathogens more likely to affect a specific population. For example, *P. neurophilia* and some *Mycobacterium* spp. are more likely to be detected in older fish (Kent et al., 2009; Murray et al., 2011; Ramsay et al., 2009). Fish with clinical signs are an obvious choice to investigate disease. Nonetheless, it is important to remember that health monitoring also aims to survey microbes in the colonies, and fish of all ages, genders, and genetic backgrounds should be sampled to support the detection of unknown pathogens.

Sentinels

The sentinel method is used to increase the chances of pathogen detection. Sentinel fish are exposed to water leaving colony fish tanks before it is filtered. This is deemed to increase the risk that prevalent pathogen will contaminate sentinel fish, compared to colony fish. Hence, prefiltration sentinel fish would be exposed to pathogens before pathogens have reached a high enough prevalence in the colonies to allow detection by a low number of sampled colony fish. The exposure of prefiltration sentinel fish to sump water is achieved by use of a water pump set in the sump, or by manual changes of prefiltration sentinel tank water three times a week (Mocho et al., 2017). Sentinel fish can be selected to control gender, age, genetic background, and exposure length. For example, the wild-type line most representative of the facility colonies can be used (e.g., AB), selecting 25 females and 25 males, 3–6 months of age. Fifty fish are set in the prefiltration sentinel tanks, and 10 fish are sampled quarterly. At the end of a 12-month period, the sentinel tank is renewed.

Environmental samples

There are various options and methods to monitor chemical and microbial environmental contaminations. For the screening of specific zebrafish pathogens, described sampling methods consist of water filtration (Crim et al., 2017), biofilm swabbing at the water surface, and sludge collection from the tank bottoms (Mocho et al., 2017). All techniques allow detection of potential bacterial pathogens like *Mycobacterium* spp. However, the presence of such commensal microbes is expected. Therefore care should be taken to identify the mycobacterial species, and to monitor any impact on fish health with morbidity and mortality records, as well as confirmation from clinical cases. Sludge analysis also allows prompt detection of parasites (or their eggs, e.g., *P. tomentosa*) by microscopy (or PCR) (Mocho et al., 2017), and of other forms of life thriving on the life support systems (Mocho, 2016). Environmental samples bring another dimension to the monitoring of fish health (Mocho, 2016), though they do not suffice as a sole screening technique, as much as a fish-only approach would be limited by its sample number. Fish and environmental samples should both be performed regularly for a complete approach. It can also be judicious to stagger samples monthly, rather than to sample all media together quarterly, so that routine and continuous surveillance is established.

Diagnostic assays

Necropsy and fresh mounts

Freshly dead animals, either by natural death or euthanasia, should be immediately necropsied or be directly processed for histopathology. Gross necropsy is a diagnostic tool for most external parasites and fungi, and can provide clues regarding infectious and noninfectious diseases that will determine other diagnostic examinations (e.g., microbiology, molecular biology, and histopathology). It should be done, along with histopathology exams, as a routine health screening test in sentinel fish and in nonroutine testing (e.g., sick or surplus fish). The first step of necropsy is an external and internal gross examination, performed under a stereoscopic dissecting microscope due to the zebrafish small dimensions, and using small dissecting instruments. Cutaneous mucous and scales, gill and fin biopsies are observed in wet mounts between coverslip and glass slide with a compound microscope. Small pieces of organs like the liver, intestine, spleen, kidney, and central nervous system tissues are observed under the microscope in squash preparations. These wet mounts and squash preparations are important for the detection of external parasites, fungi, gliding bacteria like *Flavobacterium* spp., granulomas or cysts, intestinal nematodes, and even microsporidian like *P. neurophilia* in central nervous tissue smears. These spores are better visualized by phase-contrast microscopy. Necropsy procedures are identical to what is described for larger fish and scale-related adaptations need to be done for zebrafish (Borges et al., 2016; Cartner et al., 2020; Noga, 2010; Smith, 2019; Yanong, 2003).

Histopathology

Histopathology is an essential diagnostic tool to assess pathogen load and noninfectious disease severity and morbidity. Due to the rapid onset of fish autolysis, which renders this technique practically worthless, samples should be fixed immediately after death (e.g., maximum of 10 min after cessation of opercular beats). Traditionally, various tissue samples of an animal are submitted separately for histology processing, and this can also be done with zebrafish in the case of a gross necropsy. Yet, a major advantage of zebrafish is that, due to their small size, it is possible to submit the entire animal to histology processing for a single slide preparation. To help fish body fixation, the coelomic cavity can be nicked, trying not to disrupt internal organs, and the tail can be removed by cutting through the caudal peduncle well behind the anal fin (unless the tail presents lesions) (Cartner et al., 2020; Zebrafish

International Resource Center, 2020b). The whole fish is fixed in Davidson's or Dietrich's solutions (preferred) (Fournie et al., 2000; Ramsay et al., 2009; Zebrafish International Resource Center, 2020b), or in 10% neutral buffered formalin for at least 48 h. The following decalcification with 10% formic acid, which will not interfere with an eventual Ziehl–Neelsen–Fite staining, should be done in different lengths according to fish dimensions. The next step is paraffin embedding through standard methods (Meyers, 2000). Mid-sagittal histological 3-μ sections should be done in a way to attempt observation in the same plane or section of brain, spinal cord, gills, gonads, and kidney. These sections need skills with a nonimmediate learning curve (Cartner et al., 2020). Ideally, three routine stainings should be done. Hematoxylin and eosin for general purposes, Ziehl–Neelsen–Fite for detection of acid-fast bacteria, and Luna to highlight *P. neurophilia* spores. Additional Gram staining can be performed to differentiate any visible bacteria. For cutaneous ulcers, imprint smears stained with Ziehl–Neelsen can be performed, transverse sections of the whole animal can be done, or the ulcerated skin can be carefully dissected and processed in an independent paraffin block. To optimize results, histopathology slides should be evaluated by a fish veterinary pathologist.

Polymerase chain reaction

PCR consists of detecting a piece of DNA deemed specific to a sought pathogen. The main advantage is that a wide range of material can be used as sample, although care should be taken to ensure the diagnostic laboratory has validated their PCR assay to any submitted material. Otherwise, for example, uncontrolled presence of inhibitors may induce false negatives. PCR can be performed on environmental samples (e.g., surface sump swab biofilm, sludge (Mocho et al., 2017) and water (Crim et al., 2017)). It is the recommended tool to screen found dead fish for pathogens. Fish samples can eventually be pooled to reduce cost. The main disadvantage of PCR is that it only allows detection of predetermined pathogens; it cannot be used for noninfectious disease or to survey against contamination by undefined pathogens. Moreover, PCR detects DNA, whether it is part of a live microbe or an inert DNA sequence. Thus false-positive results are a risk since PCR may reveal the former presence of a pathogen currently unable to multiply. In consequence, whenever possible, any unexpected results should be confirmed by another diagnostic assay allowing detection of live pathogens only, i.e., histopathology, culture, or microscopy.

Determining the number of samples

Defining an epidemiological unit

The first step to determine the number of fish to sample is to estimate the size of the population to screen. The population is called an EU and includes all fish exposed to the same microorganisms because cross-contamination between fish cannot be avoided. For example, fish in the same room, rack, tank, or behind the same biosecurity barrier may be part of the same EU.

Specific pathogen free for a threshold prevalence

In the SPF concept, the term "free" is misleading. The World Organization for Animal Health stipulates that "Scientific methods cannot provide absolute certainty of the absence of disease." "Instead, the aim is to provide adequate evidence (to an acceptable level of confidence) that disease, if present, is present in less than a specified proportion of the population (i.e., threshold prevalence)" (The World Organization for Animal Health (OIE), 2019). In consequence, a fish facility may only claim to be SPF for a defined threshold prevalence. The SPF concept also relies on measures to prevent contamination by the pathogens specifically screened against. Introduction of colony fish through a triage and quarantine process and biosecurity programs is therefore of particular importance when aiming for an SPF status.

Number of fish sample

There is a formula to demonstrate that if the pathogen is present at the time of sampling, it is below a chosen threshold prevalence. The formula takes into account a chosen probability level (e.g., 95% confidence) and diagnostic assay specificity and sensitivity. Application of the formula determines the number of colony fish to sample according to the EU population size (Sergeant, 2018). We propose to apply this formula and to introduce bias toward increasing chances of pathogen detection. For example, as described earlier, at least 10 of the screened fish are prefiltration sentinels. Environmental samples are added to the calculated number of fish samples. Samples are taken on a quarterly basis and staggered monthly. Table 4.1 shows examples of screening pattern for various threshold prevalence and specific zebrafish pathogens. It is deemed that at least six consecutive negative quarterly sets of screens are necessary to claim some SPF status, as long as threshold prevalence is clearly defined, and there is transparency regarding screening, import, and biosecurity programs so that collaborators can assess biosecurity risks when importing fish from the SPF-claimed EU. Quarantine and fish imported in quarantine should be screened as well. For a busy quarantine, health monitoring can follow a similar pattern to a main holding EU. Otherwise, each import should be screened (Mocho et al., 2017).

Introduction of zebrafish colonies and biosecurity

Triage of imports

To preserve the microbiological status of an EU, care should be taken when introducing fish from an external source into the animal unit. First, the risk

TABLE 4.1 Quarterly screening pattern per epidemiological unit (EU) according to threshold prevalence.

Samples	Month 1	Month 2	Month 3	Threshold prevalence	Monitored pathogens and diseases
Sump surface swab	2 samples			N/A	*Mycobacterium* spp.
Live feed cultures	2 samples				*Mycobacterium* spp.
Sludge		2 samples			*Mycobacterium* spp., *Pseudocapillaria tomentosa*
Number of fish samples according to threshold prevalence			15	20%	Any pathogens and diseases by histopathology Only specific pathogens by polymerase chain reaction (PCR)
			30	10%	
			60	5%	
			150	2%	

Environmental samples are screened for *Mycobacterium* spp. by PCR. *P. tomentosa* can be detected in sludge by PCR or by microscopy (Mocho et al., 2017). Fish samples represent prefiltration sentinel and colony fish. When PCR is used, fish samples are screened quarterly for *Mycobacterium* spp., *P. tomentosa,* and *Pseudoloma neurophilia* and other pathogens deemed a hazard for the EU due to specific fish sources or multispecies holding conditions. Other pathogens may be screened twice a year or not monitored. Histopathology allows a broad screening for pathogens and noninfectious diseases. The number of fish samples according to threshold prevalence is calculated with online software (Sergeant, 2018). Specificity and sensitivity are set to 100%, confidence at 95%, and the EU holds more than 1000 fish. For example, when the threshold prevalence is set at 20%, 15 fish samples are required quarterly. This increases to 150 fish for a threshold prevalence of 2%.

related to each specific import should be assessed. For example, adult fish represent a higher risk than fertilized embryos that could be surface disinfected at arrival in the facility. It is also relevant to inquire whether the facility of origin is able to provide some information on their health and microbiological status, and on their biosecurity process. For example, pet shops and multispecies facilities with small barriers between aquaria dedicated to species may present a higher risk than zebrafish-only laboratories. Once the risk is assessed, then a decision is taken not to import the colony, import in quarantine, or import directly into the EU. This is based on the presence in the exporting facility of pathogens excluded from the importing facility (e.g., *M. marinum*), on the need to further assess the microbiological status in quarantine (Mocho et al., 2017), or on the need to treat the fish (e.g., *P. tomentosa*) before EU introduction (Kent et al., 2019a; Erlacher-Reid, 2018; Cartner et al., 2020).

The option to import directly into an EU should only be reserved to a trusted facility demonstrating thorough health monitoring and biosecurity programs. The triage system to decide how to handle imports should be established in advance in agreement with users.

Quarantine

In zebrafish laboratories, the import process often consists of receiving fish (embryo whenever possible), and rearing and breeding imported generations in quarantine until they are culled. Only offspring embryos from imported fish are surface disinfected and introduced to the EU. The choice of disinfectant, concentration, and length of exposure (e.g., 10 min in 30 ppm unbuffered sodium hypochlorite (Mason et al., 2016), 5 min in 100 ppm unbuffered sodium hypochlorite (Kent et al., 2014), 2 min in 12.5–25 ppm buffered iodine (Cartner et al., 2020)) should be adapted to the pathogens to exclude and to the required zebrafish embryo survival rate. This practice aims at isolating quarantine microbiological status from the EU. Biosecurity barriers between quarantine and EU(s) are therefore essential to the relevance of the process. Cross-contaminations between the two areas must be avoided. Practical implications are numerous, and the set-up of reliable quarantine biocontainment should be central to the design of the facility and to the biosecurity program. The main principles are that quarantine tanks, racks, or systems should not share the same room with the main EUs, otherwise water cross-contamination would be very challenging to avoid. Personnel assuming quarantine duties should not go back on the same day to other EU rooms, ensuring a "from clean to dirty" flow. Gears and materials (e.g., maintenance tools, water test kits, pens, feeds and feeding devices, incubators, microscopes, etc.) should be dedicated to quarantine and not shared with EUs. Tanks and other dirty items should be cleaned and disinfected before they cross paths with devices from main EUs. For example, tanks can be manually cleaned and bagged in the quarantine room. As this first bag crosses the barrier to exit quarantine, it is set in a second bag which can then lie in corridors or storage for dirty items waiting for further processing. Cage or tank wash machines usually do not provide a sterilization process and introducing contaminated tanks in these machines would spread contaminants to tanks destined for the main EUs. Therefore dirty quarantine tanks are autoclaved before reaching the main tank wash process, and eventually will be reused in quarantine or EUs.

Biosecurity

The biosecurity program can be particularly challenging for multispecies facilities. The precautions described earlier for quarantine may apply between EUs deemed at risk of cross-contamination. Such strong barriers between EUs preserve each EU's microbiome as it avoids contamination by unknown microorganisms. However, such restrictions have a significant cost on daily work pattern and budget. Alternatively, barriers can be set according to prevalent and excluded pathogens in each EU, and treatment and prevention efficacy for the specific pathogens representing a hazard. When the main pathogen(s) likely to be transmitted between two EUs is eliminated by means other than autoclave sterilization, a less resource-consuming process can be adopted. For example, *P. tomentosa* eggs can be destroyed by desiccation (Kent et al., 2019b), and a drying period between EUs may therefore constitute a sufficient barrier. The biosecurity program also needs to consider entry of hazards carried by personnel, visitors, biological material, feed, etc. Specific restrictions can be set according to risks, e.g., limiting visitor entry to only one EU per day. Experience shows that reliability of barriers relies on two pillars that need to be addressed at managerial level: buy-in from users and frequency of passage at the barriers. The former point reflects on culture of care, training, stakeholders' involvement in the decision-making process, and internal communication. The latter point highlights the importance of bearing in mind biosecurity when designing animal facilities. First, flows for personnel, eggs, larvae, adult fish, feed, nets, tanks, cadavers, storage, etc must be considered. Then barriers can be set with specific details on disinfection and personal protective equipment. This is particularly important to prevent staff contamination with zoonotic pathogens. For example, at the minimum, the use of gloves seems unavoidable when hands may come in contact with water. Finally, the last pillar of the biosecurity program is the treatment of water supply. The risk of incoming water introducing pathogens can be mitigated by reverse osmosis, ultraviolet, and other filtrations. Still, water should be seen as the main pathogen carrier across the fish facility. When designing the facility water circuits, fish on the same recirculation loop or potentially exposed to the same water should therefore be considered in the same EU since they are behind the same barrier and potentially exposed to the same pathogens.

Conclusion

The EU concept is key to the design, biosecurity, and health monitoring programs of fish facilities. From there, detection of specific pathogens can be developed to address most prevalent laboratory zebrafish pathogens or other relevant microorganisms deemed a hazard due to the local research models or multispecies context. However, health monitoring must also rely on a broader approach to monitor husbandry output, screen for noninfectious disease, and survey emergence of pathogens.

Acknowledgments

The authors would like to thank Pedro Faísca, Joana Lóios, and Andreia Mindouro from the Instituto Gulbenkian de Ciência (IGC) (https://gulbenkian.pt/ciencia/), Ana Farinho, Petra

Pintado, and Fábio Valério from the Chronic Diseases Research Center at the Nova Medical School (CEDOC) (http://cedoc.unl.pt/), and Tânia Carvalho from the Champalimaud Centre for the Unknown for their input with the histopathology section and figures. The authors would like to thank IGC and CEDOC fish facilities staff. Finally, the authors would like to thank Chereen Collymore, Susan Farmer, Katy Murray, and Emmanuel Leguay for the fruitful discussion within the FELASA-AALAS Working Group on Health Monitoring for fish in research.

References

Aleström, P., D'Angelo, L., Midtlyng, P.J., Schorderet, D.F., Schulte-Merker, S., Sohm, F., Warner, S., 2019. Zebrafish: housing and husbandry recommendations. Lab. Anim. 54 (3), 213–224.

Altan, E., Kubinski, S.V., Boros, A., Reuter, G., Sadeghi, M., Deng, X., et al., 2019. A highly divergent picornavirus infecting the gut epithelia of zebrafish (*Danio rerio*) in research institutions worldwide. Zebrafish 16 (3), 291–299.

Bermudez, R., Losada, A.P., de Azevedo, A.M., Guerra-Varela, J., Pérez-Fernández, D., Sánchez, L., et al., 2018. First description of a natural infection with spleen and kidney necrosis virus in zebrafish. J. Fish. Dis. 41 (8), 1283–1294.

Binesh, C.P., 2013. Mortality due to viral nervous necrosis in zebrafish *Danio rerio* and goldfish *Carassius auratus*. Dis. Aquat. Org. 104 (3), 257–260.

Borges, A.C., Pereira, N., Franco, M., Vale, L., Pereira, M., Cunha, M.V., et al., 2016. Implementation of a zebrafish health program in a research facility: a 4-year retrospective study. Zebrafish 13 (Suppl. 1), 115–126.

Cartner, S.C., Eisen, J.S., Farmer, S.C., Guillemin, K.J., Kent, M.L., Sanders, G.E. (Eds.), 2020. The Zebrafish in Biomedical Research. Elsevier, San Diego.

Chang, C.T., Benedict, S., Whipps, C.M., 2019a. Transmission of *Mycobacterium chelonae* and *Mycobacterium marinum* in laboratory zebrafish through live feeds. J. Fish. Dis. 42 (10), 1425–1431.

Chang, C.T., Lewis, J., Whipps, C.M., 2019b. Source or sink: examining the role of biofilms in transmission of *Mycobacterium* spp. in laboratory zebrafish. Zebrafish 16 (2), 197–206.

Clark, T.S., Pandolfo, L.M., Marshall, C.M., Mitra, A.K., Schech, J.M., 2018. Body condition scoring for adult zebrafish (*Danio rerio*). J. Am. Assoc. Lab. Anim. Sci. 57 (6), 698-702.

Crim, M.J., Lawrence, C., Livingston, R.S., Rakitin, A., Hurley, S.J., Riley, L.K., 2017. Comparison of antemortem and environmental samples for zebrafish health monitoring and quarantine. J. Am. Assoc. Lab. Anim. Sci. 56 (4), 412–424.

Erlacher-Reid, C.D., 2018. Considerations for treatment of large zoologic collections: fish. Vet. Clin. North Am. Exot. Anim. Pract. 21 (2), 311–325.

European Commission, 2020. 2019 Report on the Statistics on the Use of Animals for Scientific Purposes in the Member States of the European Union in 2015–2017. Report from the commission to the European parliament and the council. Report no. COM(2020) 16 final, 5 February 2020. https://ec.europa.eu/environment/chemicals/lab_animals/reports_en.htm. (Accessed 28 October 2020).

Farm Animal Welfare Council, 2012. Five Freedoms. http://webarchive.nationalarchives.gov.uk/20121007104210/http:/www.fawc.org.uk/freedoms.htm. (Accessed 28 October 2020).

Fournie, J.W., Krol, R.M., Hawkins, W.E., 2000. Fixation of fish tissues. In: Ostrander, G.K. (Ed.), The Laboratory Fish. Academic Press, San Diego, pp. 569–578.

Gaulke, C.A., Martins, M.L., Watral, V.G., Humphreys, I.R., Spagnoli, S.T., Kent, M.L., Sharpton, T.J., 2019. A longitudinal assessment of host-microbe-parasite interactions resolves the zebrafish gut microbiome's link to Pseudocapillaria tomentosa infection and pathology. Microbiome 7 (1), 1–16.

Harper, C., Lawrence, C. (Eds.), 2011. The Laboratory Zebrafish. CRC Press, Boca Raton (FL.

Kent, M.L., Buchner, C., Barton, C., Tanguay, R.L., 2014. Toxicity of chlorine to zebrafish embryos. Dis. Aquat. Org. 107 (3), 235–240.

Kent, M.L., Buchner, C., Watral, V.G., Sanders, J.L., Ladu, J., Peterson, T.S., Tanguay, R.L., 2011. Development and maintenance of a specific pathogen-free (SPF) zebrafish research facility for *Pseudoloma neurophilia*. Dis. Aquat. Org. 95 (1), 73–79.

Kent, M.L., Feist, S.W., Harper, C., Hoogstraten-Miller, S., Law, J.M., Sánchez-Morgado, J.M., et al., 2009. Recommendations for control of pathogens and infectious diseases in fish research facilities. Comp. Biochem. Physiol. C Toxicol. Pharmacol. 149 (2), 240–248.

Kent, M.L., Watral, V., Gaulke, C.A., Sharpton, T.J., 2019a. Further evaluation of the efficacy of emamectin benzoate for treating *Pseudocapillaria tomentosa* (Dujardin 1843) in zebrafish *Danio rerio* (Hamilton 1822). J. Fish. Dis. 42 (10), 1351–1357.

Kent, M.L., Watral, V., Villegas, E.N., Gaulke, C.A., 2019b. Viability of *Pseudocapillaria tomentosa* eggs exposed to heat, ultraviolet light, chlorine, iodine, and desiccation. Zebrafish 16 (5), 460–468.

Lee, C.J., Paull, G.C., Tyler, C.R., 2019. Effects of environmental enrichment on survivorship, growth, sex ratio and behaviour in laboratory maintained zebrafish *Danio rerio*. J. Fish. Biol. 94 (1), 86–95.

Lieggi, C., Kalueff, A.V., Lawrence, C., Collymore, C., 2020. The influence of behavioral, social, and environmental factors on reproducibility and replicability in aquatic animal models. ILAR J. 60 (2), 270–288.

Mason, T., Snell, K., Mittge, E., Melancon, E., Montgomery, R., McFadden, M., et al., 2016. Strategies to mitigate a *Mycobacterium marinum* outbreak in a zebrafish research facility. Zebrafish 13 (Suppl. 1), 77–87.

Meyers, T.R. (Ed.), 2000. Fish Pathology Section Laboratory Manual. Alaska Department of Fish and Game, Juneau.

Midttun, H.L.E., Vindas, M.A., Nadler, L.E., Øverli, Ø., Johansen, I.B., 2020. Behavioural effects of the common brain-infecting parasite Pseudoloma neurophilia in laboratory zebrafish (*Danio rerio*). Sci. Rep. 10 (1), 8083.

Mocho, J.P., 2016. Three-dimensional screen: a comprehensive approach to the health monitoring of zebrafish. Zebrafish 13 (Suppl. 1), 132–137.

Mocho, J.P., Martin, D.J., Millington, M.E., Saavedra Torres, Y., 2017. Environmental screening of *Aeromonas hydrophila*, *Mycobacterium* spp., and *Pseudocapillaria tomentosa* in zebrafish systems. J. Vis. Exp. 130, 55306. https://doi.org/10.3791/55306.

Murray, K.N., Dreska, M., Nasiadka, A., Rinne, M., Matthews, J.L., Carmichael, C., et al., 2011. Transmission, diagnosis, and recommendations for control of Pseudoloma neurophilia infections in laboratory zebrafish (*Danio rerio*) facilities. Comp. Med. 61 (4), 322–329.

Murray, K.N., Peterson, T.S., 2015. Pathology in practice. *P. tomentosa* infection in zebrafish. J. Am. Vet. Med. Assoc. 246 (2), 201–203.

Noga, E.J. (Ed.), 2010. Fish Disease: Diagnosis and Treatment. Wiley-Blackwell, Ames.

Rácz, A., Dwyer, T., Killen, S.S., 2019. Overview of a disease outbreak and introduction of a step-by-step protocol for the eradication of *Mycobacterium haemophilum* in a zebrafish system. Zebrafish 16 (1), 77–86.

Ramsay, J.M., Watral, V., Schreck, C.B., Kent, M.L., 2009. *Pseudoloma neurophilia* infections in zebrafish *Danio rerio*: effects of stress on survival, growth, and reproduction. Dis. Aquat. Org. 88 (1), 69–84.

Romano, L.A., Klosterhoff, M.C., de Medeiros, A.F.F., Pedrosa, V.F., 2020. Comparative study of Ziehl neelsen's technique and the fite-faraco technique in the histopathological diagnosis of mycobacteriosis in fish. Asian J. Res. Anim. Vet. Sci. 6 (1), 27–34.

Sanders, J.L., Monteiro, J.F., Martins, S., Certal, A.C., Kent, M.L., 2020. The impact of Pseudoloma neurophilia infection on body condition of zebrafish. Zebrafish 17 (2), 139–146.

Sergeant, E.S.G., 2018. Epitools Epidemiological Calculators. Ausvet. http://epitools.ausvet.com.au. (Accessed 28 October 2020).

Smith, S.A. (Ed.), 2019. Fish Diseases and Medicine. CRC Press, Boca Raton.

Spitsbergen, J.M., Buhler, D.R., Peterson, T.S., 2012. Neoplasia and neoplasm-associated lesions in laboratory colonies of zebrafish emphasizing key influences of diet and aquaculture system design. ILAR J. 53 (2), 114–125.

The World Organisation for Animal Health (OIE), 2019. Aquatic Animal Health Code. https://www.oie.int/en/standard-setting/aquatic-code/access-online. (Accessed 28 October 2020).

Wafer, L.N., Jensen, V.B., Whitney, J.C., Gomez, T.H., Flores, R., Goodwin, B.S., 2016. Effects of environmental enrichment on the fertility and fecundity of zebrafish (*Danio rerio*). J. Am. Assoc. Lab. Anim. Sci. 55 (3), 291–294.

Yanong, R.P.E., 2003. Necropsy techniques for fish. Seminars Avian Exot. Pet Med. 12 (2), 89–105.

Zebrafish International Resource Center, 2020a. Summary of Submissions to the ZIRC Diagnostic Service. https://zebrafish.org/wiki/health/submission/report. (Accessed 28 October 2020).

Zebrafish International Resource Center, 2020b. ZIRC Health Services Sample/Specimen Preparation. https://zebrafish.org/wiki/health/submission/preperation. (Accessed 28 October 2020).

Chapter 5

The welfare of zebrafish

Paul Georg Schroeder
Department of Biomedical Services, Oxford University, United Kingdom

Introduction

In the course of devising this addition to the textbook on laboratory fishes, there was one recurring question: Why would zebrafish, but not other fish species, merit their own welfare chapter in this (already comprehensive) piece of work? The answer may be somewhat subjective as it is informed by my own and my colleagues' years working as lab animal veterinarians: no fish species is used more widely in the laboratory and is therefore more relevant when it comes to considering its welfare. With a history of more than 45 years of keeping this species in laboratories and using it for procedures, there is, or at least there should be, a wealth of knowledge and experience from which a meaningful welfare chapter could be distilled.

As with any other laboratory species, desired testing and research outcomes can limit husbandry improvements and therapeutic options in other ways than for fish exploited as food or ornamental species. When licensed research and regulatory toxicology establishments discuss welfare-relevant aspects of their projects with zebrafish, this process can be particularly marred by stakeholders adopting some of the more entrenched views on the capacity of these animals to feel pain and anxiety (e.g., Rose et al., 2012). This can be a repeated source of frustration for any clinician or technician trying to promote welfare improvements (Schroeder and Mocho, 2014).

With 388,000 (accounting for 9% of the total) regulated procedures carried out in 2019, zebrafish are the United Kingdom's second most popular research model (UK Home Office, 2020). For the European Union, the European Parliament reported the use of 1.2 million fish in regulated procedures for the first time in 2017, including 500,000 zebrafish (The Commission to the Council and the European Parliament, 2020). There are no precise figures for the United States but extrapolating from the species split in the United Kingdom and European Union as well as the overall reported total of 820,000 nonrodent, nonaquatic vertebrates (DEPARTMENT OF AGRICULTURE - Animal and Plant Health Inspection Service, 2017) used in regulated

Laboratory Fish in Biomedical Research. https://doi.org/10.1016/B978-0-12-821099-4.00010-9
Copyright © 2022 Elsevier Inc. All rights reserved.

procedures in the United States in 2016, a yearly figure of 1–2 million zebrafish could be approximated.

In the United Kingdom, like all other vertebrates used in scientific procedures, fish are protected under the Animals (Scientific Procedures) Act 1986 2012 Amendment Regulation, which was itself a national implementation of the new Directive 2010/63 EU. As a result, breeding, husbandry, and scientific use of these animals are strictly regulated. Fish are protected from the time of free feeding. For zebrafish this means that from the fifth day postfertilization, any intervention that may constitute pain, distress, suffering, or lasting harm requires full licensing as prescribed by the regulator. Over the last two decades a considerable amount of information has been gained exploring physiological and behavioral welfare indicators in fish, including zebrafish, carp, trout, and goldfish (Sneddon, 2003, 2006, 2012; Sneddon et al., 2003; Braithwaite and Boulcott, 2007; Branson, 2008; Reilly et al., 2008; Nordgreen et al., 2009).

Two major factors affect the welfare of laboratory fish. Apart from the stress that can be caused by painful or stressful scientific procedures (such as invasive surgery or behavioral work with food deprivation), inappropriate husbandry (including housing, arrangement of social groups, feeding regime) can impose a significant welfare cost on the animals held in the research facility (Carbone, 2004). This reduction in welfare through an animal's exposure to an unnatural, captive environment was coined "contingent suffering" (Richmond, 2002). This is an expansion of the concept of "suffering in the absence of disease" (Fox, 1989), where animals are well catered yet are also in poor welfare, even without the infliction of harmful procedures. In these cases, welfare is reduced by the denial of what is pleasant (Richmond, 2002) or natural (Shepherdson et al., 1998). Despite this evidence, a review of research scientists' priorities in terms of housing standards for laboratory animals showed that the actual requirements of the animals ranked behind concern over legal constraints and economic issues (Morton, 1998). Contingent suffering can lead to poor health and abnormal behavior, which undermines the validity of data gained from animal experiments (Würbel and Garner, 2007).

Measuring welfare in zebrafish

Laboratory animals face a wide range of stimuli and experiences, from repeated confrontations with humans, for some a perceived enemy species (Laule, 2010), to restraint and transport. Stressed animals display differing baseline behaviors and after experimental interventions their responses are altered compared with unstressed individuals (Richmond, 2010). As found in other vertebrates, fish react to environmental challenges (stressors) through a range of neuroendocrine and behavioral stress responses (Huntingford et al., 2006).

Behavioral stress response

Stress response, to one or several stressor challenges (Fig. 5.1) is routinely used to evaluate the welfare state of an animal, true to the concept of welfare

FIGURE 5.1 Experimental zebrafish tank with a black marker line indicating "hovering low" threshold and stressor challenge with mock heron; the latter is exacerbated by injecting alarm pheromone into the water (Schroeder, 2014).

as "the animal's state in regard to its ability to cope with its environment" (Broom, 1991), which focuses on the adaptive response to outside changes. The most widely used behavioral welfare assessments in (laboratory) fish include time in the tank bottom, activity, and ventilation rate. "Time in the tank bottom" is a test measuring time spent in the bottom third of a tank ("hovering low" Fig. 5.1), described as an indicator of stress in zebrafish (Parker et al., 2012) and pain in goldfish (Nordgreen et al., 2009), adapted from the "novel tank diving test" (Egan et al., 2009). "Activity" is another useful indicator, which is the number of direct swimming movements longer than the length of the animal (Reilly et al., 2008) in 5 min. This is a measure of swimming frequency rather than distance swum per time period. Finally, "ventilation rate" (opercular beat rate per min) can be measured by direct observation with the observer obscured behind a screen or observed through a high-definition camera when feasible (Reilly et al., 2008; Schroeder and Sneddon, 2017). This can be particularly challenging as opercular movements in zebrafish are extremely hard to detect when the animals are not stressed, whereas the very visible and high-frequency hyperventilation displayed by stressed animals can be very hard to measure when opercular beat rate increases to 300 per minute or above (Schroeder and Sneddon, 2017).

More recently, the concept of fractal dimension was introduced to assess postnociceptive zebrafish behavior, calculating 3D trajectories to determine the effect of a range of algesiogenic stimuli on the complexity of movement patterns (Deakin et al., 2019; Fig. 5.2).

Neuroendocrine stress response

Like humans, but unlike rodents who employ corticosterone, various publications have suggested cortisol as the primary stress response hormone for fish

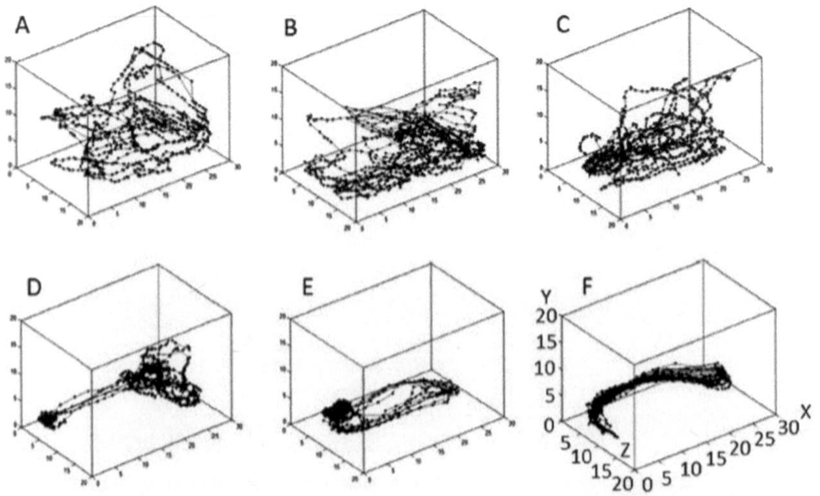

FIGURE 5.2 (A–C) Comparison between 3D movement trajectories from control animals, with high fractal dimension scores and animals subjected to three different algesiogenic stimuli (D = acetic acid injection lip, E = fin clip, F = implant of passive integrated transponder) (Deakin et al., 2019).

(Schreck, 1981; Egan et al., 2009). In the past, studies used whole body cortisol measurements (the fish is killed and homogenized to extract cortisol; in larger fish [>2 g] plasma cortisol is often measured from blood samples) to investigate the effect of common laboratory stressors such as crowding (Ramsay et al., 2006), predator exposure (Barcellos et al., 2007), and net handling (Ramsay et al., 2009) on zebrafish. While net handling stress (with a cortisol peak at 15 min postnet stress and recovery down to baseline levels after 1 h) and predator exposure lead to significant increases in whole body cortisol levels, crowding only leads to higher levels of stress hormone when modulated with fasting.

While the corticosteroid measurements listed up to this point require sacrificing the whole fish, there are noninvasive methods to measure corticosteroids from zebrafish holding water, based on the fact that these hormones are released into the tank water through the gills. For example, cortisol released to holding water correlates positively with plasma concentrations, with the ACTH challenge test demonstrating that a hike in circulating cortisol concentration can be detected in the surrounding tank water (Felix et al., 2013).

Husbandry

Environmental enrichment

The term environmental enrichment relates to a broad spectrum of husbandry improvements and includes structural enrichment as well as behavioral

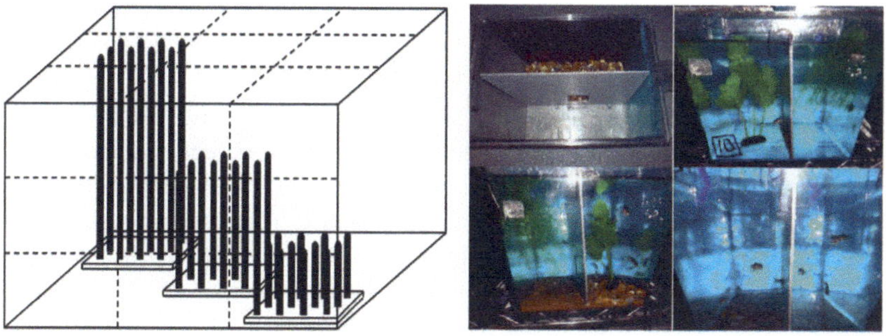

FIGURE 5.3 Tank designs for environmental enrichment zebrafish preference assays: (*left*) autoclavable glass rods with different heights (Wilkes et al., 2012); (*right*) binary preference tank with different naturalistic designs; gravel image bottom right (Schroeder et al., 2014).

engineering, social context, and live feeding. For laboratory fishes, research has focused mainly on structural enrichment.

Preference studies on stickleback (*Gasterosteus aculeatus*) showed that they preferred more complex substrate (gravel and sand) to simple substrate (just sand) although this was linked to higher food availability (Webster and Hart, 2004; Brydges and Braithwaite, 2009). More recently, studies have focused on zebrafish: fish reared in barren conditions preferred structural enrichment over standard (barren) conditions (Kistler et al., 2011); however, when fish were held in pairs this was influenced by dominance status and in groups this was influenced by gender (Schroeder et al., 2014). The natural zebrafish habitat has been extensively characterized (Engeszer et al., 2007) and tank environments designed accordingly, extrapolating from natural habitat preferences (Schroeder et al., 2014; Fig. 5.3). When the resulting designs were offered in binary choice assays, preference for enrichment elements was strongest for gravel substrate and overhanging plants but in almost equal measure, images of gravel (Fig. 5.3). This showed that captive-bred zebrafish prefer structurally enriched environments, suggesting that zebrafish may have some behavioral needs that are not met by barren conditions. These findings were used to inform a long-term study: 10-week-old zebrafish were offered the most preferred enrichments (gravel substrate, artificial plants) or barren tanks over 4 months, after which animals were assayed for poststress behavior. There was no significant behavioral variation between fish from different tank designs. Instead, behavioral stress response was much more influenced by stressor type. Water cortisol analysis did not reveal any treatment effect with regard to enrichment, nor was there significant cortisol elevation after the actual stressor challenges (Schroeder, 2014).

It is not always feasible to offer naturalistic environmental enrichment in every institution: in facilities conducting regulatory toxicology testing, any tank additions need to be inert and autoclavable. This limitation affects the range of possible enrichment items and resulting designs may bear very little biological relevance. For example, toxicology compliant structures include glass rods that neither measurably augmented nor decreased welfare of zebrafish (Wilkes et al., 2012) (Fig. 5.3).

Rather than imposing enrichment, it is vital that we provide fish in our care with choices to understand their subjective preferences. However, caution must be applied to simple preference tests as the animals will make a choice between whatever two or more items are presented to them but not necessarily shed any light on the biological relevance of these items. To address that matter, a preference order was compiled from binary choice tests by providing zebrafish with a range of items not only routinely used as enrichment in ornamental and laboratory aquaria, but also guided by evidence on wild zebrafish habitats (Schroeder et al., 2014).

In research facilities across the United Kingdom, environmental enrichment is adopted as standard for primates, rodents, and birds (Wolfensohn and Lloyd, 2013) with modified cages informed by preference tests. If the absence of comparable guidelines for fish is due to the lack of evidence about the benefits (Wilkes et al., 2012) associated with enrichment for this group of animals, then the preference study findings (Kistler et al., 2011; Schroeder et al., 2014) should inform a reevaluation of zebrafish housing. If the strong evidence of preference testing alone constitutes little incentive for facilities to add enrichments, one could also argue, with the same logic, that the addition of structural enrichment is not a confounding factor for zebrafish when recovering from pain or stress.

Strains, genetic modifications, and harmful phenotypes

Strains and genetic modifications deserve a brief mention here despite a lack of published data on their welfare cost. For mammals used in laboratories the breeding of certain strains and creation of harmful knockouts generates a series of welfare concerns (Workman et al., 2010). Similar to rodents, there are a number of fish strains with phenotypes impacting welfare. For example, leopard strain zebrafish were used as a model for anxiety (Egan et al., 2009) as they were thought to be more prone to stress. Several induced mutations create particular pathological or genetic conditions such as retinal photoreceptor degeneration (Stearns and Evangelista, 2007), mental retardation (Wu et al., 2017), liver tumors (Li et al., 2019), or cardiac ablation (Wang et al., 2013). While ecotoxicological procedures often employ strain-specific phenotypical differences, the creation of transgenic lines for certain purposes such as measuring the effect of endocrine disrupting materials (Lee and Lu, 2014) can raise difficult ethical questions.

Refinements in regulated procedures

Arguably, welfare of zebrafish as a species overwhelmingly relates to how these animals are used and housed in research facilities, while a range of interventions and refinements have the potential to attenuate these impacts. In the European Union and the United Kingdom, with very few exceptions, invasive procedures with the potential to cause pain, suffering, distress, or lasting harm can only be carried out on protected animals after these are anesthetized. Further procedural refinements mainly focus on optimizing anaesthetic protocols, for example, through the addition of premedication and/or analgesia.

Pain and analgesic considerations

Pain perception in fish is likely to differ from human experience and may not be consistent with our understanding of mammalian physiology (Stoskopf, 1994). Rather than ruling out pain on the basis of mere anatomical differences, judgments should be made based on evidence. For example, as a starting point, the understanding of pain in a particular group of animals can be based upon behavioral or physiological changes linked to painful events in other species, such as postsurgically increased respiratory rates in cats and dogs and a higher opercular beat rate for fish subjected to trauma (Brown, 1993).

Investigations of postnociceptive behaviors in rainbow trout (*Oncho-rhynchus mykiss*) (Sneddon, 2003), common carp (*Cyprinus carpio*), and zebrafish (*Danio rerio*) (Reilly et al., 2008; Maximino, 2011) established that the type and magnitude of behavioral patterns after painful events (injection of acetic acid) is species specific. For example, an increased ventilation rate was indicative of nociceptive stimuli in zebrafish and trout but not in carp. In the same study, zebrafish displayed a reduced swim rate, whereas rainbow trout showed an increase. Injection of acetic acid into the lip elicited "lip rubbing" in rainbow trout, an atypical behavior thought to be analogous to humans rubbing painful areas (although this causality has not been accepted by everyone (Rose, 2003)) but is not seen in control fish. This particular behav-ioral response has also been recorded in goldfish experiencing a painful event (Newby et al., 2009). A potentially analogous response has been observed in zebrafish, where after acid injections near the tailfin (Maximino, 2011) or tailfin clipping (Schroeder and Sneddon, 2017), animals have reacted with rapid fin movements described as "tailfin fanning."

The ongoing debate about the capacity of fish to experience pain beyond pure reflexes has dominated the Ethical Review process in several research institutions (pers. obs.) and accordingly the field of laboratory animal welfare science. With the application of the 3Rs (replacement, reduction, refinement) the concept of replacement by *less sentient species* has triggered a debate (Chandroo et al., 2004; Volpato et al., 2007) on whether it is justifiable to attribute less sentience to fish than, for example, rodents. One should argue instead that such assumptions need to be backed up by solid research findings.

As evidence on pain and pain management in zebrafish is mounting (Schroeder and Sneddon, 2017; Lopez-Luna et al., 2017; Lelek et al., 2020) it is worth discussing the practical implications of this, not least the idea of a usable protocol. To that objective, different immersion drugs were trialed with varying effect on zebrafish. Lidocaine administered at 5 mg/L has been shown to modulate postnociceptive behavior after a tailfin clip (Schroeder and Sneddon, 2017) and noxious temperature stimuli (Lopez-Luna et al., 2017). For larval zebrafish morphine (48 mg/L), immersion proved efficacious (Lopez-Luna et al., 2017), possibly due to better gill permeability for this type of molecule in juvenile fish. For cardiac cryoinjury procedures, morphine, but not lidocaine, significantly improved animal welfare post surgery, without impairing the heart regeneration process (Lelek et al., 2020), although studies on rodents have shown that opioids prevent tissue regeneration in mice through inhibition of reactive oxygen species production (Labit et al., 2018).

Anaesthetic aversiveness

The only two agents (buffered tricaine methanesulfonate and 2-phenoxyethanol) currently licensed as fish anaesthetics in the United Kingdom have been shown to be aversive to zebrafish (Readman et al., 2013; Wong et al., 2014) and the same authors suggest that nonaversive agents such as etomidate and 2,2,2-tribromoethanol should be considered as anaesthetics of choice. Yet etomidate has shown poor analgesic capacity (Flecknell, 2009), which casts doubt on its suitability for mass use as immersion anaesthetic. Furthermore, since etomidate inhibits adrenal steroidogenesis as shown for mammalian (Wagner et al., 1984) and aquatic species (Ross et al., 2008), it may be regarded as a source of interference in experimental models.

Humane endpoints

EU legislation prescribes that "death as the end-point of a procedure shall be avoided as far as possible and replaced by early and humane end-points (HEP). Where death as the end-point is unavoidable, the procedure shall be designed so as to: (a) result in the deaths of as few animals as possible; and (b) reduce the duration and intensity of suffering to the animal to the minimum possible" (European Parliament and Council of the European Union, 2010). In the context of managing animal welfare during regulated procedures, the endpoint is the timepoint when exposure to a harmful or stressful treatment ends (Wolfensohn and Lloyd, 2013). To limit such exposure, knowledge of fish physiology and fish welfare expertise needs to match the severity of procedures and stressors that these organisms are subjected to (Nordgreen et al., 2009). It is essential that clinical signs, indicating when predefined thresholds preceding the point-of-no-return to agonal suffering and mortality (Ellis and Katsiadaki, 2020) have been reached, are understood. For zebrafish used in laboratory procedures, this is aided by methodical capture of behavioral and physiological welfare parameters (details in previous section). Indexes and scoring systems based on clinically relevant observations and with robust cross-observer correlation can further aid consistency in the recognition of endpoints. Despite the notable absence of a standardized approach and set criteria for fishes (Ellis and Katsiadaki, 2020) the peer-reviewed evidence base for humane endpoints applicable to zebrafish and other common laboratory fish species is gradually increasing, e.g., with focus on body condition scores (Clark and Pandolfo, 2018) or pain response and tissue injury (Sneddon, 2009).

Welfare-focused systematic review of surgical zebrafish models for cardiac regeneration

Background

The ARRIVE guidelines (Kilkenny et al., 2010) were devised to improve transparency where studies reported results derived from animal research. Ideally, any publications adhering to these rules objectively describe the welfare cost to the animals involved. The guidelines are particularly linked to the following:

(1) Provide precise details of all procedures carried out, for example (…) anaesthesia and analgesia, (…) surgical procedure. (2) Provide details of (….) welfare related assessments and interventions that were carried out prior to, during, or after the experiment. (3) Specify the total number of animals used in each experiment.

Surgically induced cardiac injury as a popular invasive research procedure, which can be reasonably perceived to cause pain and distress, was reviewed in terms of the three ARRIVE criteria laid out here. This utilizes the ability of fish and amphibia to replace heart tissue. Myocardial infarction is simulated by surgical amputation of the heart apex (Wang et al., 2013), cryoinjury with copper probes (Chablais and Jaźwińska, 2012; González-Rosa and Mercader, 2012), or cauterization (Lafontant and Burns, 2012); notably using the closely related giant danio *Devario aequipinnatus*, rather than zebrafish. Nonsurgical alternative methods include genetic ablation (Wang et al., 2013; Zhang et al., 2013, Curado et al., 2007) and immersion in hypoxic tank water (Parente et al., 2013). In the first two cases the procedure incorporates incision into the skin and pericardial sac with microdissection scissors. This is followed by ventricle exposure through gentle abdominal pressure after which 20%–30% of the ventricle is removed with scissors/scalpel. Alternatively, a 0.2–0.3 mm Ø copper filament cooled in liquid nitrogen is placed on the ventricular surface for a few seconds, leading to similar ventricular damage.

Search strategy

The Web of Science database was used to perform a systematic review of the available literature published between January 1, 2012 and October 30, 2020. Search terms and keywords included:

"Zebrafish"
AND
"cardiac" OR "heart"
AND
"regeneration" OR "regrowth".

Study inclusion and exclusion criteria

Electronic search and title management was accomplished using a commercial software program (Endnote X8, Clarivate Analytics). Study selection was based on the following inclusion criteria:

1. Publication in the international peer-reviewed literature.
2. English language.
3. Zebrafish studies.
4. Cardiac resection surgery, cryoinjury, or cauterization.

Results

Of 47 publications that passed the selection criteria, 44 studies described the cardiac excision procedure. Five studies were dedicated to cryoinjury. Two incorporated both excision and cryoinjury.

Seven manuscripts detailed their procedure sufficiently to satisfy the first ARRIVE criterion. However, only one of these described duration of the surgical procedure, elaborating that the cryoinjury protocol takes 3–5 min once animals have reached surgical anesthesia and that subsequent recovery from anesthetic can take up to 5 min (González-Rosa and Mercader, 2012). Thirty of the remaining 39 studies, as recent as 2019 (Ben-Yair et al., 2019), referred to a resection protocol published in 2002 (Poss et al., 2002), which itself is not sufficiently detailed to satisfy the guidelines, not least because these were not published for another nine years and because there are no specifications on buffering and dosing of tricaine. Therefore this protocol could not be endorsed as suitable with today's knowledge of best practice. One study referred to a protocol (Jopling et al., 2010) that turned out to be a secondary source, and unusable to boot, citing a 20-fold tricaine overdose of 4000 mg/L (0.4%); another study (Xiao et al., 2016) erroneously referenced as source of its resection protocol a publication which had no detail of such a procedure.

Worryingly, cryoinjury protocols have been referenced to several publications at once, each with different intervention points and anesthetic regimes; this "scattergun" approach cannot be reconciled with the ARRIVE criteria (e.g. Lai et al., 2017).

Only two pieces of work (González-Rosa and Mercader, 2012; Goldmann and Kuzu, 2017) at least partially satisfied the second criterion, detailing intervention points and reporting higher mortality rates for younger fish

(González-Rosa and Mercader, 2012) and giving some detail about stopping hemorrhaging (citing the use of "Kim-Wipes" rather than sterile swabs) and aiding recovery (Goldmann and Kuzu, 2017). No other studies reported on mortality rates associated with the procedure; this could be perceived as a decline in reporting diligence compared with publications from the pre-ARRIVE era (Poss et al., 2002; Raya et al., 2004).

Finally, only three studies including (Gupta et al., 2013) presented numbers of animals used for the surgical procedure, while one further work reported on the number of hearts studied instead ("50–60", (Chen and Wang, 2016)).

Mortality rates and speciesism

From a veterinary perspective, there is a sizeable elephant in the room: the postsurgical mortality rate associated with cardiac injury protocols. As already mentioned, there is scant detail that can be gleaned from scientific manuscripts, despite their ever-increasing number. Under current interpretation of UK laboratory animal legislation, any "built-in" mortality (i.e., where protected animals are found dead, rather than euthanized after reaching an agreed humane endpoint) which is attributable to regulated procedures will always push a protocol and its associated project license into the highest ("severe") severity band, as long as this affects no more than a small percentage of the animals used. Yet with recorded mortality rates (taking into consideration animals euthanized for failing to recover and those found dead after the procedure) of typically >10% and as high as 30%–40% in some facilities (pers. obs.), one needs to question why this has not been identified as a major issue by the regulator and if procedures with similar mortality levels would have been accepted for nonaquatic vertebrates. One definition of the term "speciesism" is "assigning certain values on membership of a specific genetic clade"; this aptly sums up this particular issue as the value attributed to the survival of a fish appears several orders of magnitude lower than that of a mammalian research animal.

Use of anaesthetics

There is justified concern that, in the comparatively niche field of fish anaesthesia, new developments and palpable improvements are slow to take hold. In terms of qualitative changes to mainstream anesthetic practice, the use of tricaine (also called tricaine mesylate, tricaine methanesulfonate, or MS 222) was first described for lake trout in 1960 (Schoettger, 1967). Back then, tricaine and quinaldine sulfate were used in approximately equal measures (Schoettger and Julin, 1969); arguably the main change in the ensuing 60 years was that the latter has all but disappeared from our shelves. While any perceived lack of progress and pervasive over-reliance on tricaine may be specific to laboratory fishes, accounts from ornamental fish anaesthesia have shown a wider range of chemicals and administrative routes. Therefore there is scope to also expand the anaesthetic arsenal for laboratory fish as there is no published evidence that refinement and consideration afforded to anaesthesia for laboratory fishes are even approaching levels comparable to mammalian practice (Schroeder et al., 2021). In addition, there are some concerns about the dosing of tricaine, especially when considering published doses for recovery anaesthesia for experimental use in zebrafish cryoinjury procedures (320–1000 mg/L; González-Rosa and Mercader, 2012; Chablais and Veit, 2011) versus veterinary guidelines (100–200 mg/L; Mattews and Varga, 2012). It is conceivable that such an excessive dose, which entails the risk of medullary collapse and ensuing brain death, is frequently chosen to allow longer out-of-water procedures and more invasive protocols. It is also conceivable that these high anaesthetic doses exacerbate already high mortality rates during and after the cardiac excision procedure.

Conclusion: the state of zebrafish welfare today

The outcome of the review of cardiac injury protocols shows that attempts at quantification of the cost to zebrafish welfare are still few and far between. With the majority of peer-reviewed and published manuscripts not even describing the surgical protocol in sufficient detail, and little in the way of detailing welfare issues and intervention points, this is a poignant reminder that awareness of fish welfare issues in the field of laboratory animal research has to improve.

Of the concepts and paradigms used to assess animal welfare, the "Two Pillars" concept by Marian Stamp Dawkins is probably the most reductive. This concept is based on asking "if the animal has what it wants" coupled with an examination of its physical health (Dawkins, 2008). On the evidence of preference testing for different structural environmental enrichments, the welfare of zebrafish in barren tanks does need addressing as they are housed in an environment that they avoid when given the opportunity to locate themselves with enrichment (Schroeder et al., 2014).

Regarding the second "pillar," the practice of health screening zebrafish is still in its infancy, but on current evidence, the majority of European and North American facilities have to contend with microsporidiosis, a parasitosis caused by the spore-forming *Pseudoloma neurophilia*. The parasite infects the central nervous system and the skeletal musculature and is associated with emaciation and skeletal deformities (Sanders et al., 2012). However, many fish exhibit subclinical infections with intermittent acute morbidity and mortality triggered by environmental stressors such as pH and water temperature changes. *P. neurophilia* was identified in 74% of facilities that submitted fish to the pathology service at the Zebrafish International Resource Center in Oregon, USA (Murray and Dreska, 2011). An alternative estimate for Western Europe is 90%–95% of facilities (QM diagnostics, 2015), with 90% of all deformed or cachectic fish and 10%–30% of fish of normal appearance affected (Matthews et al., 2005). Consequently, on the evidence of these high-prevalence figures, the clinical welfare of zebrafish for the majority of facilities can be justifiably considered problematic and an appraisal based on the reductive "Two-Pillar" approach presents ample scope for improvements to the welfare of zebrafish in laboratories.

Relative scarcity of empirical data on many aspects of zebrafish welfare makes this species no less deserving of our attention. In fact, the steady rise in invasive fish procedures such as surgically induced cardiac injury should be utilized to collate postnociceptive data to improve the understanding of pain-associated behavioral patterns and develop efficacious analgesic regimes. As already seen in mouse models, future scientific challenges will require standardization of health status and welfare provisions. Veterinary surgeons specializing in that area should play a key part in the application of the 3Rs in research facilities using fish. For this they need a comprehensive understanding of the latest diagnostic technique, the animal's physiology, and welfare requirements.

The ultimate objective is to utilize the plethora of procedures carried out on zebrafish to improve the understanding of fish welfare, for example, by adapting welfare scoring systems for different experimental procedures and developing analgesic protocols to alleviate their welfare impact.

Acknowledgments

The author is grateful to the Wellcome Trust Doctoral Studentships Programme, Dr. Jean-Phillippe Mocho MRCVS, Maggie Lloyd MRCVS (Red Kite Vets), the Oxford University Biomedical Services Dept., and Dr. Lynne Sneddon (University of Gothenburg).

References

Barcellos, L.J.G., Ritter, F., et al., 2007. Whole-body cortisol increases after direct and visual contact with a predator in zebrafish, *Danio rerio*. Aquaculture 272 (1–4), 774–778.

Ben-Yair, R., Butty, V.L., et al., 2019. H3K27me3-mediated silencing of structural genes is required for zebrafish heart regeneration. Development 146, dev178632.

Braithwaite, V., Boulcott, P., 2007. Pain perception, aversion and fear in fish. Dis. Aquat. Org. 75 (2), 131.

Branson, E.J., 2008. Fish Welfare. Wiley-Blackwell.

Broom, D.M., 1991. Animal welfare: concepts and measurement. J. Anim. Sci. 69 (10), 4167–4175.

Brown, L., 1993. Aquaculture for Veterinarians. Fish Husbandry and Medicine. Pergamon Press, UK.

Brydges, N.M., Braithwaite, V.A., 2009. Does environmental enrichment affect the behaviour of fish commonly used in laboratory work? Appl. Anim. Behav. Sci. 118 (3–4), 137–143.

Carbone, L., 2004. What Animals Want: Expertise and Advocacy in Laboratory Animal Welfare Policy. Oxford University Press, USA.

Chablais, F., Jaźwińska, A., 2012. Induction of myocardial infarction in adult zebrafish using cryoinjury. J. Vis. Exp. 62.

Chablais, F., Veit, J., et al., 2011. The zebrafish heart regenerates after cryoinjury-induced myocardial infarction. BMC Dev. Biol. 11 (21).

Chandroo, K.P., Duncan, I.J., et al., 2004. Can fish suffer?: perspectives on sentience, pain, fear and stress. Appl. Anim. Behav. Sci. 86 (3), 225–250.

Chen, W.C.W., Wang, X., 2016. Decellularized zebrafish cardiac extracellular matrix induces mammalian heart regeneration. Sci. Adv. 2 (11), e1600844.

Clark, T.S., Pandolfo, L.M., et al., 2018. Body condition scoring for adult zebrafish (*Danio rerio*). J. Am. Assoc. Lab. Anim. Sci. 57, 698–702. https://doi.org/10.30802/AALAS-JAALAS-18-000045.

Commission to the Council and the European Parliament, 2020. 2019 Report on the Statistics on the Use of Animals for Scientific Purposes in the Member States of the European Union in 2015–2017. https://eur-lex.europa.eu/legal-content/EN/TXT/PDF/?uri=CELEX:52020DC0016&from=ENU.S.

Curado, S., Anderson, R.M., et al., 2007. Conditional targeted cell ablation in zebrafish: a new tool for regeneration studies. Dev. Dynam. 236, 1025–1035.

Dawkins, M.S., 2008. The science of animal suffering. Ethology 114 (10), 937–945.

Deakin, A.G., Spencer, J.W., et al., 2019. Welfare challenges influence the complexity of movement: fractal analysis of behaviour in zebrafish. Fishes 4 (1), 8.

Department of Agriculture - Animal and Plant Health Inspection Service, 2017. Annual Report Animal Usage by Fiscal Year 2016. https://www.aphis.usda.gov/animal_welfare/downloads/reports/Annual-Report-Animal-Usage-by-FY2016.pdf.

Egan, R.J., Bergner, C.L., et al., 2009. Understanding behavioral and physiological phenotypes of stress and anxiety in zebrafish. Behav. Brain Res. 205 (1), 38–44.

Ellis, T., Katsiadaki, I., 2020. Clarification of early end-points for refinement of animal experiments, with specific reference to fish. Lab. Anim. https://doi.org/10.1177/0023677220971002.

Engeszer, R.E., Patterson, L.B., et al., 2007. Zebrafish in the wild: a review of natural history and new notes from the field. Zebrafish 4 (1), 21–40.

European Parliament and Council of the European Union, 2010. European Parliament, Council of the European Union Directive 2010/63/EU of the European Parliament and of the Council of 22 September 2010 on the Protection of Animals Used for Scientific Purposes. Council of Europe, Strasbourg.

Felix, A.S., Faustino, A.I., et al., 2013. Noninvasive measurement of steroid hormones in zebrafish holding-water. Zebrafish 10 (1).

Flecknell, P., 2009. Laboratory Animal Anaesthesia. Academic Press.

Fox, M.W., 1989. Genetic engineering and animal welfare. Appl. Anim. Behav. Sci. 22 (2), 105–113.

Goldmann, J.A., Kuzu, G., et al., 2017. Resolving heart regeneration by replacement histone profiling. Dev. Cell 40 (4), 392–404.

González-Rosa, J.M., Mercader, N., 2012. Cryoinjury as a myocardial infarction model for the study of cardiac regeneration in the zebrafish. Nat. Protoc. 7 (4), 782–788.

Gupta, V., Gemderling, M., et al., 2013. An injury-responsive Gata4 program shapes the zebrafish cardiac ventricle. Curr. Biol. 23 (13), 1221–1227.

Home Office, 2020. Annual Statistics of Scientific Procedures on Living Animals Great Britain 2019. The Stationery Office, Home Office, London. In: https://assets.publishing.service.gov.uk/government/uploads/system/uploads/attachment_data/file/901224/annual-statistics-scientific-procedures-living-animals-2019.pdf.

Huntingford, F.A., Adams, C., et al., 2006. Current issues in fish welfare. J. Fish. Biol. 68 (2), 332–372.

Jopling, C., Sleep, E., et al., 2010. Zebrafish heart regeneration occurs by cardiomyocyte dedifferentiation and proliferation. Nature 464, 606–609.

Kilkenny, C., Browne, W.J., et al., 2010. Improving bioscience research reporting: the ARRIVE guidelines for reporting animal research. PLoS Biol. 8 (6), e1000412.

Kistler, C., Hegglin, D., et al., 2011. Preference for structured environment in zebrafish (*Danio rerio*) and checker barbs (*Puntius oligolepis*). Appl. Anim. Behav. Sci. 135 (4), 318–327.

Labit, E., Rabiller, L., et al., 2018. Opioids prevent regeneration in adult mammals through inhibition of ROS production. Sci. Rep. 8, 12170.

Lafontant, P., Burns, A., et al., 2012. The giant danio (D. aequipinnatus) as a model of cardiac remodeling and regeneration. Anat. Rec. 295, 234–248.

Lai, S.-L., Marín-Juez, R., et al., 2017. Reciprocal analyses in zebrafish and medaka reveal that harnessing the immune response promotes cardiac regeneration. eLife 6, e25605.

Laule, G., 2010. Positive reinforcement training for laboratory animals. In: The UFAW Handbook on the Care and Management of Laboratory and Other Research Animals, pp. 206–218.

Lee, H.C., Lu, P.N., et al., 2014. Zebrafish transgenic line huORFZ is an effective living bioindicator for detecting environmental toxicants. PLoS ONE 9 (e90160).

Lelek, S., Guedes Simoes, M., et al., 2020. Morphine alleviates pain after heart cryoinjury in zebrafish without impeding regeneration. bioRxiv. https://doi.org/10.1101/2020.10.01.322560.

Li, H., Li, Y., et al., 2019. Liver-specific androgen receptor knockout attenuates early liver tumor development in zebrafish. Sci. Rep. 9, 10645.

Lopez-Luna, J., Canty, M.N., Al-Jubouri, Q., Al-Nuaimy, W., Sneddon, L.U., 2017. Behavioural responses of fish larvae modulated by analgesic drugs after a stress exposure. Appl. Anim. Behav. 195, 115–120.

Matthews, J.L., Brown, A.M.V., et al., 2005. *Pseudoloma neurophilia* n. g., n. sp., a new microsporidium from the central nervous system of the zebrafish (*Danio rerio*). J. Eukaryotic Biol. 48 (2), 227–233.

Mattews, M., Varga, Z.M., 2012. Anesthesia and euthanasia in zebrafish. ILAR J. 53 (2), 192–204.

Maximino, C., 2011. Modulation of nociceptive-like behavior in zebrafish (*Danio rerio*) by environmental stressors. Psychol. Neurosci. 4 (1), 149–155.

Morton, D.B., 1998. Perceived and actual welfare issues - laboratory animals. In: Michell, A.R., Ewbank, R. (Eds.), Ethics, Welfare, Law and Market Forces: The Veterinary Interface. UFAW.

Murray, K.N., Dreska, M., 2011. Transmission, diagnosis, and recommendations for control of pseudoloma neurophilia infections in laboratory zebrafish (*Danio rerio*) facilities. Comp. Med. 61 (4), 322–329.

Newby, N.C., Wilkie, M.P., et al., 2009. Morphine uptake, disposition, and analgesic efficacy in the common goldfish (*Carassius auratus*). Can. J. Zool. 87 (5), 388–399.

Nordgreen, J., Garner, J.P., et al., 2009. Thermonociception in fish: effects of two different doses of morphine on thermal threshold and post-test behaviour in goldfish (*Carassius auratus*). Appl. Anim. Behav. Sci. 119 (1–2), 101–107.

Parente, V., Balasso, S., et al., 2013. Hypoxia/reoxygenation cardiac injury and regeneration in zebrafish adult heart. PloS One 8 (1), e53748.

Parker, M.O., Millington, M.E., et al., 2012. Housing conditions differentially affect physiological and behavioural stress responses of zebrafish, as well as the response to anxiolytics. PloS One 7 (4), e34992.

Poss, K.D., Wilson, L.G., et al., 2002. Heart regeneration in zebrafish. Science 298 (5601), 2188–2190.

QM Diagnostics, 2015. Parasite Screening for Zebrafish – a Conference Talk by Ronald de Boer. Fish Veterinary Society Annual Conference, Edinburgh.

Ramsay, J.M., Feist, G.W., et al., 2006. Whole-body cortisol is an indicator of crowding stress in adult zebrafish, *Danio rerio*. Aquaculture 258 (1–4), 565–574.

Ramsay, J.M., Feist, G.W., et al., 2009. Whole-body cortisol response of zebrafish to acute net handling stress. Aquaculture 297 (1–4), 157–162.

Raya, Á., Consiglio, A., et al., 2004. The zebrafish as a model of heart regeneration. Clon Stem Cell 6 (4), 345–351.

Readman, G.D., Owen, S.F., et al., 2013. Do fish perceive anaesthetics as aversive? PloS One 8 (9), e73773.

Reilly, S.C., Quinn, J.P., et al., 2008. Behavioural analysis of a nociceptive event in fish: comparisons between three species demonstrate specific responses. Appl. Anim. Behav. Sci. 114 (1), 248–259.

Richmond, J., 2002. Refinement, reduction, and replacement of animal use for regulatory testing: future improvements and implementation within the regulatory framework. ILAR J. 43 (Suppl. 1), S63–S68.

Richmond, J., 2010. The three Rs. In: The UFAW Handbook on the Care and Management of Laboratory and Other Research Animals, pp. 3–22.

Rose, J., Arlinghaus, R., et al., 2012. Can fish really feel pain?. In: Fish and Fisheries.

Rose, J.D., 2003. A critique of the paper: "Do fish have nociceptors: evidence for the evolution of a vertebrate sensory system". Proc. Roy. Soc. Biol. Sci. 270 (1520), 1115–1121, 2003 by Sneddon, Braithwaite and Gentle. Information Resources on Fish Welfare 1970–2003 (Animal Welfare Information Resources No. 20): 49–51.

Ross, L.G., Ross, B., et al., 2008. Anaesthetic and Sedative Techniques for Aquatic Animals. Wiley Online Library.

Sanders, J., Watral, V., et al., 2012. Microsporidiosis in zebrafish research facilities. ILAR J. 53 (2), 106–113.

Schoettger, R.A., 1967. Efficacy of MS-222 as an anesthetic on four salmonids. Invest. Fish Contr. 13, 1–15.

Schoettger, R.A., Julin, A.M., 1969. Efficacy of quinaldine as an anesthetic for seven species of fish. Invest. Fish Contr. 22, 3–10.

Schreck, C., 1981. Stress and compensation in teleostean fishes: response to social and physical factors. In: Stress and Fish, vol. 367. Academic Press, London, pp. 295–321.

Schroeder, P.G., 2014. Refinements in husbandry practices and scientific procedures: impact upon stress and pain in the zebrafish (*Danio rerio*) (Doctoral thesis). In: School of Veterinary Sciences. University of Nottingham. Dr Vet Med, Nottingham, pp. 98–110.

Schroeder, P.G., Jones, S., et al., 2014. What do zebrafish want? Impact of social grouping, dominance and gender on preference for enrichment. Lab. Anim 48 (4), 328–337.

Schroeder, P.G., Lloyd, R., et al., 2021. Anaesthesia of laboratory, aquaculture and ornamental fish: proceedings of the first LASA-FVS symposium. Lab. Anim. https://doi.org/10.1177/0023677221998403. (in press).

Schroeder, P.G., Sneddon, L.U., 2017. Exploring the efficacy of immersion analgesics in zebrafish using an integrative approach. Appl. Anim. Behav. Sci. 187, 93–102.

Schroeder, P.G., Mocho, J.-P., 2014. A veterinary perspective on laboratory zebrafish welfare. Fish. Vet. J. 14 (1), 34–43.

Shepherdson, D.J., Mellen, J.D., et al., 1998. Second Nature: Environmental Enrichment for Captive Animals. Smithsonian Institution Press.

Sneddon, L.U., 2003. The evidence for pain in fish: the use of morphine as an analgesic. Appl. Anim. Behav. Sci. 83 (2), 153–162.

Sneddon, L., 2006. Ethics and welfare: pain perception in fish. Bull. Eur. Assoc. Fish Pathol. 26 (1), 6.

Sneddon, L.U., 2009. Pain perception in fish: indicators and endpoints. ILAR J. 50, 338–342.

Sneddon, L.U., 2012. Clinical anesthesia and analgesia in fish. J. Exot. Pet Med. 21 (1), 32–43.

Sneddon, L.U., Braithwaite, V.A., et al., 2003. Do fishes have nociceptors? Evidence for the evolution of a vertebrate sensory system. Proc. Roy. Soc. Lond. B Biol. Sci. 270 (1520), 1115–1121.

Stearns, G., Evangelista, M., et al., 2007. A mutation in the cone-specific pde6 gene causes rapid cone photoreceptor degeneration in zebrafish. J. Neurosci. 27, 13866–13874.

Stoskopf, M.K., 1994. Pain and analgesia in birds, reptiles, amphibians, and fish. Investig. Ophthalmol. Vis. Sci. 35 (2), 775–780.

Volpato, G.L., Gonçalves-de-Freitas, E., et al., 2007. Insights into the concept of fish welfare. Dis. Aquat. Org. 75 (2), 165–171.

Wagner, R.L., White, P.F., et al., 1984. Inhibition of adrenal steroidogenesis by the anesthetic etomidate. N. Engl. J. Med. 310 (22), 1415–1421.

Wang, J., Karra, R., et al., 2013. Fibronectin is deposited by injury-activated epicardial cells and is necessary for zebrafish heart regeneration. Dev. Biol. 382 (2), 427–435.

Webster, M.M., Hart, P.J.B., 2004. Substrate discrimination and preference in foraging fish. Anim. Behav. 68 (5), 1071–1077.

Wilkes, L., Owen, S.F., et al., 2012. Does structural enrichment for toxicology studies improve zebrafish welfare? Appl. Anim. Behav. Sci. 139 (1–2), 143–150.

Wolfensohn, S., Lloyd, M., 2013. Handbook of Laboratory Animal Management and Welfare. Wiley Online Library.

Wong, D., Kayserling, M.A.C., et al., 2014. Conditioned place avoidance of zebrafish (*Danio rerio*) to three chemicals used for euthanasia and anaesthesia. PloS One 9, e88030.

Workman, P., Aboagye, E., et al., 2010. Guidelines for the welfare and use of animals in cancer research. Br. J. Cancer 102 (11), 1555–1577.

Wu, Y., Hsu, M., et al., 2017. Fragile X mental retardation-1 knockout zebrafish shows precocious development in social behavior. Zebrafish 14 (5), 438–443.

Würbel, H., Garner, J.P., 2007. Refinement of rodent research through environmental enrichment and systematic randomization. In: National Centre for the Replacement, Refinement and Reduction of Animals in Research.

Xiao, C., Gao, L., et al., 2016. Chromatin-remodelling factor Brg1 regulates myocardial proliferation and regeneration in zebrafish. Nat. Commun. 7, 13787.

Zhang, R., Han, P., et al., 2013. In vivo cardiac reprogramming contributes to zebrafish heart regeneration. Nature 498 (7455), 497–501.

Chapter 6

Analgesia, anesthesia, and euthanasia in zebrafish

Almut Köhler[1], Ana M. Valentim[2,3]

[1]*Safety and Environment, Karlsruhe Institute of Technology (KIT), Karlsruhe, Germany;* [2]*i3S - Instituto de Investigação e Inovação em Saúde, Universidade do Porto, Portugal;* [3]*Laboratory Animal Science, IBMC — Instituto de Biologia Molecular e Celular, Universidade do Porto, Porto, Portugal*

Many experimental procedures include steps where different developmental stages of zebrafish are subjected to somewhat painful or stressful procedures. These protocols may be common to most labs (like fin clipping, short-term imaging) or more specific to certain types of research (injections, implantation of electrodes, cardiac injury). Analgesia and anesthesia are in most cases used to suppress distress and/or pain sensation/perception. While anesthesia is often used mainly to immobilize animals for surgical and nonsurgical procedures, analgesia is a fairly new aspect added to zebrafish protocols to take into account the increasing awareness of pain sensation and animal welfare in fish. The application of analgesics often takes place after the procedure itself and it is often unclear what the side effects of the treatment are. Therefore the decision concerning effective, safe, and humane methods of analgesia and anesthesia that are ethically justifiable as well as scientifically reproducible is still difficult and depends on the specific situation. There are different procedures with different levels of invasiveness and duration to which analgesia and anesthesia protocols must be adapted, as there is no "one size fits all."

Often, at some part of an experiment, animals have to be sacrificed using euthanasia methods. Euthanasia or humane killing might be applied due to protocol requirements, but also to avoid more suffering, applying a final pain relief if ethical justification of a harmful state of the animal is not given. There are several euthanasia methods depending on the purpose, developmental stage of the animal, and legislation of the country/region.

Chemical substances are involved in all the referred procedures, analgesia, anesthesia, and chemical euthanasia, thus one has to pay special attention to legal considerations. While in earlier years, people working with fish as animal models in laboratory animal science often used pure chemical substances to

Laboratory Fish in Biomedical Research. https://doi.org/10.1016/B978-0-12-821099-4.00016-X
Copyright © 2022 Elsevier Inc. All rights reserved.

make their own "drugs" with lab ingredients and did not think about medical details, responsible authorities are starting to require the use of medically approved drugs and solvents. Especially with fish this leads to increasing problems requiring expert knowledge: many of the traditionally used substances are not approved for fish or only for specific fish species, and some are not for other animals, but only for humans. Also, many formerly approved drugs currently lose their approval since the approval process in earlier years missed important analysis aspects, and most pharmaceutical companies avoid the costly procedure of approving drugs for a relatively small field of use. Thus people in fish labs have to use drugs approved for mammalian species or even humans, and this so-called "rededication" often requires intensive documentation and approval by authorities.

In this chapter, we will discuss different protocols for analgesia, anesthesia, and euthanasia based on different requirements of various protocols.

Analgesia

Analgesic substances are used to relieve pain in an animal. In contrast to anesthetic substances, which suppress pain sensation and induce immobility/loss of consciousness by depression of the central nervous system (CNS), analgesic substances mainly work by action on signal transmission and modulation. While analgesia is routinely used in mammals, the level of knowledge is still low in fish. Fish have opioid receptors and show complex behaviors to flee or manage pain, among other indications (Sneddon et al., 2014) but there are still some contradictory opinions regarding fish feeling pain. Nevertheless, all agree that pain or nociception cause physiological and behavioral alterations that should be prevented. Usually, before a painful procedure, loss of pain sensation in fish is tested by the response to a pinch in the caudal fin performed by hand or forceps. Several behavioral indicators of pain in fish have been described in the literature: reduction of activity and food intake, rubbing the injured body part, increase of opercular movement rate, and bottom dwelling, but species-specific behaviors must be considered (Reilly et al., 2008; Deakin et al., 2019).

To prevent behavioral alterations related with pain, several classes of analgesic substances (overview in Table 6.1) are described in fish: opioids, nonsteroidal antiinflammatory drugs (NSAIDs), and paracetamol (aniline analgesics/derivatives). The group of local anesthetics also induces analgesia but also acts like an anaesthetic by suppressing pain sensation in a particular region. Analgesics can be injected intramuscularly (IM) or intraperitoneally (IP); however, bath application (BA) is usually preferred. This has the advantage of less direct handling and thus is less stressful for the animals with no need to use anesthetics to render the animal immobile to the injection. To avoid injection administration is particularly advantageous for small fish such as zebrafish. However, the effect of BA compared to injections is delayed, and

TABLE 6.1 Overview of analgesics indicating analgesic properties in zebrafish.

Substance	Dosage	Remarks	Literature
Opioids			
Buprenorphine	0.1 μg/mL	Larvae	Steenbergen and Bardine (2014)
	5 μM	Larvae; temperature stress test	Curtright et al. (2015)
Morphine	1.5–3 mg/L	Acetic acid bath	Currie (2014)
	6 mg/kg	Acetic acid test	Correia et al. (2011)
	48 mg/L	Fin clip	Deakin et al. (2019)
	1.5 mg/L	Cryoinjury	Lelek et al. (2020)
NSAIDs			
Aspirin	2.5 mg/L	Fin clip	Schroeder and Sneddon (2017)
Flunixin	8 mg/L	Fin clip	Deakin et al. (2019)
Local anesthetics			
Lidocaine	5 mg/L	Larvae, heat exposure	Lopez-Luna et al. (2017a)
		Larvae, acetic acid exposure	Lopez-Luna et al. (2017a)
		Larvae, CO_2 exposure	Lopez-Luna et al. (2017a), Deakin et al. (2019)
		Fin clip	Schroeder and Sneddon (2017)
Bupivacaine	0.5, 1 mg/L	Fin clip	Deakin et al. (2019)

If not stated otherwise, the mentioned dosage applies to adult zebrafish via a water bath.

greater quantities of analgesics are required, which have to be disposed of with the water in an environmentally friendly manner. Unfortunately, there is a lack of comprehensive data on the pharmacokinetics and physiological effects of analgesics, which is crucial for experimental use.

Opioids

Opioids mainly bind to opioid receptors located in the brain and spinal cord. μ-, κ-, and two δ-receptors have been demonstrated in all areas of the zebrafish adult brain (Alvarez et al., 2006; Pinal-Seoane et al., 2006; Sanchez-Simon and Rodriguez, 2008; Gonzalez-Nunez and Rodríguez, 2009; Sivalingam et al., 2020). μ- and δ-receptors are already expressed at embryonal stages, while κ-receptors start to be expressed at later stages (Sanchez-Simon and Rodriguez, 2008).

The nociceptive system of fish is similar to that of mammals, but there are differences that affect the effects of analgesics, e.g., no efferent fibers have been identified from the brain to the spinal cord so far, which is one of the main areas of action for opioid analgesics in mammals (Stevens, 2008). Furthermore, analgesics distribute more slowly in fish tissue compared to mammals due to the lower body temperature, but also seem to work for much longer (Stevens, 2008). As the opioid system is already mature in early stages, opioids can already be used at young stages for analgesia (Lopez-Luna et al., 2017a,b,c).

Morphine

Morphine mainly binds to the μ-opioid receptors to express its analgesic properties in mammals (Kieffer, 1999). The effective morphine dose for adult *Danio rerio* is between 1.5 and 3 mg/L (BA) or 6 mg/kg (IM) (Correia et al., 2011; Currie, 2014). Initial contact with morphine when applied to the water can be stressful for fish as they react similarly to the application of acids, which are often used as a pain stimulus (Lopez-Luna et al., 2017a,b,c). This reaction subsides after a short time. Morphine has analgesic potential in larvae in case of administration of acid or warm temperatures, but not of cold temperatures (Lopez-Luna et al., 2017a,b,c). It is also used as an effective pain relief drug in case of heart cryoinjury studies up to 6 h postinjury, without affecting the regeneration process in adult zebrafish (Lelek et al., 2020). When applying opioids, one should consider the addictive potential of the drugs. For morphine, Khor et al. (2011) demonstrated an increased level of anxiety when withdrawing morphine after 2 weeks' application of 1.5 mg/L. In reverse, application of morphine results in reduced anxiety (Wong et al., 2010). Pharmacokinetic data as well as reports about side effects for zebrafish are not available, but prove to be species specific in other fish species (Chatigny et al., 2018). It is still a matter of study whether morphine does not show any physiological side effects in zebrafish. As morphine is a regulated substance, special considerations are required concerning waste disposal and documentation, thus making the morphine bath a method with drawbacks in handling.

Buprenorphine

Buprenorphine is a partial opioid agonist binding mainly to μ-receptors. As it has a lower affinity to κ- and δ-receptors than morphine, usually the spectrum of side effects is smaller than for morphine in mammals, and it has fewer addictive effects (Gudin and Fudin, 2020). Its analgesic properties in larvae reduce responses to thermal or acidic stimuli (Steenbergen and Bardine, 2014; Curtright et al., 2015). There are no data available for adult zebrafish, but the results in rainbow trout are not promising (Mettam et al., 2011).

Butorphanol

Butorphanol binds agonistically with high affinity to δ- and κ-receptors and is an antagonist on μ-receptors. Use in zebrafish is currently quite limited and does not show a clear analgesic effect (Schroeder and Sneddon, 2017). So, the benefit of use in zebrafish is not clear. As butorphanol has no addictive effect in mammals and it is not a regulated substance it can be used in the laboratory.

Tramadol

Tramadol is a κ- and μ-receptor agonist and a serotonin reuptake inhibitor. Developmental toxicity (Bachour et al., 2020) and its pharmacokinetics and metabolites distribution have been studied in adult zebrafish. IM injection or BA of tramadol reduces zebrafish activity and increases their preference for the water surface (Zhuo et al., 2016). Apart from these side effects, there are no reports about its analgesic effects in zebrafish. Since it is a regulated substance, injection might be the favorable way of administration regarding waste disposal.

NSAIDs

NSAIDs produce their analgesic and antiinflammatory effects by inhibiting the cyclooxygenase enzyme, which is involved in the synthesis of prostaglandins.

Aspirin

The nonsteroidal aspirin has a good analgesic effect by decreasing adult zebrafish behaviors related with pain when administered prior to the painful stimulus (Schroeder and Sneddon, 2017). As there is no proof of the absorption of this substance at 2.5 mg/L (BA), the method of action is not clear. Aspirin is also effective in larvae (Lopez-Luna et al., 2017c).

Flunixin

Flunixin is believed to inhibit cyclooxygenases similarly to aspirin. It shows no effect for heat or acidic stimuli at 8 and 20 mg/L in zebrafish larvae (Lopez-Luna et al., 2017b) but reduces the noxious stimuli of fin clipping in adult zebrafish (Deakin et al., 2019).

Ibuprofen

Ibuprofen is considered a nonselective cyclooxygenase inhibitor and acts as an antiinflammatory drug in fish. However, ibuprofen does not reverse the sensitization for noxious temperatures in larvae when used at 400 μM (Curtright et al., 2015). No studies have been performed in adults regarding its analgesic effects.

Local anesthetics

Local anesthetics cause absence of pain sensation by preventing action potential formation through blocking mainly sodium channels, thus leading to insensitivity. They can prevent response to painful stimuli in zebrafish, indicating analgesia. In higher doses, these substances act systemically and induce anesthesia. In zebrafish, they are also used as analgesic substances for post-surgery/postprocedure treatment.

Lidocaine

A preoperative administration of the local anesthetic lidocaine (5 mg/L) shows positive effects on activity levels and opercular beat rate related to pain response of fin clipping in adult zebrafish. These effects are accompanied by pharmacokinetic validation, i.e., a dose-dependent uptake is observed (Schroeder and Sneddon, 2017). This protocol is also an effective analgesic for heat stimulus and noxious chemicals in zebrafish larvae with minor side effects (Lopez-Luna et al., 2017a,b,c; Schroeder and Sneddon, 2017). Use in muscle-affecting protocols (like heart regeneration studies) is questionable since several studies describe myotoxic effects and compromise the regeneration process in the heart (Carlson and Rainin, 1985; Hussain et al., 2018; Lelek et al., 2020).

Bupivacaine

A single study showed some potential of bupivacaine on pain reduction applied in BA before fin clipping in adult zebrafish (Deakin et al., 2019), but it needs clarification. Apart from this, the analgesic effects of this substance have only been studied in amphibians and reptiles (Chatigny et al., 2017).

Anesthesia

Anesthesia is characterized as a reversible intoxication of the CNS. Different depths of anesthesia are used depending on the procedure requirement (sedation, immobility, and/or pain relieve) and are obtained using different anesthetic concentrations or exposure times when anesthetics are administered via water immersion. There are several descriptions of the stages of anesthesia (example in Table 6.2), which may be hard to detect with certain quick-action anesthetics. However, the loss of equilibrium (in dorsal recumbence for more than 3 s), lack of response to a light stimulus (e.g., touch on the side of the fish), and painful stimulus (e.g., caudal tail pinch) can be monitored and indicate that the animal achieved surgical anesthesia (Valentim et al., 2016). As described for analgesics, BA is also used for anesthetics in most cases, especially in small fish like zebrafish. Anesthetics are mainly absorbed/inhaled via gills and excreted through the gills, kidneys, or skin. The accumulation of metabolites in water must therefore be considered if serial anesthesia is

TABLE 6.2 Representation of the stages of anesthesia.

Stages of anesthesia	Depth of anesthesia	Observable behavioral signs
0	No anesthesia	Normal swimming, responsive
1	Sedation	Decreased activity, delayed reaction to visual and/or tactile stimuli
2	Excitement phase	Equilibrium loss (total or partial) or not but hyperactivity, increased respiratory rate, and response to stimulus
3	Light anesthesia	Equilibrium loss, decrease in muscle tone and respiratory rate, no response to visual or tactile light stimulus
4	Deep/surgical anesthesia	Equilibrium loss, decrease in muscle tone, respiratory rate very slow, no response to painful stimulus
5	Nonrecovery anesthesia	Equilibrium loss, relaxed muscle tone, rare respiratory movements, no response to any stimulus
6	Overdose	Equilibrium loss, flaccid/no muscle tone, apnea or absence of respiratory movements, no response to any stimulus—(imminent) death

Adapted from Martins, T., Valentim, A.M., Pereira, N., Antunes, L.M., 2016. Anaesthesia and analgesia in laboratory adult zebrafish: a question of refinement. Lab. Anim 50 (6), 476–488 and Collymore, C., Tolwani, A., Lieggi, C., Rasmussen, S., 2014. Efficacy and safety of 5 anesthetics in adult zebrafish (Danio rerio). J. Am. Assoc. Lab. Anim. Sci. 53 (2), 198–203.

applied to fish in the same tank. A good anesthetic protocol must not induce mortality, have a quick induction (<3 min) and recovery, allow a controllable depth of anesthesia, be safe for the animal and operator, inducing no or minimal changes in fish physiology, behavior, and stress, be soluble in water in case of BA, and must be adequate for the experimental purpose (e.g., in surgery, first verify the loss of response to a painful stimulus). Care must be taken regarding the concentration and duration of anesthesia to avoid too light anesthesia or overdose and the quality of the water (temperature, pH, conductivity, oximetry, etc.) usually using the water from the system where the fish lives; consider also using the air supply for longer anesthesia. Nonpowdered gloves should be used and food restriction is required for bigger fish; this is also recommended for zebrafish to avoid food being stuck on the gills if the animal vomits, and to prevent increased levels of ammonia if the BA is performed in the fish tank. During anesthesia, it is important to monitor the operculum movements (respiratory rate), because if they are too slow or stop, the fish must be placed in clean water to recover. Gill color alterations to pale or dark red may indicate lack of oxygen or hemorrhage, respectively. Also, sensitivity to anesthetics is different depending on the health status, strain, and developmental stage of the fish. Usually, larvae have higher resistance to anesthetics because they do not have developed gills, where the anesthetics are mainly absorbed. In additon of not having gills, embryos are also envelopped in the chorion, which constitutes an additional barrier to the anesthetics (Rombough, 2007).

MS-222

Tricaine methanesulfonate (MS-222) is the most frequently used anesthetic in laboratories. It is the only drug approved by the US Food and Drug Administration for fish anesthesia in the food industry with 21 days of withdrawal and has approval as a medical drug in several countries across Europe. Tricaine is a sodium channel blocker that inhibits signal transmission from the periphery to the CNS, blocks central neurons, and acts as a hypnotic and anesthetic in BA acting systemically, although it is classified as a local anesthetic (Walsh and Schopp, 1966; Attili and Hughes, 2014). At higher doses, it leads to muscle relaxation and thus reduced cardiac activity with symptoms of hypoxia. The effect strongly depends on concentration. The depth of anesthesia increases with duration. Long duration anesthesia depresses respiration, and hypoxia and metabolic acidosis may happen, leading to hypotension and changes of heart rate (Carter et al., 2011). To avoid this, tricaine has been combined with isoflurane (Huang et al., 2010). In general, tricaine interferes with cardiovascular, endocrine, and osmoregulatory systems. Nevertheless, it seems to be indicated for quick procedures, but should be considered with caution for long-lasting or repeated anesthesia (Dang et al., 2016). The use of tricaine also creates stress during induction (Readman et al., 2013; Wong et al., 2014) and anesthesia (Carter et al., 2011; Matthews and Varga, 2012). This reaction occurs to varying degrees with all anesthetics and is not a sole problem of tricaine. Stock solution must be buffered and stored in the dark for not more than 3 months; nevertheless, the optical aspect of the solution must be evaluated before each use and legal regulations may prohibit storage of the solution since it is not part of the approval protocol as a veterinary drug. Retinotoxicity (Carter et al., 2011) is relevant for animals and researchers in the case of repeated anesthesia. Moreover, tricaine powder can be irritant for the mucosa, thus it should be prepared in a fume hood wearing gloves, and goggles are recommended (Ross and Ross, 2008).

Benzocaine

Benzocaine works similarly to tricaine, but is slower and less effective in warm water, which is a disadvantage for use in zebrafish, in addition to its aversive properties (Readman et al., 2013). Although 200 ppm of benzocaine causes a rapid depth of anesthesia and recovery in adults, it also induces 50% of mortality. However, it seems to work well in zebrafish larvae and embryos (Ehrlich et al., 2019). When used in an intermittent way, it is useful for long anesthesia duration for imaging in adult zebrafish (Wynd et al., 2017). Benzocaine must be diluted in a solvent (e.g., ethanol) to be used in a water bath. Alternatively, benzocaine hydrochloride can be used since it has good water solubility.

Clove oil

Clove oil and its components eugenol, methyleugenol, and isoeugenol probably also inhibits sodium channels. However, the analgesic effect is low (Sladky et al., 2001) and the anesthesia often incomplete. Single or frequent use of clove oil can lead to gill necrosis and other histological alterations in fish (Marina Hassan, 2015; Chance et al., 2018). In zebrafish, it induces a quick anesthesia, with longer recovery times compared with tricaine (Grush et al., 2004), but it induces mortality when exposure is too long in adult zebrafish. Nevertheless, it may be used in embryos and larvae (Ehrlich et al., 2019). Thus clove oil is often used for easier handling of animals using short periods of exposure or low concentrations; surgical procedures should be performed with more effective anesthetics. Clove oil blocks cytochrome P-450, a class of enzymes involved in steroid hormone synthesis (Cochrane, 2015). This is why it is not indicated for many experiments unless this effect is desired. Isoeugenol and methyleugenol may have carcinogenic effects (Johnson et al., 2000). As a natural substance, clove oil is subject to strong batch fluctuations, which leads to difficult experimental dosing and reproducibility. However, there are market solutions based on isoeugenol with a fixed composition (Aqui-S), which, contrary to clove oil and eugenol, is water soluble. Clove oil is often used because it is less expensive.

Propofol

Propofol is a GABAA agonist inducing quick hypnotic effects. Being a hypnotic, it does not provide analgesia, although it seems to numb painful stimuli in higher doses, which may be unsafe because it affects the respiratory and circulatory system (Valentim et al., 2016). In addition, developmental and cardiac alterations and neurotoxic effects were reported when embryos are exposed to propofol from 8 to 120 h postfertilization (hpf) (Guo et al., 2015; Luo et al., 2020). Nevertheless, propofol has the advantage of being rapidly metabolized with fewer cumulative effects (Ross and Ross, 2008). This agent has been used in combination with lidocaine providing a balanced anesthesia where low doses of both drugs can be used for achieving the desired anesthesia, increasing protocol safety (Collymore et al., 2014; Valentim et al., 2016; Martins et al., 2018). Thus the analgesic properties of lidocaine together with the hypnotic properties of propofol induce a sufficiently deep anesthesia for general surgical interventions. Note that this combination may form droplets from 30 min to 1 h after solution preparation (Valentim et al., 2016).

Etomidate and metomidate

Etomidate and metomidate are imidazole-based nonbarbiturates with similar properties, and they are hypnotics that act on GABAA receptors. They inhibit cortisol production, which should be taken into consideration for research purposes, and they do not induce analgesia (Ross and Ross, 2008). Thus these may be used to sedate/calm the animal or to render fish immobile for imaging. Aquacalm is a commercially available solution of metomidate. Metomidate induces a quick loss of equilibrium with a wide safety margin (Collymore et al., 2014), while etomidate also induces quick loss of equilibrium and a low respiratory rate (Martins et al., 2018). When etomidate is combined with lidocaine, loss of response to a painful stimulus is quickly obtained (Martins et al., 2018). Etomidate is also effective in zebrafish larvae (Du et al., 2018). Etomidate provides surgical anesthesia when combined with lidocaine but induces a lower respiratory rate than the combination propofol/lidocaine (Martins et al., 2018). Both etomidate and metomidate are less aversive for zebrafish than MS-222 (Readman et al., 2013; Wong et al., 2014).

Other agents such as 2-phenoxyethanol, isoflurane, etc. are rarely used in zebrafish (Matthews and Varga, 2012). Although 2-phenoxyethanol is often used in research (Lidster et al., 2017), there are almost no studies of its effect and efficacy in zebrafish, only some applications in zebrafish larvae (Lisse et al., 2015). It does not seem to alleviate stress and induces hemodynamic alterations in fish (Ross and Ross, 2008). Table 6.3 gives an overview of some anesthetic protocols described in the literature. Each anesthetic may have a wide range of concentrations because the effectiveness will depend on the fish characteristics, as mentioned before, and on the properties of the water such as pH, temperature, and conductivity, which will influence the bioavailability of the anesthetic in the solution. Therefore it is extremely important to do a pilot test with your laboratory conditions to test anesthetic efficacy and adequacy to the procedure prior to the experiment.

Euthanasia

Euthanasia or humane killing is a procedure where the animal is killed with minimal pain or distress. This procedure is performed whenever the animal is in great suffering, when there is the need to make postmortem analysis, when the experiment ends, or due to husbandry considerations and health issues. Thus euthanasia must induce a quick loss of consciousness, be irreversible,

TABLE 6.3 Overview of anesthetic drugs used in zebrafish.

Substance	Concentrations (mg/L)	Remarks	Literature
Tricaine	10–17	Sedation of adult fish for *N-ethyl-N-nitrosourea* mutagenesis	Trevarrow (2011)
	100–168	Tail fin pinch in adults	Collymore et al. (2014), Nordgreen et al. (2014), Valentim et al. (2016)
	168	Long duration anesthesia for adults	Xu et al. (2015)
	31–106	Sedation to anesthesia of larvae	Rombough (2007)
	200	Deep anesthesia for larvae	Ehrlich et al. (2019)
	160	Long duration immobility for embryos and larvae	Kamei & Weinstein (2005)
Tricaine + isoflurane[a]	40–80 + 100	Long duration anesthesia for adults	Dang et al. (2016)
	60–80 + 60–80	Anesthesia for adults	Huang et al. (2010)
	175 + 175–200	Repeated anesthesia for adults (TraNac transgenics)	Lockwood et al. (2017)
Benzocaine	35	Long duration anesthesia protocols using intermittent application in adults	Wynd et al. (2017)
	200	Anesthesia for larvae	Ehrlich et al. (2019),

Continued

TABLE 6.3 Overview of anesthetic drugs used in zebrafish.—cont'd

Substance	Concentrations (mg/L)	Remarks	Literature
Clove oil	45–140[a]	Short duration anesthesia (eugenol)	Grush et al. (2004), Valentim et al. (2018)
	90–120[a]	Anesthesia for larvae	Ehrlich et al. (2019)
	20	Anesthesia for adult fin clipping (isoeugenol)	Musk et al. (2020)
Etomidate	2	Light anesthesia for adults	Martins et al. (2018), Valentim et al. (2018)
	7.3–24.4	Anesthesia for larvae	Du et al. (2018)
Etomidate + lidocaine	2 + 100	Anesthesia for adults	Martins et al. (2018)
Metomidate	6–10	Light anesthesia for adults	Collymore et al. (2014)
Propofol	1.25–7.5	Light anesthesia for adults	Valentim et al. (2016), Martins et al. (2018)
	5.3–17.8 1–3	Anesthesia for larvae	Du et al. (2018) Guo et al. (2015)
Propofol + lidocaine	1.25–2.5 + 50–150	Anesthesia for adults	Valentim et al. (2016), Martins et al. (2018)
2-Phenoxyethanol	100	Anesthesia for larvae	van den Bos et al. (2020)

[a]*Diluted in ethanol at 10%.*

ideally with minimal handling, be safe to staff, and consider the consequences of the procedure chosen to the postmortem analysis as it may interfere with the results. The choice of method also depends on species, strain, age, number, and health of the animals (Leary et al., 2020). Confirmation of final death is important in fish since the absence of a heartbeat is not as good an indicator as it is in mammals. Fish heart muscle tissue contains relatively high amounts of stored glycogen. Thus the heart muscle and pacemaker cells have energy to maintain the heartbeat for quite a long time without oxygen uptake by respiration and without a functional CNS. This causes problems in interpretation. On the one hand, a missing heartbeat may be reversible when the fish is returned to its normal fish water surroundings. On the other hand, although the fish is definitely dead (like after decapitation or destruction of the brain), the heart can continue beating.

Within the European Union, under European Directive 2010/63/EU on the protection of animals used for scientific purposes, legally approved methods for euthanasia of fish include overdoses of anesthetics, concussion/percussive blow to the head, and electrical stunning. Other methods can be used in animals that do not regain consciousness before death. The concussion methodology is not practical for *Danio rerio* due to the size of the animals and is therefore not considered further; however, it would make it impossible to use the brain for further analysis. Only anesthetic overdose is adequate for use in zebrafish larvae.

Overdose of anesthetics

Anesthetic overdose kills fish by hypoxia, which can be stressful. Also, anesthetics may be aversive to the animals, causing more stress, and this, associated with being a chemical, may influence the test results. People mostly use tricaine (200–300 mg/L), rarely also benzocaine (250 mg/L) (Thurman et al., 2019), or sometimes clove oil (50–100 mg/L) (Davis et al., 2015; Ferreira et al., 2018). A 10–20-min period without opercular movement has been recommended to confirm final death (Matthews and Varga, 2012; Ferreira et al., 2018). Hence, the confirmation of death by other methods is of ethical importance. This is critical in young fish that have not developed gills (major organ of anesthetic entrance), absorbing chemicals only by diffusion, thus becoming more tolerant to anesthetics. Anesthetics are therefore insufficient for the complete euthanasia of embryos and larvae, and the concentrations used and exposure times must be quite high. Tricaine at 1633 mg/L and 730 mg/L stops the heart beating at 3 days postfertilization (dpf) and 9 dpf larvae, respectively (Rombough, 2007). However, other studies show that the heart of 144 dpf larvae stops beating with 900 mg/L tricaine, but they recover in clean water. Even 1500 μL/L of eugenol will avoid the recovery of half of the 14 dpf larvae (Strykowski and Schech, 2015). This includes a critical period from 120 hpf to approximately 14 days. Embryos are also quite resistant to anesthetics and may continue to develop during exposure (Köhler et al., 2017). In addition to the physical methods to complete euthanasia referred to at the end of the chapter, chemical methods (proteinase K, bleach solution) can also be used to ensure death.

Electrical stunning

Electrical stunning or electroeuthanasia results in cardiac fibrillation inducing cardiac arrest. The resulting cerebral hypoxia then leads to death of the animal.

The advantage of this method is the absence of chemical substances influencing subsequent analyses. Nevertheless, it is not often used for zebrafish since technical parameters have to be optimized and standardized devices are missing. Technical details to be considered are the type of electric current (smooth DC, pulse DC, although it is not recommended to use AC), power, voltage, water conductivity, and form of the tank and electrodes to guarantee a uniform electric field for a humane killing (Lines et al., 2003; Lines and Kestin, 2004, 2005). If an inadequate current is used for some time, it may cause muscle exhaustion and the fish may be mistakenly considered unconscious. A suggested protocol combines an initial smooth DC with a following pulse DC (van Krogh et al., 2019). While the smooth DC results in anodotaxis and electroanesthesia, pulse DC affects the muscles and leads to an epileptiform insult. This can lead to tonic–clonic spasms by epileptic seizures ranging from hemorrhages and injuries to broken spines. Therefore it is important to anesthetize the animals before using smooth DC so that muscles are relaxed as much as possible to avoid injuries and to start with sufficient voltage immediately to reduce time until stunning and death. Currently, no commercial devices are available and only handmade devices may be used. This is a topic that needs proper studies and protocol refinement.

Rapid chilling

Rapid chilling or rapid cooling is not accepted by the European Directive 2010/63/EU and its transposition into national law in each European country by standard. However, it can be used in small tropical fish in the United States under American Veterinary Medical Association guidelines (American Veterinary Medical Association, 2020). For this method, animals are transferred abruptly from the housing temperature into ice-cold water at 0–4°C without direct contact with any ice. This leads to spontaneous loss of all signs of life, and the reliability of the occurrence of death is 100%. Recent studies (Wilson et al., 2009; Köhler et al., 2017) meanwhile lead to the recognition of this method for *Danio rerio* and other small tropical fish species in several countries (Leary et al., 2020). For young stages, additional measures and prolonged exposure are necessary: larvae until 14 dpf need to be exposed for 12 h, while older larvae until 90 dpf need between 30 s and 5 min (Wallace et al., 2018). Larvae and embryonic stages are sometimes very resistant, so proteinase K, bleach solution, or liquid nitrogen can be used in addition.

Although rapid chilling cannot currently be used as standard in research in European countries, it is possible to ask for a special permit from the competent authority to use it (or other methods) justifying scientifically why you cannot use the accepted methods in your project and/or assuring that the method chosen will not compromise the animals' welfare (2010/63/EU, article 5, point 4).

At the end of all the euthanasia methods, death must be confirmed by loss of reflexes or clinical signs (respiratory and cardiac arrest), and by a secondary method such as decapitation or exsanguination. Usually for embryos or larvae, the secondary method is exposure to bleach, proteinase K, or even maceration. The carcass must then be refrigerated until proper disposal by a specialized company.

Conclusions

This chapter is a hub of information regarding analgesia, anesthesia, and euthanasia of zebrafish. Recommendation of a good protocol depends highly on the procedure/research and animal characteristics, as discussed. As fish comprise so many different species, it is essential that the methods are refined for each species. Zebrafish is now widely used in research, but knowledge of these topics is still lacking compared with the other widely used models, i.e., mice. Thus research needs to continue in these areas, as some challenges need to be overcome, such as protocols for repeated anesthesia, clarification of the potentially aversive profile of the anesthetics, influence of substance-specific side effects on different protocols, and implications of rapid cooling on animal welfare.

References

Alvarez, F.A., Rodriguez-Martin, I., Gonzalez-Nuñez, V., Marrón Fernández de Velasco, E., Gonzalez Sarmiento, R., Rodríguez, R.E., 2006. New kappa opioid receptor from zebrafish *Danio rerio*. Neurosci. Lett. 405 (1–2), 94–99.

American Veterinary Medical Association, 2020. AVMA Guidelines for the Euthanasia of Animals. Available from: https://www.avma.org/sites/default/files/2020-01/2020-Euthanasia-Final-1-17-20.pdf. (Accessed 5 November 2020).

Attili, S., Hughes, S.M., 2014. Anaesthetic tricaine acts preferentially on neural voltage-gated sodium channels and fails to block directly evoked muscle contraction. PLoS One 9 (8), e103751.

Bachour, R.-L., Golovko, O., Kellner, M., Pohl, J., 2020. Behavioral effects of citalopram, tramadol, and binary mixture in zebrafish (*Danio rerio*) larvae. Chemosphere 238, 124587.

Carlson, B.M., Rainin, E.A., 1985. Rat extraocular muscle regeneration. Repair of local anesthetic-induced damage. Arch. Ophthalmol. 103 (9), 1373–1377.

Carter, K.M., Woodley, C.M., Brown, R.S., 2011. A review of tricaine methanesulfonate for anesthesia of fish. Rev. Fish Biol. Fish. 21 (1), 51–59.

Chance, R.J., Cameron, G.A., Fordyce, M., Noguera, P., Wang, T., Collins, C., Secombes, C.J., Collet, B., 2018. Effects of repeated anaesthesia on gill and general health of Atlantic salmon, *Salmo salar*. J. Fish. Biol. 93 (6), 1069–1081.

Chatigny, F., Kamunde, C., Creighton, C.M., Stevens, E.D., 2017. Uses and doses of local anesthetics in fish, amphibians, and reptiles. J. Am. Assoc. Lab. Anim. Sci. 56 (3), 244–253.

Chatigny, F., Creighton, C.M., Stevens, E.D., 2018. Updated review of fish analgesia. J. Am. Assoc. Lab. Anim. Sci. 57 (1), 5–12.

Cochrane, M.L., 2015. Inhibition of Cytochrome P450 2C9 by Essential Oils, UNCG NC DOCKS (North Carolina Digital Online Collection of Knowledge and Scholarship). Master-Thesis. The University of North Carolina, Greensboro.

Collymore, C., Tolwani, A., Lieggi, C., Rasmussen, S., 2014. Efficacy and safety of 5 anesthetics in adult zebrafish (*Danio rerio*). J. Am. Assoc. Lab. Anim. Sci. 53 (2), 198–203.

Correia, A.D., Cunha, S.R., Scholze, M., Stevens, E.D., 2011. A novel behavioral fish model of nociception for testing analgesics. Pharmaceuticals 4 (4), 665–680.

Currie, A.D., 2014. Toward a novel model of pain in zebrafish: exposure to water containing dilute concentrations of acetic acid. In: Psychology Honors Projects. Macalester College, St. Paul, Minnesota.

Curtright, A., Rosser, M., Goh, S., Keown, B., Wagner, E., Sharifi, J., Raible, D.W., Dhaka, A., 2015. Modeling nociception in zebrafish: a way forward for unbiased analgesic discovery. PLoS One 10 (1), e0116766.

Dang, M., Henderson, R.E., Garraway, L.A., Zon, L.I., 2016. Long-term drug administration in the adult zebrafish using oral gavage for cancer preclinical studies. Dis. Models Mech. 9 (7), 811–820.

Davis, D.J., Klug, J., Hankins, M., Doerr, H.M., Monticelli, S.R., Song, A., Gillespie, C.H., Bryda, E.C., 2015. Effects of clove oil as a euthanasia agent on blood collection efficiency and serum cortisol levels in *Danio rerio*. J. Am. Assoc. Lab. Anim. Sci. 54 (5), 564–567.

Deakin, A.G., Buckley, J., AlZu'bi, H.S., Cossins, A.R., Spencer, J.W., Al'Nuaimy, W., Young, I.S., Thomson, J.S., Sneddon, L.U., 2019. Automated monitoring of behaviour in zebrafish after invasive procedures. Sci. Rep. 9 (1), 9042.

Du, W.-J., Zhang, R.-W., Li, J., Zhang, B.-B., Peng, X.-L., Cao, S., Yuan, J., Yuan, C.-D., Yu, T., Du, J.-L., 2018. The locus coeruleus modulates intravenous general anesthesia of zebrafish via a cooperative mechanism. Cell Rep. 24 (12), 3146–3155.e3.

Ehrlich, O., Karamalakis, A., Krylov, A.J., Dudczig, S., Hassell, K.L., Jusuf, P.R., 2019. Clove oil and AQUI-S efficacy for zebrafish embryo, larva, and adult anesthesia. Zebrafish 16 (5), 451–459.

Ferreira, J.M., Olsson, I.A.S., Valentim, A.M., 2018. Adult zebrafish euthanasia: efficacy of anaesthesia overdose versus rapid cooling. Preprint. https://doi.org/10.7287/peerj.preprints.27432v1. (Accessed 5 November 2020).

Gonzalez-Nunez, V., Rodríguez, R.E., 2009. The zebrafish: a model to study the endogenous mechanisms of pain. ILAR J. 50 (4), 373–386.

Grush, J., Noakes, D.L.G., Moccia, R.D., 2004. The efficacy of clove oil as an anesthetic for the zebrafish, *Danio rerio* (Hamilton). Zebrafish 1 (1), 46–53.

Gudin, J., Fudin, J., 2020. A narrative pharmacological review of buprenorphine: a unique opioid for the treatment of chronic pain. Pain Ther. 9 (1), 41–54.

Guo, P., Huang, Z., Tao, T., Chen, X., Zhang, W., Zhang, Y., Lin, C., 2015. Zebrafish as a model for studying the developmental neurotoxicity of propofol. J. Appl. Toxicol. 35 (12), 1511–1519.

Hassan, M., 2015. Toxicity induced historical changes in gills of Oreochromis sp. by exposed to clove oil. J. FisheriesSciences.com 9 (4).

Huang, W.-C., Hsieh, Y.-S., Chen, I.-H., Wang, C.-H., Chang, H.-W., Yang, C.-C., Ku, T.-H., Yeh, S.-R., Chuang, Y.-J., 2010. Combined use of MS-222 (tricaine) and isoflurane extends anesthesia time and minimizes cardiac rhythm side effects in adult zebrafish. Zebrafish 7 (3), 297–304.

Hussain, N., McCartney, C.J.L., Neal, J.M., Chippor, J., Banfield, L., Abdallah, F.W., 2018. Local anaesthetic-induced myotoxicity in regional anaesthesia: a systematic review and empirical analysis. Br. J. Anaesth. 121 (4), 822–841.

Johnson, J.D., Ryan, M.J., Toft JD, I.I., Graves, S.W., Hejtmancik, M.R., Cunningham, M.L., Herbert, R., Abdo, K.M., 2000. Two-year toxicity and carcinogenicity study of methyleugenol in F344/N rats and B6C3F(1) mice. J. Agric. Food Chem. 48 (8), 3620–3632.

Kamei, M., Weinstein, B.M., 2005. Long-term time-lapse fluorescence imaging of developing zebrafish. Zebrafish 2 (2), 113–123.

Khor, B.-S., Jamil, M.F.A., Adenan, M.I., Shu-Chien, A.C., 2011. Mitragynine attenuates withdrawal syndrome in morphine-withdrawn zebrafish. PLoS One 6 (12), e28340.

Kieffer, B.L., 1999. Opioids: first lessons from knockout mice. Trends Pharmacol. Sci. 20 (1), 19–26.

Köhler, A., Collymore, C., Finger-Baier, K., Geisler, R., Kaufmann, L., Pounder, K.C., Schulte-Merker, S., Valentim, A., Varga, Z.M., Weiss, J., Strähle, U., 2017. Report of workshop on euthanasia for zebrafish-a matter of welfare and science. Zebrafish 14 (6), 547–551.

Leary, S., Underwood, W., Anthony, R., Cartner, S., Grandin, T., Greenacre, C., Gwaltney-Brant, S., et al., 2020. AVMA Guidelines for the Euthanasia of Animals: 2020 Edition, 1 October.

Lelek, S., Simões, M.G., Hu, B., Alameldeen, A.M., Czajkowski, M.T., Meyer, A.M., Ferrara, F., et al., 2020. Morphine alleviates pain after heart cryoinjury in zebrafish without impeding regeneration. Preprint, bioRxiv. https://doi.org/10.1101/2020.10.01.322560. (Accessed 5 November 2020).

Lidster, K., Readman, G.D., Prescott, M.J., Owen, S.F., 2017. International survey on the use and welfare of zebrafish *Danio rerio* in research. J. Fish. Biol. 90 (5), 1891–1905.

Lines, J., Kestin, S., 2004. Electrical stunning of fish: the relationship between the electric field strength and water conductivity. Aquaculture 241 (1–4), 219–234.

Lines, J., Kestin, S., 2005. Electric stunning of trout: power reduction using a two-stage stun. Aquacult. Eng. 32 (3–4), 483–491.

Lines, J.A., Robb, D.H., Kestin, S.C., Crook, S.C., Benson, T., 2003. Electric stunning: a humane slaughter method for trout. Aquacult. Eng. 28 (3–4), 141–154.

Lisse, T.S., Brochu, E.A., Rieger, S., 2015. Capturing tissue repair in zebrafish larvae with time-lapse brightfield stereomicroscopy. J. Vis. Exp. (95).

Lockwood, N., Parker, J., Wilson, C., Frankel, P., 2017. Optimal anesthetic regime for motionless three-dimensional image acquisition during longitudinal studies of adult nonpigmented zebrafish. Zebrafish 14 (2), 133–139.

Lopez-Luna, J., Al-Jubouri, Q., Al-Nuaimy, W., Sneddon, L.U., 2017a. Impact of analgesic drugs on the behavioural responses of larval zebrafish to potentially noxious temperatures. Appl. Anim. Behav. Sci. 188, 97–105.

Lopez-Luna, J., Al-Jubouri, Q., Al-Nuaimy, W., Sneddon, L.U., 2017b. Reduction in activity by noxious chemical stimulation is ameliorated by immersion in analgesic drugs in zebrafish. J. Exp. Biol. 220 (Pt 8), 1451–1458.

Lopez-Luna, J., Canty, M.N., Al-Jubouri, Q., Al-Nuaimy, W., Sneddon, L.U., 2017c. Behavioural responses of fish larvae modulated by analgesic drugs after a stress exposure. Appl. Anim. Behav. Sci. 195, 115–120.

Luo, X., Chen, L., Zhang, Y., Liu, J., Xie, H., 2020. Developmental and cardiac toxicities of propofol in zebrafish larvae. Comp. Biochem. Phys. Toxicol. Pharmacol. 237, 108838.

Martins, T., Valentim, A.M., Pereira, N., Antunes, L.M., 2016. Anaesthesia and analgesia in laboratory adult zebrafish: a question of refinement. Lab. Anim 50 (6), 476–488.

Martins, T., Diniz, E., Félix, L.M., Antunes, L., 2018. Evaluation of anaesthetic protocols for laboratory adult zebrafish (*Danio rerio*). PLoS One 13 (5), e0197846.

Matthews, M., Varga, Z.M., 2012. Anesthesia and euthanasia in zebrafish. ILAR J. 53 (2), 192–204.

Mettam, J.J., Oulton, L.J., McCrohan, C.R., Sneddon, L.U., 2011. The efficacy of three types of analgesic drugs in reducing pain in the rainbow trout, *Oncorhynchus mykiss*. Appl. Anim. Behav. Sci. 133 (3–4), 265–274.

Musk, G.C., Ezzy, B.J., Kenchington, L.M., Hopper, W.A., Callahan, L.M., 2020. A comparison of buffered tricaine methanesulfonate (MS-222) and isoeugenol anesthesia for caudal fin clipping in zebrafish (*Danio rerio*). J. Am. Assoc. Lab. Anim. Sci. 59.

Nordgreen, J., Tahamtani, F.M., Janczak, A.M., Horsberg, T.E., 2014. Behavioural effects of the commonly used fish anaesthetic tricaine methanesulfonate (MS-222) on zebrafish (*Danio rerio*) and its relevance for the acetic acid pain test. PLoS One 9 (3), e92116.

Pinal-Seoane, N., Martin, I.R., Gonzalez-Nuñez, V., Marron Fernandez de Velasco, E., Alvarez, F.A., Sarmiento, R.G., Rodriguez, R.E., 2006. Characterization of a new duplicate delta-opioid receptor from zebrafish. J. Mol. Endocrinol. 37 (3), 391–403.

Readman, G.D., Owen, S.F., Murrell, J.C., Knowles, T.G., 2013. Do fish perceive anaesthetics as aversive? PLoS One 8 (9), e73773.

Reilly, S.C., Quinn, J.P., Cossins, A.R., Sneddon, L.U., 2008. Behavioural analysis of a nociceptive event in fish: comparisons between three species demonstrate specific responses. Appl. Anim. Behav. Sci. 114 (1–2), 248–259.

Rombough, P.J., 2007. Ontogenetic changes in the toxicity and efficacy of the anaesthetic MS222 (tricaine methanesulfonate) in zebrafish (*Danio rerio*) larvae. Comp. Biochem. Physiol. Mol. Integr. Physiol. 148 (2), 463–469.

Ross, L.G., Ross, B., 2008. Anaesthetic and Sedative Techniques for Aquatic Animals, third ed. Blackwell, Oxford, Ames, Iowa.

Sanchez-Simon, F.M., Rodriguez, R.E., 2008. Developmental expression and distribution of opioid receptors in zebrafish. Neuroscience 151 (1), 129–137.

Schroeder, P.G., Sneddon, L.U., 2017. Exploring the efficacy of immersion analgesics in zebrafish using an integrative approach. Appl. Anim. Behav. Sci. 187, 93–102.

Sivalingam, M., Ogawa, S., Parhar, I.S., 2020. Mapping of morphine-induced OPRM1 gene expression pattern in the adult zebrafish brain. Front. Neuroanat. 14, 5.

Sladky, K.K., Swanson, C.R., Stoskopf, M.K., Loomis, M.R., Lewbart, G.A., 2001. Comparative efficacy of tricaine methanesulfonate and clove oil for use as anesthetics in red pacu (Piaractus brachypomus). Am. J. Vet. Res. 62 (3), 337–342.

Sneddon, L., Elwood, R., Adamo, S., Leach, M., 2014. Defining and assessing animal pain. Anim. Behav. 97, 201–212.

Steenbergen, P.J., Bardine, N., 2014. Antinociceptive effects of buprenorphine in zebrafish larvae: an alternative for rodent models to study pain and nociception? Appl. Anim. Behav. Sci. 152, 92–99.

Stevens, D., 2008. "Pain" and analgesia in fish: what we know, what we do not know, and what we need to know, before using analgesics in fish. In: ANZCCART (Ed.), Blue Sky to Deep Water: The Reality and the Promise. ANZCCART Conference, Auckland, pp. 115–124.

Strykowski, J.L., Schech, J.M., 2015. Effectiveness of recommended euthanasia methods in larval zebrafish (*Danio rerio*). J. Am. Assoc. Lab. Anim. Sci. 54 (1), 81–84.

Thurman, C.E., Rasmussen, S., Prestia, K.A., 2019. Effect of 3 euthanasia methods on serum yield and serum cortisol concentration in zebrafish (*Danio rerio*). J. Am. Assoc. Lab. Anim. Sci. 58 (6), 823–828.

Trevarrow, B., 2011. Techniques for optimizing the creation of mutations in zebrafish using N-ethyl-N-nitrosourea. Lab. Anim. 40 (11), 353–361.

Valentim, A.M., Félix, L.M., Carvalho, L., Diniz, E., Antunes, L.M., 2016. A new anaesthetic protocol for adult zebrafish (*Danio rerio*): propofol combined with lidocaine. PLoS One 11 (1), e0147747.

Valentim, A.M., Ferreira, J., Olsson, A., 2018. Protocols' comparison to zebrafish anaesthesia and euthanasia. In: Karolinska Institutet (Ed.), Nordic Zebrafish and Husbandry Meeting, Nordic Zebrafish and Husbandry Meeting, 7-9.11.2018, Stockholm, pp. 16–17.

van den Bos, R., Cromwijk, S., Tschigg, K., Althuizen, J., Zethof, J., Whelan, R., Flik, G., Schaaf, M., 2020. Early life glucocorticoid exposure modulates immune function in zebrafish (*Danio rerio*) larvae. Front. Immunol. 11, 727.

van Krogh, K., Jencic, V., Lundegaard, P.R., McKimm, R., Mocho, J.-P., Ramos Blasco, J., 2019. Update from the FELASA working group on methods of humane killing of laboratory fish. In: Abstracts of 14th FELASA Congress 2019, 14th FELASA Congress, Prague, pp. 82–83.

Wallace, C.K., Bright, L.A., Marx, J.O., Andersen, R.P., Mullins, M.C., Carty, A.J., 2018. Effectiveness of rapid cooling as a method of euthanasia for young zebrafish (*Danio rerio*). J. Am. Assoc. Lab. Anim. Sci. 57 (1), 58–63.

Walsh, R.R., Schopp, R.T., 1966. Frequency-depressing action of ethylaminobenzoate isomers on a mormyrid electric fish. Am. J. Physiol. 211 (1), 1–5.

Wilson, J.M., Bunte, R.M., Carty, A.J., 2009. Evaluation of rapid cooling and tricaine methanesulfonate (MS222) as methods of euthanasia in zebrafish (*Danio rerio*). J. Am. Assoc. Lab. Anim. Sci. 48 (6), 785–789.

Wong, K., Elegante, M., Bartels, B., Elkhayat, S., Tien, D., Roy, S., Goodspeed, J., Suciu, C., Tan, J., Grimes, C., Chung, A., Rosenberg, M., Gaikwad, S., Denmark, A., Jackson, A., Kadri, F., Chung, K.M., Stewart, A., Gilder, T., Beeson, E., Zapolsky, I., Wu, N., Cachat, J., Kalueff, A.V., 2010. Analyzing habituation responses to novelty in zebrafish (*Danio rerio*). Behav. Brain Res. 208 (2), 450–457.

Wong, D., Keyserlingk, M.A.G. von, Richards, J.G., Weary, D.M., 2014. Conditioned place avoidance of zebrafish (*Danio rerio*) to three chemicals used for euthanasia and anaesthesia. PLoS One 9 (2), e88030.

Wynd, B.M., Watson, C.J., Patil, K., Sanders, G.E., Kwon, R.Y., 2017. A dynamic anesthesia system for long-term imaging in adult zebrafish. Zebrafish 14 (1), 1–7.

Xu, C., Volkery, S., Siekmann, A.F., 2015. Intubation-based anesthesia for long-term time-lapse imaging of adult zebrafish. Nat. Protoc. 10 (12), 2064–2073.

Zhuo, H., Jin, H., Peng, H., Huang, H., 2016. Distribution, pharmacokinetics and primary metabolism model of tramadol in zebrafish. Mol. Med. Rep. 14 (6), 5644–5652.

Chapter 7

Transgenesis, mutagenesis, knockdown, and genetic colony management

Claire Allen[1], Francesco Argenton[2], Fredericus J. Van Eeden[1], Natascia Tiso[2]

[1]*Bateson Centre, University of Sheffield, Sheffield, United Kingdom;* [2]*Dipartimento di Biologia, Università degli Studi di Padova, Padua, Italy*

Introduction

Genetic engineering techniques have rapidly developed since the first large-scale forward genetics screen was developed in 1996, which generated a spectrum of phenotypic zebrafish lines with mutations in many essential genes for normal development (Haffter et al., 1996). Management of this original collection was relatively straightforward and embryonic analysis was the focus. Adults were heterozygous in their genotype, and phenotypes were primarily within homozygous embryos. Homozygous fish were rarely viable and reporter lines were in their molecular infancy. Genotyping methods were not available since extensive molecular investigation was required to piece together the pathways involved. Mapping and genetic characterization of mutations was propelled forward in the early 2000s by the full sequencing of the *Danio rerio* genome (Howe et al., 2013).

This chapter aims to provide a comprehensive overview of the ethical issues in zebrafish (*Danio rerio*) handling and the most recent practices for the generation and management of zebrafish transgenic, mutant, and mosaic lines.

The discussion will take into account recent legislation, international publications, and standard guidelines applied in world reference centers such as the Zebrafish International Resource Center (ZIRC) and the European Zebrafish Resource Center (EZRC), as well as our personal experience in the management of national-level facilities.

A set of schemes will be included to provide the reader with an immediate overview of the main operational phases in the different workflows.

Laboratory Fish in Biomedical Research. https://doi.org/10.1016/B978-0-12-821099-4.00006-7
Copyright © 2022 Elsevier Inc. All rights reserved.

Ethical issues in genetic manipulation of laboratory fish

Being a vertebrate with a high degree of similarity with humans, the zebrafish is often used to assess the function of genes involved in diseases. The advantages of using this model in quantitative biology is mostly due to its fecundity, genetic manipulability, transparency of its embryos, and easy husbandry, allowing rapid screens of drugs or genetic modifications. As for all animal models in research, while the number of zebrafish used and their suffering has to be minimized, the amount of information gathered in experiments needs to be maximized.

Several national legislations on laboratory animal welfare use as reference the OIE Terrestrial Animal Health Code and the EU Directive 2010/63 (Ogden et al., 2016). Considering the EU example, the aim of the Directive 2010/63/EU, which replaced the 86/609/EEC, is to improve the welfare of those animals still needed in research and to deliver the 3R principles (replacement, reduction, and refinement). While the final goal of the directive is the full replacement of procedures on live animals, it also states that "the use of live animals continues to be necessary to protect human and animal health and the environment" and that "the methods selected should use the minimum number of animals that would provide reliable results and require the use of species with the lowest capacity to experience pain, suffering, distress or lasting harm that are optimal for extrapolation into target species."

Hence, to minimize the suffering of animals, genetically manipulated embryos, injected with oligos, mRNAs, or DNA, should be screened for deleterious morphological and behavioral phenotypes at 5 days postfertilization (dpf) just before they become protected animals; embryos with significant phenotypes will be excluded unless explicitly allowed by the relevant animal license. We have developed a set of criteria that we use to judge whether it is justified to keep fish alive under "standard circumstances," that is in the absence of pathological signs (Westerfield, 2007).

Although the recent clustered regularly interspaced short palindromic repeats (CRISPR)/Cas9 technology-mediated mutagenesis does not involve foreign DNA, it is important to note that the European Court has surprisingly classified CRISPR/Cas9-edited plants under the same regulatory rules as genetically modified organisms (GMOs) that contain foreign DNA. It will be important to check local GMO regulations to ascertain how CRISPR mosaics (CRISPants) and CRISPR mutants need to be licensed and documented (see https://curia.europa.eu/jcms/upload/docs/application/pdf/2018-07/cp180111en.pdf).

In this chapter we take for granted that all experiments will be conducted under the 3R principles and their implementation in the legislation of each country.

Fish management during generation of mutants

Soon after the random mutagenesis era, large consortia, such as ZF-Health, implemented mutant creation by utilizing the genomic sequence through the technique of Targeting Induced Local Lesions in Genomes (TILLING) (Moens et al., 2008). This was the first time that reverse genetics could efficiently be applied in fish, although exact mutation and position were still unpredictable. Preparation of germline mutants requires a difficult balance between (1) the direct and indirect deleterious effects of mutagenesis on animal health and (2) the efficiency in targeting the right portion of the genome. As nowadays most mutagenesis screens, either small or large scale, are performed by CRISPR/Cas9, in this chapter we will focus on the management of fish stocks during CRISPR/Cas9-mediated genome editing. Notably, researchers can now design and tailor-make specific gene changes using the CRISPR technology (Rafferty and Quinn, 2018), recently improved using modified oligonucleotides, such as end-phosphorothioated long oligonucleotides (~127 bp) (Renaud et al., 2016).

The first step in a CRISPR/Cas9-mediated mutagenic effort is the analysis of the efficiency of either the single guide RNAs (sgRNAs) or the CRISPR RNA (crRNA) in combination with a trans-activating crRNA (tracrRNA) (Hoshijima et al., 2019). This can be evaluated using a simple polymerase chain reaction (PCR) and melting analysis on the genes targeted in F0 embryos injected with Cas9 and the different sg- or tracrRNA/crRNAs pairs. Different approaches can be used to evaluate quickly the annealing efficiency of mutagenized alleles once they are amplified by PCR: agarose gel analysis using the T7E1 endonuclease mismatch test (Ablain et al., 2015), heteroduplex mutant analysis of melted DNA in polyacrylamide gels (Zhu et al., 2014), or high-resolution analysis of the annealing dynamics of real-time melted DNA (Samarut et al., 2016). The analysis of single injected F0 embryos will reveal the global and individual efficiency of each sgRNA or tracrRNA/crRNA.

As an alternative to annealing-based methods, high-resolution fragment analysis (Varshney et al., 2016) or, if designed in a suitable sequence, restriction digests can be employed to assess guide efficiency (Keatinge et al., 2020). Once selected for a proper mutagenic efficiency, mutagenic RNAs can be used for the production and isolation of germline mutations. However, to isolate F0 carriers of new allelic variants generated by CRISPR/Cas9, homozygosity for null alleles of essential genes should be avoided. Hence, if the purpose of the research is to produce a complete knockout, a global efficiency of mutagenesis around 80% is desirable.

When preparing a targeted "knock-in," the efficiency of the donor also needs to be tested. Nowadays, the strategy for an efficient substitution of specific nucleotides in the zebrafish genome takes advantage of long asymmetric synthetic single-strand DNA donors noncomplementary to the crRNA or sgRNA and containing the mutation to be introduced, a mutated protospacer adjacent motif, and a genotyping bar to recover the mutated genome with a simple restriction analysis (Prykhozhij et al., 2018).

To rear as few animals as possible, the efficiency of mutagenesis will be evaluated in a portion of the embryos before they reach the 6-day stage. F0 embryos from clutches efficiently mutagenized are selected for the absence of anatomical and behavioral defects before day 6 and grown to adulthood.

Adult F0 are outcrossed with wild-type animals for the analysis of germline transmission of mutated alleles: a fraction of the F1 embryos resulting from each outcross is analyzed at 48 h postfertilization (hpf) using PCR amplification and melting/size/restriction analysis of the targeted regions, as reported earlier. These methods do not establish the precise nature of the allele. Sequencing of F1 PCR products on both strands can be used and often allows deconvolution, either manually or using specific software (Dehairs et al., 2016), to determine if it is worth raising the allele in question. Subsequently, F1 individuals from transmitting founders are genotyped by noninvasive methods within 6 dpf and heterozygous individuals are grown to obtain F1 animals that, when adult, will be fin clipped, analyzed on both strands to confirm the sequence of the mutant allele, and outcrossed for its isolation.

To determine the possible deleterious effects of the mutant allele in the F2 (or F3 generation if maternal effects are expected), the different F1 (or F2) heterozygotes will be incrossed to generate null homozygotes (or compound null homozygotes) or homozygous knock-in animals to be analyzed from embryonic stages up to the adult; for zebrafish, a reference value of 20 larvae out of at least two clutches of eggs is recommended. Evaluation at the 6-day stage will consider the morphology, structure, and behavior of the larva (edema, swim bladder, body shape, swimming behavior). A second detailed evaluation will take place at the adult stage (3 months). During the characterization of the lines, primary attention will be given to the occurrence of signs of suffering, pain, or distress. Therefore once the larval stage has been passed, if a severe phenotype occurs, i.e., the animal is unable to move and feed sufficiently (the humane endpoint is reached), it will be suppressed after annotation.

Only the heterozygous animals shall be kept for maintaining the line when the homozygous displays a deleterious phenotype.

Taking all these elements into consideration, it is believed that a total of about 40 adults (10 F0 and 30 F1 heterozygotes) will be sufficient to produce and isolate a reasonable number of null alleles of an essential gene. In addition, a total of 40 animals from two different families should be used to assess the level of the deleterious effect in homozygotes, either null or knock-in (Fig. 7.1). To breed as few animals as possible, genotyping should ideally be carried out on individual larvae at the 4–6 days stage by means of a PCR of the water in which the larvae are grown individually. Only larvae with a genotype of interest will be grown to the adult stage. To breed as few animals as possible, sperm from the best F0 and F1 carriers shall be frozen.

A total of about 40 fish (10 F0 injected and 10 F1 in three different families) are grown to isolate a reasonable number of mutants. Forty more F2 (or F3 in case a maternal phenotype is expected) are needed for the assessment of deleterious phenotypes.

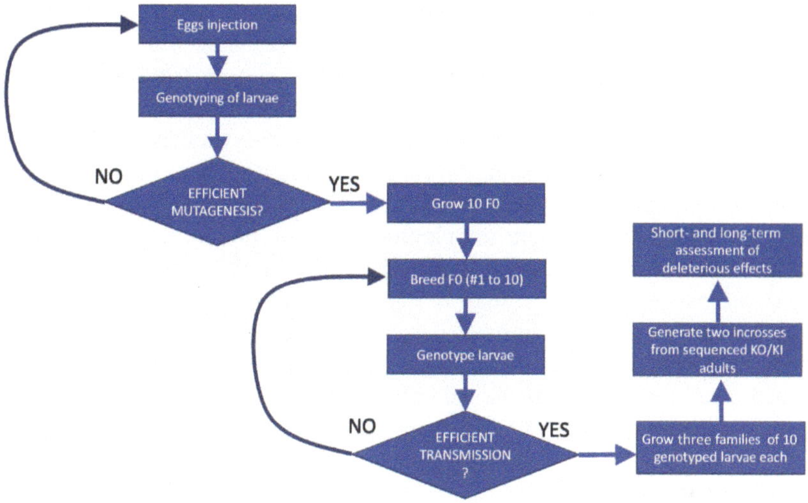

FIGURE 7.1 Flowchart for the generation of KO (knockout) and KI (knock-in) mutants.

Mutant fish management

In-depth analysis of gene function often requires specialized genetic backgrounds that provide enhanced readouts of gene function, either by sensitizing the fish to the mutation in question or by providing relevant fluorescent readouts. This has led to extensive interbreeding of multiple mutants and transgenes. Nowadays, it is not unheard of for researchers to breed over six genetic elements into one line. For lines with multiple alleles, sufficient backup stocks of each component line are required. Indeed, all genetically modified zebrafish lines require careful management to preserve their distinct genotype. The zebrafish is very susceptible to inbreeding depressions after successive incrosses, which rapidly leads to viability, fertility, and phenotypic problems. Careful genetic tracking of parentage and the alternate in- and outcrossing of both homozygotes (when viable and fertile) and heterozygotes in different combinations, with a constant eye on the potential evidence of inbreeding, is required. For newly generated lines, particularly by reverse genetics (if justified over CRISPR interference - CRISPRi - methods), several heterozygote generations originating from outcrossed parents will need to show consistent phenotypes, only then can you attribute these defects to the genetic modification. For CRISPR, we generally create two alleles initially, and then transheterozygotes; this minimizes the chance of other homozygous mutations contributing to a phenotype. Line management must consider and mitigate for the consequences of breeding a homogeneous population, particularly in an attempt to deliver research data as quickly as possible. If homozygous animals are viable, then populations of both homozygous and heterozygous carriers need to be alternately in- and outcrossed, with significant introduction of fresh wild-type genomes, to maintain health within the background. Genetically homogeneous tractable strains are known to be difficult in the zebrafish and tend to result in nonspecific phenotypes, biased sex ratios, increased mortality, and reduced fecundity (Franěk et al., 2020). The absence of sex chromosomes in common zebrafish strains may contribute to this, as homozygous populations will tend toward one of the two sexes. This contrasts with medaka where sex chromosomes are present and inbred lines are available. None of these attributes are conducive to robust research findings and experimental potential is only realized through optimal mutant management of many additional factors; husbandry knowledge; a health monitoring program; quarantine provision and procedures; optimal environmental conditions, nourishment and enrichment; and consideration of species-specific experimental design (Aleström et al., 2020).

Husbandry

The zebrafish is a communal shoaling fish in the wild (Engeszer et al., 2007), and develops a shoaling preference based on visual early experience (Engeszer et al., 2004). The identification of adverse effects within a shoal or individual fish is paramount to manage mutant lines. An understanding of their response to sickness, aging, hierarchy, environment, stress, and fear will help manage even the most vulnerable lines. Maintaining solitary or pairs of zebrafish is deemed to be an unnatural environment for this species and for this reason should be avoided, and visual contact to conspecifics is warranted; nevertheless, isolation studies have not detected strong effects on relevant neurotransmitters related to depression unless fish were additionally stressed (Fulcher et al., 2017). In our experience, keeping fish in pairs often produces a strongly dominant and subordinate fish; the latter tends to languish and receive less food. If these situations are perpetuated without resolution for the fish, then their wellbeing will be diminished and subsequent breeding is unlikely.

Dominant and subordinate fish can emerge in any tank scenario, which deteriorates the health and wellbeing of the oppressed fish. As a prey species, individuals within a shoal will attempt to hide afflictions from internal and external causes for as long as possible. This culminates in significantly impaired fish falling from the shoal, often close to death. To help identify sick fish, two significant publications detail the signs and symptoms that you should be looking out for daily (Goodwin et al., 2016; Clark et al., 2018).

Reverse genetics has resulted in many subtle, apparently asymptomatic mutant lines that are being maintained. These lines may be vulnerable to subtle impacts upon their wellbeing and need to be carefully monitored and, once characterized, known phenotypes can be assumed to affect all genotyped fish within the family.

Subsequent interline breeding to create a combination of traits can offer new challenges in adverse effect management with potential for unexpected severity and welfare issues. To support and validate findings, researchers are increasingly maintaining, and housing separately, a control shoal made up of nonmutant sibling fish.

Critical thinking and knowledge of the variables within the life-course of the fish help research to develop a more succinct knowledge of each line as having individual characteristics and needs.

Health monitoring

Mutant fish need to be monitored for their health status throughout their life and it is important to know the risks from subclinical infections, clinical diseases (Collymore et al., 2016), and noninfectious diseases (Kent et al., 2020) (Mocho and Pereira, Chapter 4), taking the expected function of the mutated gene into consideration. As previously mentioned, daily visual assessment of a population using both the Standardized Welfare Terms for the Zebrafish Community (Goodwin et al., 2016) and the Body Condition Scoring for Adult Zebrafish (*Danio rerio*; Clark et al., 2018) enables a picture of health to emerge within a facility. A carefully considered health monitoring program is also essential for all facilities, which includes both PCR panels and histopathology (Collymore et al., 2016). This must be conducted with some degree of regularity, on a significant number of fish, from a selection of candidate individuals and pooled cohorts, perhaps pooled in age- or location-specific sets.

Quarantine provision and procedures

The international movement of zebrafish mutant lines is a key attribute of this research model and sharing of lines is commonplace. Usually, shipments of 24-hpf embryos are couriered (taking <48 h door to door) in temperature-controlled parcels. Regardless of the collaboration and knowledge of the origin's health monitoring program, all external sources of fish require a barrier upon entry to prevent the introduction of infections. This barrier should be at least one, if not two, quarantine provisions that are completely separate from the main facility. Surface cleaning with one or both "bleach" and iodine (Chang et al., 2016) is required prior to transportation and analysis of sacrificed subsets of imported fish to build knowledge of their pathogenic load (Westerfield, 2007).

Genotyping mutant lines to manage stocks

All mutant lines can be managed by a selection of genotyping techniques in the absence or alongside any phenotypic characteristics. The classic method of fin clipping is a robust method of obtaining sufficient genetic material to identify carriers. New methods have been developed: (1) to reduce the number of fish grown past the point of protection for subsequent fin clipping (once large enough), and (2) to refine the technique in light of finding that fish feel pain during the fin-clip procedure (Deakin et al., 2019). These alternative methods include larval fin clipping (Wilkinson et al., 2013), swabbing (Tilley et al., 2020), and embryonic/larval DNA extraction (Lambert et al., 2018).

Postresearch management for future opportunities and collaboration: cryopreservation and stock centers

Once a mutant line has been established and bred sufficiently for research, it may be advisable to preserve the line for perpetuity either in-house by in vitro fertilization or by shipping it to one of the three nonprofit resource centers: ZIRC, EZRC, and China Zebrafish Resource Center.

An in-house provision to cryopreserve any genetically modified zebrafish line is important for both contingency line management and to preserve potential research options when research funding is unstable. Several publications and reference centers can guide you through setting up the method in-house, with minimal technical effort (see, for instance, Carmichael et al., 2009, and more in detail Khosla et al., Chapter 8). Hindsight tells us that if a zebrafish mutant fails to produce results in one avenue of investigation, research should certainly value unknown potential, within the unique genetic modification, and ensure the line is preserved for future unprecedented collaborations (see, for instance, ZebraShare at the Zebrafish Information Network [ZFIN]). However, an important consideration is the presence of (always increasing) welfare and customs information requirements. We are getting to a situation where recreating mutant alleles using CRISPR may be preferable to the hassle and cost of importing a line. CRISPRs can be used directly in the background of interest and avoid a lengthy quarantine.

Data management

The research administration associated with zebrafish mutant management requires the use of a secure database. Robust live and historic data storage, with the ability to track and trace family lineage, facilitates workflow and reporting (see, for instance, Labtracks at http://www.locustechnology.com/labtracks.html; ARMIS at https://www.armis.co.uk; Zebrabase at https://zebrabase.org). A public resource for zebrafish genetic and genomic data is stored on ZFIN by biocurators and offers integration, transparency, and research links into parallel human disease information (www.zfin.org). ZFIN is also home to an encyclopedic style glossary on anatomy, development, genetics, and bioinformatics.

Fish management during generation of transgenics

The extraordinary optical clarity and genetic manipulability of zebrafish make this animal one of the best vertebrate organisms on which to apply a plethora of different transgenesis techniques. The simplest method, based on the microinjection of bacterial artificial chromosomes or DNA constructs into zebrafish zygotes, has a reported yield (germline transmission) around 1%, which can be raised to around 30% by coinjecting rare-cutter enzymes (e.g., *I-SceI* meganuclease). Currently, most laboratories apply retroviral or transposon-based methods, such as *Tol2* or Sleeping Beauty, as comprehensively reviewed in Teh et al. (2005). Zebrafish transgenic lines have been produced with a wide range of reporter proteins, such as classic, enhanced, destabilized, and organelle-directed green/red/yellow/cyan fluorescent proteins (GFP, RFP, YFP, CFP), or with other variants (such as mCherry, mOrange, Venus, Cerulean, dTomato, and so on) (Piatkevich and Verkhusha, 2011), driven by tissue-specific or pathway-activated promoters (Teh et al., 2005; Moro et al., 2013). Moreover, the zebrafish has proved to be suitable for gene-trapping/promoter-trapping or conditional transgenesis, including site-specific recombination, induced cell ablation, and heat-shock- and binary system-mediated (Gal4-VP16/UAS) gene expression (Scheer and Campos-Ortega, 1999; Teh et al., 2005; Davison et al., 2007; Curado et al., 2008; Hans et al., 2009).

Most of these methods imply the preparation of a transgenic construct, nowadays often adopting the Gateway technology, a multistep recombinational cloning (Kwan et al., 2007). After injecting zebrafish zygotes with the construct and suitable coreagents for integration, mosaic expression can be tracked by checking the reporter fluorescence, possibly assisted by linked markers (e.g., "green-heart"/*cmlc2:GFP* or "blue-eye"/*cryaa:CFP* transgenes) and/or specific strategies for multiple reporters, such as the internal ribosome entry site technology or the "ribosomal skipping" mechanism of the viral T2A peptide (Kim et al., 2011). After excluding insertional events potentially toxic for zebrafish larvae (euthanized within 1 week postfertilization if abnormal or transgene negative), normally developing individuals, transgene positive, are grown to adulthood. Putative F0 founders are then outcrossed to verify the germline transmission of the randomly integrated transgene and, in case of multiple insertions, to progressively dilute the transgene copies until obtaining positive offspring with Mendelian ratios. The incross of a transgenic line is expected to produce transgenic individuals in homozygous conditions. This may simplify further crosses and, sometimes, improve the transgene visualization (unless homozygosity has deleterious effects). However, repeated inbreeding for maintaining such lines can lead to inbreeding depression, therefore it is advisable to outcross homozygotes every other generation to rederive "fresh" homozygotes.

Recent transgenesis strategies apply the CRISPR/Cas9 technology; in this case, a site-directed insertion of the transgene can be obtained, avoiding random and potentially disrupting integration events. For instance, Luo and colleagues set up in zebrafish a CRISPR/Cas9-based integration system able to exploit endogenous promoters while maintaining the integrity of the targeted genes (Luo et al., 2018).

Transgenic fish management

After producing or obtaining a transgenic zebrafish line, the strain can be cryopreserved (https://zebrafish.org/wiki/protocols/cryo) or maintained as a livestock in the facility.

Maintenance of transgenic lines follows standard procedures, essentially like those applied for wild-type and mutant fish lines at the ZIRC and EZRC reference centers (Westerfield 2007; Geisler et al., 2016).

Criteria to propagate a specific transgenic line may vary but usually depends on: (1) internal and external requests; (2) low number (fewer than 10 individuals per tank); and (3) age of the livestock (we schedule new crosses every 6–12 months).

After intercross or outcross of transgenic carriers, eggs are collected and monitored for their fluorescence, taking care of the stage for the initial transgene expression, which can differ based on the specific line.

To check for possible cross-contaminations among adjacent tanks, fluorescent embryos/larvae are always checked under different filter/illumination conditions, keeping individuals with the expected pattern, and discarding those bearing wrong transgenes.

Some transgenes show an inherently unstable expression level, perhaps due to epigenetic silencing; therefore continuous selection for the brightest patterns is made at every generation. Should this strategy be insufficient, the generation of a new transgenic line, reinjecting the original transgene constructs/reagents, may be required.

Specific attention is paid to maternally activated transgenes. For instance, in case of a heterozygous mother, maternal fluorescence can mask negative genotypes in the offspring. In this case, we suggest monitoring and reevaluating the offspring after 2–4 dpf, usually sufficient for the fading of the maternal signal.

Conditional transgenes may require additional strategies for their visualization. For binary systems (e.g., GAL4/UAS), we separately maintain each activator or responsive line, annually intercrossing them for carrier identification and system validation. Some lines can allow a direct (e.g., fluorescence-based) screen, while others may require specific (e.g., PCR-based) genotyping. Drug- or heat-shock-induced transgenic lines are instead chemically or thermally treated, based on line-related published protocols. Identified carriers are then inter- or outcrossed to produce a new fish generation.

Generation and management of morphants and CRISPants

To study gene function in a process of interest, analysis of a true genetic mutant is still the gold standard. However, often gene function needs to be studied in a sophisticated genetic "assay background," containing other mutations/transgenes to allow measurement of a relevant biological variable. Creating and crossing mutations in such a background can be costly in terms of fish usage, time, and money. Often two generations and extensive genotyping and selection are needed to identify the desired parents. In the case of recessive (lethal) mutants, bringing many of these together in the homozygous state means that only a tiny fraction of the resulting embryos is useful. This makes screening approaches where a larger set of genes is tested impracticable and often prohibitively expensive. Such screens are becoming more and more desirable as many proteomics, transcriptomics, and genome-wide association studies create lists of candidate genes that require some form of in vivo verification.

Targeted nucleases, most importantly, CRISPR/Cas9 technology (Cong et al., 2013; Jinek et al., 2012, 2013), hold great promise as a way to circumvent elaborate breeding schemes. The technology is now rapidly maturing, and the availability of commercial sources of guides and Cas9 protein means this technology is accessible to many labs, and technical or cost barriers to adoption are extremely low. CRISPR/Cas9 is robust, often highly efficient, and therefore capable of creating a mosaic mutant by microinjection into freshly laid eggs: a CRISPant (Burger et al., 2016; Wu et al., 2018). This is now providing a viable alternative (albeit no full replacement) to stable mutants for studying gene function. Importantly, as such injections can be done in any mutant/transgenic background, this obviates breeding schemes. Furthermore, all embryos injected are bound to have loss of function, rather than 25% of the embryos resulting from classical heterozygous incross.

CRISPants are most reliably introduced by injection of a combination of guides. Theoretically, after injection of a mutagenic guide, a third of all induced insertions/deletion (indels) will preserve the reading frame of the targeted gene; such indels are unlikely to affect gene function (Wu et al., 2018). There are two copies of a gene per cell, therefore even a 100% mutagenic CRISPR is expected to leave ~44% of the cells with one "poor-quality" mutant allele. For this reason, combinations of guides should be injected, and most recent reports suggest use of two to three per gene (Hoshijima et al., 2019; Keatinge et al., 2020; Kroll et al., 2020). Furthermore, although sgRNAs were often used initially (Hruscha et al., 2013), most are now reverting to the original two-component crRNA + tracrRNA system. Shorter RNAs can be chemically synthesized more reliably and cheaply, and are clearly superior to in vitro transcribed guides (Hoshijima et al., 2019; Keatinge et al., 2020; Kroll et al., 2020). The use of CRISPants is rising rapidly; in BioRXiv the number of manuscripts referring to CRISPants as of May 2021 is 74.

An issue with the method is the difficulty of finding the appropriate controls. There is no one-size-fits-all method to determine the precise level of knockdown that is easy to apply. Next generation sequencing (NGS) can be used to sequence the target region to determine the relative number of intact and mutated alleles but is laborious and the required PCRs can lead to bias. As negative controls, just Cas9, guides that are not expected to cut in the genome, or guides that act against a gene that is not expected to be involved in the process of interest have been used. This can give some reassurance on specificity of results (Buglo et al., 2020; Dasyani et al., 2020; Marchi et al., 2020; Neiswender and LeMosy, 2020). It is important to remember that guides will inevitably induce DNA damage. Cells in the embryo will detect and respond to this, therefore active Cas9 complexes may be more comprehensive controls. We have used *lamb1a* as a standard control guide for some experiments where the glucocorticoid pathway was studied; we knew from earlier experiments that these guides were active and would similarly result in DNA damage (Marchi et al., 2020). It would be good to adapt and take note of the recommendations that were drawn up for the use of morphant phenotypes (Stainier et al., 2017) to avoid the temptation of "Inject, until you see what you expect."

An alternative to CRISPant is a morphant produced by injection of morpholinos. These oligomers block splicing or translational initiation, and were pioneered in zebrafish at the start of the millennium (Nasevicius and Ekker, 2000). An advantage of morpholinos is that if appropriately chosen they may be less prone to elicit a compensatory response of the genome that has been observed in certain genes as a result of mutations that lead to decay of the mutant messenger RNA (El-Brolosy et al., 2019; El-Brolosy and Stainier, 2017; Rossi et al., 2015). It is important to remember that theoretically, such compensatory responses are likely to occur in CRISPants. The main issue with morpholinos is that often multiple sequences need testing before efficient and specific morpholinos are found; also, mass spectrometry is not available in every lab to check the chemical quality and real sequence of the reagents. These tests can be costly compared to CRISPR. Another advantage of the latter is that when guides are designed for a CRISPant, the same guides can be reused to create a genetic mutant when required.

Management of CRISPants/morphants

In general, it is unlikely that embryos from morpholino experiments will be kept beyond 5 dpf as the effect of morpholinos declines strongly after 2–3 days. In CRISPants the block in gene function is permanent, therefore there may be more scientific incentive to raise such animals. Concerning their welfare, the same caution should be used as would be applied to raising homozygous mutants for that gene for the first time.

Fig. 7.2 summarizes the main steps for the generation and management of morphants and CRISPants, in comparison with mutants and transgenics.

Conclusions

The maintenance of zebrafish transgenic, mutant, and mosaic lines, although sharing common procedures with regard to aquaculture and health monitoring,

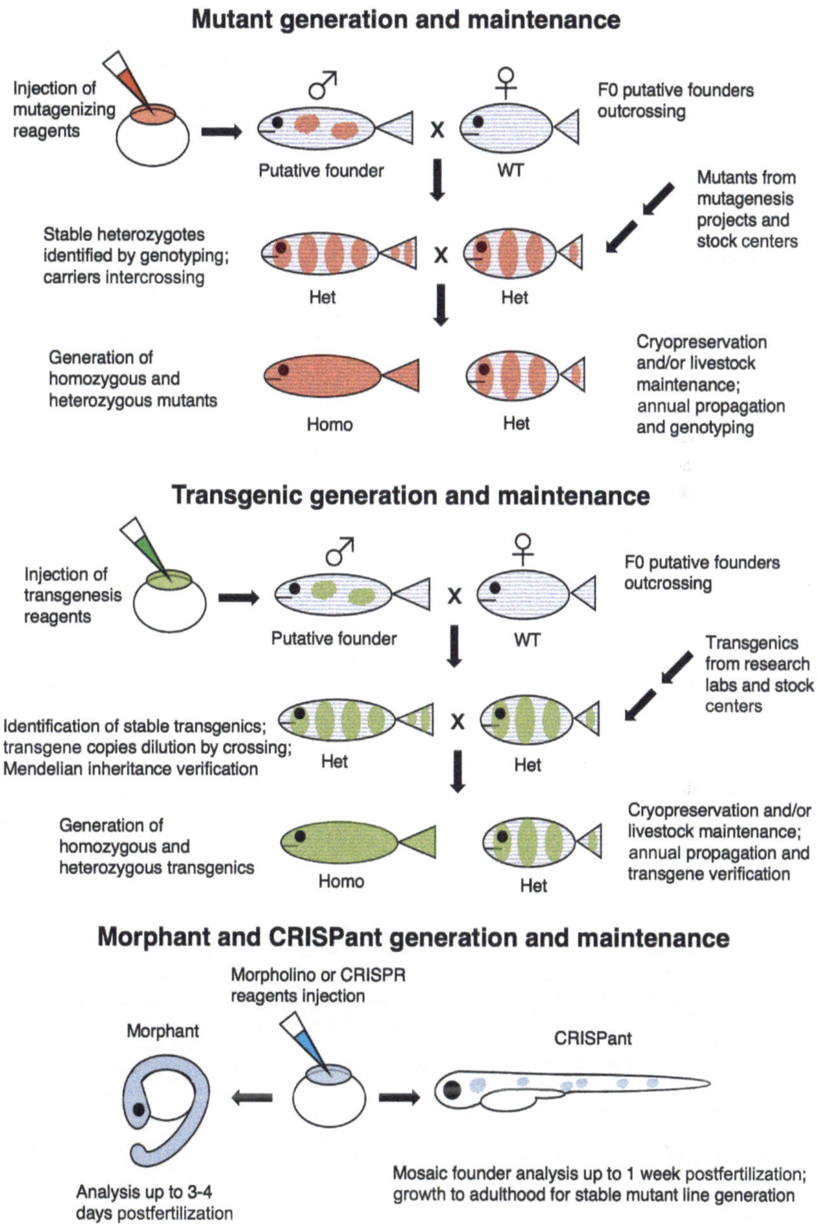

FIGURE 7.2 Schematic representation of the main steps in the generation and maintenance of zebrafish mutants, transgenics, morphants, and CRISPants. *WT*, wild type.

requires distinct procedures in terms of generation, genotyping, carrier identification, and line propagation. In the light of current literature, international standard procedures applied in the main reference centers (ZIRC and EZRC), and based on our experience, we hope to have provided an overview as comprehensive as possible on the best practices for managing zebrafish lines in small- to medium-sized facilities.

Acknowledgments

F.A. and N.T. were supported by the AIRC project IG-2017-19928 to F.A. and the Italian Telethon project GGP19287 to F.A. and N.T. F.v.E. was supported by BBSRC grant BB/R015457/1.

References

Ablain, J., Durand, E.M., Yang, S., Zhou, Y., Zon, L.I., 2015. A CRISPR/Cas9 vector system for tissue-specific gene disruption in zebrafish. Dev. Cell 32 (6), 756–764.

Aleström, P., D'Angelo, L., Midtlyng, P.J., Schorderet, D.F., Schulte-Merker, S., Sohm, F., Warner, S., 2020. Zebrafish: housing and husbandry recommendations. Lab. Anim 54 (3), 213–224.

Buglo, E., Sarmiento, E., Martuscelli, N.B., Sant, D.W., Danzi, M.C., Abrams, A.J., Dallman, J.E., Zuchner, S., 2020. Genetic compensation in a stable slc25a46 mutant zebrafish: a case for using F0 CRISPR mutagenesis to study phenotypes caused by inherited disease. PLoS One 15, e0230566.

Burger, A., Lindsay, H., Felker, A., Hess, C., Anders, C., Chiavacci, E., Zaugg, J., Weber, L.M., Catena, R., Jinek, M., Robinson, M.D., Mosimann, C., 2016. Maximizing mutagenesis with solubilized CRISPR-Cas9 ribonucleoprotein complexes. Development 143, 2025–2037.

Carmichael, C., Westerfield, M., Varga, Z.M., 2009. Cryopreservation and in vitro fertilization at the zebrafish international resource center. Methods Mol. Biol. 546, 45–65.

Chang, C.T., Amack, J.D., Whipps, C.M., 2016. Zebrafish embryo disinfection with povidone-iodine: evaluating an alternative to chlorine bleach. Zebrafish 13 (S1), S96–S101.

Clark, T.S., Pandolfo, L.M., Marshall, C.M., Mitra, A.K., Schech, J.M., 2018. Body condition scoring for adult zebrafish (*Danio rerio*). J. Am. Assoc. Lab. Anim. Sci. 57 (6), 698–702.

Collymore, C., Crim, M.J., Lieggi, C., 2016. Recommendations for health monitoring and reporting for zebrafish research facilities. Zebrafish 13 (S1), S138–S148.

Cong, L., Ran, F.A., Cox, D., Lin, S., Barretto, R., Habib, N., Hsu, P.D., Wu, X., Jiang, W., Marraffini, L.A., Zhang, F., 2013. Multiplex genome engineering using CRISPR/Cas systems. Science 339, 819–823.

Curado, S., Stainier, D.Y., Anderson, R.M., 2008. Nitroreductase-mediated cell/tissue ablation in zebrafish: a spatially and temporally controlled ablation method with applications in developmental and regeneration studies. Nat. Protoc. 3 (6), 948–954.

Dasyani, M., Gao, F., Xu, Q., Fossan, D.V., Zhang, E., Pinto, A.F.M., Saghatelian, A., Skowronska-Krawczyk, D., Chao, D.L., 2020. ELOVL2 Is Required for Robust Visual Function in Zebrafish. bioRxiv, 2020.2010.2013.336974.

Davison, J.M., Akitake, C.M., Goll, M.G., Rhee, J.M., Gosse, N., Baier, H., Halpern, M.E., Leach, S.D., Parsons, M.J., 2007. Transactivation from Gal4-VP16 transgenic insertions for tissue-specific cell labeling and ablation in zebrafish. Dev. Biol. 304 (2), 811–824.

Deakin, A.G., Buckley, J., AlZu'bi, H.S., et al., 2019. Automated monitoring of behaviour in zebrafish after invasive procedures. Sci Rep 9042. https://doi.org/10.1038/s41598-019-45464-w.

Dehairs, J., Talebi, A., Cherifi, Y., Swinnen, J.V., 2016. CRISP-ID: decoding CRISPR mediated indels by Sanger sequencing. Sci. Rep. 6, 28973.

El-Brolosy, M.A., Stainier, D.Y.R., 2017. Genetic compensation: a phenomenon in search of mechanisms. PLoS Genet. 13 (7), e1006780.

El-Brolosy, M.A., Kontarakis, Z., Rossi, A., Kuenne, C., Gunther, S., Fukuda, N., Kikhi, K., Boezio, G.L.M., Takacs, C.M., Lai, S.L., Fukuda, R., Gerri, C., Giraldez, A.J., Stainier, D.Y.R., 2019. Genetic compensation triggered by mutant mRNA degradation. Nature 568, 193–197.

Engeszer, R.E., Ryan, M.J., Parichy, D.M., 2004. Learned social preference in zebrafish. Curr. Biol. 14 (10), 881–884.

Engeszer, R.E., Patterson, L.B., Rao, A.A., Parichy, D.M., 2007. Zebrafish in the wild: a review of natural history and new notes from the field. Zebrafish 4 (1), 21–40.

Franěk, R., Baloch, A.R., Kašpar, V., Saito, T., Fujimoto, T., Arai, K., Pšenička, M., 2020. Isogenic lines in fish – a critical review. Rev. Aquacult. 12, 1412–1434.

Fulcher, N., Tran, S., Shams, S., Chatterjee, D., Gerlai, R., 2017. Following developmental isolation: the zebrafish as a model for major depression. Zebrafish 14 (1), 23–34.

Geisler, R., Borel, N., Ferg, M., Maier, J.V., Strähle, U., 2016. Maintenance of zebrafish lines at the European zebrafish resource center. Zebrafish 13 (S1), S19–S23.

Goodwin, N., Karp, N.A., Blackledge, S., Clark, B., Keeble, R., Kovacs, C., Murray, K.N., Price, M., Thompson, P., Bussell, J., 2016. Zebrafish 13 (S1). S-164–S-168.

Haffter, P., Granato, M., Brand, M., Mullins, M.C., Hammerschmidt, M., Kane, D.A., Odenthal, J., van Eeden, F.J., Jiang, Y.J., Heisenberg, C.P., Kelsh, R.N., Furutani-Seiki, M., Vogelsang, E., Beuchle, D., Schach, U., Fabian, C., Nüsslein-Volhard, C., 1996. The identification of genes with unique and essential functions in the development of the zebrafish, *Danio rerio*. Development 123, 1–36.

Hans, S., Kaslin, J., Freudenreich, D., Brand, M., 2009. Temporally-controlled site-specific recombination in zebrafish. PLoS One 4 (2), e4640.

Hoshijima, K., Jurynec, M.J., Klatt Shaw, D., Jacobi, A.M., Behlke, M.A., Grunwald, D.J., 2019. Highly efficient CRISPR-Cas9-based methods for generating deletion mutations and F0 embryos that lack gene function in zebrafish. Dev. Cell 51 (5), 645–657.e4.

Howe, K., Clark, M.D., Torroja, C.F., Torrance, J., Berthelot, C., Muffato, M., Collins, J.E., Humphray, S., McLaren, K., Matthews, L., McLaren, S., Sealy, I., Caccamo, M., Churcher, C., Scott, C., Barrett, J.C., Koch, R., Rauch, G.J., White, S., Chow, W., et al., 2013. The zebrafish reference genome sequence and its relationship to the human genome. Nature 496 (7446), 498–503.

Hruscha, A., Krawitz, P., Rechenberg, A., Heinrich, V., Hecht, J., Haass, C., Schmid, B., 2013. Efficient CRISPR/Cas9 genome editing with low off-target effects in zebrafish. Development 140, 4982–4987.

Jinek, M., Chylinski, K., Fonfara, I., Hauer, M., Doudna, J.A., Charpentier, E., 2012. A programmable dual-RNA-guided DNA endonuclease in adaptive bacterial immunity. Science 337, 816–821.

Jinek, M., East, A., Cheng, A., Lin, S., Ma, E., Doudna, J., 2013. RNA-programmed genome editing in human cells. eLife 2, e00471.

Keatinge, M., Tsarouchas, T.M., Munir, T., Larraz, J., Gianni, D., Tsai, H.-H., Becker, C.G., Lyons, D.A., Becker, T., 2020. Phenotypic Screening Using Synthetic CRISPR gRNAs Reveals Pro-regenerative Genes in Spinal Cord Injury. bioRxiv, 2020.2004.2003.023119.

Kent, M.L., Sanders, J.L., Spagnoli, S., Al-Samarrie, C.E., Murray, K.N., 2020. Review of diseases and health management in zebrafish *Danio rerio* (Hamilton 1822) in research facilities. J. Fish. Dis. 43 (6), 637–650.

Kim, J.H., Lee, S.R., Li, L.H., Park, H.J., Park, J.H., Lee, K.Y., Kim, M.K., Shin, B.A., Choi, S.Y., 2011. High cleavage efficiency of a 2A peptide derived from porcine teschovirus-1 in human cell lines, zebrafish and mice. PLoS One 6 (4), e18556.

Kroll, F., Powell, G.T., Ghosh, M., Antinucci, P., Hearn, T.J., Tunbak, H., Lim, S., Dennis, H.W., Fernandez, J.M., Hoffman, E.J., Whitmore, D., Dreosti, E., Wilson, S.W., Rihel, J., 2020. A Simple and Effective F0 Knockout Method for Rapid Screening of Behaviour and Other Complex Phenotypes. bioRxiv, 2020.2006.2004.133462.

Kwan, K.M., Fujimoto, E., Grabher, C., Mangum, B.D., Hardy, M.E., Campbell, D.S., Parant, J.M., Yost, H.J., Kanki, J.P., Chien, C.B., 2007. The Tol2kit: a multisite gateway-based construction kit for Tol2 transposon transgenesis constructs. Dev. Dynam. 236 (11), 3088–3099.

Lambert, C.J., Freshner, B.C., Chung, A., Stevenson, T.J., Bowles, D.M., Samuel, R., Gale, B.K., Bonkowsky, J.L., 2018. An automated system for rapid cellular extraction from live zebrafish embryos and larvae: development and application to genotyping. PLoS One 13 (3), e0193180.

Luo, J.J., Bian, W.P., Liu, Y., Huang, H.Y., Yin, Q., Yang, X.J., Pei, D.S., 2018. CRISPR/Cas9-based genome engineering of zebrafish using a seamless integration strategy. FASEB J. 32 (9), 5132–5142.

Marchi, D., Santhakumar, K., Markham, E., Li, N., Storbeck, K.H., Krone, N., Cunliffe, V.T., van Eeden, F.J.M., 2020. Bidirectional crosstalk between Hypoxia-Inducible Factor and glucocorticoid signalling in zebrafish larvae. PLoS Genet. 16 (5), e1008757.

Moens, C.B., Donn, T.M., Wolf-Saxon, E.R., Ma, T.P., 2008. Reverse genetics in zebrafish by TILLING. Briefings Funct. Genomics Proteomics 7 (6), 454–459.

Moro, E., Vettori, A., Porazzi, P., Schiavone, M., Rampazzo, E., Casari, A., Ek, O., Facchinello, N., Astone, M., Zancan, I., Milanetto, M., Tiso, N., Argenton, F., 2013. Generation and application of signaling pathway reporter lines in zebrafish. Mol. Genet. Genom. 288 (5–6), 231–242.

Nasevicius, A., Ekker, S.C., 2000. Effective targeted gene 'knockdown' in zebrafish. Nat. Genet. 26 (2), 216–220.

Neiswender, H., LeMosy, E.K., 2020. Acute Knockdown of Extracellular Matrix Protein Tinagl1 Disrupts Heart Laterality and Pronephric Cilia in Zebrafish Embryonic Development. bioRxiv, 2020.2006.2005.136747.

Ogden, B.E., Pang William, W., Agui, T., Lee, B.H., 2016. Laboratory animal laws, regulations, guidelines and standards in China Mainland, Japan, and Korea. ILAR J. 57 (3), 301–311.

Piatkevich, K.D., Verkhusha, V.V., 2011. Guide to red fluorescent proteins and biosensors for flow cytometry. Methods Cell Biol. 102, 431–461.

Prykhozhij, S.V., Fuller, C., Steele, S.L., Veinotte, C.J., Razaghi, B., Robitaille, J.M., McMaster, C.R., Shlien, A., Malkin, D., Berman, J.N., 2018. Optimized knock-in of point mutations in zebrafish using CRISPR/Cas9. Nucleic Acids Res. 46 (17), e102.

Rafferty, S.A., Quinn, T.A., 2018. A beginner's guide to understanding and implementing the genetic modification of zebrafish. Prog. Biophys. Mol. Biol. 138, 3–19.

Renaud, J.B., Boix, C., Charpentier, M., De Cian, A., Cochennec, J., Duvernois-Berthet, E., Perrouault, L., Tesson, L., Edouard, J., Thinard, R., Cherifi, Y., Menoret, S., Fontanière, S., de Crozé, N., Fraichard, A., Sohm, F., Anegon, I., Concordet, J.P., Giovannangeli, C., 2016. Improved genome editing efficiency and flexibility using modified oligonucleotides with TALEN and CRISPR-Cas9 nucleases. Cell Rep. 14 (9), 2263–2272.

Rossi, A., Kontarakis, Z., Gerri, C., Nolte, H., Holper, S., Kruger, M., Stainier, D.Y., 2015. Genetic compensation induced by deleterious mutations but not gene knockdowns. Nature 524, 230–233.

Samarut, É., Lissouba, A., Drapeau, P., 2016. A simplified method for identifying early CRISPR-induced indels in zebrafish embryos using High Resolution Melting analysis. BMC Genomics 17, 547.

Scheer, N., Campos-Ortega, J.A., 1999. Use of the Gal4-UAS technique for targeted gene expression in the zebrafish. Mech. Dev. 80 (2), 153–158.

Stainier, D.Y.R., Raz, E., Lawson, N.D., Ekker, S.C., Burdine, R.D., Eisen, J.S., Ingham, P.W., Schulte-Merker, S., Yelon, D., Weinstein, B.M., Mullins, M.C., Wilson, S.W., Ramakrishnan, L., Amacher, S.L., Neuhauss, S.C.F., Meng, A., Mochizuki, N., Panula, P., Moens, C.B., 2017. Guidelines for morpholino use in zebrafish. PLoS Genet. 13 (10), e1007000.

Teh, C., Parinov, S., Korzh, V., 2005. New ways to admire zebrafish: progress in functional genomics research methodology. Biotechniques 38 (6), 897–906.

Tilley, C.A., Carreño Gutierrez, H., Sebire, M., Obasaju, O., Reichmann, F., Katsiadaki, I., Barber, I., Norton, W.H.J., 2020. Skin swabbing is a refined technique to collect DNA from model fish species. Sci. Rep. 10 (1), 18212.

Varshney, G.K., Carrington, B., Pei, W., Bishop, K., Chen, Z., Fan, C., Xu, L., Jones, M., LaFave, M.C., Ledin, J., Sood, R., Burgess, S.M., 2016. A high-throughput functional genomics workflow based on CRISPR/Cas9-mediated targeted mutagenesis in zebrafish. Nat. Protoc. 11 (12), 2357–2375.

Westerfield, M., 2007. The Zebrafish Book: A Guide for the Laboratory Use of Zebrafish (*Danio rerio*), fifth ed. University of Oregon Press, Eugene.

Wilkinson, R.N., Elworthy, S., Ingham, P.W., van Eeden, F.J., 2013. A method for high-throughput PCR-based genotyping of larval zebrafish tail biopsies. Biotechniques 55 (6), 314–316.

Wu, R.S., Lam, I.I., Clay, H., Duong, D.N., Deo, R.C., Coughlin, S.R., 2018. A rapid method for directed gene knockout for screening in G0 zebrafish. Dev. Cell 46 (1), 112–125.e4.

Zhu, X., Xu, Y., Yu, S., Lu, L., Ding, M., Cheng, J., Song, G., Gao, X., Yao, L., Fan, D., Meng, S., Zhang, X., Hu, S., Tian, Y., 2014. An efficient genotyping method for genome-modified animals and human cells generated with CRISPR/Cas9 system. Sci. Rep. 4, 6420.

Chapter 8

Sperm cryopreservation, in vitro fertilization, and embryo freezing

Kanav Khosla[1,4], **John Bischof**[1,2,4], **Zoltán M. Varga**[3]
[1]*Department of Mechanical Engineering, University of Minnesota Twin Cities, Minneapolis, MN, United States;* [2]*Department of Biomedical Engineering, University of Minnesota Twin Cities, Minneapolis, MN, United States;* [3]*Zebrafish International Resource Center, University of Oregon, Eugene, OR, United States;* [4]*NSF Center for Advanced Technologies for the Preservation of Biological Systems (ATP-Bio)*

Introduction

Genetic resources are generated and used at an increasing pace and represent billions of dollars in research and industry investments. While it is possible to recreate virtually any genetic modification these days, the storage, availability, and accessibility of previously developed genetic resources prevents the unnecessary duplication of expenditures and effort and ensures that genetic materials are available for reproducible future research.

The number of genetically modified zebrafish lines has increased considerably after several large-scale mutagenic screens generated thousands of novel genetically modified lines (Haffter et al., 1996; Driever et al., 1996; Amsterdam et al., 1999; Kettleborough et al., 2013; Varshney et al., 2013). With the advent of genome editing technologies (Varshney et al., 2015a,b; Auer and Del Bene, 2014), the ability to generate even more lines in efficient and targeted ways has now increased dramatically. In addition, transgenesis is a well-established method for the tissue-, organ-, and cell-specific observation of reporter genes (Lin et al., 1994; Stuart et al., 1990; Long et al., 1997; Yin et al., 2015), adding even more lines that are actively studied in the research community. Frequently, more fish lines than can be actively studied are maintained in laboratories, and some stocks may have to be kept alive for months to years before they are used in an investigation. However, the maintenance of live stocks requires sufficiently large facilities with space for water filtration equipment, racks, and aquaria. Aquatic life support systems require routine cleaning, maintenance, and repair, and animals require

Laboratory Fish in Biomedical Research. https://doi.org/10.1016/B978-0-12-821099-4.00007-9
Copyright © 2022 Elsevier Inc. All rights reserved.

appropriate feeding, health monitoring, water conditioning, and filtration, all of which have associated expenses for electricity, chemicals, water, and personnel. Indeed, personnel costs represent the largest expense for ensuring proper animal care (Varga, 2016; Alestrom et al., 2019).

To achieve a state of suspended animation, cryopreservation relies on the ability to prevent damage from intracellular ice formation during the transition from room temperature to cryogenic temperatures and vice versa. For successful cryopreservation, some intracellular water must be replaced with viscous cryoprotective agents (CPAs) (Mazur, 1977, 1984). Once a cell is partially dehydrated and loaded with CPA, it can better withstand the stress of cooling and low-temperature storage ($< -130°C$) as long as the cooling and rewarming rates of specimens are faster than the rate of ice formation. These "critical cooling" and "critical warming" rates are inversely related to the concentration of CPA used. For example, the higher the CPA concentration is, the lower the critical cooling and warming rates need to be. Previously, researchers found that the warming rates and not the cooling rates were the most important in determining overall success of cryopreservation protocols (Seki and Mazur, 2009). In addition, it is well understood that the warming rates need to be faster than the cooling rates by at least an order of magnitude (Seki and Mazur, 2009; Han and Bischof, 2020; Peyridieu et al., 1996).

To mitigate cell injury during the cryopreservation of zebrafish sperm, several cryoprotectants, including methanol (Harvey et al., 1982), dimethylacetamide (DMA) (Morris et al., 2003), and dimethyl formamide (Zheng and Zhang, 2012), have been tested. Systematic parallel comparisons of methanol, DMA, dimethyl sulfoxide, and glycerol (Yang et al., 2007) revealed that methanol is best suited for sperm cryopreservation, whereas glycerol, the first cryoprotectant discovered for fowl sperm cryopreservation (Polge et al., 1949), is toxic to zebrafish sperm. Through several iterations of the initial zebrafish sperm cryopreservation protocol, methanol has been considered the most suitable cryoprotectant. Based on the postthaw motility and fertilization success, methanol has been used in six (Harvey et al., 1982; Yang et al., 2007; Berghmans et al., 2004; Draper et al., 2004; Carmichael et al., 2009; Matthews et al., 2018) out of eight previous studies. A methanol concentration of 8.3% (Harvey et al., 1982) has been utilized based on postfertilization outcomes. However, because male-to-male and batch-to-batch variations in postthaw sperm motility and fertility were regularly observed (Torres et al., 2017), further evaluation of methanol at different concentrations was required. Membrane analysis by flow cytometry provides a tool for monitoring sperm cell membrane integrity and protocol-induced changes. We recently compared fresh sperm motility with postthaw motility and fertility at various methanol concentrations and exposure periods and found that 5% methanol was the most suitable concentration for cryoprotecting zebrafish sperm (Yang et al., 2020).

In spite of the manifold cellular challenges and upfront investments for equipment, such as liquid nitrogen (LN_2) freezers, bulk LN_2 storage tanks, racks, and safety equipment, the cryopreservation of sperm cells is still the most cost-effective method for managing fish strains (Varga and Tiersch, 2012). There may also be construction costs associated with running vacuum-jacketed pipelines and stainless-steel manifolds if several freezers are used. Additionally, time investment is necessary to train personnel to become proficient with cryopreservation methods. Then, sperm has to be obtained and several samples frozen for each individual fish line. However, these initial investments for cryogenic storage, training, and line preservation are rather negligible compared to the long-term cost savings cryopreserved samples generate when compared with live husbandry expenses (Varga and Tiersch, 2012). The process and effort of reactivating a cryopreserved line is also minimal. Obtaining eggs from females is a straightforward, well-established process in virtually all zebrafish laboratories (Westerfield, 2007). The thawing of sperm samples and in vitro fertilization (IVF) take but a few minutes, as we will show later. In sum, the long-term cost and overall effort for maintaining frozen lines are far less than those of live maintenance, and the short-term effort for preserving and recovering lines is negligible.

While the successful cryopreservation of sperm (haploid genetic material) has progressed significantly, the preservation of the diploid genome, in the form of somatic cells with nuclear transfer (Prukudom et al., 2019; Siripattarapravat et al., 2016) and embryo cryopreservation (Khosla et al., 2017), has remained elusive for the past 25 years. With further development, embryo cryopreservation can be an ideal way to preserve the diploid genome and achieve easier breeding line regeneration and proliferation (Robles et al., 2009). Stock centers that rely solely on sperm cryopreservation need to maintain females of the corresponding strain or wild-type strains in their animal colonies. This requirement makes the banking of species with limited populations (i.e., threatened or novel lines) impossible. In 2018, an expert panel organized by the NIH recognized the need for the cryopreservation of whole embryos to preserve the diploid genome and further recommended it as a key strategy for preserving transgenic lines (Hagedorn et al., 2019).

The successful cryopreservation of zebrafish embryos presents a unique set of challenges in comparison to the cryopreservation of mammalian embryos. These challenges were first enumerated by Hagedorn et al. (Hagedorn et al., 1996, 1997, 1998) to be (1) the relatively large size of the embryo (1000 times bigger in volume than mammalian gametes), resulting in a low surface-to-volume ratio that impedes water and CPA efflux/influx; (2) the presence of multiple compartments, such as the blastoderm and yolk, with differing permeability properties (especially the yolk syncytial membrane, which is a barrier to most cryoprotectants); (3) the large yolk, which increases the likelihood of membrane disruption by intracellular ice formation upon cooling and ice crystal growth upon warming; and (4) a higher susceptibility to chilling injury. Zebrafish oocytes present similar challenges to achieve successful cryopreservation, with some researchers preferring oocytes over embryos due to their smaller size, absence of the yolk syncytial membrane, and relatively better resilience to chilling injury (Martínez-Páramo et al., 2017). Oocytes also offer the flexibility of selecting a sperm donor after rewarming, which is not possible with embryos. However, a major challenge associated with implementing oocyte cryopreservation is the development of in vitro oocyte maturation protocols postthawing. Seki and colleagues developed a successful protocol for the in vitro maturation of stage III zebrafish oocytes and demonstrated the ability to fertilize and hatch embryos after IVF (Seki et al., 2011).

Since embryos are highly susceptible to injury from exposure to near-zero and subzero temperatures, researchers have mostly ruled out the use of controlled rate (<1000°C/min) freezing. For controlled rate or slow freezing to be successful, specimens need to be able to withstand extreme dehydration and extended exposure to near-zero or subzero temperatures (e.g., 30 min−2 h) as cooling occurs (Saragusty and Arav, 2011). To determine if slow freezing protocols would work for zebrafish embryos, Hagedorn et al. evaluated both the extracellular and intracellular ice nucleation temperatures within embryos at later epiboly stages using cryomicroscopy (Hagedorn et al., 2004). The researchers reported that zebrafish embryos were not good candidates for slow freezing since they had higher ice nucleation temperatures than expected along with increased mortality from chilling due to long exposure to near-zero or subzero temperatures (Hagedorn et al., 2004). Rapid cooling to achieve vitrification (i.e., amorphous glassy phase) was suggested as a better way to reduce chilling injury since the time of exposure to sub- or near-zero temperatures was much shorter (a few seconds).

Researchers have attributed the lack of success in cryopreservation to the inability of CPAs to penetrate the yolk due to the presence of the yolk syncytial layer, which creates a permeability barrier (Hagedorn et al., 1998). Therefore past efforts have focused on improving CPA penetration by either dechorionation, artificial expression of aquaporin-3 to induce hydraulic channels within cell membranes (Hagedorn et al., 2002), microinjection of CPA (Janik et al., 2000), or the use of ultrasound (Wang et al., 2008) and lasers (Kohli et al., 2007). Table 8.1 summarizes studies on the cryopreservation of sperm, embryos, oocytes, and somatic cells that are not extensively discussed in this chapter. Few studies have shown the ability to revive embryos, with this failure attributed to the inability to prevent intracellular ice formation during warming. For instance, Janik et al. used microinjection to overcome the permeability barrier between compartments by directly introducing up to 5 M CPA into the yolk (Janik et al., 2000). However, the study was not able to successfully cryopreserve embryos since the warming rates achieved with a water bath (i.e., convective warming, ~20,000°C/min) were below the critical warming rate (Janik et al., 2000). In 2017, Khosla et al. showed that zebrafish embryos could be safely microinjected with 2 M propylene glycol (PG) and then rapidly cooled (20,000°C/min) by plunging them into LN_2. However, combining this

TABLE 8.1 Key studies on the cryopreservation of sperm, oocytes, embryos, and somatic cells of zebrafish in the last 25 years.

Cryo material	Ploidy	Stage	Cryoprotectants used	Cooling device	Postwarming key finding	References
Sperm	Haploid	n/a	10% MeOH	CR freezer	52% motility	Harvey et al. (1982)
Sperm	Haploid	n/a	8% DMA and 8% MeOH	CR freezer	78% motility	Yang et al. (2007)
Oocytes	Haploid	Stage I/II	1.5M MeOH and 4.5M PG	Metal container	74% intact	Marques et al. (2015)
Oocytes	Haploid	Stage III	2 M MeOH, 2 M PG	Cryohook (10 μL)	31% survival	Guan et al. (2010)
Caudal fin	Diploid	n/a	20% EG and 20% DMSO	0.25 mL straws	60% cell attachment	Cardona-Costa et al. (2006)
Blastomeres	Diploid	High cell	5 M PG and 5 M EG	Nylon loop (0.25 μL)	20% cell survival	Cardona-Costa and García-Ximénez (2007)
Blastomeres	Diploid	5 somite	5 M DMSO, 1 M EG, 10% sucrose	0.5 mL straws	50% cell survival	Martínez-Páramo et al. (2009)
PGC	Diploid	64 cell—20 somite	3 M EG and 3 M PG	Nylon mesh (0.5 μL)	25% cell survival	Higaki et al. (2009)
PGC	Diploid	14—18 somite	3 M EG and 2.5 M DMSO	Nylon mesh (0.5 μL)	81% cell survival	Higaki et al. (2013)
Embryo	Diploid	6 somite	2 M DMSO	0.25 mL straws	32% intact	Zhang and Rawson (1996)
Embryo	Diploid	100% epiboly	2 M PG and 6 M PG	0.25 mL straws	No viable embryos	Hagedorn et al. (1998)
Embryo	Diploid	High cell—5 somite	5 M DMSO, 2 M MeOH,1 M EG, and 10% Suc	0.5 mL straws	12% enzymatic activity	Robles et al. (2004)

DMSO, Dimethyl sulfoxide; *EG,* ethylene glycol; *MeOH,* methanol; *Raff,* raffinose; *PG,* propylene glycol; *PGC,* primordial germ cell; *Suc,* sucrose.

approach with traditional convective warming still led to ice formation within the embryos upon warming, resulting in 0% survival (Khosla et al., 2017). In short, even with faster cooling and CPA permeation via microinjection, higher warming rates (10–100× the cooling rates) remain essential to "outrun" ice crystal growth, especially during low-concentration CPA conditions (Khosla et al., 2018a; Zhan et al., 2021).

Previously, researchers estimated that rates over 10 million °C/min are necessary to prevent ice crystallization in droplets of vitrified CPA (~2–3 M) (Khosla et al., 2018a; Zhan et al., 2021). Many groups have proposed the use of a powerful millisecond laser pulse to achieve such ultrarapid rates. Peter Mazur's group was the first to implement and demonstrate the successful laser rewarming of cryopreserved specimens (Jin et al., 2014; Kleinhans and Mazur, 2015). Using Indian ink as an external absorbing dye, their study reported that 90% of mice oocytes survived warming even when less than the optimal amount of CPA was used (Jin et al., 2014). Since zebrafish embryos are ~1000× larger in volume than mice oocytes, laser warming protocols were modified to load biocompatible intracellular absorbers such as polyethylene glycol (PEG)-coated gold nanorods (GNR) into embryos (Khosla et al., 2017, 2018). Extensive modeling and imaging studies have shown that the distribution of laser absorbers within each compartment of the embryo is essential for preventing the formation of damaging temperature gradients, which lead to cracking (Khosla et al., 2017; Liu et al., 2020). By microinjecting GNR along with CPA into the embryo, researchers are able to spread the heat induced by the laser throughout the different compartments of the embryo. In 2017, researchers successfully laser warmed microinjected embryos and demonstrated a 10% survival rate beyond 24 h postfertilization. However, none of the cryopreserved and laser-warmed embryos hatched naturally or survived beyond 5 days postfertilization (Khosla et al., 2017).

In a 2020 follow-up study, researchers improved their protocol such that rewarmed cryopreserved zebrafish embryos were capable of growing into adult fish that bred normally (Khosla et al., 2020). This new protocol, which will be presented later, employed a single injection of CPA and GNR into the yolk and immersion in a precooling bath to dehydrate the perivitelline space. Then, embryos were encapsulated within droplets containing CPA and GNR, plunged into LN_2, cryogenically stabilized, and rewarmed with a laser pulse. The results from the new protocol showed a threefold improvement in survival rates over the 2017 protocol (Khosla et al., 2020). Approximately 9% of the embryos that were laser warmed hatched after 48 h, and 3% began actively swimming after Day 5. This study also showed that the laser-warmed fish that survived until Day 5 could grow to adulthood and spawn, yielding survival rates comparable to those of noncryopreserved fish (Khosla et al., 2020). Despite its low efficacy, microinjection of CPA combined with laser warming is one of the only protocols to achieve consistent embryo survival beyond 24 h, followed by hatching, development to adulthood, and normal breeding.

On the following pages, we provide detailed methods for both zebrafish sperm and embryo cryopreservation. For Zebrafish International Resource Center (ZIRC)'s operations, the cryopreservation of sperm has become a resource management tool for importing, storing, and distributing the ever-growing number of zebrafish lines (Freeman et al., 2019). Here, we describe one of the cryopreservation and recovery processes we use, including recent modifications that further refine our previously described method (Matthews et al., 2018), providing a step-by-step guide for (1) obtaining pooled sperm by stripping males, (2) determining sperm cell density with a spectrophotometer and diluting sperm for optimal storage and recovery, (3) freezing samples, (4) thawing frozen samples, and (5) performing IVF of eggs. We also present the best protocol currently available for the successful cryopreservation of zebrafish embryos using CPA and GNR microinjection and laser warming. Together, these methods will provide laboratories with the ability to better plan their research programs and save funds. Once novel lines have been generated or acquired, they can be cryopreserved and backed up for later reactivation as needed.

Materials and methods

Protocol for sperm cryopreservation and in vitro fertilization

Collection and pooling of sperm

1. Prepare a 0.5 mL microcentrifuge tube with E400G for each fish line (see Note[1] for E400G starting volume considerations). Use a separate micro-centrifuge tube for each fish line. To make 1 L of E400G, combine the following in 800 mL dH_2O: 9.70 g KCl, 2.92 g NaCl, 2.0 mL of 1.0 M $CaCl_2$ (or 0.29 g $CaCl_2 \cdot 2H_2O$), 1.0 mL of 1.0 M $MgSO_4$ (or 0.25 g $MgSO_4 \cdot 7H_2O$), 1.8 g D-(+)-Glucose, 7.15 g HEPES, and 1 g gelatin from cold water fish skin (Sigma #G7041). Add dry components first, stir to dissolve, add liquid components, and stir. Adjust the pH to 7.9 with 5 M KOH and bring the final volume to 1000 mL with dH_2O. It is advisable to check the osmolality of E400, which should be close to 400 mmol/kg. The solution should be filter sterilized and stored at 4 °C. The resulting E400G contains 130 mM KCl, 50 mM NaCl, 2 mM $CaCl_2$, 1 mM $MgSO_4$, 10 mM D-(+)-Glucose, 30 mM HEPES-KOH (pH 7.9), and 0.1 % w/v gelatin.

---

1. A few general guidelines for selecting the starting volume of E400G into which sperm is collected are provided. The use of an appropriate starting volume will ensure that the sperm concentration is as optimal as possible in the resulting samples without activating the cells. The starting volume of E400G depends on: (a) the number of available males (short-fin fish typically give more sperm than long-fin fish); (b) the number of desired samples. To calculate a conservative starting volume of E400G, use the following equations. $\text{short-fin fish}: (\#\text{ males} - 2) \times 5\ \mu l$ and $\text{long-fin fish}: (\#\text{ males} - 4) \times 5\ \mu l$

2. Keep E400G and the collected sperm at room temperature throughout the collection and prefreezing procedures.
3. Sedate or preanesthetize males as needed for at least 10 min before the procedure (see Note[2]). Prepare and store a tricaine (MS-222) stock solution as follows: 4 g/L in dH_2O, pH adjusted to 7.0 with Tris-HCl (pH 9.0). To prepare the MS-222 preanesthesia solution of 48 mg/L, dilute 12 mL MS-222 stock solution in 1000 mL fish water.
4. For anesthesia, dilute 4.2 mL tricaine stock solution in 100 mL fish water (168 mg/L). Anesthetize two to three males in 168 mg/L MS-222.
5. Briefly rinse a male in phosphate buffered saline (PBS) isotonic fish rinse, dry it by rolling gently on a soft and absorbent paper towel (see Note[3]), and place it belly up in a dampened sponge/foam holder (in a 35 × 10 mm Petri dish). The isotonic PBS fish rinse is prepared from PBS (pH 7.4) powder packets (Sigma #P3813) that are dissolved in 870 mL dH_2O. The final osmolality should be approximately 315–325 mmol/kg. Place the sponge holder under a dissecting microscope with incident lighting.
6. Place the end of a calibrated 10 μL borosilicate glass microcapillary (Drummond #2-000-010) on the urogenital opening.
7. Use rubber-tipped Millipore forceps (Millipore #XX6200006P; rubber tips made from heat-shrink tubing) to apply *gentle* abdominal pressure to the sides of the male. Move the forceps gently from anterior to posterior toward the urogenital opening. Collect sperm into the microcapillary.
8. Expel sperm immediately into the E400G solution in the microcentrifuge collection tube (Fig. 8.1(1)).
9. Transfer fish into fresh system water for recovery from anesthesia.
10. Continue collecting sperm from all males from the same family/stock and pool into the same E400G microcentrifuge tube. Repeat steps 4 – 8 for each male (Fig. 8.1A(1)).

---

2. Preanesthesia/sedation of males is recommended for lines sensitive to bleeding from the gills during anesthesia. Males should be placed in preanesthesia tricaine solution (48 mg/L MS-222 in fish water) for 10–30 min prior to anesthesia.
3. Eggs and sperm are activated by water, and it is virtually impossible to dry fish completely. Therefore it is important to rinse anesthetized males and females with isotonic PBS solution before gamete collection to avoid premature activation of sperm or unfertilized eggs.

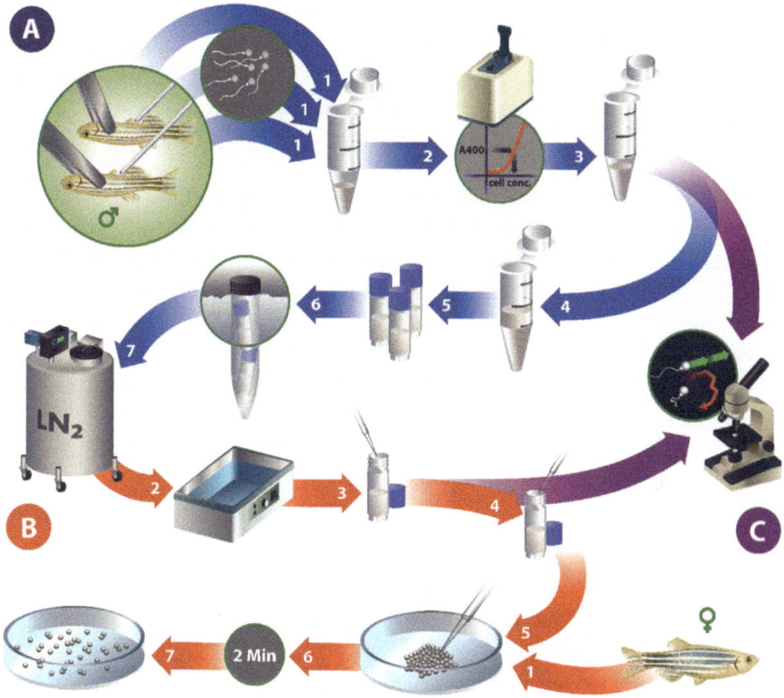

FIGURE 8.1 Overview of sperm cryopreservation, thawing, and in vitro fertilization (IVF). (A) (1) Sperm is collected from multiple males into E400G extender. (2) Cell density is estimated using a spectrophotometer and cell concentration is adjusted (3) to yield circa 4.0×10^8 cells/mL (2.0×10^6 cells/sample). (4) Three times the sperm sample volume of RMMB cryoprotectant is added, mixed, and (5) 20 μL samples are aliquoted and frozen (6) in dry ice snow. (7) Samples are transferred from dry ice to a liquid nitrogen freezer for long-term storage. (B) (1) Eggs are collected from females and stored for up to 5 min in a humid chamber (Petri dish) until a sperm sample has been thawed. Several egg clutches can be combined before IVF as needed. (2) A sperm sample(s) is retrieved from liquid nitrogen and thawed in a water bath at 38°C for 10–15 s. (3) SS300 with or without milk is added to the vial. (4) To activate sperm, water is added to the sample vial and the mixed sperm is used to fertilize eggs in vitro (5). (C) Evaluation of sperm motility by microscope, with or without computer-assisted sperm analysis, is an optional assessment of sample quality before and/or after freezing.

Estimation of sperm cell densities and sample dilution

At ZIRC, we use a NanoDrop 2000 spectrophotometer to determine the cell density by light absorption at 400 nm (see Note[4]). The worksheet we use to determine NanoDrop cell counts is available in the Cryopreservation and IVF section on the ZIRC site: https://zebrafish.org/wiki/protocols/cryo (see Note[5]).

1. Estimate the volume of pooled sperm in E400G in the microcentrifuge tube for each line using a Pipetman. Draw the sperm into the tip and adjust the pipette volume until all of the solution fills the tip. Record the estimated volume for later use. While measuring, gently pipette the sperm to mix completely.
2. Prepare a 1:10 dilution of the sperm solution in E400G for a density measurement. If the collected sample appears less opaque (i.e., less concentrated), a 1:5 dilution can be used. Pipette 9 μL (or 4 μL for a 1:5 dilution) E400G into a 0.6 mL microcentrifuge tube and add 1 μL of the sperm suspension.
3. Mix the diluted sample by flicking the tube and keep the tube at room temperature. Optional: Use separate (colored) microcentrifuge tubes for each line (see Note[6]).
4. Calibrate A_{400} absorption of a blank E400G sample using the NanoDrop's Cell Cultures menu. Set the Absorbance Cursor to 400 nm and add 1.5 μL E400G to the spectrophotometer. Read the E400G blank sample to verify

---

4. The density of sperm cells in the extender and cryoprotective medium plays a crucial role in optimal cryopreservation, thawing, and reactivation. While satisfying results can be achieved without cell density estimation and adjustment, we strongly recommend measuring concentrations prior to freezing to obtain the best possible recovery results. Too few cells (or too many) in relation to the available solution ingredients will not freeze as well as a well-adjusted cell density. There are two key benefits to adjusting cell densities: (1) further dilution of sperm (if possible) generates more samples and a longer lasting resource, and (2) an optimal ratio of cells to available salt and buffer molecules ensures that the solution toxicity and osmotic shock are reduced, and as a result, cells suffer less damage during freezing and thawing.
5. This calculator helps to estimate and adjust cell densities in most cases. However, because several variables can influence the readout, it is recommended to develop a calibration curve and formula like this with your own equipment. Download the complete ZIRC cryopreservation and IVF protocol (PDF) and Excel sperm density calculator files from the ZIRC website: https://zebrafish.org/wiki/protocols/cryo. This worksheet/calculator has been generated with ZIRC equipment, and it should generate results within the correct order of magnitude. It may or may not be necessary to calibrate a new A_{400} absorption curve for your own spectrophotometer for a more accurate conversion. To generate a curve specific to your equipment, use a hemocytometer or Makler chamber to determine cell densities microscopically. Determine the A_{400} of a corresponding dilution series with your spectrophotometer and generate a curve from the dilution series. Calculate a best-fit curve (or line) from your absorption data and use the formula in a worksheet to determine the cell density that corresponds to the A_{400} of unknown cell densities.
6. Using a designated colored microcentrifuge tube for the NanoDrop dilution makes it easily distinguishable from samples of other stocks.

that it has set the instrument's readings close to 0. If a blank sample reads higher than 0.005, recalibrate with a fresh 1.5 μL aliquot of E400G. When prompted, save the NanoDrop data with the date of the freeze event and line name(s).

5. To measure the absorption of a sample, enter the sample ID on the Measure Cell Cultures page.
6. Mix the sample well using a vortex mixer set at intermediate speed (~1300 rpm).
7. Immediately load 1.5 μL of the diluted sperm and read the cursor absorbance (A_{OD400}; select the measure button in the left-hand corner). Readings above 0.3 are acceptable, and you can proceed with the dilution. Any reading at or below 0.2 should be diluted 1:5 and repeated.
8. Measure each sample three to five times and calculate the average A_{400} for all samples. A minimum of three successful readings is optimal (Fig. 8.1A(2)).
9. Clean the NanoDrop using a clean Kimwipe and deionized water. Wipe the top arm (mirror) and the bottom lens with a moistened Kimwipe and then dry completely with a Kimwipe before closing the arm.
10. Calculate the cell density for each sample with the averaged A_{400} using the Excel cell density calculator.
11. Dilute the sperm with E400G according to the desired number of samples or sperm concentration. Optimal sperm cell densities should be between 4.0×10^8 and 1.6×10^9 cells/mL. This density range will result in samples with 2.0×10^6–8.0×10^6 cells/sample (see Note[7]; Fig. 8.1A(3)).

Pooled sperm freezing

1. Prepare RMMB cryoprotective medium (100 mL) in the following order: Combine 20.0 g D-(+)-raffinose pentahydrate (Sigma R7630 or 83400) and 70 mL dH_2O in a 250 mL beaker. Then, place the beaker in an evaporating dish (Pyrex 3140) or large beaker containing hot water (~70 °C) on a stir plate. Stir the mixture until the raffinose is completely dissolved.

---

7. If your priority is to achieve a particular number of samples, you can dilute the sample as far as 4.0×10^8 cells/mL (2.0×10^6 cells/sample). If your priority is to ensure that a line is reactivated with only a single thaw, you may want to leave the concentration higher and add less E400G for a lower total volume. Cell densities should not be lower than 4.0×10^8 cells/mL for best results. If more than 12 samples are going to be frozen, divide the final sperm volume into two aliquots. In addition, 0.5–1 μL of the final diluted sperm suspension can be removed for an optional prefreeze motility assay. Computer-assisted sperm analysis or microscopic visualization can be used before and/or after freezing samples to assess sperm cell motility. The complete absence of motile sperm cells indicates that cells may have been irreversibly damaged. In this case, IVF can be used to determine whether a pool of samples fertilizes at all. Low fertilization rates can sometimes be obtained from low-motility/nonmotile samples.

Add 2.5 g skim milk (Difco #232100) and stir until it is completely dissolved. Let the solution cool to room temperature and add 3 mL of 1 M Bicine-NaOH (8.0). Add 6.67 mL absolute methanol (acetone-free, absolute, Certified ACS Reagent Grade), transfer the mixture to a 100 mL volumetric flask, adjust the final volume to 100 mL with dH_2O, and mix by inversion three to four times. Transfer the solution to two 50 mL conical tubes and centrifuge at 15,000 × g for 20 min at 25°C. Transfer the clear supernatant into a clean beaker and aliquot into 1.5 mL microfuge tubes, 1 mL each, or a different convenient volume for daily use. Store the RMMB solution frozen at −20 or −80 °C until use. The resulting RMMB cryoprotective solution will contain 20 % (w/v) D-(+)-**R**affinose pentahydrate, 2.5 % (w/v), Difco skim **M**ilk, 6.67 % (v/v), **M**ethanol, and 30 mM **B**icine-NaOH.

2. Fill a Styrofoam container or cooler (≈30 cm × 30 cm) with powdered dry ice made from liquid CO_2 (Carmichael et al., 2009) (see Note[8]).
3. Prepare 15 mL Falcon tubes (Falcon 352,096) with an empty Matrix cryovial tube (0.5 mL Matrix Screw Top Storage Tubes, Thermo Scientific, Item #3745-BR or 2 mL Corning vials, Item #430488) functioning as a spacer (Fig. 8.1A) in each along with the Falcon tube caps so that the tubes are ready to hold samples immediately after the cryovials have been capped.
4. Prepare labeled sample cryovials as needed prior to freezing (all samples should be labeled with the freeze date and allele number and/or line identification number). Additional vial color coders (colored caps or cap inserts specific for vial type) are helpful to distinguish freeze events or different stocks.
5. If you are using a multicapper, prepare it with the appropriate caps already attached before adding cryoprotectant (RMMB) to sperm.
6. For each sample, determine the volume of RMMB (RMMB volume = 3× the sperm volume, see Note[9]).
7. Add RMMB to sperm and mix by pipetting (Fig. 8.1A(4)).

---

8. A cooling rate between 10 and 15°C per minute was determined to be the optimal range for the materials and solutions used in this protocol. Using powdered dry ice with stacked cryovials in a 15 mL Falcon tube will achieve a cooling rate of 10–15°C/min. If possible, avoid changing the materials specified in this protocol. If the specified materials are not available (for example, the 15 mL vials or cryovials with the same wall material and thickness as specified here), you need to determine which freezing rate provides the optimal postthaw fertilization rates with the replacements. You may also have to adjust the thawing protocol to match the new freezing rate. The description for making powdered dry ice is included in the online protocol: https://zebrafish.org/wiki/protocols/cryo.
9. Before use, thaw RMMB aliquots in a water bath or heating block at 45–50°C. Raffinose precipitates if the RMMB solution is kept on ice. If precipitation occurs, heat the solution slightly to get it back into solution prior to use. Cool the solution to room temperature before mixing with sperm. Keep diluted sperm and RMMB at room temperature.

8. Immediately aliquot 20 μL into the prelabeled 0.5 mL cryovials (Fig. 8.1A(5)).
9. Without delay, cap the cryovials (use an automated capper for rows of eight tubes, if available) and place the cryovials into the 15 mL conical tubes (containing a Matrix cryovial spacer).
10. Cap the conical tubes and drive the tubes into the dry ice until the caps are flush with the surface (Fig. 8.1A(6) and Note[10]).
11. Freeze the samples in dry ice for 20–40 min and then quickly transfer them to a cryogenic freezer box submerged in LN_2 (Fig. 8.1A(7)).

Sperm motility assessment

A good way to assess the quality of sperm is to observe its motility (Fig. 8.1C). Computer-assisted sperm analysis software systems provide an objective and comprehensive quantification of the density and motility parameters, but a manual, subjective assessment is feasible in most labs and is sufficient for most sperm-freezing applications. A compound microscope with a 10× or 20× objective and DIC or dark field illumination is ideal. Osmolality affects both the speed and duration of sperm motility. Fresh (prefreeze) sperm will be faster and have a higher percentage of motile cells than postthaw samples.

To estimate the prefreeze motility, place 6 μL dH_2O on a microscope slide. Add 0.5–1 μL of the final sperm dilution (in E400G) to the drop and then mix the drop and spread it quickly with the pipette tip. Observe immediately (see Note[11]).

To estimate the postthaw sperm motility, thaw a sperm sample in a water bath as described later, add 150 μL of SS300 solution to the thawed sperm, and mix gently. Remove 10–20 μL of the sample for motility assessment (see Note[12]).

---

10. Because the RMMB cryoprotective medium is toxic to sperm cells, steps 6–9 should be carried out swiftly with as little delay as possible once the RMMB is mixed with the sperm. However, our research suggests that cell viability is not affected significantly when sperm is exposed to cryoprotective medium for up to 5 min before freezing. If more than 12 samples are to be frozen, divide the final sperm volume into two smaller aliquots and freeze them in two rounds (steps 6–9) to ensure the rapid transfer of the mixture into dry ice after mixing with the RMMB cryoprotective medium.
11. Examining the remainder of the NanoDrop dilution, if performed, is a good use of the sperm. Checking the prefreeze motility confirms the viability and concentration of the sperm being frozen.
12. At this point, a small portion (10–20 μL) of the sample can be removed and held on ice for motility assessment. The remainder of the sample can be used for IVF as described earlier. It is best to view the motility as soon as possible after thawing, but samples are typically stable on ice for 10–20 min.

Place 5 μL dH_2O on a slide and add 4.25 μL of thawed sperm/SS300 solution (see Note[13]). Mix briefly with the pipette tip on the slide and observe immediately.

Egg collection

1. Place females in a tank with preanesthesia solution at least 10 min before full anesthesia. However, fish can be held in a preanesthesia solution for several hours (see Note [14]).
2. Anesthetize females in tricaine/MS-222 solution.
3. Rinse fish in isotonic PBS and blot dry by gently rolling on a paper towel.
4. Place fish on its side in a small Petri dish (35 or 60 mm).
5. Dampen fingers with PBS fish rinse.
6. Expel eggs by applying gentle (light) finger pressure on the ventral abdomen and move your finger from anterior to posterior. Eggs will be expelled readily if the female is ready (see Note[15]).
7. Transfer the female to a recovery tank.
8. Combine several clutches of eggs if needed by gently moving eggs to another dish with a fine-tipped paint brush dampened with isotonic PBS (see Note[16]).
9. Maintain pooled eggs in a closed dish in a moisture chamber at room temperature no longer than 5–10 min before IVF (Fig. 8.1B(1)).

Sperm sample thawing

1. To prepare Sperm Solution SS300 (1 L), combine 8.2 g NaCl, 5 mL of 1 M KCl (or 0.37 g KCL), 1 mL of 1 M $CaCl_2$ (or 0.15 g $CaCl_2{-}2H_2O$), 1 mL of 1 M $MgSO_4$ (or 0.25 g $MgSO_4 \cdot 7H_2O$), 1.8 g D-(+)-Glucose, and 20 mL of 1 M Tris-Cl (pH 8.0) in 800 mL dH_2O. Add all dry ingredients first and stir until they are dissolved. Then add liquid ingredients and mix. Bring the final volume to 1000 mL with dH_2O and check the osmolality (should be approximately 300 mmol/kg). Filter sterilize the solution and

13. For postthaw motility, activating the sperm on a slide in the same relative proportions as in the IVF procedure gives a consistent method and provides a good sense of sperm concentration and motility as it is applied to the eggs.
14. Egg collection and IVF of eggs should be performed early in the morning. Egg quality is optimal the first few hours after facility lights have been turned on. The resorption of eggs begins 2–3 h after daybreak.
15. Good eggs will be golden in color and have little fluid and no opaque or white eggs intermixed (Fig. 8.1B–D). Eggs can be moved away from the fish using a fine water-color brush.
16. If females do not provide large enough clutches, or to maximize the number of embryos produced by IVF, it is helpful to combine several clutches of eggs. A soft water-color brush is an effective and gentle tool for moving and mixing eggs. Dampen the fingers and paint brush with the isotonic PBS solution before squeezing females or manipulating eggs.

store it at 4 °C. The resulting SS300 solution will contain 140 mM NaCl, 5 mM KCl, 1 mM $CaCl_2$, 1 mM $MgSO_4$, 10 mM D-(+)-Glucose, and 20 mM Tris-Cl (8.0).

To prepare Sperm Solution SS300 with 2 mg/mL Difco Skim Milk (SS300 + milk), add 100 mg Difco Skim Milk to 50 mL SS300 and stir or vortex to dissolve. Aliquot the solution into microcentrifuge tubes and store frozen at −20 °C. Thaw and use at room temperature. The milk helps to prevent the sperm tails from sticking and tangling and is thought to contain antioxidants that protect against oxidative damage during cryopreservation and thawing (Grażyna et al., 2017; Khan et al., 2019).

2. Remove the cryovial from the LN_2 and quickly open the cap to vent any LN_2 in the vial (see Note[17] (Fig. 8.1B(2)).
3. Thaw the cryovial in a 38 °C water bath until the frozen pellet is less than 3 mm in diameter (takes approximately 10–15 s; Fig. 8.1B(2)).
4. Immediately add 150 μL room-temperature SS300 solution to the cryovial (Fig. 8.1B(3)). If you are thawing sperm that was frozen without milk (see Note[18]), add 2 mg/mL Difco Skim Milk (Difco #232100) to the SS300 solution. See Note[19] for an optional postthaw motility assessment.

In vitro fertilization

1. Add 200 μL dH_2O to the cryovial to activate the sperm (Fig. 8.1B(4)). Gently mix the sperm one to two times with a micropipette and transfer the sample to the eggs: slide the pipette tip sideways along the bottom of the Petri dish from the edge of the pile of eggs into the center. Expel the activated sperm into the mass of eggs (not on top of the eggs (Fig. 8.1B(5)).
2. Start a 2-min countdown timer (Fig. 8.1B(6)).
3. Do not move, mix, or swirl the dish; let it sit completely undisturbed for 2 min, then flood the dish with embryo medium (Fig. 8.1B(7)).

---

17. Before squeezing females, sperm samples can be retrieved from the LN_2 freezer and maintained in a tray of LN_2 within a small Styrofoam cooler until eggs are available.
18. SS300 + milk is used for thawing sperm samples that were frozen with a cryopreservation medium not containing skim milk powder (other protocols). The milk helps to prevent the sperm tails from sticking and tangling and is thought to contain antioxidants that mitigate oxidative damage.
19. For the postthaw motility assessment, a small portion (10–20 μL) of the sperm/SS300 mix can be removed and held in a microcentrifuge tube at room temperature if assessed immediately or on ice if stored for later (see the "Motility assessment" section).

4. Determine the fertilization rate 2–4 h postfertilization or as soon as cell divisions are recognizable. Count embryos and remove unfertilized eggs (see Note[20]).

Protocol for embryo cryopreservation

Fig. 8.2 shows a step-by-step overview of the microinjection and laser warming approach. The results from this new protocol demonstrated the first

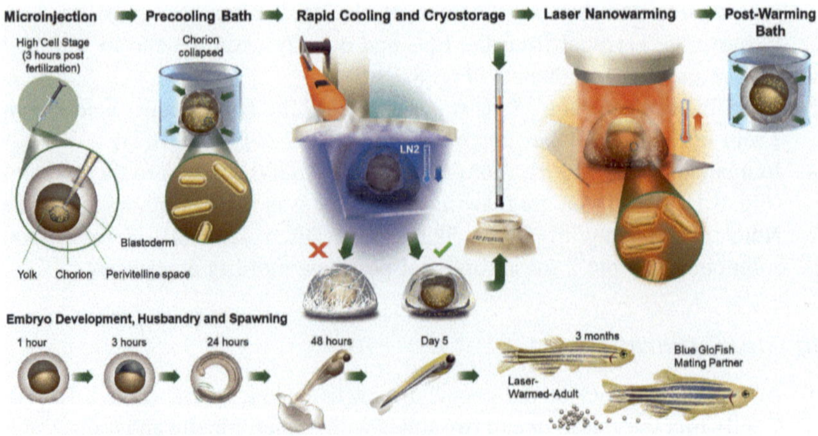

FIGURE 8.2 Step-by-step overview of the microinjection, rapid cooling, and laser warming protocol for cryopreserving zebrafish embryos. The protocol begins with the microinjection of cryoprotective agent (CPA) (propylene glycol and MeOH) and gold nanorods into the yolk of a zebrafish embryo reaching the high cell stage. After a recovery period of 2 h, the injected embryos are placed in a precooling bath for 5 min to dehydrate the perivitelline fluid. Immediately after, the embryos are placed on a polypropylene strip called a cryotop and plunged into liquid nitrogen. After equilibration to −196°C, the embryo is brought under the laser for warming. After warming, the embryo is placed in a postwarming washout bath for rehydration and CPA removal from the perivitelline fluid, after which the embryo is placed in embryo medium and stored in an incubator at 28°C. Embryo development is monitored regularly up to 5 days. Select embryos surviving to Day 5 are transferred to separate housing, where they reach spawning age. Once sexual maturation is reached, the adult zebrafish can be spawned with the corresponding transgenic or wild-type zebrafish adults. *Original source: Khosla, K., Kangas, J., Liu, Y., Zhan, L., Daly, J., Hagedorn, M., Bischof, J. Cryopreservation and laser nanowarming of zebrafish embryos followed by hatching and spawning. Adv. Biosyst., 2000138. Copyright Wiley-VCH Verlag GmbH & Co. KGaA. Reproduced with permission.*

20. A fertilized egg will be visible by the swelling of the first embryonic cell (zygote) and the chorion or the first cell divisions. Because fertilization and the emergence of the first cell and its division are usually not perfectly synchronized, the number of fertilized eggs can be more conveniently determined during the blastula stages after several rounds of cell divisions. The fertilization rate can be obtained by dividing the number of embryos with proliferating cells (fertilized) by the number of total viable eggs (×100).

instance of the hatching of cryopreserved fish embryos, the growth of the fish to adulthood, and the successful spawning of normal offspring (Khosla et al., 2020). While the protocol outlined in the following section produced the best results to date, we expect further optimization and refinement of the process and its steps in the future.

Zebrafish care and embryo collection

The Zebrafish Book by Monte Westerfield provides detailed protocols for feeding, housing, and spawning of embryos (Westerfield, 2007). Embryos should be cultured for up to 5 days in standard embryo medium (recipe in *The Zebrafish Book* (Westerfield, 2007)) at 28°C. Larvae will require feeding if they are grown beyond 5 days. To spawn embryos, place a female and two males overnight together in a tank with a collection net underneath. As soon as the lights turn on the next morning, the fish will begin spawning. Within 30 min of the lights turning on, collect embryos, wash them with embryo medium, and sort into treatment groups.

Microinjection of CPA and GNR

The microinjection of solutions into zebrafish embryos is a well-established practice (Kettleborough et al., 2013; Bill et al., 2009). In addition to online tutorials, prospective users can consult the step-by-step videos published in the *Journal of Visualized Experiments* for injecting mRNA into zebrafish embryos between 1−4 cell stages (Rosen et al., 2009). The microinjection protocol for injections of CPA and GNR into fish embryos at the high cell stage was initially adapted from Janik et al. (2000). It is recommended that the chorion should not be removed prior to injection to protect the embryo during handling and culturing. The high cell stage was shown to be the most robust of the early developmental stages during which yolk syncytial layer formation is minimal and cytoplasmic streaming allows for the rapid distribution of the GNR and CPA (Khosla et al., 2017) into the blastoderm. Injection needles should be prepared by pulling borosilicate glass capillaries with a micropipette puller (Flaming/Brown, Model P87, Sutter Instruments Inc). The needle puller's instrument settings (heat, pull, velocity, etc.) have to be customized for the specific instrument and type of glass capillary used. To create the needle opening, it is recommended that a needle beveler (EG 401, Narishige) be used to create a sharp beveled tip. The needles are backfilled with a premixed injection solution and attached to an air pressure injector (PLI-100, Harvard Apparatus) to begin injection volume calibration. The goal of calibration is to find the appropriate injector settings (i.e., injection time and pressure) needed to dispense the necessary volume from the needle. The recommended method for calibrating the injected volume is to use a stage micrometer to measure the diameter of a droplet made by injecting the CPA and GNR solution into a larger droplet of mineral oil. The injection solution should be prepared by mixing MeOH, PG, and GNRs to obtain a concentration of 10.6 M PG, 4.8 M MeOH, and 9.3×10^{11} np/mL GNR. Previous studies have found PEG-coated GNR to be more biocompatible than traditionally made GNR coated with cetyltrimethylammonium bromide (Khosla et al., 2017; Liu et al., 2015). The PEG-coated GNR used in cryopreservation and laser nanowarming studies are manufactured and sold by nanoComposix Inc. based in San Diego, California.[21]

Once embryos reach the high cell stage, they should be lined up in the ridges of a custom-made agarose-covered injection tray. Bill et al. (2009) provided a recipe for preparing the injection trays with 1% molten agarose (Bill et al., 2009). All cryopreservation and laser warming studies inject 9 nL into the yolk resulting in a concentration of 1.6 M PG, 0.7 M MeOH, and 1.4×10^{11} np/mL GNR. This concentration was estimated by using the total volume (88 nL) and water fraction (59%) of the yolk (Hagedorn et al., 1998; Janik et al., 2000). The solution must be injected into the yolk without damaging the blastoderm. The injected solution (CPA and GNR) will diffuse evenly throughout the yolk within 30 min (Janik et al., 2000). Any embryos that show leakage from the injection site should be discarded.

Use of a precooling bath

It is recommended that injected embryos be cultured for at least 2 h prior to the application of the precooling bath to allow time for recovery and resealing of the yolk membrane. Since CPA and GNR solution were not injected in the perivitelline fluid, the main purpose of the precooling bath is to remove the perivitelline fluid via water transport through a mostly permeable chorion. The embryo should appear to quickly shrink as soon as it is placed in the bath solution due to the dehydration of the perivitelline space. The bath concentration can be modified as long as 1 μL droplets of the chosen concentration can be vitrified by plunging them into LN_2. We recommend using a 5 min immersion period in the precooling bath (2 M PG, 1.2 M MeOH, and 0.5 M Tre), as this duration has previously been found to be effective at reducing ice formation in the outer embryo compartments (Khosla et al., 2020). Once removed from the precooling bath, individual embryos can be placed on the tip of a custom-made cryotop ($3.0 \times 2.0 \times 0.08$ mm, polypropylene) and encapsulated in a 1 μL droplet (2 M PG, 1.2 M MeOH, 0.5 M Tre) with a GNR concentration of 4.2×10^{11} np/mL. Finite element multiphysics simulations showed a three times higher GNR concentration in the encapsulating droplet than in the solution that is injected in the yolk was needed to generate the necessary warming rates (Khosla et al., 2020).

21. In addition to UV-VIS spectroscopy, potential users can apply the photothermal conversion and nanoparticle stability techniques described in Liu et al. (2020) to extensively study nanoparticles.

Rapid cooling and laser warming

A custom automated device was built to plunge the cryotops into LN_2, rapidly position the embryo under the laser beam, and trigger the laser once the embryo is in position (Fig. 8.2). The automated system[22] is placed on top of a Styrofoam container containing LN_2. It uses a 3D-printed plastic holder to mount the cryotop onto a servo motor. The x–y stage of the automated system allows for fine adjustments of the cryotop. After plunging the cryotop in LN_2, the embryo should be held at −196°C for at least 2 min to allow for equilibration. We recommend purchasing the millisecond pulsed 1064 nm lasers sold by LaserStar Technologies (iWeld 980 Series, 80J).[23] When injected GNR concentrations are between 0.8 and 4.8×10^{11} np/mL, we recommend using a fluence rate of 1.1×10^9 W/m^2 (300 V power, 1 ms pulse time) to generate the necessary warming rates over 10 million °C/min. These lasers are also equipped with an overhead camera that should be used to record video of the embryo during laser nanowarming to isolate failure modes such as extensive ice formation within embryos or laser misfires. After laser warming is accomplished, the rewarmed embryos should be delicately transferred to a postwarming rehydration bath (half the strength of the precooling bath) for 20 min. Once the chorion regains its original shape, the embryo should be washed three times in standard embryo medium.

Survival assessment

Observing developmental changes in zebrafish embryos through Day 5 is commonly used for understanding mutagenetic changes in response to various treatments (van Eeden et al., 1998). Table 8.2 presents examples of live and dead embryos evaluated according to our developmental criteria for selected time points at 1 h, 3 h, 24 h, 48 h, and 5 days after a specific treatment (i.e., injection, bath, or laser warming). Embryos should be considered viable if they are intact after 1 h, continue to develop by 3 h, and show movement (i.e., tail twitching) past 24 h. By the 48 h time point, we expect embryos to begin hatching. By Day 5, an embryo is considered viable if the hatched larva has proper cardiac development, tail musculature, fins, swim bladder, and ability to swim. Any fish that do not match these criteria are considered abnormal and not counted among the "live" fish. Based on previous studies, expected survival thresholds can be defined for each specific treatment (Khosla et al., 2017, 2020). We expect the minimum average Day 5 survival fraction for embryos

---

22. Users can design their own version or replicate other designs used in previous laser warming studies (Jin et al., 2014; Daly et al., 2018; Panhwar et al., 2018). Please contact the authors for more information regarding replicating the automated device design.
23. The energy per pulse can be varied by changing the user settings, such as the input voltage and pulse time. Previous studies have extensively characterized the available laser fluence rates for different laser input settings for the iWeld 980 series model (Khosla et al., 2018a,b).

TABLE 8.2 Survival criteria for zebrafish embryos after injection and laser nanowarming.

Time	Survival Criteria	Dead	Live
Post injection (~30min)	No leaking yolk	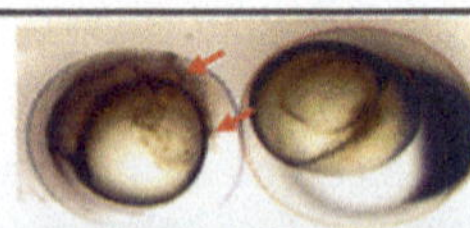	
1 hr	Intact, no membrane damage	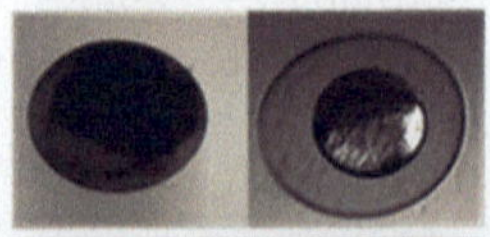	
3 hr	Developing, ex: epiboly to somite	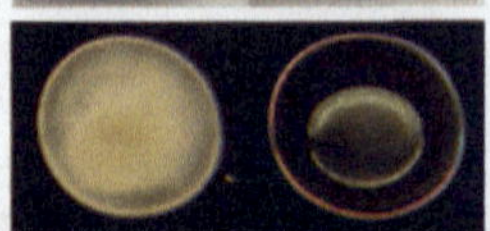	
24 hr	Movement, tail twitching	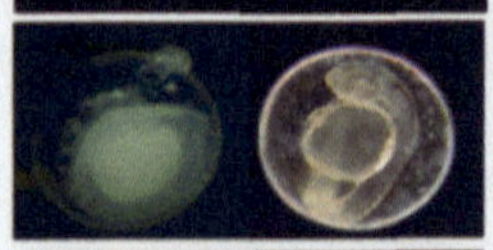	
48 hr	Eye pigmentation, hatching begins	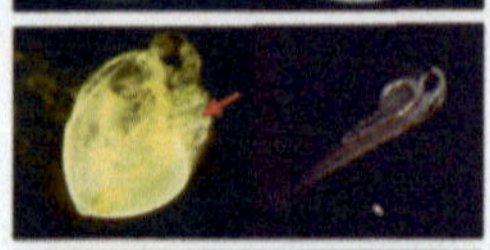	
Day 5	Swimming, no malformations		

Note: These survival criteria should be used to determine the impact of various treatments applied to zebrafish embryos. After a specific treatment such as microinjection and/or laser warming, the development of embryos should be monitored up to 5 days while the embryos are kept at 28°C. The injection protocol and survival metrics were adapted from Janik et al. (2000) and Khosla et al. (2017), respectively.
Original source: Khosla, K., Kangas, J., Liu, Y., Zhan, L., Daly, J., Hagedorn, M., Bischof, J. Cryopreservation and laser nanowarming of zebrafish embryos followed by hatching and spawning. Adv. Biosyst., 2000138. Copyright Wiley-VCH Verlag GmbH & Co. KGaA. Reproduced with permission.

exposed to (1) no treatments (i.e., not injected) to be 80%; (2) injection only to be 60%; and (3) injection and bath treatments to be 50%. Additionally, we expect that postlaser nanowarming, 40% of embryos will remain intact after 1 h, 11% will show movement after 24 h, 9% will begin hatching after 48 h, and 3% will show active swimming after Day 5.

Conclusion

The cryopreservation of zebrafish gametes enables the cost-effective long-term storage of valuable research lines without incurring the cost of maintaining large colonies. In this chapter, we shared the protocol developed by ZIRC to cryopreserve zebrafish sperm in conjunction with in vitro fertilization. In addition, we also presented the latest protocol available for the successful cryopreservation of zebrafish embryos using CPA microinjection and laser warming. The ability to cryopreserve the entire diploid genome will be critical for many applications. Though the current survival rates for embryo cryopreservation protocol remain low, optimization studies to increase the efficacy and translation to embryos of other fish species are under way.

Acknowledgments

ZIRC is grateful for continued NIH support from the Office of Research Infrastructure Programs (ORIP, P40OD011021) in collaboration with the Eunice Kennedy Shriver National Institute of Child Health and Human Development (NICHD). We thank the ZIRC staff for daily help and discussion, and we gratefully acknowledge Huiping Yang, Leticia Torres, Yue Liu, Mary Hagedorn, and Terry Tiersch for their advice and collaboration during the research and development phases of this protocol.

Research on the cryopreservation of zebrafish embryos was funded by NIH SBIR Phase I (1R41OD024430-01) and Phase II (9R44MH122118-02) grants. We would like to thank Mr. Marc Tye and Dr. Mark Masino of the University of Minnesota Zebrafish Core for their support of our zebrafish studies. The protocol detailed in this chapter was first published in the journal *Advanced Biosystems* in September 2020 and can be found here:https://doi.org/10.1002/adbi.202000138.

References

Alestrom, P., D'Angelo, L., Midtlyng, P.J., Schorderet, D.F., Schulte-Merker, S., Sohm, F., Warner, S., 2019. Zebrafish: housing and husbandry recommendations. Lab. Anim. 54 (3), 213–224. https://doi.org/10.1177/0023677219869037. 31510859.

Amsterdam, A., Burgess, S., Golling, G., Chen, W., Sun, Z., Townsend, K., Farrington, S., Haldi, M., Hopkins, N., 1999. A large-scale insertional mutagenesis screen in zebrafish. Genes Dev. 13 (20), 2713–2724.

Auer, T.O., Del Bene, F., 2014. CRISPR/Cas9 and TALEN-mediated knock-in approaches in zebrafish. Methods 69 (2), 142–150.

Berghmans, S., Morris, J.P., Kanki, J.P., Look, A.T., 2004. Zebrafish sperm cryopreservation. Zebrafish 77, 645–659.

Bill, B.R., Petzold, A.M., Clark, K.J., Schimmenti, L.A., Ekker, S.C., 2009. A primer for morpholino use in zebrafish. Zebrafish 6 (1), 69–77.

Cardona-Costa, J., García-Ximénez, F., 2007. Vitrification of zebrafish embryo blastomeres in microvolumes. Cryoletters 28 (4), 303–309.

Cardona-Costa, J., Roig, J., Perez-Camps, M., Garcia-Ximenez, F., 2006. Vitrification of caudal fin explants from zebrafish adult specimens. Cryoletters 27 (5), 329–332.

Carmichael, C., Westerfield, M., Varga, Z.M., 2009. Cryopreservation and in vitro fertilization at the zebrafish international resource center. Methods Mol. Biol. 546, 45–65.

Daly, J., Zuchowicz, N., Lendo, C.I.N., Khosla, K., Lager, C., Henley, E.M., Bischof, J., Kleinhans, F.W., Lin, C., Peters, E.C., 2018. Successful cryopreservation of coral larvae using vitrification and laser warming. Sci. Rep. 8 (1), 15714.

Draper, B.W., McCallum, C.M., Scout, J.L., Slade, A.J., Moens, C.B., 2004. A high-throughput method for identifying N-Ethyl-N-Nitrosourea (ENU)-induced point mutations in zebrafish. In: Detrich III, H.W., Westerfield, M., Zon, L.I. (Eds.), The Zebrafish Genomics, and Informatics, Volume 77 of Methods in Cell Biology. Elsevier Press, San Diego, pp. 91–112.

Driever, W., Solnica-Krezel, L., Schier, A.F., Neuhauss, S.C., Malicki, J., Stemple, D.L., Stainier, D.Y., Zwartkruis, F., Abdelilah, S., Rangini, Z., Belak, J., Boggs, C., 1996. A genetic screen for mutations affecting embryogenesis in zebrafish. Development 123, 37–46.

Freeman, A., Holland, R., Hwang-Shum, J.-J., Lains, D., Matthews, J., Murray, K.N., Nasiadka, A., Quinn, E., Varga, Z.M., Westerfield, M., 2019. The zebrafish international resource center. In: Jarret, R.L., McCluskey, K. (Eds.), The Biological Resources of Model Organisms. CRC Press, Taylor & Francis Group, Boca Raton, London, New York, pp. 113–143.

Grażyna, C., Hanna, C., Adam, A., Magdalena, B.M., 2017. Natural antioxidants in milk and dairy products. Int. J. Dairy Technol. 70 (2), 165–178.

Guan, M., Rawson, D., Zhang, T., 2010. Cryopreservation of zebrafish (*Danio rerio*) oocytes by vitrification. CryoLetters 31 (3), 230–238.

Haffter, P., Granato, M., Brand, M., Mullins, M.C., Hammerschmidt, M., Kane, D.A., Odenthal, J., van Eeden, F.J., Jiang, Y.J., Heisenberg, C.P., Kelsh, R.N., Furutani-Seiki, M., Vogelsang, E., Beuchle, D., Schach, U., Fabian, C., Nüsslein-Volhard, C., 1996. The identification of genes with unique and essential functions in the development of the zebrafish, *Danio rerio*. Development 123, 1–36.

Hagedorn, M., Hsu, E.W., Pilatus, U., Wildt, D.E., Rall, W., Blackband, S.J., 1996. Magnetic resonance microscopy and spectroscopy reveal kinetics of cryoprotectant permeation in a multicompartmental biological system. Proceedings of the National Academy of Sciences 93 (15), 7454–7459.

Hagedorn, M., Kleinhans, F., Wildt, D., Rall, W., 1997. Chill sensitivity and cryoprotectant permeability of dechorionated zebrafish embryos, *Brachydanio rerio*. Cryobiology 34 (3), 251–263.

Hagedorn, M., Kleinhans, F., Artemov, D., Pilatus, U., 1998. Characterization of a major permeability barrier in the zebrafish embryo. Biol. Reprod. 59 (5), 1240–1250.

Hagedorn, M., Lance, S., Fonseca, D.M., Kleinhans, F., Artimov, D., Fleischer, R., Hoque, A., Hamilton, M., Pukazhenthi, B.S., 2002. Altering fish embryos with aquaporin-3: an essential step toward successful cryopreservation. Biol. Reprod. 67 (3), 961–966.

Hagedorn, M., Peterson, A., Mazur, P., Kleinhans, F., 2004. High ice nucleation temperature of zebrafish embryos: slow-freezing is not an option. Cryobiology 49 (2), 181–189.

Hagedorn, M., Varga, Z., Walter, R.B., Tiersch, T.R., 2019. Workshop report: cryopreservation of aquatic biomedical models. Cryobiology 86, 120–129.

Han, Z., Bischof, J.C., 2020. Critical cooling and warming rates as a function of CPA concentration. CryoLetters 41 (4), 185–193.

Harvey, B., Norman, K.R., Ashwood-Smith, M.J., 1982. Cryopreservation of zebrafish spermatozoa using methanol. Can. J. Zool. 60, 1867–1870.

Higaki, S., Mochizuki, K., Baba, H., Akashi, Y., Yamaha, E., Katagiri, S., Takahashi, Y., 2009. Feasibility of cryopreservation of zebrafish (*Danio rerio*) primordial germ cells by whole embryo freezing. Jpn. J. Vet. Res. 57 (2), 119–128.

Higaki, S., Kawakami, Y., Eto, Y., Yamaha, E., Nagano, M., Katagiri, S., Takada, T., Takahashi, Y., 2013. Cryopreservation of zebrafish (*Danio rerio*) primordial germ cells by vitrification of yolk-intact and yolk-depleted embryos using various cryoprotectant solutions. Cryobiology 67 (3), 374–382.

Janik, M., Kleinhans, F., Hagedorn, M., 2000. Overcoming a permeability barrier by microinjecting cryoprotectants into zebrafish embryos (*Brachydanio rerio*). Cryobiology 41 (1), 25–34.

Jin, B., Kleinhans, F., Mazur, P., 2014. Survivals of mouse oocytes approach 100% after vitrification in 3-fold diluted media and ultra-rapid warming by an IR laser pulse. Cryobiology 68 (3), 419–430.

Kettleborough, R.N., Busch-Nentwich, E.M., Harvey, S.A., Dooley, C.M., de Bruijn, E., van Eeden, F., Sealy, I., White, R.J., Herd, C., Nijman, I.J., Fenyes, F., Mehroke, S., Scahill, C., Gibbons, R., Wali, N., Carruthers, S., Hall, A., Yen, J., Cuppen, E., Stemple, D.L., 2013. A systematic genome-wide analysis of zebrafish protein-coding gene function. Nature 496 (7446), 494–497.

Khan, I.T., Nadeem, M., Imran, M., Ullah, R., Ajmal, M., Jaspal, M.H., 2019. Antioxidant properties of milk and dairy products: a comprehensive review of the current knowledge. Lipids Health Dis. 18 (1), 41.

Khosla, K., Wang, Y., Hagedorn, M., Qin, Z., Bischof, J., 2017. Gold nanorod induced warming of embryos from the cryogenic state enhances viability. ACS Nano 11 (8), 7869–7878.

Khosla, K., Zhan, L., Bhati, A., Carley-Clopton, A., Hagedorn, M., Bischof, J., 2018a. Characterization of laser gold Nanowarming: a platform for millimeter-scale cryopreservation. Langmuir 35 (23), 7364–7375.

Khosla, K., Wang, Y., Hagedorn, M., Qin, Z., Bischof, J., 2018b. Gold nanorod induced warming of frozen embryos enhances viability. Cryobiology 81, 228.

Khosla, K., Kangas, J., Liu, Y., Zhan, L., Daly, J., Hagedorn, M., Bischof, J., 2020. Cryopreservation and laser nanowarming of zebrafish embryos followed by hatching and spawning. Adv. Biosyst. 2000138.

Kleinhans, F., Mazur, P., 2015. Physical parameters, modeling, and methodological details in using IR laser pulses to warm frozen or vitrified cells ultra-rapidly. Cryobiology 70 (2), 195–203.

Kohli, V., Robles, V., Cancela, M.L., Acker, J.P., Waskiewicz, A.J., Elezzabi, A.Y., 2007. An alternative method for delivering exogenous material into developing zebrafish embryos. Biotechnol. Bioeng. 98 (6), 1230–1241.

Lin, S., Yang, S., Hopkins, N., 1994. lacZ expression in germline transgenic zebrafish can be detected in living embryos. Dev. Biol. 161 (1), 77–83.

Liu, K., Zheng, Y., Lu, X., Thai, T., Lee, N.A., Bach, U., Gooding, J.J., 2015. Biocompatible gold nanorods: one-step surface functionalization, highly colloidal stability, and low cytotoxicity. Langmuir 31 (17), 4973–4980. https://doi.org/10.1021/acs.langmuir.5b00666. 25874503.

Liu, Y., Kangas, J., Wang, Y., Khosla, K., Pasek-Allen, J., Saunders, A., Oldenburg, S., Bischof, J., 2020. Photothermal conversion of gold nanoparticles for uniform pulsed laser warming of vitrified biomaterials. Nanoscale 12 (23), 12346–12356.

Long, Q., Meng, A., Wang, H., Jessen, J., Farrell, M., Lin, S., 1997. GATA-1 expression pattern can be recapitulated in living transgenic zebrafish using GFP reporter gene. Development 124 (20), 4105–4111.

Marques, L.S., Bos-Mikich, A., Godoy, L.C., Silva, L.A., Maschio, D., Zhang, T., Streit Jr., D.P., 2015. Viability of zebrafish (*Danio rerio*) ovarian follicles after vitrification in a metal container. Cryobiology 71 (3), 367–373.

Martínez-Páramo, S., Barbosa, V., Pérez-Cerezales, S., Robles, V., Herráez, M., 2009. Cryoprotective effects of antifreeze proteins delivered into zebrafish embryos. Cryobiology 58 (2), 128–133.

Martínez-Páramo, S., Horváth, Á., Labbé, C., Zhang, T., Robles, V., Herráez, P., Suquet, M., Adams, S., Viveiros, A., Tiersch, T.R., 2017. Cryobanking of aquatic species. Aquaculture 472, 156–177.

Matthews, J.L., Murphy, J.M., Carmichael, C., Yang, H., Tiersch, T., Westerfield, M., Varga, Z.M., 2018. Changes to extender, cryoprotective medium, and in vitro fertilization improve zebrafish sperm cryopreservation. Zebrafish 15 (3), 279–290.

Mazur, P., 1977. The role of intracellular freezing in the death of cells cooled at supraoptimal rates. Cryobiology 14 (3), 251–272.

Mazur, P., 1984. Freezing of living cells: mechanisms and implications. Am. J. Physiol. Cell Physiol. 247 (3), C125–C142.

Morris, J.P., Berghmans, S., Zahrieh, D., Neuberg, D.S., Kanki, J.P., Look, A.T., 2003. Zebrafish sperm cryopreservation with N,N-dimethylacetamide. Biotechniques 35 (5), 956–968.

Panhwar, F., Chen, Z., Hossain, S.C., Wang, M., Haider, Z., Memon, K., Chen, P., Zhao, G., 2018. Near-infrared laser mediated modulation of ice crystallization by two-dimensional nanosheets enables high-survival recovery of biological cells from cryogenic temperatures. Nanoscale 10 (25), 11760–11774.

Peyridieu, J., Baudot, A., Boutron, P., Mazuer, J., Odin, J., Ray, A., Chapelier, E., Payen, E., Descotes, J., 1996. Critical cooling and warming rates to avoid ice crystallization in small pieces of mammalian organs permeated with cryoprotective agents. Cryobiology 33 (4), 436–446.

Polge, C., Smith, A.U., Parkes, A.S., 1949. Revival of spermatozoa after vitrification and dehydration at low temperatures. Nature 164 (4172), 666.

Prukudom, S., Siripattarapravat, K., Poulos, W., Cibelli, J.B., 2019. Optimized protocol of zebrafish somatic cell nuclear transfer (SCNT). In: Vertebrate Embryogenesis. Springer, pp. 353–375.

Robles, V., Cabrita, E., De Paz, P., Cuñado, S., Anel, L., Herráez, M., 2004. Effect of a vitrification protocol on the lactate dehydrogenase and glucose-6-phosphate dehydrogenase activities and the hatching rates of zebrafish (*Danio rerio*) and Turbot (*Scophthalmus maximus*) embryos. Theriogenology 61 (7–8), 1367–1379.

Robles, V., Cabrita, E., Paz Herráez, M., 2009. Germplasm cryobanking in zebrafish and other aquarium model species. Zebrafish 6 (3), 281–293.

Rosen, J.N., Sweeney, M.F., Mably, J.D., 2009. Microinjection of zebrafish embryos to analyze gene function. J. Vis. Exp. (25), e1115.

Saragusty, J., Arav, A., 2011. Current progress in oocyte and embryo cryopreservation by slow freezing and vitrification. Reproduction 141 (1), 1.

Seki, S., Mazur, P., 2009. The dominance of warming rate over cooling rate in the survival of mouse oocytes subjected to a vitrification procedure. Cryobiology 59 (1), 75–82.

Seki, S., Kouya, T., Tsuchiya, R., Valdez Jr., D.M., Jin, B., Koshimoto, C., Kasai, M., Edashige, K., 2011. Cryobiological properties of immature zebrafish oocytes assessed by their ability to be fertilized and develop into hatching embryos. Cryobiology 62 (1), 8–14.

Siripattarapravat, K., Prukudom, S., Cibelli, J., 2016. Method for somatic cell nuclear transfer in zebrafish. Methods Cell Biol. 135, 245–257.

Stuart, G.S., Vielkind, J.R., McMurray, J.V., Westerfield, M., 1990. Stable lines of transgenic zebrafish exhibiting reproducible tissue-specific transgene expression patterns. Development 109, 577–584.

Torres, L., Liu, Y., Guitreau, A., Yang, H.P., Tiersch, T.R., 2017. Challenges in development of sperm repositories for biomedical fishes: quality control in small-bodied species. Zebrafish 14 (6), 552–560.

van Eeden, F.J., Granato, M., Odenthal, J., Haffter, P., 1998. Developmental mutant screens in the zebrafish. Methods Cell Biol. 60, 21–41.

Varga, Z.M., 2016. Aquaculture, husbandry, and shipping at the zebrafish international resource center. Methods Cell Biol. 135, 509–534.

Varga, Z.M., Tiersch, T.R., 2012. Workshop and panel discussion: high-throughput cryopreservation of germplasm as an exchange currency for genetic resources. Comp. Biochem. Physiol. C Toxicol. Pharmacol. 155 (1), 167–168.

Varshney, G.K., Lu, J., Gildea, D.E., Huang, H., Pei, W., Yang, Z., Huang, S.C., Schoenfeld, D., Pho, N.H., Casero, D., Hirase, T., Mosbrook-Davis, D., Zhang, S., Jao, L.E., Zhang, B., Woods, I.G., Zimmerman, S., Schier, A.F., Wolfsberg, T.G., Pellegrini, M., Burgess, S.M., Lin, S., 2013. A large-scale zebrafish gene knockout resource for the genome-wide study of gene function. Genome Res. 23 (4), 727–735.

Varshney, G.K., Pei, W., LaFave, M.C., Idol, J., Xu, L., Gallardo, V., Carrington, B., Bishop, K., Jones, M., Li, M., Harper, U., Huang, S.C., Prakash, A., Chen, W., Sood, R., Ledin, J., Burgess, S.M., 2015a. High-throughput gene targeting and phenotyping in zebrafish using CRISPR/Cas9. Genome Res. 25 (7), 1030–1042.

Varshney, G.K., Sood, R., Burgess, S.M., 2015b. Understanding and editing the zebrafish genome. Adv. Genet. 92, 1–52.

Wang, R.Y., Guan, M., Rawson, D.M., Zhang, T., 2008. Ultrasound enhanced methanol penetration of zebrafish (*Danio rerio*) embryos measured by permittivity changes using impedance spectroscopy. Eur. Biophys. J. 37 (6), 1039–1044.

Westerfield, M., 2007. The Zebrafish Book: A Guide for the Laboratory Use of Zebrafish (*Danio rerio*), 5th. University of Oregon Press, Eugene, OR.

Yang, H., Carmichael, C., Varga, Z.M., Tiersch, T.R., 2007. Development of a simplified and standardized protocol with potential for high-throughput for sperm cryopreservation in zebrafish *Danio rerio*. Theriogenology 68 (2), 128–136.

Yang, H., Hu, E., Tiersch, T., Carmichael, C., Matthews, J., Varga, Z.M., 2020. Temporal and concentration effects of methanol on cryopreservation of zebrafish (*Danio rerio*) sperm. Zebrafish 17 (4), 233–242.

Yin, L., Maddison, L.A., Li, M., Kara, N., LaFave, M.C., Varshney, G.K., Burgess, S.M., Patton, J.G., Chen, W., 2015. Multiplex conditional mutagenesis using transgenic expression of Cas9 and sgRNAs. Genetics 200 (2), 431–441.

Zhan, L., Guo, S.-Z., Kangas, J., Shao, Q., Maple, S., Khosla, K., Low, W., McApline, M., Bischof, J., 2021. Conduction cooling and plasmonic heating dramatically increase droplet vitrification volumes for cell cryopreservation. Adv. Sci. 2004605. https://doi.org/10.1002/advs.202004605.

Zhang, T., Rawson, D.M., 1996. Feasibility studies on vitrification of intact zebrafish (*Brachydanio rerio*) embryos. Cryobiology 33 (1), 1–13.

Zheng, N., Zhang, B., 2012. A brief protocol for sperm cryopreservation and revival in zebrafish. Yi Chuan 34 (9), 1211–1216.

Section 2

Other small freshwater fish

Chapter 9

Medaka as a model teleost: characteristics and approaches of genetic modification

Tokiro Ishikawa[1], Yu Murakami[2], Chika Fujimori[3], Masato Kinoshita[4], Kiyoshi Naruse[5], Shinji Kanda[3]

[1]*Department of Biophysics, Graduate School of Science, Kyoto University, Sakyo-ku, Kyoto, Japan;* [2]*Department of Fisheries, Graduate School of Agriculture, Kindai University, Nara-shi, Nara, Japan;* [3]*Laboratory of Physiology, Atmosphere and Ocean Research Institute, The University of Tokyo, Kashiwa, Chiba, Japan;* [4]*Division of Applied Biosciences, Graduate School of Agriculture, Kyoto University, Sakyo-ku, Kyoto, Japan;* [5]*Laboratory of Bioresources, National Institute for Basic Biology, Okazaki, Aichi, Japan*

Medaka

History of the medaka

The medaka has been a popular pet in Japan

The medaka (*Oryzias latipes*) is a small freshwater teleost that lives in ponds and rivers in Japan. A fish encyclopedia published in 1835 had already described two different medaka body-color mutants (orange-red and white) as well as a wild type (brown), which indicates that medaka had been popularly bred as a pet at that time. Because they are easy to keep and breed, they are used for educational purposes in Japanese elementary schools, and thus virtually all Japanese are familiar with medaka. Although medaka are not as popular outside of Japan, their prolific traits and tolerance to various environments make them good experimental animals in various scientific fields.

In medaka, it has been well known that there are genetic variations associated with their geographical distribution, and these variations are referred to as northern and southern populations (Sakaizumi et al., 1983; Takehana et al., 2016). Recently, it was proposed that they be treated as distinct species: northern medaka (*Oryzias sakaizumii*) and Japanese medaka (*O. latipes*), respectively (Asai et al., 2011). Since most of the strains used in experiments (e.g., drR, Hd-rR, and Cab) are derived from the southern population, whose

Laboratory Fish in Biomedical Research. https://doi.org/10.1016/B978-0-12-821099-4.00008-0
Copyright © 2022 Elsevier Inc. All rights reserved.

scientific names would not be changed, most existing research would not need to be amended if this proposed change were enacted.

History of the medaka as an experimental fish

Because of their availability, especially in Japan, the medaka has been used as an experimental animal for many years. The first historical series of experiments performed by Toyama and Ishikawa dates back to the 1910s, in which they proved that Mendel's law of inheritance is applicable to vertebrates (Japanese literature). Thus for more than 100 years, medaka have contributed to the understanding of basic science, such as endocrinology (Yamamoto, 1953) and genetics (for details, Hori, 2011).

Later, once genetic manipulation was developed in the 1980s, many studies applied transgenesis to visualize specific cells/organs by using green fluorescent protein (GFP) in medaka. Unlike mammals, the fertilized eggs of medaka develop outside the maternal body, which simplifies transgenesis because a microinjection into a one-cell stage embryo followed by simple incubation can generate transgenic animals. By using transgenic and other conventional techniques, the medaka, with the advantage of being an oviparous animal, has contributed to developmental studies. Moreover, the genome sequence of medaka was made available as early as 2003 (Kasahara et al., 2007). Furthermore, the recent development of genome editing using transcription activator-like effector nuclease/clustered regularly interspaced short palindromic repeats (TALEN/CRISPR) made it possible to produce knockout medaka more efficiently than ever (Ansai and Kinoshita, 2014, 2017; Ansai et al., 2013). Because of these advantages, the medaka has been a useful model teleost.

With merits similar to those of the small teleosts just described, the zebrafish is widely used in various study areas. Compared with studies involving zebrafish, the number of those involving medaka and the number of associated publications are much smaller (zebrafish, 39,527; medaka, 3653; search results of PubMed, April 2020). Therefore medaka may be less utilized than zebrafish in terms of research population, but there are some merits to using medaka when compared to zebrafish. First, a female medaka can spawn every day under laboratory conditions, and their eggs can be stored at 4°C for several hours, which enables cell division to be delayed prior to microinjection. Additionally, medaka have high tolerance to various water conditions. They can live at a temperature of 4°C in winter and 37°C in summer, whereas zebrafish require temperatures above 20°C.

Additionally, they are nearly euryhaline, which provides researchers with the opportunity to study their adaptation to seawater (Inoue and Takei, 2003; Miyanishi et al., 2016). Moreover, because the history of medaka as experimental animals is far longer than that of zebrafish, researchers may find classical but important studies from the early 1900s, which is rarely an option for other experimental model fishes. Furthermore, medaka show high tolerance to inbreeding, and several inbred lines have been established (Naruse, 2011), which provide a good platform for genetic studies such as QTL analysis. In addition, the genome size of medaka (800 Mb) is approximately half that of zebrafish, which can be advantageous in the application of molecular genetic tools because enhancers may be packed in shorter regions. Recently, new gene models based on long-read RNA-seq, gene expression profiling by short-read RNA-seq, and chromatin accessibility by ATAC-seq during embryogenesis have become available (Li et al., 2020).

Last but not the least, medaka (Euteleostei) and zebrafish (Otocephala) branched ~250 million years ago (mya), which was a relatively early stage of teleost evolution (~283 mya) (Betancur et al., 2013). This indicates that medaka and zebrafish are evolutionary distant species within teleosts, and the understanding of both is important for understanding the generality in teleosts (Fig. 9.1).

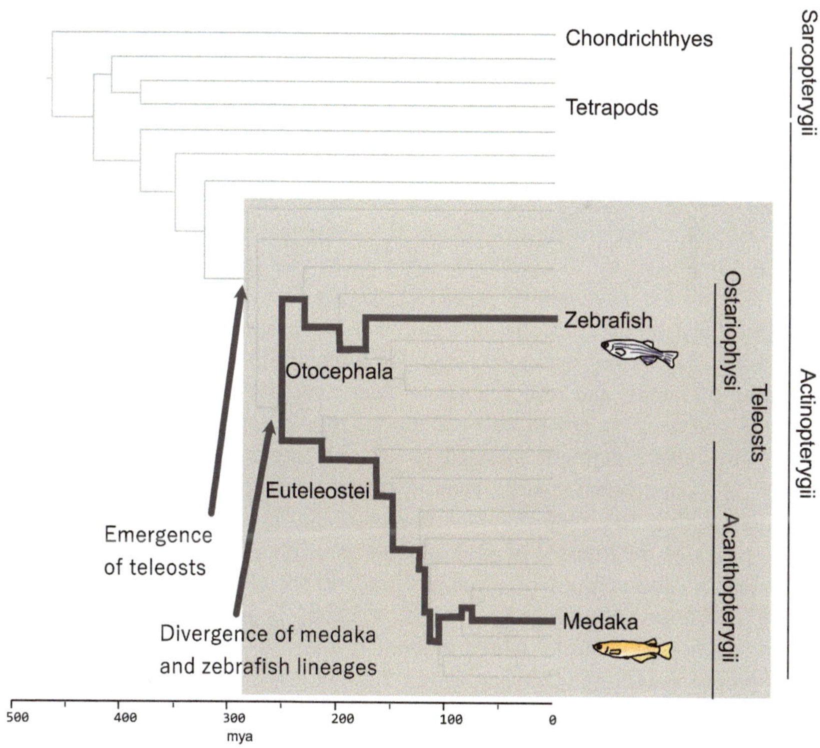

FIGURE 9.1 Phylogenetic tree of teleosts and other species in vertebrates. Medaka and zebrafish diverged ~250 million years ago (mya), which can be regarded as a relatively early stage of teleost lineage (~283 mya).

Based on these merits, medaka are now widely used for studies that require genetic manipulation. In this chapter, we will introduce classical and recently developed approaches of genetic modification in medaka, including the generation of transgenic, knockout, and knock-in varieties.

Husbandry and breeding under laboratory conditions

Animal husbandry

As expected from their natural habitat, medaka can be easily kept in dechlorinated water at room temperature. Usually, researchers keep them at 27°C in the long day condition (14L, 10D). To avoid laborious cleaning of hundreds of tanks, many researchers use a rack system with water circulation for small fishes, which are commonly used in zebrafish. However, as medaka live in a slow stream of water in rivers and ponds, they should be kept in a mild flow of water in the laboratory condition as well. Since medaka require relatively stronger light than zebrafish do to activate their reproductive functions, medaka facilities need lights over the tanks. As the density of fish affects water quality and growth/reproduction, a density of less than 5–10 fishes in 3 L tank is recommended. They can be kept in a small plastic tank just filled with dechlorinated water, but lower density is recommended in such conditions. Although they can be kept with flake food, live artemia is preferred because their growth and breeding are dependent on their nutritional conditions.

Breeding

They mature approximately 2–3 months after hatching. One can easily identify their sex by morphology since their anal fins change this shape dependent on the androgen (Ogino et al., 2014).

Once they mature, male/female paired medaka spontaneously lay 10–30 eggs every day. After spawning, eggs are attached to the female body with attaching filaments, and researchers can gently remove them from the female body. If you want to obtain one cell-stage embryo for microinjection or for other purposes, it is recommended to separate the male and female 1 day before the experiment. After removing the partition, they usually start spawning within 30 minutes.

Empirically, the number of eggs and the occurrence of spawning are highly dependent on the amount of available food. Especially, the nutritional state during the 2-day period before the experiment is important, because the luteinizing hormone surge, which induces ovulation, occurs in the afternoon 1 day before spawning (Ogiwara et al., 2013; Karigo et al., 2012) when the accumulation of yolk should be complete. For the detailed mechanism of reproductive regulation, refer to the next chapter and the following review papers (Kanda, 2019; Karigo and Oka, 2013).

Husbandry during early larval stages

Fertilized eggs can be kept in dechlorinated water with methylene blue (~0.0001%) until they hatch. Hatching occurs approximately 7–10 days after fertilization. Although the larva can be fed with only fine powder food (smaller than the mouth size of larval fish), paramecia, or rotifer during the initial several days, they grow up to be able to eat artemia afterward. Unlike zebrafish, medaka hatch out with their mouths open, and they can eat food immediately after hatching. It is suggested in teleosts that the thickness of the egg envelope and the time necessary before hatching are related (Sano et al., 2017). Medaka has a thick and stout egg envelope, which probably allows a longer time for embryogenesis in the egg. Thus the newborn medaka larva shows more developed stages compared to zebrafish.

Larvae/juvenile medaka that have grown large enough to feed on artemia can be reared like adults. Since medaka are not aggressive, medaka at different developmental stages/size can be kept in the same tank unless they differ tremendously in size. Two months after hatching, we can observe sexual dimorphism in the anal fins as described earlier. This should be a sign that they are ready to breed.

Basic approaches of generation of transgenic/knockout medaka

Conventional ways of introducing transgenes: plasmid/BAC-based constructs—their merits and demerits

At the end of the 20th century, a transgenic technique using short plasmids was introduced to zebrafish (reviewed in Lin, 2000). Later, the concept and technique of the generation of transgenic animals was also introduced to medaka (Ozato et al., 1986). This experiment was the first example of successful transgenesis in fish. The most widely used transgene is probably the GFP of *Aequorea coerulescens*, which is used to label specific cells. By labeling specific cells in living organisms or tissues, researchers can observe cellular migration, dynamics, and interactions. Moreover, such labeling has also been used to collect specifically labeled living cells for gene expression analysis and primary cell culture of specific cellular populations. Modified GFP variants can also be used (Lin and Schnitzer, 2016). GCaMP (Nakai et al., 2001) and its modified variants (Chen et al., 2013; Ohkura et al., 2012) are used for Ca^{2+} imaging, which enables the visualization of possible neurotransmitter/hormone release (detailed in the next chapter). GFP with PEST degradation sequence, which shortens the half-life of GFP in the cell, is used for the real-time observation of expressional changes in higher-time resolution (Li et al., 1998). In addition to GFP and its variants, several fluorescent proteins whose excitation/emission wavelength differ from that of GFP have been characterized/developed, such as red fluorescent protein (RFP), including dsRed and its improved variant, mCherry, derived from *Discosoma* sea anemones (Shaner et al., 2004). In combination with GFP, these fluorescent proteins can be used for the analyses of interactions among multiple cellular populations. Moreover, photoconvertible fluorescent proteins are also available. Kaede (Ando et al., 2002) later developed Kikume Green-Red (Tsutsui et al., 2005), which initially indicates green fluorescence but after irradiation with UV indicates red fluorescence. These photoconvertible fluorescent proteins can be useful for labeling specific cells within cellular populations during live imaging. These are only some of the variants of fluorescent proteins, and many other fluorescent proteins can be used in transgenic medaka.

For driving the expression of such functional proteins in specific cells or tissues, specific enhancers are necessary. However, to date, it is virtually impossible to generate target cell-specific enhancers computationally. Instead, we biologists often use long flanking sequences of the gene of interest that may include enhancements by replacing the protein-coding region with GFP or other proteins.

For such target cell-specific expressional activation, PCR-amplified 3–10 kb fragments can be included in a plasmid vector. Frequently, a 5′ flanking region of 3–4 kb is sufficient to induce cell-specific expression. To increase the probability of successful expression, enhancers can be predicted using microhomology among species of the possible essential regulatory sequences. Percent identity plot helps to find such microhomologies (http://pipmaker.bx.psu.edu/pipmaker/), and mVISTA is also useful in finding microhomologies (http://genome.lbl.gov/vista/index.shtml).

However, enhancers sometimes localize in the genomic area more than 10 kb from the open reading frame (ORF), intron, or the 3′ flanking region. In such cases, BAC or fosmid-based constructs (Nakamura et al., 2008) may be used, and they theoretically provide more accurate labeling. One can find a 50–200 kb-long BAC or fosmid that contains the gene of interest from the genome database (http://viewer.shigen.info/medakavw/mapview/mapView), which can also be obtained from the National BioResource Project Medaka(https://shigen.nig.ac.jp/medaka/top/top.jsp) (Sasado et al., 2010). The ORF of the gene of interest can be replaced with the GFP gene by homologous recombination using *Escherichia coli* strains that express recombinase, such as DY380 and SW102 (Warming, 2005).

After the generation of a transgenic line, it is also important to examine whether the target cells are specifically labeled by using dual labeling of the transgene and intrinsic mRNA (Karigo et al., 2014) or proteins (Karigo et al., 2014; Kanda et al., 2010).

Generation of knockout using TALEN/CRISPR

TALEN and CRISPR have only recently been developed (Cermak et al., 2011; Jinek et al., 2012), but they have already been widely applied to medaka

(Ansai and Kinoshita, 2014, 2017; Ansai et al., 2013). Before this breakthrough, as there were no sequence-specific gene disruption tools, accidentally discovered natural mutants and mutants that were screened from a TILLING library (Taniguchi et al., 2006) had been used. Because of the high efficiency of targeted gene disruption of TALEN/CRISPR, these techniques are mostly used in the generation of knockouts today. After breakage of the double-strand DNA genome, nonhomologous end joining (NHEJ) occurs, which is an intrinsic repair mechanism for repairing double-strand breaks (DSBs). During this process, a cell often makes deletions and/or insertions of several to hundreds of base pairs, which may cause a frameshift of ORF or a deletion of a coding sequence of important domains.

Both transgene introduction using plasmid and gene disruption using CRISPR occurs by chance and requires screening of the F1 generation. From our experience, such gene disruption much more frequently occurs compared to transgene introduction with a similar protocol involving microinjection. Thus gene knockout, which had been very difficult in teleosts with little chance of accessing embryonic stem cell lines, has become quite easy to accomplish because of the innovation of TALEN/CRISPR.

Double promoter methods for the efficient screening of transgenic lines

As described earlier, transgenic techniques using plasmids or BACs/fosmids enable us to theoretically label specific cells of the medaka. However, the successful integration of a transgene into the host genome after microinjection is usually very rare (a few percent; Rembold et al., 2006). Thus to obtain transgenic medaka efficiently, it is important to screen the embryos in which the transgene is integrated into the genome. When the construct contains a fluorescent protein such as GFP under a promoter of genes expressed during embryogenesis, we can determine whether the transgene is integrated into the genome by checking their fluorescent signals during embryogenesis. However, it is very hard to screen the embryos harboring a nonfluorescent transgene or a fluorescent protein under an adult tissue-specific promoter. In these cases, we can screen through genomic PCR of the transgene using lysates from part of the tail fin of injected medaka, but this screening method is inefficient compared with the simple observation of fluorescence at embryogenesis.

This problem can be overcome by using a double-promoter method, which employs an additional promoter and a fluorescent reporter gene for screening. Because of this additional promoter, the fluorescent reporter gene is expressed in specific tissues at the embryonic stage, which allows us to easily find transgene-positive embryos. The promoters/enhancers of genes available for screening at the embryonic stage are shown in Table 9.1. Schematics of constructs and resulting expression patterns are shown in Fig. 9.2. Note that globin b4 is an embryo-specific globin and that the reporter gene expressed

TABLE 9.1 Enhancer/promoter of genes available for screening at the embryonic stage.

Gene	Expressed tissue	References	Note
Zebrafish cardiac myosin light chain2 (zcmlc2)	Cardiac muscle	Kirchmaier et al. (2013)	Transcriptional activity to both upstream and downstream promoter
Globin b4	Hemocyte (embryo specific)	Maruyama et al. (2002), Matsuzaki et al. (2013)	
Crystallin	Lens	Vopalensky et al. (2010)	

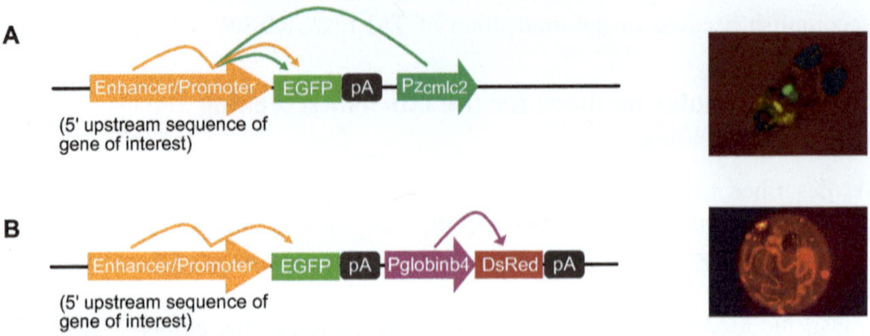

FIGURE 9.2 Examples of the double-promoter method, which increases the screening efficiency of transgenic medaka. When the enhancers of zcmlc2 and globin b4 are utilized for screening, fluorescent signals can be observed at the embryonic stage in cardiac muscle (A) and hemocytes (B), respectively. The enhancer/promoter in these examples is a 3∼5 kb DNA fragment of the 5′ flanking region of the gene of interest. Note that zcmlc enhancer has bidirectional activity, which can induce upstream enhanced green fluorescent protein (EGFP) by the activation of upstream promoter contained in the 5′ flanking region of the gene of interest.

under this promoter is not supposed to be expressed in adult tissues. Since there is toxicity in the abundantly expressed fluorescent proteins, we need to carefully choose the promoter and reporter gene for screening. In our case, when DsRed was expressed by zebrafish cardiac myosin light chain (zcmlc) enhancer, the survival rate was much lower than that in cases of enhanced GFP (EGFP), which may be due to the toxicity of DsRed (Tao et al., 2007). mCherry and tdTomato showed a much better survival ratio than

DsRedExpress2 did (Kanda and Fujimori, personal communication). Interestingly, zcmlc enhancer (Kirchmaier et al., 2013) has bidirectional transcriptional activity, which can activate promoters located either upstream or downstream of it (Fig. 9.2A). Thus in most cases we do not have to add an additional fluorescent protein if the construct contains fluorescent protein to be expressed under the enhancer/promoter of interest. On the other hand, it should be recognized that the target transgene might also be expressed in cardiac muscle cells.

It is also important to note that ectopic expression of the reporter gene sometimes occurs, which may be due to the genomic locus of transgene insertions (known as position effects). Therefore it is recommended that multiple transgenic lines from a different F0 founder for a single transgene be obtained.

Introduction of new transgenic methods

Knockout lines provide information on the overall function of the disrupted gene, whereas reporter transgenic strains are useful for the analysis of the mechanism. Thus a combination of these genetic engineering methods will enhance the understanding of the biological systems and are increasing the importance of such methods in studies of small fishes. As described in the previous sections, genome editing techniques such as Cas9/CRISPR and TALEN have made it possible to generate knockout strains in a much shorter time and with greater efficiency than before. Under these circumstances, efficient and rapid techniques for generating transgenic strains have become more important.

This section presents two efficient transgenic methods: the use of Ac/Ds transposase and the use of phiC31 integrase. Also, we present a Cas9/CRISPR-based knock-in method that can integrate genes into the allele of interest.

Transgenesis with DNA transposase

DNA transposon

DNA transposon is a DNA sequence that moves through genomic DNA. It is composed of a gene encoding an enzyme termed transposase and flanking inverted terminal repeat (ITR) sequences. The transposase expressed from the transposon gene recognizes the ITR sequences, cuts out the entire transposon from the genome, and inserts the transposon at another genomic location, allowing the transposon to move through the genome.

Many DNA transposons are isolated from a variety of species, such as tol1 and tol2 from medaka, piggyBac from silkworm, sleeping beauty from salmon, and Ac/Ds from maize.

Transposase improves transgenic efficiency

As transposase works on artificially induced DNA fragments, it can be used to increase transgenic efficiency. Practically, a donor vector containing a gene of interest flanked by ITR sequences is injected into the embryo along with mRNA encoding a transposase. Induced transposase cuts out the gene of interest between ITR sequences in the donor vector and transfers it to the host genome. This reaction achieves a much more efficient gene integration than a simple conventional injection method would.

In addition, it is noteworthy that transposon introduces a single copy of the transgene into the genome, unlike a simple conventional injection, which usually introduces multiple copies of injected DNA fragments into a single allele (single-allele multicopy). When transposase is used, the gene is inserted into multiple alleles one by one (multiallele single copy). Thus with transposase, it is possible to establish strains with a single copy of the transgene by crossing them.

In zebrafish, Kawakami's pioneering work has made the tol2 transposase method a standard transgenic method (Kawakami, 2005).

In medaka, it is reported that sleeping beauty and Ac/Ds function and improve transgenesis efficiency (Grabher et al., 2003; Froschauer et al., 2011; Boon Ng and Gong, 2011).

We have examined the efficiency of some transposases on transgenesis and discovered that the Ac/Ds system has a higher efficiency than the others (Fig. 9.3). We also discovered that about 26% of injected embryos expressed the marker gene ubiquitously, and 65% of screened G0 with marker genes are transgenic founders. Additionally, we determined that 8 bp duplications often observed after transposition (Rubin et al., 2001) were generated at both ends of integrated DNA in the established strains, indicating that this integration was not due to random integration but instead resulted from Ac transposase activity. For these reasons, we recommend Ac/Ds for transgenesis in medaka fish.

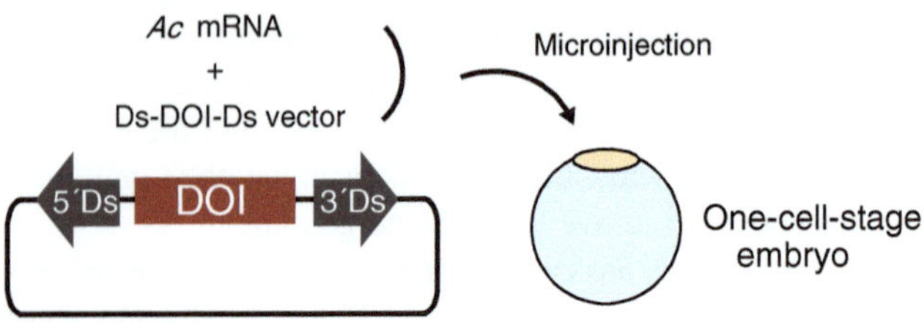

FIGURE 9.3 Ac transposase-mediated transgenesis. Coinjection of Ac mRNA and donor vector induces integration of the transgene. *DOI*, DNA of interest. *Figure adapted, under license CCBY4.0, from Ishikawa, T., Ansai, S., Kinoshita, M., Mori, K., 2018. A collection of transgenic medaka strains for efficient site-directed transgenesis mediated by phiC31 integrase. G3 (Bethesda) 8, 2585–2593.*

phiC31 integrase-mediated transgenesis

Regarding phiC31

The phiC31 recombinase is an enzyme that was originally found in the phage phiC31 genome. It catalyzes site-directed and unidirectional recombination between the two 34 bp motifs, attB (attachment site bacterium) and attP (attachment site phage) (Thorpe and Smith, 1998; Groth et al., 2000) (Fig. 9.4).

phiC31 integrase improves transgenic efficiency

Based on this principle, phiC31 integrase can be used to increase the efficiency of transgenesis. For the use of this system, we first have to generate or obtain the transgenic strain harboring a single attP motif in a chromosome as a landing site (attP-landing strain). Then, phiC31 integrase mRNA and a donor vector containing an attB DNA motif are injected into the one-cell-stage eggs of the attP-landing strain. In the embryo, the phiC31 integrase catalyzes recombination between attP and attB motifs in the host genome and donor vector, respectively. From this reaction, the donor vector is integrated into the host genome.

This phiC31-mediated integration occurs with high efficiency among various species. Pioneering studies have established attP-landing strains and shown that this method is useful in *Drosophila*, zebrafish, and medaka (Groth et al., 2004; Mosimann et al., 2013; Kirchmaier et al., 2013).

Advantage of phiC31 integrase

As described in the previous section, with conventional or transposase-mediated transgenesis, the transgenes are integrated randomly into the host genome, and their copy numbers vary. Moreover, the genomic location of the transgene also affects its expression. Thus the expression level and pattern of

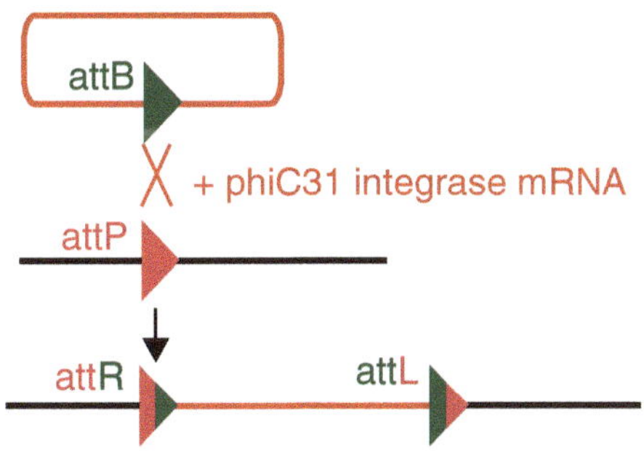

FIGURE 9.4 The phiC31 recombinase. phiC31 catalyzes the recombination between an attP and attB motif to generate an attL and attR. The vector sequence indicated in red is integrated.

integrated genes differ even among strains harboring the same vector. Therefore to obtain a reliable result, we must establish at least two strains and compare their phenotypes.

In this regard, transgenesis with phiC31 integrase offers a significant advantage (Groth et al., 2004). As phiC31 integrase always integrates a single copy of the transgene into the identical attP-landing allele in the host genome, the transgene's expression level becomes stable and comparable to that of other transgenic strains derived from the same attP-landing strains.

In our experience, using the same attP-landing strain, the transgenes' expression levels and patterns among two or more different transgenic medaka strains are quite similar.

The collection of attP-landing medaka strains

The collection and use of attP-landing strains with a single attP DNA motif in various chromosomes is desirable for the use of phiC31-mediated transgenesis because it is theoretically impossible to generate more than a double heterozygote transgenic line by crossing different transgenic lines that have respective transgenes in the same locus.

Things are more critical when the transgenic line used in combination with the knockout gene that is in close locus to the integration site since it usually requires homo knockout. With the attP-landing strain collection, we can choose a suitable strain, avoiding chromosome overlapping. For *Drosophila melanogaster*, such a collection has already been built and distributed (Bischof et al., 2007).

In medaka, three attP-landing strains were previously established by Wittbrodt et al. (Kirchmaier et al., 2013). In addition, Ishikawa et al. recently established 12 independent transgenic strains harboring a single copy of the attP motif in at least 11 of the 24 medaka chromosomes (Ishikawa et al., 2018) (Fig. 9.5). This collection is available upon request (https://shigen.nig.ac.jp/medaka/) (TG1091-1103).

Genetical features of attP-landing strains in the collection

Ishikawa et al. established multiple attP-landing strains harboring 5′DS ITR, tandem chicken insulator (2xcHS4), an attP DNA motif, Pzhsp70-mCherry-pA, an FRT DNA motif, and 3′DS ITR in a 5′ to 3′ order on the various genomic locations (Fig. 9.5A). The locations are shown in Fig. 9.5B.

The strains, except for Chr22, can be maintained as homozygous, suggesting that the integrated attP-landing site does not disrupt any essential genes.

In these strains, it is possible to transpose the attP-landing site using 5′D and 3′D ITR sequences and Ac transposase (see the "Transgenesis with DNA transposase" section). These strains also express mCherry in the lens to identify attP-landing strains. The mCherry will be ubiquitously expressed upon heat shock.

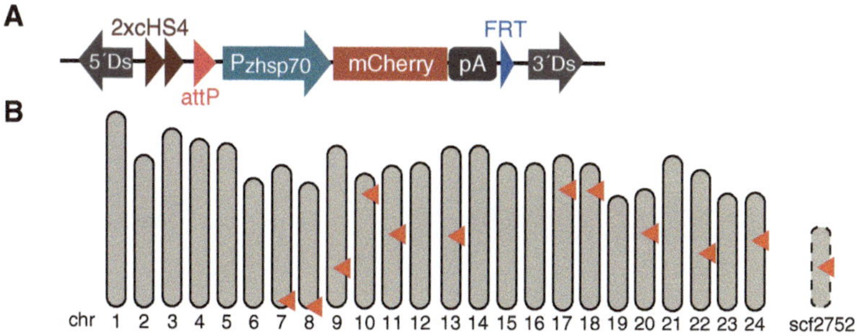

Allele	Integration site (version 2.2.4)	Gene adjacent to the attP-landing site		Phenotype of homozygotes
chr7	Chr7: 27,763,996-27,764,003	ENSORLG00000016806	40kb upstream	–
chr8	Chr8: 25,152,030-25,152,037	*carm1*	15kb upstream	–
chr9	Chr9 :24,958,590-24,958,583	*zgc:65851*	2kb upstream	–
chr10	Chr10: 8,675,165-8,675,172	*glra4a*	250kb upstream	–
chr11	Chr11: 14,688,303-14,688,296	si:ch211-232m10.6	0.7kb upstream	–
chr13	Chr13: 18,585,077-18,585,070	*acsl3a*	4kb upstream	–
chr17	Chr17: 7,100,452-7,100,445	ENSORLG00000005611	intron	–
chr18	Chr18: 6,137,116-6,137,123	*tmem256*	intron	–
chr20	Chr20: 11,133,136-11,133,143	*col11a1a*	intron	–
chr22	Chr22: 17,934,472-17,934,465	ENSORLG00000018396	intron	lethal
chr24	Chr24: 10,109,527-10,109,531	ENSORLG00000015040	intron	–
scf2752	000850F: 71,471-71,478	no Ensembl gene	–	–

FIGURE 9.5 A collection of attP-landing medaka strains. (A) Schematic representation of the transgene included in the attP-landing site of each strains. (B) Schematic representation of the locations of the attP-landing site indicated by arrowheads. Detailed integration positions are shown in the table. *Figure adapted, under license CCBY4.0, from Ishikawa, T., Ansai, S., Kinoshita, M., Mori, K., 2018. A collection of transgenic medaka strains for efficient site-directed transgenesis mediated by phiC31 integrase. G3 (Bethesda) 8, 2585–2593.*

attB-targeting vector

For a donor vector containing an attB motif, Ishikawa et al. designed the attB-targeting vector, which works best with the attP-landing strains described in Fig. 9.5. This vector contains a multicloning site for the cloning of DNA of interest (DOI), FRT DNA motif, and tandem chicken insulator, in this order (Fig. 9.6). When this vector is recombined into the genome of the attP-landing strain, the allele after recombination contains a tandem chicken insulator, DOI, a tandem chicken insulator, FRT DNA motif, vector backbone, pzhsp70-mCherry-pA, and FRT DNA motif in a 5′ to 3′ order.

Since the two tandem chicken insulators flank the DOI, it is less susceptible to the potent enhancers around the integration site (Shimizu and Shimizu, 2012).

A vector backbone and an mCherry transgenic marker gene can be removed by FLP recombinase-mediated recombination between two FRT DNA motifs that flank the sequences, as described next.

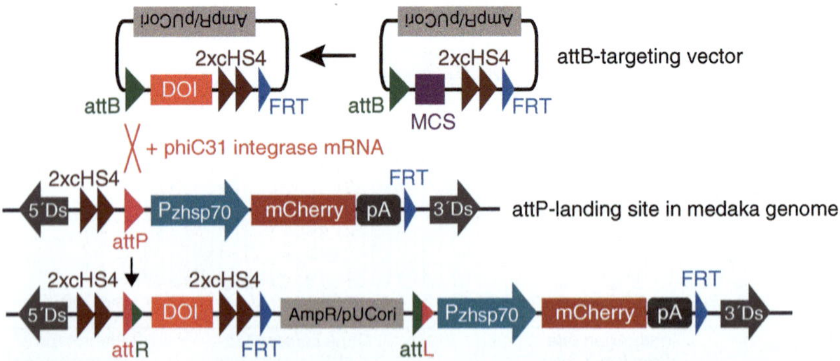

FIGURE 9.6 The attB-targeting vector. Schematic representation of the attB-targeting vector and genomic structure after recombination with phiC31 integrase is shown. *DOI*, DNA of interest. *Figure adapted, under license CCBY4.0, from Ishikawa, T., Ansai, S., Kinoshita, M., Mori, K., 2018. A collection of transgenic medaka strains for efficient site-directed transgenesis mediated by phiC31 integrase. G3 (Bethesda) 8, 2585–2593.*

Advanced attB-targeting vector containing fluorescent protein for efficient screening in embryos

Ishikawa et al. also designed an advanced attB-targeting vector. As described in the previous section, this is a dual-promoter vector, and it adds the Pzcmlc2-tagCFP gene to the just-described attB-targeting vector, making it easier to identify the transgenic founder, especially when the DOI does not contain a fluorescent protein or expresses fluorescent protein at a lower level in embryonic stages (Fig. 9.7A).

When the advanced attB-targeting vector is integrated, the tagCFP gene is expressed in the heart through the zebrafish cmlc2 promoter (Pzcmlc2) activity, as described in the previous section. Furthermore, when this vector is recombined into an attP-landing site, the Pzcmlc2-tagCFP fragment will be integrated right next to Pzhsp70-mCherry. Because the zhsp70 promoter can be influenced by a nearby enhancer (ENCODE Project Consortium, 2012), it is driven by the zcmlc2 enhancer, and thus mCherry will be expressed in the heart. This feature makes it easy to determine whether the transgene was integrated to the correct attP site, expressing both tagCFP and mCherry in the heart, or integrated to another locus, expressing only the tagCFP (Fig. 9.7B).

Efficiency of phiC31-mediated transgenesis

To conduct phiC31-mediated transgenesis, we recommend low concentrations of phiC31 mRNA (10 ng/μL) because of its toxicity. At a concentration of 40 ng/μL, most of the injected embryos died before hatching.

In our experience, about 20% of injected embryos are well integrated and can be screened as transgenic founder candidates, and more than 50% of the candidates are transgenic founders. Without screening, at least 10% of injected G0s are transgenic founders.

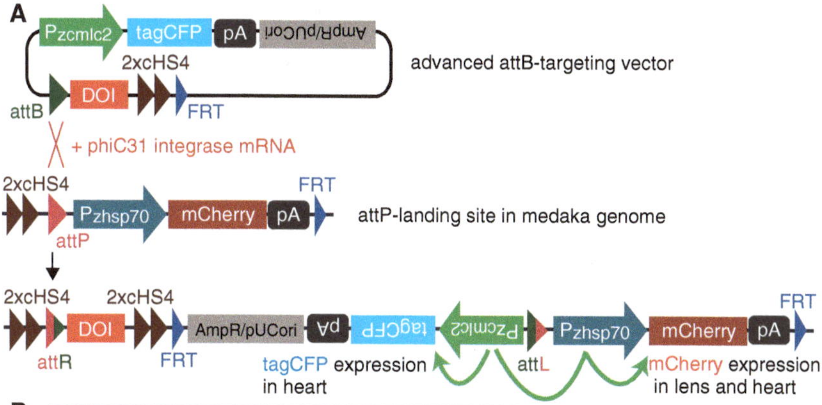

B

	random integration	PhiC31 mediated recombination
tagCFP	heart	heart
mcherry	lens	lens, heart

FIGURE 9.7 The advanced attB-targeting vector. (A) Schematic representation of the advanced attB-targeting vector and genomic structure after recombination with phiC31 integrase is shown. (B) The expected expression pattern of tagCFP and mCherry in transgenic embryos produced by random integration or phiC31 mediated the advanced attB-targeting vector's recombination. *DOI*, DNA of interest. *Figure adapted, under license CCBY4.0, from Ishikawa, T., Ansai, S., Kinoshita, M., Mori, K., 2018. A collection of transgenic medaka strains for efficient site-directed transgenesis mediated by phiC31 integrase. G3 (Bethesda) 8, 2585–2593.*

After phiC31 recombination, 5′D and 3′D ITR sequences are still intact. Thus the transgene can be transposed by inducing Ac expression.

FLP recombinase-mediated locus cleanup

As described earlier, unnecessary sequences, such as the vector backbone, which may affect DOI expression (Tasic et al., 2011), and the transgenic marker mCherry gene are flanked by an FRT DNA motif after recombination (Fig. 9.8A). Thus these sequences can be removed by FLP recombinase-mediated recombination between two FRT motifs.

In our experience, FLP mRNA has low toxicity. We can inject high concentrations (100 ng/μL) of FLP mRNA, and FLP-mediated recombination is very efficient. As shown in Fig. 9.8B, FLP mRNA injection decreases the expression of mCherry without a decrease of DOI (EGFP) expression in the G0.

Primer sequences to confirm integration by phiC31 integrase

To confirm the locus generated by phiC31-mediated recombination, attL motif, genomic PCR can be performed. Primers 5′-GTAGGTCACGGTCTCGAAGC-3′ binding to the attL sequence and 5′-TTACCCTGGTGCCATGAAAT-3′ binding to the zhsp70 promoter will generate a 398 bp PCR product.

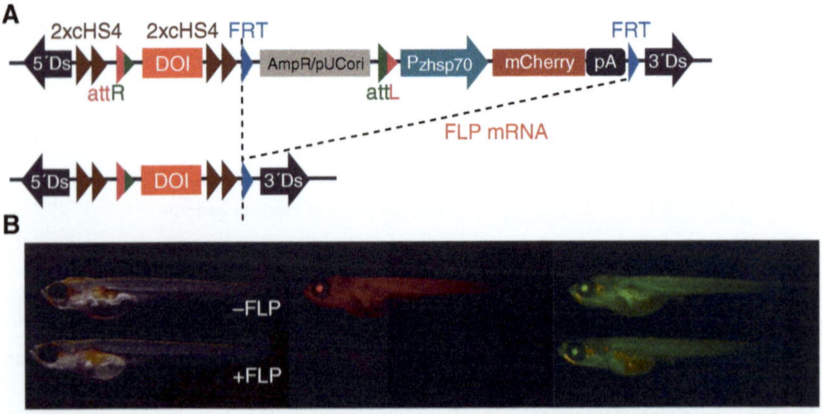

FIGURE 9.8 FLP recombinase-mediated locus cleanup. (A) Schematic representation of FLP-mediated locus cleanup. (B) By FLP mRNA injection, the mCherry marker's expression is decreased, whereas that of the DNA of interest (DOI), Pzhsp70-EGFP, maintains its level of expression. *Figure adapted, under license CCBY4.0, from Ishikawa, T., Ansai, S., Kinoshita, M., Mori, K., 2018. A collection of transgenic medaka strains for efficient site-directed transgenesis mediated by phiC31 integrase. G3 (Bethesda) 8, 2585–2593.*

phiC31 integrase can be applicable to BAC/fosmid-based transgenesis

As described in the prior sections, BAC and fosmid vectors can contain a large genomic DNA fragment of 200 and 40 kb, respectively.

In *D. melanogaster*, BAC–vector integration has been achieved by phiC31 integrase-mediated transgenesis (Venken et al., 2006). In the current study, we attempted to integrate the fosmid vector into the medaka genome and confirmed that phiC31 integrase could integrate it as efficiently as in the plasmid case.

Application of the highly efficient phiC31 integrase-based transgenesis for analysis of G0 generation

Due to the high efficiency of phiC31 integrase-mediated transgenesis, 20% of injected fish contain the transgene in most somatic cells. Therefore we can analyze the phenotype even in the G0 generation.

For example, we can assess the promoter activity of specific DNA sequences at the G0 embryo (Kirchmaier et al., 2013). This technique allows for a much faster analysis than the traditional method involving the F1 generation embryo.

An advantage and a basic principle of CRISPR/Cas-based knock-in approaches in medaka

Until recently, in species that do not have embryonic stem cell lines, it has been impossible to insert DNA fragments into the target location using what is called the knock-in technique. However, the development of genome-editing technologies such as TALEN and the CRISPR/Cas system ushered in a new era of genetic engineering and biotechnology. Unlike the transgenic approaches involving the transposases or integrases, the genome editing tool-based knock-in technique enables researchers to precisely integrate transgenes into the selected target sites on the host genome. With its superior simplicity and robust accuracy, the CRISPR/Cas system has been applied to a number of knock-in/out experiments using cultured cells, plants, and animals. In medaka, shortly after the CRISPR-mediated knockout without transgene insertions became a routine procedure, several studies reported successful knock-in of the *GFP* gene using this tool (Stemmer et al., 2015; Murakami et al., 2017; Watakabe et al., 2018; Gutierrez-Triana et al., 2018). Although the CRISPR-assisted knock-in technology has currently achieved only such reporter tagging in medaka, future studies would undoubtedly prove its suitability for other types of genetic alterations such as exon replacements, introduction of point mutations, and conditional knockout combined with the FLP/FRT or Cre/LoxP systems. These reverse genetic approaches hold promise for clarifying the molecular basis of complex life phenomena in medaka.

Currently, the combination of the CRISPR/Cas9 system with a donor plasmid has become an effective strategy for the generation of knock-in medaka. The knock-in events induced by this strategy are as follows: the target site in the host genome and that of donor plasmid, the latter of which is called the "bait," are simultaneously cleaved by the complex of Cas9 and guide RNA (gRNA) in the cells. Then, the resultant DNA DSBs on the genome trigger several repair mechanisms, which can mediate the integration of insert fragments derived from the donor plasmid into the target site. To successfully achieve an efficient gene knock-in, it is important to understand not only the mechanisms of the DNA repair pathways but also the roles of the components of the plasmid. Thus we will mainly focus on such key factors and introduce the performance and limitations of the current state of the knock-in method.

An efficient knock-in using a donor plasmid with bait sequences

Previous studies have demonstrated that simultaneous cleavage of both target sites on the genome and the donor plasmid by targetable nucleases can promote knock-in events in zebrafish (*Danio rerio*) and sea urchin (*Hemicentrotus pulcherrimus*) embryos (Irion et al., 2014; Ochiai et al., 2012) although the detailed mechanism of this process remains unclear. The effect of bait sequences on knock-in efficiency has also been examined in medaka; the integration efficiencies of a donor plasmid with or without bait sequences were compared according to the expression area of GFP green fluorescence in the embryos (Murakami et al., 2017). The comparisons in somatic level revealed that a donor plasmid with bait sequences yielded fivefold higher knock-in efficiencies than did the other plasmid without bait sequences. Furthermore, the same experimental scheme showed that knock-in efficiencies were considerably varied among eight bait sequences screened and that the knock-in system with BaitD achieved the highest integration efficiency among them. Based on these results, we suggest that a donor plasmid with a BaitD sequence (GATCTTCGGCCTAGACTGCGAGG) is suitable for an efficient generation of knock-in medaka (Fig. 9.9A and B). An in silico analysis using CCTop (Stemmer et al., 2015) nominated two potential off-target candidates against BaitD in the reference genome of medaka, which includes four mismatches in the target sequence of 20 bp for gRNA-BaitD. However, we confirmed that the knock-in system with BaitD did not induce any off-target mutagenesis at either loci (unpublished data). Additionally, BaitD has been reported to contain few off-target candidates in the reference genomes of 11 other fish species, including zebrafish, Nile tilapia (*Oreochromis niloticus*), and red sea bream (*Pagrus major*) (Murakami et al., 2017), which may help to establish a new knock-in system with minimized off-target effects in these fish species. The reporter plasmid with BaitD was distributed from the RIKEN Bioresource Center DNA Bank (http://dna.brc.riken.jp/).

Pathways of DNA double-stranded break repair and their use for producing knock-in medaka

With the use of the CRISPR/Cas9 system, DSBs are induced at the genomic target sites, which activates several DNA repair processes, via NHEJ, microhomology-mediated end joining (MMEJ), or homology-directed repair (HDR) pathways (Symington and Gautier, 2011). These repair mechanisms have been exploited for the development of various targeting strategies (He et al., 2018). In this section, after introducing the feature of the three knock-in methods based on the foregoing three pathways, we will explain the application of each method in medaka.

It is well known that NHEJ is frequently utilized to produce small insertions/deletions (indels) in the gene-coding region for the generation of knockouts. NHEJ is generally active through the overall cell cycle except for mitosis (Lieber, 2008); thus its high activity is also effective for an efficient knock-in generation. The NHEJ-mediated gene knock-in can directly join both ends of the genomic DNA and the insert gene at the target site (Fig. 9.10A). This method has one advantage: there is no requirement for homology arms (HAs) flanking the DSB sites. Therefore the donor plasmid can be more simply constructed than that of other plasmids for a homology-dependent knock-in such as HDR and MMEJ. In contrast, one clear disadvantage of this method is that it cannot control the direction of the insert gene, which randomly generates both forward- and reverse-directional integrations of transgenes into the target sites (Watakabe et al., 2018). Moreover, the NHEJ repair process is often accompanied by small indels at the junctions between the target site and the insert gene, making its use in seamless knock-in

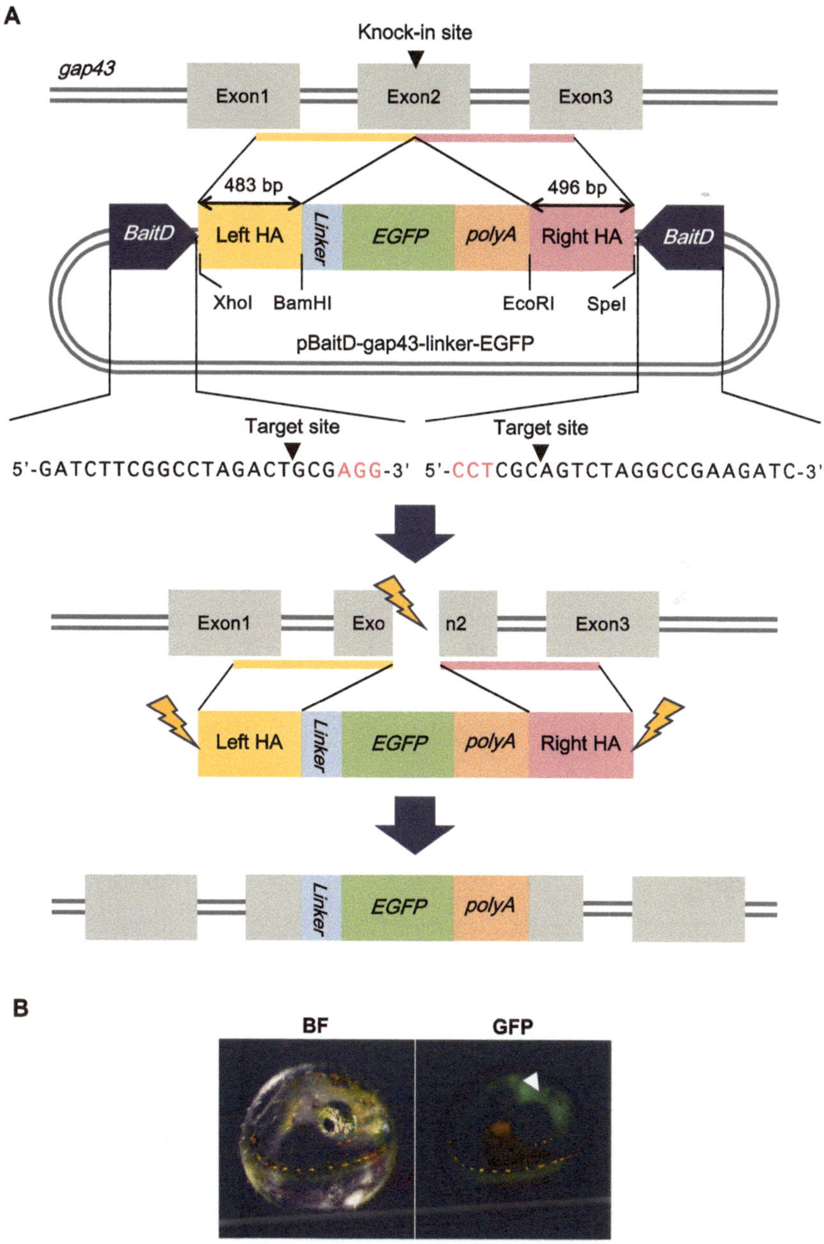

FIGURE 9.9 Schematic overview of the homology-directed repair (HDR)-mediated knock-in system with BaitD. (A) Schematic design of a donor plasmid for precise integration of *EGFP* into the *gap43* locus (Ensembl transcript number ENSORLT00000015837). A linker sequence is located between a left homology arm (HA) and *EGFP* to increase the flexibility of the fusion protein, the endogenous GAP43, and EGFP. The 3' terminal AGG and the 5' terminal CCT (red letters) indicate a protospacer adjacent motif (PAM) sequence in BaitD. After the Cas9-mediated cleavage at BaitD of the donor plasmid and the genomic target site, which is shown as lightning bolt signs, the green fluorescent protein (GFP) reporter cassette is seamlessly integrated into the second exon of the *gap43* locus via HDR. (B) GFP expression in the injected embryos at 4 days postfertilization. Embryos injected with the donor plasmid expressed GFP in their central nervous system (CNS). White arrowheads show fluorescence in the CNS. *HAs*, homology arms.

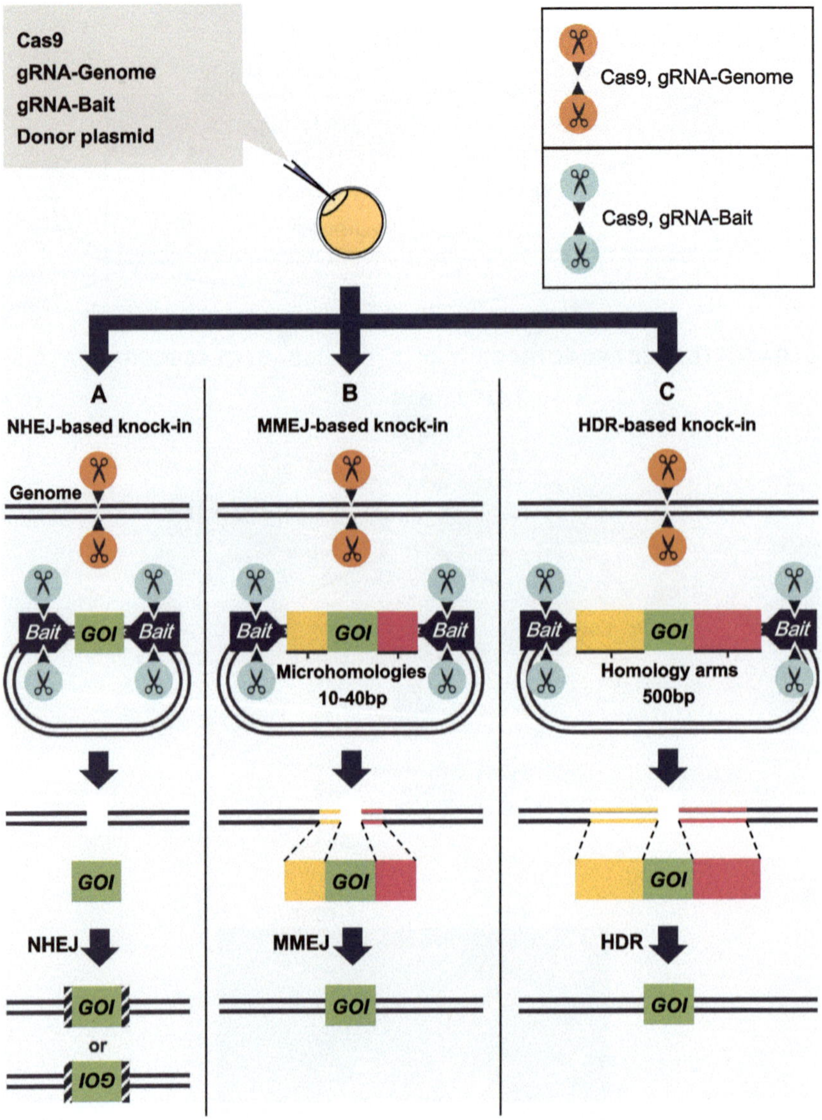

FIGURE 9.10 Schematic overview of nonhomologous end joining (NHEJ-), microhomology-mediated end joining (MMEJ-), and homology-directed repair (HDR)-mediated gene knock-in. (A) The NHEJ-based knock-in. No homology arms (HAs) are required, but various types of indel mutations are often generated (shown as striped boxes). (B) The MMEJ-based knock-in. Microhomologies (usually less than 40 bp) are required. (C) The HDR-based method. Long HAs (~500 bp) are required. *GOI*, gene of interest.

challenging. Taking into account that NHEJ is highly error prone, it is better to integrate the donor templates with promoter into noncoding or untranslated genome regions.

MMEJ was identified as a new DNA repair pathway later than NHEJ and HDR, and, in this method, the short homologous sequences (microhomology: 10–40 bases) around the DNA DSB region are annealed and the injured DNA strands are repaired (McVey and Lee, 2008). As shown in Fig. 9.10B, the microhomologies in the donor plasmid are annealed with the same sequences flanking the DSB site, and the DNA fragment edged by the microhomologies is then integrated into the chromosome. Although the MMEJ-mediated knock-in requires a molecular cloning step for insertion of the microhomologies into both edges of the knock-in fragment, the length of the microhomologies is extremely short. Therefore they can simply be added by a single PCR without amplification from genomic DNA. In contrast to the NHEJ-mediated knock-in, the advantage of the MMEJ-mediated knock-in is that it allows us to regulate the direction of the insert fragment. Thus the use of the MMEJ-mediated gene knock-in has expanded in cultured cell lines, zebrafish, and frogs (*Xenopus laevis*) as a simple gene knock-in method, which is known as the precise integration into the target chromosome (PITCh) system (Nakade et al., 2014; Sakuma et al., 2016).

HDR is an error-free DNA repair pathway that utilizes the homologous sequences of the intact sister chromatid as the template for DSB healing, which is activated in an independent manner from NHEJ and MMEJ. Because of its accuracy, HDR (previously termed HR) has been used for the conventional gene knock-in/out in embryonic stem cells (Tong et al., 2010; Davis et al., 2009). The donor plasmid suited for HDR carries HAs that lead to seamless and accurate integration of the desired fragments into the DSB region (Fig. 9.9A), which is common in the MMEJ-mediated knock-in method. However, the HDR-mediated knock-in method requires much longer HAs (~500 bp) than those of MMEJ-mediated knock-in (10–40 bp), indicating that time-consuming effort is required to construct the donor plasmid used for HDR.

The CRISPR-based knock-in technology diversifies quite frequently, which has both advantages and disadvantages. Therefore researchers should select the current appropriate technology for the efficient generation of their desired knock-in medaka. The choice of method largely depends on whether in-frame knock-in such as fusion of a tag to the ORF of interest or introduction of a point mutation into the coding region is required. When these homology-dependent insertions are not needed, the NHEJ-mediated knock-in is a better choice due to its ease of donor plasmid construction. In contrast, in the case of in-frame knock-in, the used of methods via MMEJ or HDR are recommended. For efficient knock-in in medaka, HDR would be more suitable than MMEJ because it was reported that the MMEJ-based PITCh system induced knock-in with significantly lower efficiency in medaka compared to those reported in cultured cell lines, zebrafish, and frogs (Nakade et al., 2014; Sakuma et al., 2016; Murakami et al., 2017). However, the HDR-mediated knock-in method has not become popular in medaka compared to the NHEJ-mediated targeted mutagenesis because of the laborious donor template construction. A large portion of DSBs is generally repaired via NHEJ; thus a slight modification is needed to overcome this problem. Recently, successful trials have been reported for switching the DSB repair pathway from NHEJ to HDR by suppressing key NHEJ molecules such as KU70, KU80, and DNA ligase IV in rats and cell lines (Chu et al., 2015; Ma et al., 2016). Future research may reveal that the combination of these molecules with the HDR-mediated gene knock-in system can facilitate seamless gene modifications in medaka.

Advantages, difficulties, and prospects for research on transgenic and genome editing in medaka

As indicated earlier, powerful tools such as transgenic and genome editing are available for medaka studies. Furthermore, by taking advantage of the characteristics of medaka as shown next, we can apply unique approaches that would be difficult to utilize with other experimental animals. In the following paragraphs, we will briefly introduce the advantages of medaka for research and potential future prospects.

The first advantage is that medaka have small amounts of tissue, allowing us to analyze the entire tissue easily, such as the pituitary for neural projection analysis by immunohistochemistry (Nakajo et al., 2018; Hodne et al., 2019) and the retina for clonal analysis (Kromm et al., 2016). By using light-sheet microscopy (selective plane illumination microscopy or digital single lens mirrorless) and confocal microscopy, it is possible to scan a large portion of the tissue without disrupting its structure. In combination with optical clearing such as Sca*l*e (Hama et al., 2011) or CUBIC (Susaki et al., 2014), light-sheet microscopy allows for a deeper observation of the tissue, which helps us view the broader picture. In most cases, the use of medaka strains that genetically lack chromatophores (Wakamatsu et al., 2001) may provide better results. Recently, methods for enhanced *in vivo* imaging in medaka by optimized anesthesia and the removal of pigmentation have been reported (Lischik et al., 2019). The methods described here can allow for clearer observation of existing GFP/RFP transgenic lines. On the other hand, for detailed analysis at the cellular level, the preparation of tissue sections is necessary, as described in the next chapter. The small size of the tissue also makes it easier to perform *in vitro* imaging. For instance, Ca^{2+} imaging using the whole brain (discussed in the next chapter) and live imaging of germ cells using the ovarian epithelial tissue (Nakamura et al., 2010) have been reported.

The second advantage is that organ cultures and primary culture protocols have been established for some organs. For example, the entire pituitary or isolated cells can be cultured (Kayo et al., 2019; Ager-Wick et al., 2018). The entire ovary or isolated ovarian follicles can be cultured in a medium, and the cultured follicles are physiological enough to exhibit ovulation after gonadotropin stimulation (Pendergrass and Schroeder, 1976; Ogiwara et al., 2010). It is reported that testicular tissue can be cultured *in vitro* and that spermatogenesis can be induced (Saiki et al., 1997). By combining these organ culture techniques with transgenic and genome-editing technologies, the medaka can provide a convenient experimental model for understanding molecular mechanisms of the biological phenomena at both tissue and organ levels.

The third advantage is that the behavior can be observed easily and clearly. In particular, the patterns of sexual behavior are well defined (Ono and Uematsu, 1957; Walter and Hamilton, 1970), and several papers that combine a genetic approach and behavior have been published (Okuyama et al., 2014; Yokoi et al., 2015, 2020). In addition, through the evaluation of anxiety and social behaviors of knockout medaka, the functions of neurotransmitters have been examined (Ansai et al., 2017). Moreover, semi-automated behavioral analysis systems that can be applied to medaka makes behavioral analyses easier (Tomihara et al., 2021).

On the other hand, some experimental techniques have not been fully applied due to the small tissue size. Fiber optic techniques such as fiber photometry and optogenetics are difficult to use at present because the small size of the medaka brain makes it difficult to accurately locate optical fibers.

For more precise temporal/spatial expressional regulation, conditional genetic modification using site-specific recombination systems such as the Cre-loxP and FLP/FRT systems are widely used in mice and other model organisms. Recent studies revealed that they can also be used in medaka (Fig. 9.8; Ishikawa et al., 2018; Kishimoto et al., 2016). In addition, expressional induction using a heat-shock promoter can be used in the embryo and larvae of medaka (Kirchmaier et al., 2013; Okuyama et al., 2013; Shimada et al., 2013). However, fewer such findings have been reported in adults (Kobayashi et al., 2013). On the other hand, in our case, nitroreductase/metronidazole, which is used to ablate specific cells in the presence of drugs (Bridgewater et al., 1995), did not work in the neurons or pituitary cells of adult medaka (Kanda and Takahashi, personal communication). However, other conditional genetic tools have been developed and used in other model species such as Gal4/UAS and Tet On/Off systems. By making the most of the techniques described in this chapter and those under development, the medaka will become an ever more useful model animal.

Acknowledgment

This work was supported by NBRP medaka.

References

Ager-Wick, E., Hodne, K., Fontaine, R., Von Krogh, K., Haug, T.M., Weltzien, F.-A., 2018. Preparation of a high-quality primary cell culture from fish pituitaries. JoVE e58159.

Ando, R., Hama, H., Yamamoto-Hino, M., Mizuno, H., Miyawaki, A., 2002. An optical marker based on the UV-induced green-to-red photoconversion of a fluorescent protein. Proc. Natl. Acad. Sci. U. S. A. 99, 12651–12656.

Ansai, S., Kinoshita, M., 2014. Targeted mutagenesis using CRISPR/Cas system in medaka. Biol. Open 3, 362–371.

Ansai, S., Kinoshita, M., 2017. Genome editing of medaka. Methods Mol. Biol. 1630, 175–188.

Ansai, S., Hosokawa, H., Maegawa, S., Naruse, K., Washio, Y., Sato, K., Kinoshita, M., 2017. Deficiency of serotonin in raphe neurons and altered behavioral responses in tryptophan hydroxylase 2-knockout medaka (*Oryzias latipes*). Zebrafish 14, 495–507.

Ansai, S., Sakuma, T., Yamamoto, T., Ariga, H., Uemura, N., Takahashi, R., Kinoshita, M., 2013. Efficient targeted mutagenesis in medaka using custom-designed transcription activator-like effector nucleases. Genetics 193, 739–749.

Asai, T., Senou, H., Hosoya, K., 2011. *Oryzias sakaizumii*, a new ricefish from northern Japan (Teleostei: Adrianichthyidae). Ichthyol. Explor. Freshw. 22, 289–299.

Betancur, R.R., Broughton, R.E., Wiley, E.O., Carpenter, K., Lopez, J.A., Li, C., Holcroft, N.I., Arcila, D., Sanciangco, M., Cureton Ii, J.C., Zhang, F., Buser, T., Campbell, M.A., Ballesteros, J.A., Roa-Varon, A., Willis, S., Borden, W.C., Rowley, T., Reneau, P.C., Hough, D.J., Lu, G., Grande, T., Arratia, G., Orti, G., 2013. The tree of life and a new classification of bony fishes. PLoS Curr. 5.

Bischof, J., Maeda, R.K., Hediger, M., Karch, F., Basler, K., 2007. An optimized transgenesis system for *Drosophila* using germ-line-specific phiC31 integrases. Proc. Natl. Acad. Sci. U. S. A. 104, 3312–3317.

Boon Ng, G.H., Gong, Z., 2011. Maize Ac/Ds transposon system leads to highly efficient germline transmission of transgenes in medaka (*Oryzias latipes*). Biochimie 93, 1858–1864.

Bridgewater, J.A., Springer, C.J., Knox, R.J., Minton, N.P., Michael, N.P., Collins, M.K., 1995. Expression of the bacterial nitroreductase enzyme in mammalian cells renders them selectively sensitive to killing by the prodrug CB1954. Eur. J. Cancer 31a, 2362–2370.

Cermak, T., Doyle, E.L., Christian, M., Wang, L., Zhang, Y., Schmidt, C., Baller, J.A., Somia, N.V., Bogdanove, A.J., Voytas, D.F., 2011. Efficient design and assembly of custom TALEN and other TAL effector-based constructs for DNA targeting. Nucleic Acids Res. 39, e82.

Chen, T.W., Wardill, T.J., Sun, Y., Pulver, S.R., Renninger, S.L., Baohan, A., Schreiter, E.R., Kerr, R.A., Orger, M.B., Jayaraman, V., Looger, L.L., Svoboda, K., Kim, D.S., 2013. Ultrasensitive fluorescent proteins for imaging neuronal activity. Nature 499, 295.

Chu, V.T., Weber, T., Wefers, B., et al., 2015. Increasing the efficiency of homology-directed repair for CRISPR-Cas9-induced precise gene editing in mammalian cells. Nat. Biotechnol. 33, 543–548.

Davis, R.P., Grandela, C., Sourris, K., et al., 2009. Generation of human embryonic stem cell reporter knock-in lines by homologous recombination. Curr. Protoc. Stem Cell Biol. 11, 5B.1.1–5B.1.34.

Encode Project Consortium, 2012. An integrated encyclopedia of DNA elements in the human genome. Nature 489, 57–74.

Froschauer, A., Sprott, D., Gerwien, F., Henker, Y., Rudolph, F., et al., 2011. Effective generation of transgenic reporter and gene trap lines of the medaka (*Oryzias latipes*) using the Ac/Ds transposon system. Transgenic Res. 21, 149–162.

Grabher, C., Henrich, T., Sasado, T., Arenz, A., Wittbrodt, J., et al., 2003. Transposon-mediated enhancer trapping in medaka. Gene 322, 57–66.

Groth, A.C., Olivares, E.C., Thyagarajan, B., Calos, M.P., 2000. A phage integrase directs efficient site-specific integration in human cells. Proc. Natl. Acad. Sci. U. S. A. 97, 5995–6000.

Groth, A.C., Fish, M., Nusse, R., Calos, M.P., 2004. Construction of transgenic Drosophila by using the site-specific integrase from phage phiC31. Genetics 166, 1775–1782.

Gutierrez-Triana, J.A., Tavhelidse, T., Thumberger, T., et al., 2018. Efficient single-copy HDR by 5' modified long dsDNA donors. eLife 7.

Hama, H., Kurokawa, H., Kawano, H., Ando, R., Shimogori, T., Noda, H., Fukami, K., Sakaue-Sawano, A., Miyawaki, A., 2011. Scale: a chemical approach for fluorescence imaging and reconstruction of transparent mouse brain. Nat. Neurosci. 14, 1481–1488.

He, X., Li, Y.X., Feng, B., 2018. New turns for high efficiency knock-in of large DNA in human pluripotent stem cells. Stem Cell. Int. 2018, 9465028.

Hodne, K., Fontaine, R., Ager-Wick, E., Weltzien, F.-A., 2019. Gnrh1-Induced responses are indirect in female medaka fsh cells, generated through cellular networks. Endocrinology 160, 3018–3032.

Hori, H., 2011. A glance at the past of medaka fish biology. In: Medaka: A Model for Organogenesis, Human Disease, and Evolution, pp. 1–16.

Inoue, K., Takei, Y., 2003. Asian medaka fishes offer new models for studying mechanisms of seawater adaptation. Comp. Biochem. Physiol. B Biochem. Mol. Biol. 136, 635–645.

Irion, U., Krauss, J., Nüsslein-Volhard, C., 2014. Precise and efficient genome editing in zebrafish using the CRISPR/Cas9 system. Development 141, 4827–4830.

Ishikawa, T., Ansai, S., Kinoshita, M., Mori, K., 2018. A collection of transgenic medaka strains for efficient site-directed transgenesis mediated by phiC31 integrase. G3 (Bethesda) 8, 2585–2593.

Jinek, M., Chylinski, K., Fonfara, I., Hauer, M., Doudna, J.A., Charpentier, E., 2012. A programmable dual-RNA-guided DNA endonuclease in adaptive bacterial immunity. Science 337, 816–821.

Kanda, S., 2019. Evolution of the regulatory mechanisms for the hypothalamic-pituitary-gonadal axis in vertebrates-hypothesis from a comparative view. Gen. Comp. Endocrinol. 284, 113075.

Kanda, S., Nishikawa, K., Karigo, T., Okubo, K., Isomae, S., Abe, H., Kobayashi, D., Oka, Y., 2010. Regular pacemaker activity characterizes gonadotropin-releasing hormone 2 neurons recorded from green fluorescent protein-transgenic medaka. Endocrinology 151, 695–701.

Karigo, T., Aikawa, M., Kondo, C., Abe, H., Kanda, S., Oka, Y., 2014. Whole brain-pituitary in vitro preparation of the transgenic medaka (*Oryzias latipes*) as a tool for analyzing the differential regulatory mechanisms of LH and FSH release. Endocrinology 155, 536–547.

Karigo, T., Kanda, S., Takahashi, A., Abe, H., Okubo, K., Oka, Y., 2012. Time-of-Day-Dependent changes in GnRH1 neuronal activities and gonadotropin mRNA expression in a daily spawning fish, medaka. Endocrinology 153, 3394–3404.

Karigo, T., Oka, Y., 2013. Neurobiological study of fish brains gives insights into the nature of gonadotropin-releasing hormone 1-3 neurons. Front. Endocrinol. 4, 177.

Kasahara, M., Naruse, K., Sasaki, S., Nakatani, Y., Qu, W., Ahsan, B., Yamada, T., Nagayasu, Y., Doi, K., Kasai, Y., Jindo, T., Kobayashi, D., Shimada, A., Toyoda, A., Kuroki, Y.,

Fujiyama, A., Sasaki, T., Shimizu, A., Asakawa, S., Shimizu, N., Hashimoto, S., Yang, J., Lee, Y., Matsushima, K., Sugano, S., Sakaizumi, M., Narita, T., Ohishi, K., Haga, S., Ohta, F., Nomoto, H., Nogata, K., Morishita, T., Endo, T., Shin, I.T., Takeda, H., Morishita, S., Kohara, Y., 2007. The medaka draft genome and insights into vertebrate genome evolution. Nature 447, 714–719.

Kawakami, K., 2005. Transposon tools and methods in zebrafish. Dev. Dynam. 234, 244–254.

Kayo, D., Zempo, B., Tomihara, S., Oka, Y., Kanda, S., 2019. Gene knockout analysis reveals essentiality of estrogen receptor β1 (Esr2a) for female reproduction in medaka. Sci. Rep. 9, 8868.

Kirchmaier, S., Höckendorf, B., Möller, E.K., Bornhorst, D., Spitz, F., Wittbrodt, J., 2013. Efficient site-specific transgenesis and enhancer activity tests in medaka using PhiC31 integrase. Development 140, 4287.

Kishimoto, K., Nakayama, M., Kinoshita, M., 2016. In vivo recombination efficiency of two site-specific recombination systems, VCre/VloxP and SCre/SloxP, in medaka (*Oryzias latipes*). Dev. Growth Differ. 58, 516–521.

Kobayashi, K., Kamei, Y., Kinoshita, M., Czerny, T., Tanaka, M., 2013. A heat-inducible CRE/LOXP gene induction system in medaka. Genesis 51, 59–67.

Kromm, D., Thumberger, T., Wittbrodt, J., 2016. Chapter 5 - An eye on light-sheet microscopy. In: Detrich, H.W., Westerfield, M., Zon, L.I. (Eds.), Methods in Cell Biology. Academic Press.

Li, X., Zhao, X., Fang, Y., Jiang, X., Duong, T., Fan, C., Huang, C.C., Kain, S.R., 1998. Generation of destabilized green fluorescent protein as a transcription reporter. J. Biol. Chem. 273, 34970–34975.

Li, Y., Liu, Y., Yang, H., Zhang, T., Naruse, K., Tu, Q., 2020. Dynamic transcriptional and chromatin accessibility landscape of medaka embryogenesis. Genome Res. 30 (6), 924–937.

Lieber, M.R., 2008. The mechanism of human nonhomologous DNA end joining. J. Biol. Chem. 283, 1–5.

Lin, M.Z., Schnitzer, M.J., 2016. Genetically encoded indicators of neuronal activity. Nat. Neurosci. 19, 1142–1153.

Lin, S., 2000. Transgenic zebrafish. Methods Mol. Biol. 136, 375–383.

Lischik, C.Q., Adelmann, L., Wittbrodt, J., 2019. Enhanced in vivo-imaging in medaka by optimized anaesthesia, fluorescent protein selection and removal of pigmentation. PLoS One 14 (3), e0212956.

Ma, Y., Chen, W., Zhang, X., et al., 2016. Increasing the efficiency of CRISPR/Cas9-mediated precise genome editing in rats by inhibiting NHEJ and using Cas9 protein. RNA Biol. 13, 605–612.

Maruyama, K., Yasumasu, S., Iuchi, I., 2002. Characterization and expression of embryonic and adult globins of the teleost *Oryzias latipes* (Medaka)1. J. Biochem. 132, 581–589.

Matsuzaki, Y., Hosokai, H., Mizuguchi, Y., Fukamachi, S., Shimizu, A., Saya, H., 2013. Establishment of HRASG12V transgenic medaka as a stable tumor model for in vivo screening of anticancer drugs. PLoS One 8, e54424.

Mcvey, M., Lee, S.E., 2008. MMEJ repair of double-strand breaks (director's cut): deleted sequences and alternative endings. Trends Genet. 24, 529–538.

Miyanishi, H., Inokuchi, M., Nobata, S., Kaneko, T., 2016. Past seawater experience enhances seawater adaptability in medaka, *Oryzias latipes*. Zool. Lett. 2, 12.

Mosimann, C., Puller, A.-C., Lawson, K.L., Tschopp, P., Amsterdam, A., et al., 2013. Site-directed zebrafish transgenesis into single landing sites with the phiC31 integrase system. Dev. Dynam. 242, 949–963.

Murakami, Y., Ansai, S., Yonemura, A., Kinoshita, M., 2017. An efficient system for homology-dependent targeted gene integration in medaka (*Oryzias latipes*). Zool. Lett. 3, 10.

Nakade, S., Tsubota, T., Sakane, Y., et al., 2014. Microhomology-mediated end-joining-dependent integration of donor DNA in cells and animals using TALENs and CRISPR/Cas9. Nat. Commun. 5, 5560.

Nakai, J., Ohkura, M., Imoto, K., 2001. A high signal-to-noise Ca^{2+} probe composed of a single green fluorescent protein. Nat. Biotechnol. 19, 137–141.

Nakajo, M., Kanda, S., Karigo, T., Takahashi, A., Akazome, Y., Uenoyama, Y., Kobayashi, M., Oka, Y., 2018. Evolutionally conserved function of kisspeptin neuronal system is nonreproductive regulation as revealed by nonmammalian study. Endocrinology 159, 163–183.

Nakamura, S., Kobayashi, K., Nishimura, T., Higashijima, S., Tanaka, M., 2010. Identification of germline stem cells in the ovary of the teleost medaka. Science 328, 1561–1563.

Nakamura, S., Saito, D., Tanaka, M., 2008. Generation of transgenic medaka using modified bacterial artificial chromosome. Dev. Growth Differ. 50, 415–419.

Naruse, K., 2011. Genetics, genomics, and biological resources in the medaka, *Oryzias latipes*. In: Medaka: A Model for Organogenesis, Human Disease, and Evolution, pp. 19–37.

Ochiai, H., Sakamoto, N., Fujita, K., et al., 2012. Zinc-finger nuclease-mediated targeted insertion of reporter genes for quantitative imaging of gene expression in sea urchin embryos. Proc. Natl. Acad. Sci. U. S. A. 109, 10915–10920.

Ogino, Y., Hirakawa, I., Inohaya, K., Sumiya, E., Miyagawa, S., Denslow, N., Yamada, G., Tatarazako, N., Iguchi, T., 2014. Bmp7 and Lef1 are the downstream effectors of androgen signaling in androgen-induced sex characteristics development in medaka. Endocrinology 155, 449–462.

Ogiwara, K., Fujimori, C., Rajapakse, S., Takahashi, T., 2013. Characterization of luteinizing hormone and luteinizing hormone receptor and their indispensable role in the ovulatory process of the medaka. PLoS One 8, e54482.

Ogiwara, K., Ikeda, T., Takahashi, T., 2010. A new in vitro ovulation model for medaka based on whole ovary culture. Zool. Sci. 27, 762–767.

Ohkura, M., Sasaki, T., Sadakari, J., Gengyo-Ando, K., Kagawa-Nagamura, Y., Kobayashi, C., Ikegaya, Y., Nakai, J., 2012. Genetically encoded green fluorescent Ca^{2+} indicators with improved detectability for neuronal Ca^{2+} signals. PLoS One 7.

Okuyama, T., Isoe, Y., Hoki, M., Suehiro, Y., Yamagishi, G., Naruse, K., Kinoshita, M., Kamei, Y., Shimizu, A., Kubo, T., Takeuchi, H., 2013. Controlled Cre/loxP site-specific recombination in the developing brain in medaka fish, *Oryzias latipes*. PLoS One 8, e66597.

Okuyama, T., Yokoi, S., Abe, H., Isoe, Y., Suehiro, Y., Imada, H., Tanaka, M., Kawasaki, T., Yuba, S., Taniguchi, Y., Kamei, Y., Okubo, K., Shimada, A., Naruse, K., Takeda, H., Oka, Y., Kubo, T., Takeuchi, H., 2014. A neural mechanism underlying mating preferences for familiar individuals in medaka fish. Science 343, 91–94.

Ono, Y., Uematsu, T., 1957. Mating Ethogram in *Oryzias latipes* (With 1 Text-Figure).

Ozato, K., Kondoh, H., Inohara, H., Iwamatsu, T., Wakamatsu, Y., Okada, T.S., 1986. Production of transgenic fish: introduction and expression of chicken delta-crystallin gene in medaka embryos. Cell Differ. 19, 237–244.

Pendergrass, P., Schroeder, P., 1976. The ultrastructure of the thecal cell of the teleost, *Oryzias latipes*, during ovulation *in vitro*. Reproduction 47, 229.

Rembold, M., Lahiri, K., Foulkes, N.S., Wittbrodt, J., 2006. Transgenesis in fish: efficient selection of transgenic fish by co-injection with a fluorescent reporter construct. Nat. Protoc. 1, 1133–1139.

Rubin, E., Lithwick, G., Levy, A.A., 2001. Structure and evolution of the hAT transposon superfamily. Genetics 158, 949–957.

Saiki, A., Tamura, M., Matsumoto, M., Katowgi, J., Watanabe, A., Onitake, K., 1997. Establishment of in vitro spermatogenesis from spermatocytes in the medaka, *Oryzias latipes*. Dev. Growth Differ. 39, 337–344.

Sakaizumi, M., Moriwaki, K., Egami, N., 1983. Allozymic variation and regional differentiation in wild populations of the fish *Oryzias latipes*. Copeia 311–318.

Sakuma, T., Nakade, S., Sakane, Y., Suzuki, K.T., Yamamoto, T., 2016. MMEJ-assisted gene knock-in using TALENs and CRISPR-Cas9 with the PITCh systems. Nat. Protoc. 11, 118–133.

Sano, K., Kawaguchi, M., Katano, K., Tomita, K., Inokuchi, M., Nagasawa, T., Hiroi, J., Kaneko, T., Kitagawa, T., Fujimoto, T., Arai, K., Tanaka, M., Yasumasu, S., 2017. Comparison of egg envelope thickness in teleosts and its relationship to the sites of ZP protein synthesis. J. Exp. Zool. B Mol. Dev. Evol. 328, 240–258.

Sasado, T., Tanaka, M., Kobayashi, K., Sato, T., Sakaizumi, M., Naruse, K., 2010. The National BioResource Project Medaka (NBRP Medaka): an integrated bioresource for biological and biomedical sciences. Exp. Anim. 59, 13–23.

Shaner, N.C., Campbell, R.E., Steinbach, P.A., Giepmans, B.N., Palmer, A.E., Tsien, R.Y., 2004. Improved monomeric red, orange and yellow fluorescent proteins derived from *Discosoma* sp. red fluorescent protein. Nat. Biotechnol. 22, 1567–1572.

Shimada, A., Kawanishi, T., Kaneko, T., Yoshihara, H., Yano, T., Inohaya, K., Kinoshita, M., Kamei, Y., Tamura, K., Takeda, H., 2013. Trunk exoskeleton in teleosts is mesodermal in origin. Nat. Commun. 4, 1639.

Shimizu, A., Shimizu, N., 2012. Dual promoter expression system with insulator ensures a stringent tissue-specific regulation of two reporter genes in the transgenic fish. Transgenic Res. 22, 435–444.

Stemmer, M., Thumberger, T., Del Sol Keyer, M., Wittbrodt, J., Mateo, J.L., 2015. CCTop: an intuitive, flexible and reliable CRISPR/Cas9 target prediction tool. PLoS One 10, e0124633.

Susaki, E.A., Tainaka, K., Perrin, D., Kishino, F., Tawara, T., Watanabe, T.M., Yokoyama, C., Onoe, H., Eguchi, M., Yamaguchi, S., Abe, T., Kiyonari, H., Shimizu, Y., Miyawaki, A., Yokota, H., Ueda, H.R., 2014. Whole-brain imaging with single-cell resolution using chemical cocktails and computational analysis. Cell 157, 726–739.

Symington, L.S., Gautier, J., 2011. Double-strand break end resection and repair pathway choice. Annu. Rev. Genet. 45, 247–271.

Takehana, Y., Sakai, M., Narita, T., Sato, T., Naruse, K., Sakaizumi, M., 2016. Origin of boundary populations in medaka (*Oryzias latipes* species complex). Zool. Sci. 33, 125–131.

Taniguchi, Y., Takeda, S., Furutani-Seiki, M., Kamei, Y., Todo, T., Sasado, T., Deguchi, T., Kondoh, H., Mudde, J., Yamazoe, M., Hidaka, M., Mitani, H., Toyoda, A., Sakaki, Y., Plasterk, R.H., Cuppen, E., 2006. Generation of medaka gene knockout models by target-selected mutagenesis. Genome Biol. 7, R116.

Tao, W., Evans, B.G., Yao, J., Cooper, S., Cornetta, K., Ballas, C.B., Hangoc, G., Broxmeyer, H.E., 2007. Enhanced green fluorescent protein is a nearly ideal long-term expression tracer for hematopoietic stem cells, whereas DsRed-express fluorescent protein is not. Stem Cells 25, 670–678.

Tasic, B., Hippenmeyer, S., Wang, C., Gamboa, M., Zong, H., et al., 2011. Site-specific integrase-mediated transgenesis in mice via pronuclear injection. Proc. Natl. Acad. Sci. U. S. A. 108, 7902–7907.

Thorpe, H.M., Smith, M.C., 1998. In vitro site-specific integration of bacteriophage DNA catalyzed by a recombinase of the resolvase/invertase family. Proc. Natl. Acad. Sci. U. S. A. 95, 5505–5510.

Tomihara, S., Oka, Y., Kanda, S., 2021. Establishment of open-source semi-automated behavioral analysis system and quantification of the difference of sexual motivation between laboratory and wild strains. Sci. Rep. 11, 10894.

Tong, C., Li, P., Wu, N.L., Yan, Y., Ying, Q.L., 2010. Production of p53 gene knockout rats by homologous recombination in embryonic stem cells. Nature 467, 211–213.

Tsutsui, H., Karasawa, S., Shimizu, H., Nukina, N., Miyawaki, A., 2005. Semi-rational engineering of a coral fluorescent protein into an efficient highlighter. EMBO Rep. 6, 233–238.

Venken, K.J.T., He, Y., Hoskins, R.A., Bellen, H.J., 2006. P[acman]: a BAC transgenic platform for targeted insertion of large DNA fragments in *D. melanogaster.* Science 314, 1747–1751.

Vopalensky, P., Ruzickova, J., Pavlu, B., Kozmik, Z., 2010. A lens-specific co-injection marker for medaka transgenesis. Biotechniques 48, 235–236.

Wakamatsu, Y., Pristyazhnyuk, S., Kinoshita, M., Tanaka, M., Ozato, K., 2001. The see-through medaka: a fish model that is transparent throughout life. Proc. Natl. Acad. Sci. U. S. A. 98, 10046–10050.

Walter, R.O., Hamilton, J.B., 1970. Head-up movements as an indicator of sexual unreceptivity in female medaka, *Oryzias latipes*. Anim. Behav. 18, 125–127.

Warming, S., 2005. Simple and highly efficient BAC recombineering using galK selection. Nucleic Acids Res. 33, e36.

Watakabe, I., Hashimoto, H., Kimura, Y., Yokoi, S., Naruse, K., Higashijima, S.I., 2018. Highly efficient generation of knock-in transgenic medaka by CRISPR/Cas9-mediated genome engineering. Zool. Lett. 4, 3.

Yamamoto, T., 1953. Artificially induced sex-reversal in genotypic males of the medaka (*Oryzias latipes*). J. Exp. Zool. 123, 571–594.

Yokoi, S., Naruse, K., Kamei, Y., Ansai, S., Kinoshita, M., Mito, M., Iwasaki, S., Inoue, S., et al., 2020. Sexually dimorphic role of oxytocin in medaka mate choice. Proc. Natl. Acad. Sci. U. S. A. 117 (9), 4802–4808.

Yokoi, S., Okuyama, T., Kamei, Y., Naruse, K., Taniguchi, Y., Ansai, S., Kinoshita, M., Young, L.J., et al., 2015. An essential role of the arginine vasotocin system in mate-guarding behaviors in triadic relationships of medaka fish (*Oryzias latipes*). PLoS Genet. 11 (2), e1005009.

Chapter 10

Integrated analyses using medaka as a powerful model animal toward understanding various aspects of reproductive regulation

Chie Umatani[1,*], Mikoto Nakajo[1,2,*], Daichi Kayo[3], Yoshitaka Oka[1], Shinji Kanda[4]

[1]*Department of Biological Sciences, Graduate School of Science, The University of Tokyo, Tokyo, Japan;* [2]*Department of Physiology, Division of Life Sciences, Faculty of Medicine, Osaka Medical and Pharmaceutical University, Takatsuki, Osaka, Japan;* [3]*Department of Aquatic Bioscience, Graduate School of Agricultural and Life Sciences, The University of Tokyo, Tokyo, Japan;* [4]*Laboratory of Physiology, Atmosphere and Ocean Research Institute, The University of Tokyo, Kashiwa, Chiba, Japan*

Medaka as a powerful model animal for research fields of reproduction

As we mentioned in the previous chapter, medaka is one of the most useful model species in terms of genetic manipulation due to its resourceful genome database and biological features for the application of gene-editing tools. In this chapter, we will describe how medaka is suitable for research in neuroendocrinology, especially in the study area of reproduction. As indicated in the previous chapter, medaka have a short life cycle (hatching within 2 weeks after fertilization) and grow up to be reproductive within 2–3 months when raised in long daylight conditions (14 h of light/10 h of dark), which means they are regarded as seasonal breeders. Their reproductive status is also controlled at ~27°C by daylight conditions (long day: reproductive, short day: nonreproductive). Moreover, medaka show very typical patterns of sexual behavior or courtship behavior and they spawn 10–20 fertilized eggs every day (normally in the beginning of the light period). Each component of their sexual behavior

*CU and MN contributed equally to this work.

Laboratory Fish in Biomedical Research. https://doi.org/10.1016/B978-0-12-821099-4.00002-X
Copyright © 2022 Elsevier Inc. All rights reserved.

sequence is as follows. First, a sexually mature male medaka follows or chases a female individual and represents courtship behavior by quickly circling in front of the female (quick circle). If the female is responsive to the male's courtship, the male holds the female's body by its anal fin (clasping), which is followed by simultaneous spawning (Ono and Uematsu, 1957; Walter and Hamilton, 1970). This stereotypical pattern of sexual behavior enables us to analyze the timing and duration of each step of their reproductive behavior, success rate, and even sexual motivation of both males and females (Ikawa et al., 2017; Hiraki-Kajiyama et al., 2019). Therefore, by using medaka, we can easily manipulate their reproductive states, obtain their eggs on a daily basis, and quantitatively analyze their sexual behavior or courtship.

In contrast, the interspawning period of zebrafish is approximately 1.9–6 days, and we cannot control their reproductive state, suggesting that they are opportunistic breeders (Lawrence, 2007; Uusi-Heikkila et al., 2012). Taken together, these features related to reproduction indicate that medaka has an advantage over zebrafish or other teleosts as a useful model species in the study of reproductive systems, the so-called hypothalamic–pituitary–gonadal (HPG) axis regulation in coordination with sexual behavior.

Mechanisms of reproductive systems clarified by histological and physiological analyses

For an understanding of neuronal systems, there are three important factors to consider: localization of cell bodies, axonal projection, and synaptic signaling. A neuron is localized in some specific nucleus (nuclei) in the central nervous system, and it projects its neuronal fiber(s) (axons or dendrites) to target cell(s). In addition, the neuron and its target cells express ligands and corresponding receptors. When the neuron is excited, its ligands are released as neurotransmitters from its terminal to the target cells (synaptic signaling). Considering these complex features of the neuronal system, it is necessary for neuroscientists to use various experimental methods to reveal each aspect of the system and to understand the entire structure of neuronal circuits. For example, to analyze axonal projection and localization of cell bodies, histological techniques such as immunohistochemistry (IHC) and in situ hybridization (ISH) are useful. In these methods, medaka brain samples are fixed so that we can analyze "static" structure in a "dead" state. On the other hand, physiological methods such as the patch clamp technique and Ca^{2+} imaging are necessary to analyze neuronal activities in a "live" state. In addition, hormonal assay or treatment normally provides data at an individual level. Moreover, for further study to clarify neuronal systems and their functions, it is very effective to apply knockout (KO) lines, which lack specific genes coding neurotransmitters or their receptors of interest, and transgenic (Tg) lines, which express green fluorescent protein (GFP) (fluorescence label) or Ca^{2+} indicator in the target neurons. With a combination of these methods that

TABLE 10.1 Categories of experimental methods for medaka described in this chapter.

Method	Histology	Hormonal assay	Neurophysiology
Individual level	○ (larva/embryo)	○	–
Intercellular level	○	–	○
Cellular level	○	–	○
Live sample?	No	Yes	Yes

Circles indicate that the method covers the biological level. Dashes indicate that the method is usually not suitable. Histology enables us to analyze cellular localization and intercellular connection such as neuronal projection. When we use larvae or embryos, we can obtain the data at the individual level. Hormonal assay mainly represents phenomena at the individual level. For the analyses at cellular-level functions, we dissect some specific tissue before the application of the following techniques. Neurophysiology provides us with data of cellular-to-intercellular activities in live cells/tissues. However, methods for the individual level such as in vivo imaging have not yet been applied to medaka. Future study will provide us a novel neurophysiological strategy at this level.

covers different biological levels (Table 10.1), we can acquire the entire picture of a neuronal system. In the following sections, we will show powerful histological tools, hormonal assays, and physiological techniques that can be applied to medaka.

Histological studies using medaka

Histological techniques such as IHC, ISH, and neuronal tracing have been established and are commonly used in neurobiological fields. In addition, recent establishment of Tg animals whose neurons of interest are specifically labeled by fluorescent proteins such as GFP and red fluorescent protein (RFP) provides us various insights into neuronal pathways. For medaka, almost all histological methods established in mammals can be applied with some minor modifications, and we use them for a variety of purposes. Here, we will introduce the histological methods we usually employ in medaka, and their features, merits, and demerits will be described (see also Fig. 10.1).

From histological data, we can obtain information on the localization of neurons of interest, colocalization of other genes in the neurons of interest, projection sites of the neurons, and interaction between the neurons and their target cells in the brain. This is also the case with endocrine cells in the pituitary, so it is possible to analyze the localization of each pituitary-hormone-releasing cell and the axonal projection to them from the brain.

For most histological analyses, we often use perfusion fixation and cryosection. Perfusion fixation enables us to quickly and evenly fix the tissue and to reduce the noise caused by hematocytes and serum. Here, we introduce a general protocol from perfusion fixation to cryosectioning.

Info. \ Method	A. Immunohisto-chemistry	B. in situ hybridization	C. Single-neuron labeling	D. GFP-labeled transgenic
Localization (cell body)	○	○	○	○
Projection (fibers)	△	×	○	○
Specificity	△	○	○	△
Easy to apply?	○	○	×	△
Representative photo	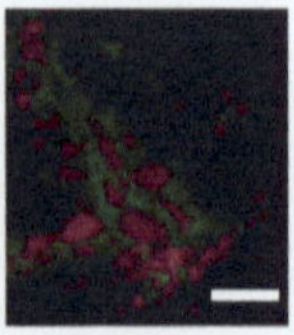	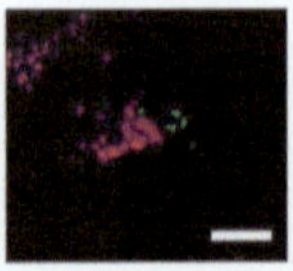	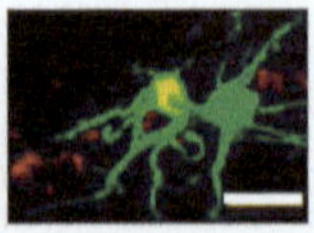	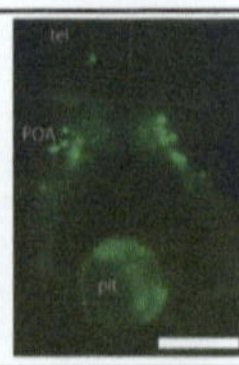

FIGURE 10.1 Comparison of histological methods for medaka described in this section. A circle indicates that the method is applicable, while a cross indicates that the method is not applicable. Triangles indicate that under certain conditions the method can detect the cellular structures or can be suitable for the condition. In the lowest line, the representative photos involving each method are shown. (A) Dual immunohistochemistry of green fluorescent protein (GFP) and isotocin (bar: 10 μm; Nakajo et al., 2018). (B) Dual in situ hybridization of *gnrh1* and *gpr54-1* (bar: 50 μm; Kanda et al., 2013). (C) Single neuron labeling by biocytin in GnRH2-ir neurons in dwarf gourami (bar: 20 μm; Kanda et al., 2010). (D) Raw GFP fluorescence of *gnrh1*:gfp transgenic medaka (bar: 100 μm; modified from Takahashi et al., 2015). Note that single neuronal labeling can also be used in medaka, but, generally, it is much easier and more effective to detect the neurons of interest using GFP-labeled transgenic medaka. *pit*, Pituitary; *POA*, preoptic area; *Tel*, telencephalon. *Photographs in (A) and (C) are reprinted with permission of Oxford University Press, (B) is reprinted with permission of PLOS, and (D) is reprinted with permission of John Wiley & Sons, Inc.*

Medaka are deeply anesthetized by MS-222 (Sigma–Aldrich, St. Louis, MO), and are pinned on styrofoam. Then, the hearts are carefully exposed by forceps on the chamber for fixation. We use a glass needle prepared from a glass capillary (G-1.5; Narishige, Tokyo, Japan) with a micropipette puller (P-97; Sutter Instruments, Novato, CA). It is important to break the tip to increase the flow of fixative before use. The needle is connected to a syringe with an ~80-cm-long silicone tube, and about 5 mL of fixative containing 4% paraformaldehyde in phosphate-buffered saline (PBS) is loaded into the syringe (Fig. 10.2A). After air is expelled from the syringe, the needle is inserted into the ventricle of the heart (Fig. 10.2B). After application of slight pressure, the ventricle is inflated (Fig. 10.2C). With this pressure, the needle is brought forward to the bulbus arteriosus, and the bulbus arteriosus is inflated (Fig. 10.2D). Finally, we apply some more pressure to let the fixative infuse throughout the body (Fig. 10.2E). We hold the needle until

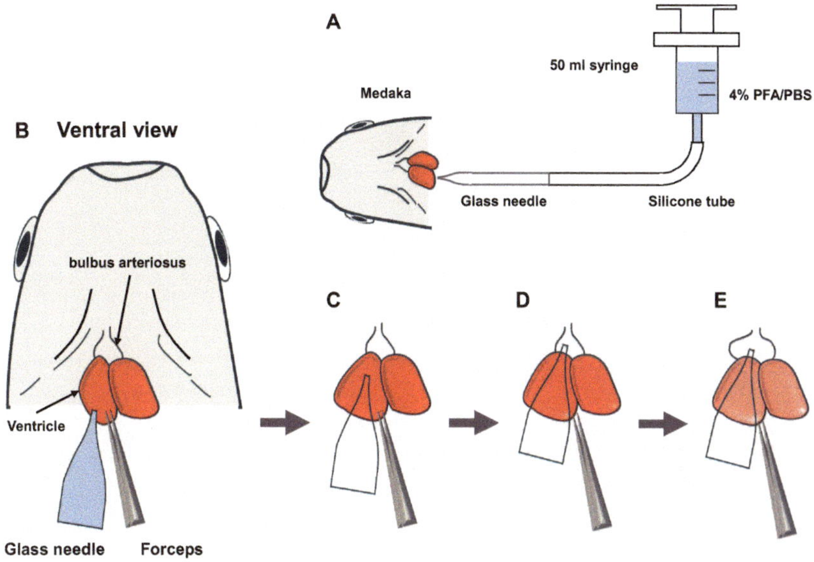

FIGURE 10.2 **Protocol of perfusion fixation for medaka and other small-sized fish. (A) Setup of perfusion equipment. A 50-mL syringe connected to a silicone tube and glass capillary needle with its tip broken are used for this protocol. (B–E) Schematic illustration of the method for inserting the pipette and perfusing fixative solution into the heart. First, the medaka is oriented ventral side up with its heart exposed by forceps and the ventricle is pierced with the glass capillary needle while holding the edge of the ventricle with forceps (B). Next, slight pressure is applied by pushing the plunger of the syringe and the ventricle is then inflated (C). After inflation of the ventricle, the pipette is moved gently toward the bulbus arteriosus and its slight inflation is confirmed (D). With some additional pressure, the bulbus arteriosus is inflated until it is balloon shaped. The needle is held with continuous pressure for ~30 s so that the whole body is fixed (E). Fixation level is confirmed when the gill or tail turns pale. *PBS*, phosphate-buffered saline; *PFA*, paraformaldehyde.**

the entire body is fixated (~30 s), as determined from the color change of the gill or tail (blood vessels usually turn pale white). Next, the entire brain is excised in the tube filled with the same fixative for ~10 min. For cryosectioning, we next put the fixated sample in 30% sucrose in PBS (2–6 h at 4°C) and embed it in 5% ultralow gelling temperature agarose (Sigma–Aldrich)/20% sucrose/PBS in the small plastic mold. By embedding in this agarose instead of a commercially available medium, we can easily adjust the position of the sample in the mold. In addition, as the brains of fishes are softer than those of mammals, this agarose mounting medium is superior for sectioning compared to a mounting medium optimized to mammalian tissues. The agarose blocks are rapidly frozen by dipping them in the chilled *n*-hexane at −80°C and storing them until cryosectioning. After sections are prepared as described, they are used in the following histological experiments (Fig. 10.3A).

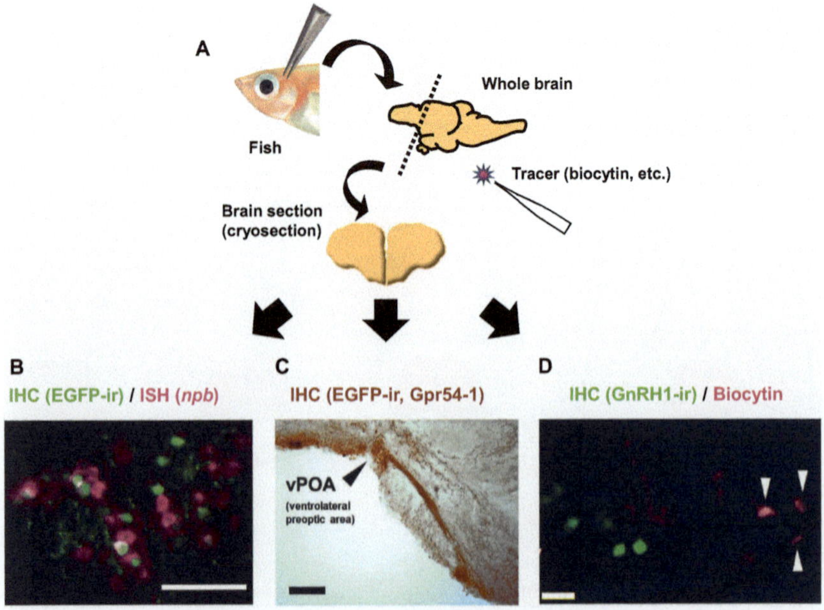

FIGURE 10.3 An overview of histological methods in medaka. (A) Workflow for preparing medaka brain samples. First, the whole brain is excised from the fish. Next, using cryostat, frozen sections of whole brain are made. The sample can be applied for various histological methods. For anterograde/retrograde neuronal tracing, the tracer is inserted into the neurons of interest or part of the brain or pituitary. (B–D) Representative photos using combinations of histological methods. (B) Dual labeling of immunohistochemistry (IHC) for EGFP-ir and in situ hybridization for *neuropeptide B* using *gpr54-1*:gfp transgenic medaka brain (bar: 50 μm; modified from Nakajo et al., 2018). (C) IHC using *gpr54-1*:gfp transgenic medaka clearly shows that EGFP-ir (*gpr54-1*-expressing) neurons in the ventrolateral preoptic area (vPOA) create a bundle of their axons and heavily project to the pituitary (bar: 100 μm; modified from Nakajo et al., 2018). Note that only for the photograph in (C), sagittal sections were used while for others frontal sections were used. (D) Retrograde tracing from pituitary to vPOA of the brain clearly shows that some GnRH1-ir neurons are biocytin labeled (*arrowheads*), suggesting that GnRH1 neurons project to the pituitary (bar: 25 μm; Karigo et al., 2012). *Photographs in (B–D) are reprinted with permission of Oxford University Press.*

Immunohistochemistry

In the IHC protocol, specific antibodies (antiserum) for the target peptide sequence are used. The target proteins are labeled not only in the cell body but also in their neuronal fibers because, in general, proteins are translated in the cell body and then transported throughout the fibers. Thus using the IHC technique, we can understand both cellular localization and the projection sites of the focused-upon neurons. Also, subcellular localization of the neurotransmitters and other molecules can be observed. We can apply both chromogenic and fluorescent signals for detection, and fluorescent signals are useful for analyses of colocalization and functional connectivity.

In IHC analysis, the critical point is the specificity of antibodies against the epitope. Without this verification, an antibody may experience cross-reactivity with similar but different epitopes. When we first obtain or prepare some antibodies (with the exception of the specificity of those that have already been reported in the same species), we have to scrutinize the specificity. The easiest way to accomplish this in medaka is to compare the immunoreactivity in KO medaka with that in wild type. The other way is through an examination using double labeling with ISH to check the colocalization of mRNA and immunoreactivity. After verifying the antibody's specificity, IHC will be effective in analyzing the neuronal pathways.

Likewise, when we generate a GFP (or other marker)-labeled Tg line, we also have to check through dual labeling of IHC/ISH to determine whether GFP-positive neurons are specifically colocalized with the signals of the target gene (mRNA) (Fig. 10.3B; Takahashi et al., 2015; Nakajo et al., 2018). After validating GFP (RFP)-labeled Tg lines, we can use commercially available anti-GFP (RFP) antibodies (anti-enhanced [E]GFP antibody raised in rat, catalog no. GF090R, Nacalai Tesque, Kyoto, Japan; anti-EGFP antibody raised in rabbit, A-11122, Thermo Fisher Scientific, Waltham MA; anti-EGFP antibody raised in rabbit #598, Medical and Biological Laboratories Co., Ltd., Nagoya Japan; anti-RFP antibody raised in rabbit, code no. PM005, Medical and Biological Laboratories Co., Ltd.). Since GFP (RFP) passively diffuses throughout the labeled neurons, GFP (RFP) signals enhanced by GFP (RFP)-IHC provide us with detailed morphological information, including their projection sites (Fig. 10.3C).

In situ hybridization

Here, we describe the use of and some tips associated with the application of ISH in medaka. For general protocols and principles of ISH, please refer to the other sources (textbooks, websites, and papers; for example, see Mezache et al., 2015). The first step of ISH is constructing the antisense RNA probe that hybridizes to mRNA coding the target gene. In our protocol, DIG/fluorsecein-labelled RNA probes transcribed from templates of PCR products or plasmids are designed to be approximately 1000 bases in length. If the length of the probe is shorter, the intensity of labeling becomes weaker. On the other hand, longer probes may decrease their accessibility in the cells. Thus we sometimes mix the probes that hybridize with different sites of the same target gene to acquire stronger signals. ISH with ~1000 bp antisense RNA probes is generally more reliable than IHC without scrutinization because it has less risk of cross-reactivity, except for cases in which the gene has a paralog whose sequence shows high homology. Moreover, the sensitivity of ISH is generally higher than that of IHC in our cases. However, as mRNA signals can be observed only in the cell body of neurons, it does not provide information on axonal projection of those neurons. Given these properties, ISH is useful for the analysis of (co-)localization of gene(s) at the cell-body level.

Single-cell labeling and tract tracing using axonal transport

Based on IHC results, multiple axonal fibers can be labeled so densely that we cannot distinguish fibers originating from a single neuron. In such cases, small molecular weight substances such as Neurobiotin (Vector Laboratories, Burlingame, CA), biocytin (Sigma−Aldrich), Lucifer yellow (Thermo Fisher Scientific, Waltham, MA), and cobaltic lysine (see Bazer and Ebbesson, 1984 for details) enable analyses of neuronal morphology and axonal projection at the single neuron level. For example, we often inject these molecules through the glass micropipette to label the entire morphology and a single neuron after electrophysiological recordings. Before the establishment of GFP-labeled Tg medaka lines, single-cell labeling in combination with IHC used to be a common method for *post hoc* identification of recorded cells by the patch clamp technique (see later) (e.g., Oka and Matsushima, 1993; Kanda et al., 2010). Besides, Neurobiotin, biocytin, and cobaltic lysine can also be used for anterograde and retrograde tracing. Although, to the best of our knowledge, anterograde tracing of neurons using medaka has not yet been reported, there are some reports using large fish (Oka et al., 1986; Shiga et al., 1989; Ito et al., 2003). For instance, Shiga et al. demonstrated retinal projections in the brain of hime salmon by removing an eyeball and injecting cobaltic lysine into the cut end of optic nerves. On the other hand, for retrograde tracing of cells of origin projecting to the pituitary in medaka, a biocytin crystal was inserted into the pituitary, which was retrogradely transported to label hypophysiotropic neurons and their axons (Fig. 10.3D; Karigo et al., 2012).

Conventional histological staining

In addition to the aforementioned gene/protein-specific labeling methods, conventional histological staining is also widely used. For instance, hematoxylin and eosin (HE) staining is used to observe the morphology of various tissues. To date, histology of the body organs and nomenclature of nuclei in the brain have been intensively studied and documented (http://theses.vet-alfort.fr/Th_multimedia/medaka/htm/intro/introa/fishes.htm, https://shigen.nig.ac.jp/medaka/medaka_atlas/index.html).

To examine the sexual maturation of the individual or to compare maturation level between wild-type and some KO medaka, we perform HE staining of the gonads of each fish using the following method. The gonads are fixed with Bouin's solution and dehydrated with a series of ethanol soluitions with ascending concentrations and cleared with xylene, followed by paraffin embedding and sectioning. Then, the sections are stained by HE. After that, gonadal-stage or germ-cell development is identified by comparing the HE-stained sections of gonads and the foregoing reference (Takahashi et al., 2016; Kayo et al., 2019; Nakajo et al., 2018).

The present scheme of HPG axis regulation clarified by multidisciplinary analyses in vertebrates

As an example of a study using adult medaka, we will introduce how the mechanism of the HPG axis regulation has been revealed mainly by histological and physiological approaches in combination with genetic tools. Since Geoffrey Harris first put forward the conception that the hypothalamus controls the pituitary gland by some neurohumoral transmission at the end of 1940s, the search for a "hypothalamic releasing factor" was under way (Harris, 1948; Watts, 2015). After the discovery of gonadotropin-releasing hormone (GnRH) at the beginning of the 1970s (Amoss et al., 1971; Schally et al., 1971; Matsuo et al., 1971), HPG axis regulation has been intensively studied. Thus this neuroendocrine system for regulation of gonadal maturation is so far considered to be common throughout vertebrates. GnRH, which consists of 10 amino acids (decapeptide), is released from GnRH neurons localized in the basal hypothalamic and/or ventral preoptic area (POA). Many previous studies have shown that GnRH neurons in mammals project to the median eminence and release GnRH peptide which is then transported via the portal vessels, while those in teleosts project directly to the pituitary (hypophysiotropic) and release GnRH there, and GnRH peptide facilitates the release of two types of gonadotropins, follicle-stimulating hormone (FSH) and luteinizing hormone (LH), from the pituitary (Pineda et al., 2010; Karigo et al., 2014; Juntti and Fernald, 2016; Herbison, 2016). To date, it is generally agreed that GnRH neurons form the final common pathway for HPG axis regulation throughout vertebrates, although a recent analysis using KO medaka of *gnrh*, *fshb*, and *lhb* showed there are some differences between mammals and teleosts in terms of FSH/LH functions and GnRH regulation on them (Kanda, 2019).

Since the GnRH neurons lack a subtype of estrogen receptor (Esr) that is essential for feedback regulation (Esr1), the existence of neurons that express Esr1 and regulate GnRH neurons has been assumed in mammals for some time. In the early 2000s, a novel neuropeptide, kisspeptin, which is the endogenous ligand of an orphan G protein-coupled receptor, Gpr54, was identified, and there has been a growing body of evidence showing that it plays a crucial role in the HPG axis regulation in mammalian species (Kotani et al., 2001; Ohtaki et al., 2001; Han et al., 2005; Pineda et al., 2010; Herbison, 2016). In mammals, hypothalamic kisspeptin-expressing neurons (Kiss1 neurons) coexpress Esr1 and show clear estrogen sensitivity (Franceschini et al., 2006; Smith et al., 2006; Adachi et al., 2007; Dungan et al., 2007; Frazao et al., 2013). Kiss1 neurons release Kiss1 peptide and regulate GnRH neurons directly via kisspeptin receptor Gpr54, which is expressed in GnRH neurons in an estrogen-dependent manner. Consistent with these histological and physiological findings, genetic studies have demonstrated that loss of function of *Kiss1* or *Gpr54* in mammals, including humans, causes severe infertility due to deficiency of gonadal development derived from the lack of expression and/or release of FSH/LH (de Roux et al., 2003; Funes et al., 2003; Seminara et al., 2003; Franceschini et al., 2006; Smith et al., 2006; Dungan et al., 2007; D'anglemont de Tassigny et al., 2007; Frazao et al., 2013). Furthermore, it has been revealed that the kisspeptin-related genes, which consist of *kiss1/2* and *gpr54-1/2*, are highly conserved (at least either paralog) among almost all vertebrates except birds. In addition, estrogen sensitivity of kisspeptin neurons are also highly conserved (Lee et al., 2009; Akazome et al., 2010; Um et al., 2010; Servili et al., 2011; Tena-Sempere et al., 2012; Kim et al., 2012; Kanda and Oka, 2012, 2013; Kanda et al., 2012).

Differences between mammals and teleosts in terms of function of the kisspeptin neuronal system and HPG axis regulation

Against this background, many researchers using nonmammalian species, especially teleosts, have tried examining possible kisspeptin functions on HPG axis regulation. However, many of these studies have shown different results from those in mammals. Recent ISH studies using different teleost species such as the African cichlid (*Astatotilapia burtoni*) (Grone et al., 2010), the European sea bass (*Dicentrarchus labrax*) (Escobar et al., 2013), and medaka (Kanda et al., 2013) have clearly demonstrated that GnRH1 neurons lack expression of *gpr54* mRNA, which is one of the essential properties of GnRH neurons to be controlled by kisspeptin neurons in mammals. Moreover, recent studies using gene-editing tools have shown that kisspeptin-related gene KO zebrafish (*Danio rerio*) and medaka are fertile (Tang et al., 2015; Nakajo et al., 2018). Furthermore, our previous study confirmed that in vivo kisspeptin administration via intracerebroventricular and intraperitoneal pathways on wild-type goldfish did not alter serum LH concentration (Nakajo et al., 2018). Taken together, a growing body of evidence has demonstrated that the kisspeptin neuronal system is not involved in HPG axis regulation in teleosts (Fig. 10.4).

Another difference in the kisspeptin neuronal system and HPG axis regulation is the existence/absence of pulsatile release of GnRH and LH and the contribution of kisspeptin neurons. It has been reported that pulsatile release of GnRH and LH is required for folliculogenesis in mammalian species. Because kisspeptin-related gene KO mice exhibited disrupted LH pulse and normal folliculogenesis, kisspeptin is considered to be essential for LH pulse generation in mammals (Smith et al., 2006; Adachi et al., 2007; Dungan et al., 2007; Ohkura et al., 2009; Okamura et al., 2013). In contrast, recent studies regarding the HPG axis regulation in medaka and zebrafish have

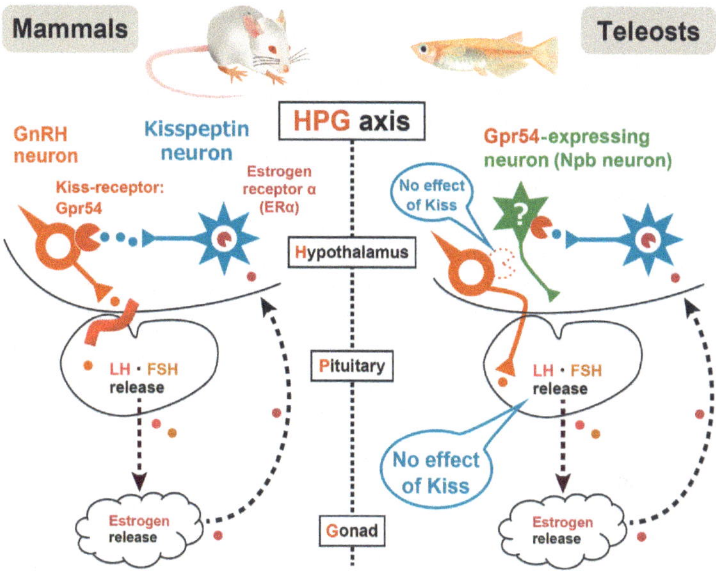

FIGURE 10.4 **Schematic illustrations for the hypothalamic–pituitary–gonadal (HPG) axis regulation and kisspeptin neuronal system in mammals and teleosts. In mammalian species (*left*), it has been generally accepted that hypothalamic kisspeptin (Kiss1) neurons, which possess estrogen sensitivity to receive gonadal estrogen signals by coexpression of estrogen receptor α (ERα), directly stimulate gonadotropin-releasing hormone (GnRH) neurons via a kisspeptin receptor, Gpr54. In turn, GnRH neurons stimulate the release of two types of gonadotropins, follicle-stimulating hormone (FSH) and luteinizing hormone (LH), in the pituitary. On the other hand, in teleost species (*right*), the kisspeptin neuronal system is involved in the regulation of neither hypophysiotropic GnRH neurons nor gonadotrophs. Nevertheless, hypothalamic estrogen-sensitive Kiss1 neurons are common between mammals and teleosts, suggesting some important common roles of Gpr54-expressing neurons. Here, to find the significance of the Kiss1 neural system in nonmammals, our recent study analyzed the Kiss1-Gpr54-1 system in medaka by various methods and suggested that some of the Gpr54-1-expressing neurons are neuropeptide B (Npb) neurons.**

suggested that GnRH and LH are essential for ovulation but not for folliculogenesis (Takahashi et al., 2016; Liu et al., 2017). Furthermore, the occurrence of LH pulses has not been reported in any nonmammalian species, and LH has been demonstrated to be dispensable for folliculogenesis in teleosts (Takahashi et al., 2016; Liu et al., 2017). Therefore, animals that do not show LH pulse may not require kisspeptin regulation of GnRH neurons, which is consistent with the fact above that GnRH neurons lack Gpr54 in teleosts (Grone et al., 2010; Escobar et al., 2013; Kanda et al., 2013). Considering these findings and the fact that avian species have survived without the kisspeptin neuronal system, it should be natural to consider that kisspeptin effects on GnRH neurons are only limited to mammals.

Various novel functions of the kisspeptin neuronal system in teleosts

Unlike in mammals, the involvement of the kisspeptin neuronal system in HPG axis regulation seems to be absent in nonmammalian species, at least in teleosts (Kanda, 2019). Therefore a recent study involving medaka searched for the possible biological functions of kisspeptin neurons in teleosts (Nakajo et al., 2018). In this study, the histological and physiological analysis of the Tg medaka whose *gpr54-1*-expressing neurons are specifically labeled by GFP demonstrated the following novel functions of the kisspeptin neuronal system. First, it was demonstrated that the Kiss1 neuronal system is involved in the regulation of the neurons that express neuropeptide B (Npb), another neuropeptide. RNA sequencing of *gpr54-1*-expressing cells suggested that those neurons express *npb*, and this result was confirmed by dual labeling of IHC for GFP (Gpr54-1) and ISH (*npb*). Furthermore, dual IHC data showed that Npb/Gpr54-1-expressing neurons in POA heavily project to the pituitary, and their axons are intertwined with those of isotocin/vasotocin (orthologous products of mammalian oxytocin and vasopressin, respectively) releasing neurons at the pituitary. These results suggested the possible involvement of the Kiss1 system in the regulation of isotocin and vasotocin release via Npb. Another previous study involving sea bass reported that the kisspeptin neuronal system may regulate neurotransmitters such as neuropeptide Y, somatostatin, and dopamine, suggesting that there is a variety of functions of the kisspeptin neuronal system (Escobar et al., 2013).

In conclusion, for understanding the kisspeptin neuronal system, in spite of the differences between mammals and nonmammals as for HPG axis regulation, medaka made a substantial contribution as a powerful nonmammalian model animal. The medaka is not only useful for histological and neurophysiological studies but also for hormonal assays as described in the following sections.

Endocrine studies using medaka

In animals, hormones play an important role in communication among organs. They are produced mainly by the endocrine glands, released into the bloodstream, and received by distant target organs as signal molecules. Thus, concentration of the hormones in the blood has been measured for improving physiological and pathological understanding. Moreover, artificial control of hormones involving the surgical removal of the endocrine glands and/or administration of hormones has also been widely used to analyze the physiological role of hormones. However, to date, such techniques have been used mainly in large animals. Here, we introduce techniques of hormonal measurement and manipulation in medaka, which, unlike large fishes, is also amenable to genetic tools.

Surgery and blood sampling

Surgical removal of the endocrine glands is a classical method of experimental biology. To our knowledge, the first study that applied surgery to medaka dates back to over 80 years ago (Nagata, 1936), in which the ovary was surgically removed (ovariectomy, OVX). In this paper, the abdominal incision was not surgically sutured, and only 15% of the OVX medaka survived post-operatively. Today, the survival rate after OVX has been dramatically improved (90%–100%) by suturing the abdomen with very thin nylon thread (Kanda et al., 2008). The methods of OVX as an example of surgeries in medaka are as follows.

OVX involves three sequential steps: abdominal opening, removal of the ovary, and surgical suturing. First, a medaka is anesthetized by 0.02% MS-222, and its abdomen is incised using a razor. Next, the ovary is pinched and removed by fine forceps with super-thin tips (e.g., Dumont Forceps #5; Manufactures D'Outils Dumont SA, Switzerland). Finally, the abdomen is sutured using very thin nylon thread (diameter, 20–30 μm). For at least 2 days postoperatively, the medaka should be kept in tank water containing 0.9% NaCl, with an osmolarity close to that of medaka's bodily fluids.

Blood is one of the most commonly used samples for hormonal assays. To collect the blood, we use puncture needles made from glass capillaries (1.0–1.5 mm outer diameter and 0.6–0.9 mm inner diameter), which are similar to the glass micropipettes used for microinjection. To allow smooth flow of the collecting blood, the needle's tip should be moderately broken with forceps. Next, this glass needle is connected to a commercially available aspirator tube assembly for microcapillary pipettes (Fig. 10.5). Note that this assembly can be replaced with a silicone tube that fits the outer diameter of the glass capillary. For blood sampling, a medaka is anesthetized and placed under a dissecting microscope. The inside of the glass needle prepared as described here is coated with an anticoagulant such as heparin sodium solution and inserted into the caudal vein while applying slight negative pressure by suction from the mouthpiece or the end of the silicone tube (Fig. 10.6). When the needle reaches the vein, blood starts to flow into the glass needle. Collected blood retrieved from the glass needle can be collected quantitatively using a capillary micropipette. In many assays, serum will be isolated to reduce the effect of hematocytes. If the volume of the collected blood is small, it is possible to use whole blood in the following assay instead of serum after confirming that the blood does not cause artifacts in the following assays. Quantitative analyses of hormones can be performed with a suitable method such as enzyme-linked immunosorbent assay (ELISA) or radioimmunoassay. Also, there are several methods for measuring small molecules such as glucose (Hasebe et al., 2016).

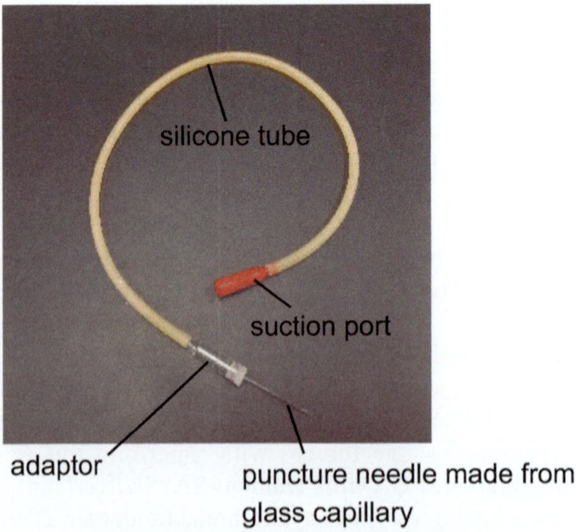

FIGURE 10.5 Blood collection instrument.

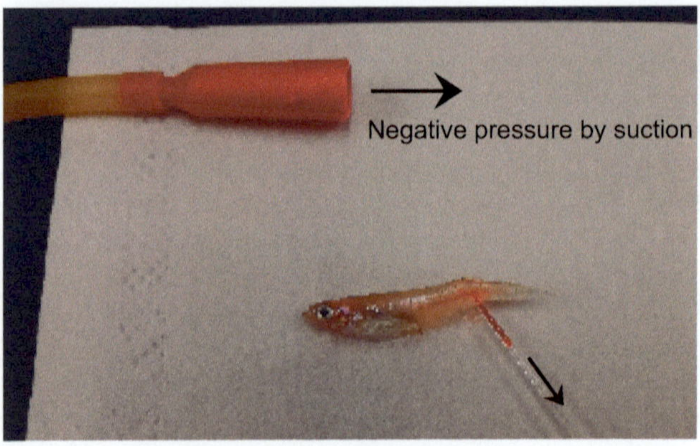

FIGURE 10.6 Blood collection steps.

Hormonal analysis of sex steroid hormones

Sex steroid hormones are typically made in the gonads and regulate sexual characteristics and the reproductive system. These hormones have a common molecular structure among species, and thus we can quantitate the amount of sex steroids in medaka using commercial products such as ELISA kits. Here, we introduce a study that analyzed the concentration of 17β-estradiol (E2), which is one of the major sex steroid hormones in vertebrates, in the blood using the techniques described previously (i.e., blood sampling and OVX) (Kayo et al., 2020). Blood E2 concentration of the sexually mature female medaka was measured at 4–13 ng/mL while that of the male medaka was 1–2 ng/mL. According to the previous studies using other teleosts, some teleosts in the breeding season showed blood E2 concentration close to that of the female medaka under breeding conditions: rainbow trout, 6–30 ng/mL (Zohar et al., 1986); Nile tilapia, 5–25 ng/mL (Tacon et al., 2000); and bitterling, 5–12 ng/mL (Shimizu et al., 1985). In contrast, others showed relatively low E2 concentrations of around 5 ng/mL: goldfish (Kobayashi et al., 1988) and goby (Kaneko et al., 1986).

Kayo et al. also reported that, depending on the method of E2 administration, the kinetics of blood E2 concentration in medaka are different. Some previous studies involving medaka used three methods of E2 administration: feeding (100 ng/day), water exposure (10 nM), and intraperitoneal implantation (25 μg in L1*W1*H1 [mm] silicon block/individual) (Kanda et al., 2011; Hasebe et al., 2014; Karigo et al., 2012). According to these studies, Kayo et al. performed those three methods and demonstrated that all three groups of E2-administrated medaka showed a transient increase in E2 concentration (see Kayo et al., 2020 for details). In E2 feeding, the average blood E2 concentration showed values approximately 2.5–8.35 times higher than that in intact female medaka after 4 h (33.4 ± 19.3 ng/mL) that then decreased at 24 h (2.2 ± 0.9 ng/mL). Surprisingly, medaka exposed to the water containing 10 nM (2.72 ng/mL) E2 for 4 h showed much higher concentration of blood E2 (100.1 ± 33.4 ng/mL) than the surrounding water did. These results may suggest bioconcentration, which indicates that some substance in the body could have absorbed and accumulated E2. The medaka implanted with silicon blocks containing E2 also showed a higher concentration of blood E2 compared to normal physiological levels, and the E2 concentration remained high 24 h after administration under this condition. Although it may be possible to mimic the physiological concentration of E2 in female medaka by using a lower dose of E2 silicon, it is difficult to imitate the diurnal fluctuation of E2 in female medaka. If one implants an E2 block in fish twice or more, these surgeries may cause severe physical damage to the experimental fish. Thus in our experience, the feeding administration of E2 may be the most convenient and appropriate method for manipulating the circulating E2 in female medaka.

Finally, to analyze the time course of the decrease in endogenous blood E2 after OVX, the concentration of blood E2 in OVX female medaka was measured (Fig. 10.7). Female medaka one or more days after OVX showed a concentration of 0.3–0.5 ng/mL of blood E2, which is significantly lower than that of intact females (~7 ng/mL). These results demonstrate that endogenous E2 was almost completely removed within 24 h.

These findings may help future analyses using OVX and E2 administration with medaka in combination with genetic modifications. For instance, some studies using both methods have been reported (Takahashi et al., 2016; Kayo et al., 2019). Moreover, recent work shows that castration of the male medaka drastically decrease endogenous 11-ketotestosterone, which is the most potent

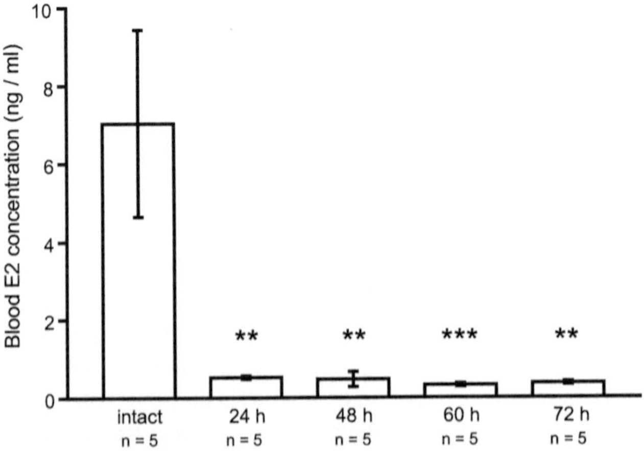

FIGURE 10.7 **Most of endogenous 17β-estradiol (E2) has been cleared in 24 h after ovariectomy (OVX) in female medaka. Time changes in blood E2 concentration after OVX. More than 24 h after OVX, blood E2 concentrations were as low as the average for males (Dunnett's test, compared with intact; **, $P < .01$; ***, $P < .001$). Data are shown as mean ± SEM ($n = 5$).** *Modified from Kayo, D., Oka, Y., Kanda, S., 2020. Examination of methods for manipulating serum 17beta-Estradiol (E2) levels by analysis of blood E2 concentration in medaka (*Oryzias latipes*). Gen. Comp. Endocrinol. 285, 113272, with permission of Elsevier.*

and prevalent androgen in fish, within 24 h (Royan et al., 2020). These methods and previous data will help us utilize medaka for endocrinological analyses and research that aims to measure and/or manipulate other hormones in the blood.

Neurophysiological studies using medaka: patch clamp recording and calcium imaging

Introduction to the neurophysiological analysis

Neurophysiological analysis provides important information on living neurons and endocrine cells. In in vitro neurophysiological studies, brains or ganglions are dissected from animal bodies, and their cellular activities are recorded in artificial spinal cord fluid. Unlike experiments using fixed tissues, this technique enables a direct analysis of living neuronal circuits. Moreover, in a comparative analysis of recordings from animals under different physiological or environmental conditions, prolonged changes in neural activities due to these differences may be observed.

Here, we introduce two widely used neurophysiological methods in medaka: electrophysiology and imaging with useful indicators. Electrophysiology indicates electrical states (membrane potential, excitability, etc.) of cells. One of the electrophysiological techniques, patch clamp recording, enables us to record fast neural activity of a single cell (in ms) (Table 10.2, left). Since the duration of the action potential is a few ms, this technique is

TABLE 10.2 **The difference between patch clamp and Ca^{2+} imaging.**

Type of recording	Patch clamp	Ca^{2+} imaging
Information	Action potential Neurotransmission Activity of ion channel	Neuropeptide/hormone release
Time scale	>milliseconds	>tens of milliseconds
Multicellular recording	No	Yes
in vitro/in vivo	in vitro	in vitro/in vivo (larval zebrafish)

suitable for precise analysis of the changes of membrane potential and/or occurrence of firings (action potentials). Furthermore, a patch clamp can record the specific ion current flowing through a cell membrane. On the other hand, imaging shows changes of intracellular concentration of the "target molecule (Ca^{2+}, cAMP, etc.)" (Odaka et al., 2014; Tsien, 1989) or can indicate pH or membrane potential of a cell (Knopfel and Song, 2019; Miesenbock et al., 1998) by using specific fluorescence probes. Among various imaging methods, Ca^{2+} imaging has been widely used. Because the release of most neurotransmitters and hormones occurs in a Ca^{2+}-dependent manner, intracellular concentration of Ca^{2+} is a good indicator of neurotransmitter/hormone release. One big advantage of Ca^{2+} imaging over a patch clamp is that it records multiple cells within the field of the imaging camera at once (Table 10.2, right). Moreover, since, in principle, it is relatively resistant to the animal movements compared to a patch clamp, in vivo Ca^{2+} imaging is often used in studies of larval zebrafish (Barker and Baier, 2015; Wee et al., 2019). Although Ca^{2+} imaging offers those advantages, the speed of Ca^{2+} imaging is limited by the speed of the camera or the time required for conformational changes of the Ca^{2+} indicator. Thus, it is not suitable for observing events in the ms order. Another obvious disadvantage of Ca^{2+} imaging is that it usually represents relative changes during the experiments, which means that we cannot compare the absolute Ca^{2+} concentrations among different individuals. For instance, we can evaluate the effects of drugs on neurons of interest, but it is difficult to compare their spontaneous activities between breeding and nonbreeding seasons through Ca^{2+} imaging.

Neural activity analyzed by patch clamp recordings

A patch clamp is used to record electrical activities of a single cell through a glass pipette with a tip diameter of $\sim 1\ \mu m$. The pipette is attached to the target cell for recording under a microscope. Infrared differential interference contrast microscopy enables researchers to find the cell located deep in the brain or brain slices. When GFP Tg animals are used, the cell of interest can be found easily by fluorescence microscopy.

Among several modes of patch clamp recordings, we often use the loose-cell patch clamp mode and the whole-cell mode in the neuroendocrinological studies of medaka. The loose-cell patch clamp recording only provides the timing of the firings, but it is less invasive to the cells. On the other hand, the whole-cell patch clamp mode provides more detailed electrophysiological information on a single cell, such as absolute membrane potential and inputs of small excitatory/inhibitory synaptic potential. Furthermore, we can isolate specific ion currents by arranging internal/external solution and membrane potentials for detailed electrophysiological analyses.

High-frequency firing of GnRH1 neuron in the afternoon stimulates LH cells

Next, we will explain how patch clamp recordings in medaka have contributed to the understanding of the HPG axis. Since medaka, a daily spawner, was suggested to possess a diurnal rhythm of hormonal release in the HPG axis, Karigo et al. examined time-of-day change of firing of GnRH1 neurons (Karigo et al., 2012), which is supposed to form the final common pathway for HPG axis in vertebrates (Knobil and Neill, 1994). They made a whole brain in vitro preparation of *gnrh1*:egfp female medaka at six time points (4-h intervals) of the day and recorded activities of GnRH1 neurons in loose-cell patch clamp mode. They found that GnRH1 neurons show a high-frequency firing in the afternoon, which may be similar to a phenomenon observed in mammals (Christian et al., 2005). In peptidergic neurons, including GnRH1 neurons, high-frequency firing is suggested to induce neuropeptide release (van den Pol, 2012). Given that the axons of GnRH1 neurons terminate in the pituitary, it was suggested that GnRH1 neurons release GnRH1 peptides in the pituitary in the afternoon. In addition, Strandabø et al. demonstrated that LH cells change their firing activities after the application of a GnRH agonist (Strandabø et al., 2013). Taken together, high-frequency firing of GnRH1 neurons may trigger a large amount of LH release, an LH surge, in the afternoon in medaka.

Application of patch clamp recording to nonhypophysiotropic GnRH neurons

Patch clamp recording also advanced the understanding of nonhypophysiotropic GnRH neurons. Terminal nerve GnRH3 (TN-GnRH3) neurons, which express one of the *gnrh* paralogs, *gnrh3*, are suggested to modulate sexual behaviors via neuropeptide release (neuromodulation) (reviewed in Umatani and Oka, 2019). These neurons in the adult animals almost always show regular pacemaker activities (Fig. 10.8A; Oka and Matsushima, 1993; Kanda et al., 2010), and their pacemaker firings are also modulated by various auto/paracrine or synaptic

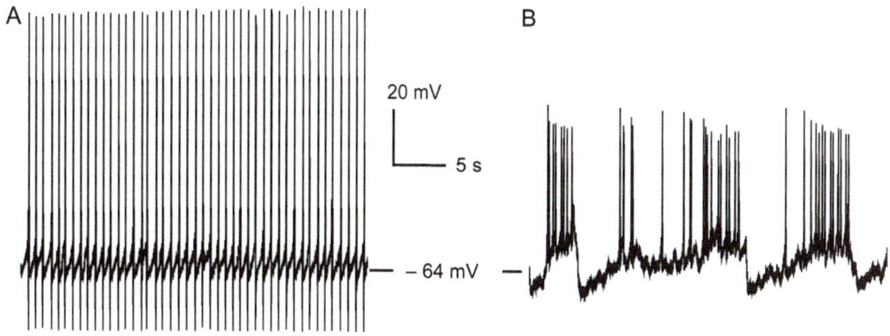

FIGURE 10.8 **Terminal nerve gonadotropin-releasing hormone (GnRH) neurons show different firing patterns between juveniles and adults. (A) A representative spontaneous activity in adult medaka. Almost all of the terminal GnRH neurons show pacemaker activity (~1 Hz). (B) A representative spontaneous activity in juvenile medaka. About 60% of terminal nerve GnRH neurons show burst firing at 3 weeks after fertilization.**

inputs (Abe and Oka, 2000; Kiya and Oka, 2003; Haneda and Oka, 2004; Wayne and Kuwahara, 2007; Nakane and Oka, 2010; Saito et al., 2010; Umatani et al., 2013). Most of these experiments could be performed without using Tg fish because TN-GnRH3 neurons were easily identified by their characteristic location and large cell bodies compared with surrounding cells. On the other hand, with the advent of Tg techniques, we were able to identify TN-GnRH3 neurons even in the embryo and to analyze their activities with the aid of their GFP fluorescence. A study using *gnrh3:EMD* (modified GFP) zebrafish demonstrated that TN-GnRH3 neurons in a 2–3-day postfertilization embryo show various firing patterns (Ramakrishnan et al., 2010). In addition, another study reported that TN-GnRH3 neurons in juvenile medaka show spontaneous high-frequency (burst) firing more frequently than do those in adults by using whole-cell patch clamp recording (Fig. 10.8B; Umatani and Oka, 2018). Because the frequency of this juvenile-specific burst firing was high enough to release neuropeptides, juvenile TN-GnRH3 neurons are suggested to release neuropeptides more frequently than the adult ones do.

Hormone/neuropeptide release clarified by Ca^{2+} imaging

Ca^{2+} imaging, another neurophysiological technique, enables us to capture the change of intracellular Ca^{2+} concentration ($[Ca^{2+}]_i$) as an increase/decrease in the intensity of a fluorescence probe. Because hormone or neuropeptide release mediated by exocytosis is triggered by Ca^{2+} (Purves, 2008), the increase in $[Ca^{2+}]_i$ indicates the occurrence of secretion. Before the development of the Tg techniques, fluorescent $[Ca^{2+}]_i$ indicators based on Ca^{2+} chelators, such as Quin2, Fluo3/4, Rhod2, and Fura2 generated by R. Y. Tisen (Tsien and Rink, 1980; Grynkiewicz et al., 1985; Minta et al., 1989), had been the only indicators of the $[Ca^{2+}]_i$. Among them, Fura2 is a ratiometric fluorescent dye, which

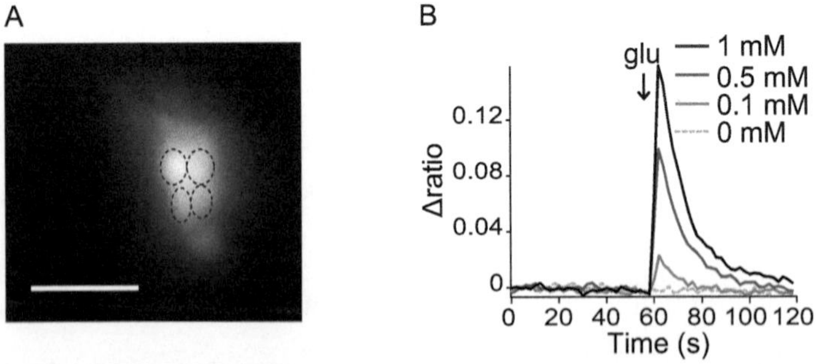

FIGURE 10.9 **Imaging of specific neurons by using green fluorescent protein (GFP) transgenic medaka and Fura2. (A) The cell bodies of juvenile terminal nerve GnRH3 neurons (4 weeks after fertilization). Fura2 was loaded into the whole brain, and the regions of interest (ROI, *dotted circle*) were selected while referring to GFP. (B) The ratio (λ340/λ380) of Fura2-fluorescence intensity was transiently increased after puffer application of glutamate (each trace represents the average of four neurons). Glutamate was applied using a puffer pipette at the point indicated by the arrow. Scale bar: 20 μm.** *Modified from Umatani, C., Oka, Y., 2018. Juvenile-specific burst firing of terminal nerve GnRH3 neurons suggests novel functions in addition to neuromodulation. Endocrinology 159, 1678–1689, with permission of Oxford University Press.*

changes the exitation spectrum after the binding of Ca^{2+} rather than its intensity. With this property, it minimizes the effects of uneven loading, focus drift, photobleaching, and varying cell thickness. With the advent of Tg techniques, it became possible to load Fura2 in the tissue of a GFP Tg medaka to perform Ca^{2+} imaging of GFP-labeled cells because the excitation wavelength of Fura2 is different from that of GFP (Fig. 10.9; Hasebe and Oka, 2017; Umatani and Oka, 2018). In addition, a more recent development of protein-based $[Ca^{2+}]_i$ indicators (inverse pericam [IP], GCaMPs, RCaMP, etc. (Nagai et al., 2001; Tian et al., 2009; Chen et al., 2013; Inoue et al., 2015)) enables a more precise analysis of genetically defined cells. On top of that, these genetically encoded indicators are less invasive than fluorescent dyes, since they do not require a loading process with detergent. Thus, in vivo imaging of hormone or neurotransmitter release may be possible in the future.

In addition to the electrophysiological approaches, Ca^{2+} imaging in medaka has also contributed to the study of HPG axis regulation. By using *lhb*:IP medaka, Karigo et al. showed that GnRH1 application induced $[Ca^{2+}]_i$ increase in LH cells (Karigo et al., 2014). In addition, electrical stimulation of axonal fibers of GnRH1 neurons or glutamate-induced high-frequency firing of GnRH1 neurons increased $[Ca^{2+}]_i$ in LH cells (Karigo et al., 2014; Hasebe and Oka, 2017). Thus it is suggested that high-frequency firing of GnRH1 neurons releases GnRH1 peptides and that these peptide releases induce an LH surge. On the other hand, the relationship between GnRH1 neurons and FSH release in teleosts was different from that in mammals. Although GnRH1 application increased $[Ca^{2+}]_i$ in FSH cells in the experiments using *fshb*:IP medaka, this increase was slower and lower than that in LH cells. In addition, a recent study reported that FSH is released indirectly by GnRH1 application (Hodne et al., 2019). Because the uncaging of Ca^{2+} in an LH cell induced hyperpolarization of the membrane potential in the neighboring FSH cell, it was suggested that FSH cells respond to GnRH1 indirectly through gap junctions between LH and FSH cells. However, *gnrh1*$^{-/-}$ medaka showed a normal expression of FSH and had mature follicles (Takahashi et al., 2016). These results suggested that *gnrh1*$^{-/-}$ medaka release FSH at a sufficient quantity for folliculogenesis. In summary, Ca^{2+} imaging of gonadotrophs played an important role in the analysis of the mechanism underlying LH surge by GnRH1 release although the meaning of the $[Ca^{2+}]_i$ increase in FSH cells by GnRH1 remains a mystery.

Neurophysiology advanced the understanding of neuroendocrinological regulation of reproduction

Neurophysiological studies using patch clamp recording and Ca^{2+} imaging clarified the cellular mechanism of LH surge induction. The techniques introduced here can be utilized on the other neurons and endocrine cells as well as cells involved in the HPG axis regulation. For instance, they have been applied to the analysis of neurons involved in behaviors or homeostasis (Ramakrishnan and Wayne, 2009; Kanda et al., 2010; Umatani and Oka, 2018; Nakajo et al., 2018; Kikuchi et al., 2019). As described earlier, we can easily conduct cell-specific neurophysiological experiments using Tg medaka. Therefore medaka may be a suitable animal for neurophysiological approaches that contribute more to the study of neuroendocrinology.

Medaka provides clues to the understanding of general mechanisms of reproduction in teleosts and even in vertebrates

Understanding of mechanism of HPG axis regulation in medaka and teleosts

In this chapter, we introduced several methods used in medaka from the cellular to the whole-body level. Since such multidisciplinary analyses could not have been used in other nonmammalian species, the results and insights from medaka have provided novel information on the ongoing research in other teleosts. In addition, the understanding of the mechanisms in one clade also provides us with an opportunity to compare the mechanism with that of the other clades of species. In our case, the understanding of the HPG axis regulation in a teleost, medaka, made it possible to compare it to that of mammals as described earlier. Such a comparison may help us understand not only the evolution of the mechanism, but also why human beings have a system that is unlike that of teleosts.

Difference of reproductive strategies and their underlying mechanisms—prolific teleosts and less prolific mammals

Here, we hypothesize that the difference in the mechanism of HPG axis regulation between teleosts and mammals underlies their degree of prolificity (greater in teleosts). Interestingly, hypothalamic and pituitary components in the HPG axis regulation, such as kisspeptin, GnRH, and FSH/LH, are mostly conserved as molecules/genes in vertebrates. In spite of this solid conservation as molecules/genes, kisspeptin does not play a role in the regulation of GnRH in teleosts, and even gonadotropins play slightly different roles in teleosts and mammals. It is important to note that kisspeptin neurons, which contribute to reproductive regulation only in mammals, play important roles in the generation of LH pulse, which is observed only in mammals as well. In other words, mammals have acquired generation mechanisms of pulsatile LH release, which contributes to folliculogenesis in addition to FSH. This pulsatile release of LH is directly regulated by kisspeptin neurons and GnRH neurons; it may enable fine tuning of folliculogenesis by modulating the neuronal activity of kisspeptin neurons in the hypothalamic nucleus, arcuate nucleus. In fact, the firing activity of arcuate kisspeptin neurons, which has been proved to be the essential component of the pulse generator, is modulated by many factors such as stress and nutrition conditions in mammals (Castellano and Tena-Sempere, 2013; Grachev et al., 2013). This fine tuning of folliculogenesis may be advantageous in the less prolific system of mammals.

Interestingly, mammals show cellular colocalization of FSH and LH in the pituitary, which retains the situation immediately after the gene duplication of *fshb* and *lhb* in the early vertebrate lineage. On the other hand, FSH and LH are expressed in separate cell populations in teleosts, which enables the completely separate regulation of FSH and LH release (Nozaki et al., 1990; Miranda et al., 2001; Yaron et al., 2001, 2003; Kanda et al., 2011; Fontaine et al., 2020). Such separate release of FSH and LH may contribute to the separate regulation of folliculogenesis and ovulation in teleosts, respectively, which may be advantageous to the regulation of prolific gonads. Although the relationship between the reproductive strategies and the underlying regulatory mechanisms is still hypothetical, further studies using medaka may enable an explanation of the merits of their systems in the prolific strategy.

Toward understanding of a general mechanism of neuroendocrine systems in vertebrates by taking advantage of medaka

Needless to say, experimental results from a single species of animal may not be applied to the entire clade to which it belongs. To lead to a consensus in teleosts, findings from medaka should be compared with the results of other teleosts. As described in the previous chapter, the accumulating information in zebrafish is important because medaka and zebrafish are phylogenetically distant among teleosts: zebrafish and medaka branched around 250 million years ago (mya), while teleosts emerged 283 mya (Betancur et al., 2013).

Especially in the field of neuroendocrinology, there are many advantages of medaka when compared to other teleosts as described earlier. However, many of these advantages are not limited to the field of neuroendocrinology, and the techniques established for medaka may also be applicable to other small fishes. It is our hope that much more researchers will take advantage of adult medaka and established techniques in the near future.

Acknowledgment

This work was supported by NBRP medaka.

References

Abe, H., Oka, Y., 2000. Modulation of pacemaker activity by salmon gonadotropin-releasing hormone (sGnRH) in terminal nerve (TN)-GnRH neurons. J. Neurophysiol. 83, 3196–3200.

Adachi, S., Yamada, S., Takatsu, Y., Matsui, H., Kinoshita, M., Takase, K., Sugiura, H., Ohtaki, T., Matsumoto, H., Uenoyama, Y., Tsukamura, H., Inoue, K., Maeda, K., 2007. Involvement of anteroventral periventricular metastin/kisspeptin neurons in estrogen positive feedback action on luteinizing hormone release in female rats. J. Reprod. Dev. 53, 367–378.

Akazome, Y., Kanda, S., Okubo, K., Oka, Y., 2010. Functional and evolutionary insights into vertebrate kisspeptin systems from studies of fish brain. J. Fish. Biol. 76, 161–182.

Amoss, M., Burgus, R., Blackwell, R., Vale, W., Fellows, R., Guillemin, R., 1971. Purification, amino acid composition and N-terminus of the hypothalamic luteinizing hormone releasing factor (LRF) of ovine origin. Biochem. Biophys. Res. Commun. 44, 205–210.

Barker, A.J., Baier, H., 2015. Sensorimotor decision making in the zebrafish tectum. Curr. Biol. 25, 2804–2814.

Bazer, G.T., Ebbesson, S.O., 1984. A simplified cobalt-lysine method for tracing axon trajectories in the central nervous system of vertebrates. Neurosci. Lett. 51, 315–318.

Betancur, R.R., Broughton, R.E., Wiley, E.O., Carpenter, K., Lopez, J.A., Li, C., Holcroft, N.I., Arcila, D., Sanciangco, M., Cureton Ii, J.C., Zhang, F., Buser, T., Campbell, M.A., Ballesteros, J.A., Roa-Varon, A., Willis, S., Borden, W.C., Rowley, T., Reneau, P.C., Hough, D.J., Lu, G., Grande, T., Arratia, G., Orti, G., 2013. The tree of life and a new classification of bony fishes. PLoS Curr. 5.

Castellano, J.M., Tena-Sempere, M., 2013. Metabolic regulation of kisspeptin. Adv. Exp. Med. Biol. 784, 363–383.

Chen, T.W., Wardill, T.J., Sun, Y., Pulver, S.R., Renninger, S.L., Baohan, A., Schreiter, E.R., Kerr, R.A., Orger, M.B., Jayaraman, V., Looger, L.L., Svoboda, K., Kim, D.S., 2013. Ultrasensitive fluorescent proteins for imaging neuronal activity. Nature 499, 295–300.

Christian, C.A., Mobley, J.L., Moenter, S.M., 2005. Diurnal and estradiol-dependent changes in gonadotropin-releasing hormone neuron firing activity. Proc. Natl. Acad. Sci. U. S. A. 102, 15682–15687.

D'anglemont de Tassigny, X., Fagg, L.A., Dixon, J.P., Day, K., Leitch, H.G., Hendrick, A.G., Zahn, D., Franceschini, I., Caraty, A., Carlton, M.B., Aparicio, S.A., Colledge, W.H., 2007. Hypogonadotropic hypogonadism in mice lacking a functional Kiss1 gene. Proc. Natl. Acad. Sci. U. S. A. 104, 10714–10719.

de Roux, N., Genin, E., Carel, J.C., Matsuda, F., Chaussain, J.L., Milgrom, E., 2003. Hypogonadotropic hypogonadism due to loss of function of the KiSS1-derived peptide receptor GPR54. Proc. Natl. Acad. Sci. U. S. A. 100, 10972–10976.

Dungan, H.M., Gottsch, M.L., Zeng, H., Gragerov, A., Bergmann, J.E., Vassilatis, D.K., Clifton, D.K., Steiner, R.A., 2007. The role of kisspeptin-GPR54 signaling in the tonic regulation and surge release of gonadotropin-releasing hormone/luteinizing hormone. J. Neurosci. 27, 12088–12095.

Escobar, S., Servili, A., Espigares, F., Gueguen, M.M., Brocal, I., Felip, A., Gomez, A., Carrillo, M., Zanuy, S., Kah, O., 2013. Expression of kisspeptins and kiss receptors suggests a large range of functions for kisspeptin systems in the brain of the European sea bass. PLoS One 8, e70177.

Fontaine, R., Ager-Wick, E., Hodne, K., Weltzien, F.A., 2020. Plasticity in medaka gonadotropes via cell proliferation and phenotypic conversion. J. Endocrinol. 245, 21–37.

Franceschini, I., Lomet, D., Cateau, M., Delsol, G., Tillet, Y., Caraty, A., 2006. Kisspeptin immunoreactive cells of the ovine preoptic area and arcuate nucleus co-express estrogen receptor alpha. Neurosci. Lett. 401, 225–230.

Frazao, R., Cravo, R.M., Donato JR., J., Ratra, D.V., Clegg, D.J., Elmquist, J.K., Zigman, J.M., Williams, K.W., Elias, C.F., 2013. Shift in Kiss1 cell activity requires estrogen receptor alpha. J. Neurosci. 33, 2807–2820.

Funes, S., Hedrick, J.A., Vassileva, G., Markowitz, L., Abbondanzo, S., Golovko, A., Yang, S.J., Monsma, F.J., Gustafson, E.L., 2003. The KiSS-1 receptor GPR54 is essential for the development of the murine reproductive system. Biochem. Biophys. Res. Commun. 312, 1357–1363.

Grachev, P., Li, X.F., O'byrne, K., 2013. Stress regulation of kisspeptin in the modulation of reproductive function. Adv. Exp. Med. Biol. 784, 431–454.

Grone, B.P., Maruska, K.P., Korzan, W.J., Fernald, R.D., 2010. Social status regulates kisspeptin receptor mRNA in the brain of *Astatotilapia burtoni*. Gen. Comp. Endocrinol. 169, 98–107.

Grynkiewicz, G., Poenie, M., Tsien, R.Y., 1985. A new generation of Ca^{2+} indicators with greatly improved fluorescence properties. J. Biol. Chem. 260, 3440–3450.

Han, S.K., Gottsch, M.L., Lee, K.J., Popa, S.M., Smith, J.T., Jakawich, S.K., Clifton, D.K., Steiner, R.A., Herbison, A.E., 2005. Activation of gonadotropin-releasing hormone neurons by kisspeptin as a neuroendocrine switch for the onset of puberty. J. Neurosci. 25, 11349–11356.

Haneda, K., Oka, Y., 2004. Selective modulation of voltage-gated calcium channels in the terminal nerve gonadotropin-releasing hormone neurons of a teleost, the dwarf gourami (*Colisa lalia*). Endocrinology 145, 4489–4499.

Harris, G.W., 1948. Neural control of the pituitary gland. Physiol. Rev. 28, 139–179.

Hasebe, M., Kanda, S., Oka, Y., 2016. Female-specific glucose sensitivity of GnRH1 neurons leads to sexually dimorphic inhibition of reproduction in medaka. Endocrinology 157, 4318–4329.

Hasebe, M., Kanda, S., Shimada, H., Akazome, Y., Abe, H., Oka, Y., 2014. Kiss1 neurons drastically change their firing activity in accordance with the reproductive state: insights from a seasonal breeder. Endocrinology 155, 4868–4880.

Hasebe, M., Oka, Y., 2017. High-frequency firing activity of GnRH1 neurons in female medaka induces the release of GnRH1 peptide from their nerve terminals in the pituitary. Endocrinology 158, 2603–2617.

Herbison, A.E., 2016. Control of puberty onset and fertility by gonadotropin-releasing hormone neurons. Nat. Rev. Endocrinol. 12, 452−466.

Hiraki-Kajiyama, T., Yamashita, J., Yokoyama, K., Kikuchi, Y., Nakajo, M., Miyazoe, D., Nishiike, Y., Ishikawa, K., Hosono, K., Kawabata-Sakata, Y., Ansai, S., Kinoshita, M., Nagahama, Y., Okubo, K., 2019. Neuropeptide B mediates female sexual receptivity in medaka fish, acting in a female-specific but reversible manner. eLife 8, e39495.

Hodne, K., Fontaine, R., Ager-Wick, E., Weltzien, F.A., 2019. Gnrh1-Induced responses are indirect in female medaka fsh cells, generated through cellular networks. Endocrinology 160, 3018−3032.

Ikawa, M., Ohya, E., Shimada, H., Kamijo, M., Fukamachi, S., 2017. Establishment and maintenance of sexual preferences that cause a reproductive isolation between medaka strains in close association. Biol. Open 6, 244−251.

Inoue, M., Takeuchi, A., Horigane, S., Ohkura, M., Gengyo-Ando, K., Fujii, H., Kamijo, S., Takemoto-Kimura, S., Kano, M., Nakai, J., Kitamura, K., Bito, H., 2015. Rational design of a high-affinity, fast, red calcium indicator R-CaMP2. Nat. Methods 12, 64−70.

Ito, H., Yamamoto, N., Yoshimoto, M., Sawai, N., Yang, C.Y., Xue, H.G., Imura, K., 2003. Fiber connections of the torus longitudinalis in a teleost: *Cyprinus carpio* re-examined. J. Comp. Neurol. 457, 202−211.

Juntti, S.A., Fernald, R.D., 2016. Timing reproduction in teleost fish: cues and mechanisms. Curr. Opin. Neurobiol. 38, 57−62.

Kanda, S., 2019. Evolution of the regulatory mechanisms for the hypothalamic-pituitary-gonadal axis in vertebrates-hypothesis from a comparative view. Gen. Comp. Endocrinol. 284, 113075.

Kanda, S., Akazome, Y., Matsunaga, T., Yamamoto, N., Yamada, S., Tsukamura, H., Maeda, K., Oka, Y., 2008. Identification of KiSS-1 product kisspeptin and steroid-sensitive sexually dimorphic kisspeptin neurons in medaka (*oryzias latipes*). Endocrinology 149, 2467−2476.

Kanda, S., Akazome, Y., Mitani, Y., Okubo, K., Oka, Y., 2013. Neuroanatomical evidence that kisspeptin directly regulates isotocin and vasotocin neurons. PLoS One 8, e62776.

Kanda, S., Karigo, T., Oka, Y., 2012. Steroid sensitive kiss2 neurones in the goldfish: evolutionary insights into the duplicate kisspeptin gene-expressing neurones. J Neuroendocrinol 24 (6), 897−906.

Kanda, S., Nishikawa, K., Karigo, T., Okubo, K., Isomae, S., Abe, H., Kobayashi, D., Oka, Y., 2010. Regular pacemaker activity characterizes gonadotropin-releasing hormone 2 neurons recorded from green fluorescent protein-transgenic medaka. Endocrinology 151, 695−701.

Kanda, S., Oka, Y., 2012. Evolutionary insights into the steroid sensitive kiss1 and kiss2 neurons in the vertebrate brain. Front. Endocrinol. 3, 28.

Kanda, S., Oka, Y., 2013. Structure, synthesis, and phylogeny of kisspeptin and its receptor. Adv. Exp. Med. Biol. 784, 9−26.

Kanda, S., Okubo, K., Oka, Y., 2011. Differential regulation of the luteinizing hormone genes in teleosts and tetrapods due to their distinct genomic environments - insights into gonadotropin beta subunit evolution. Gen. Comp. Endocrinol. 173, 253−258.

Kaneko, T., Aida, K., Hanyu, I., 1986. Studies on the reproductive rhythms of fishes. 20. Changes in ovarian activity and fine-structure of pituitary gonadotrophs during spawning cycle of the chichibu-goby *Tridentiger obscurus*. Bull. Jpn. Soc. Sci. Fish. 52, 1923−1928.

Karigo, T., Aikawa, M., Kondo, C., Abe, H., Kanda, S., Oka, Y., 2014. Whole brain-pituitary in vitro preparation of the transgenic medaka (*Oryzias latipes*) as a tool for analyzing the differential regulatory mechanisms of LH and FSH release. Endocrinology 155, 536−547.

Karigo, T., Kanda, S., Takahashi, A., Abe, H., Okubo, K., Oka, Y., 2012. Time-of-day-dependent changes in GnRH1 neuronal activities and gonadotropin mRNA expression in a daily spawning fish, medaka. Endocrinology 153, 3394−3404.

Kayo, D., Oka, Y., Kanda, S., 2020. Examination of methods for manipulating serum 17beta-Estradiol (E2) levels by analysis of blood E2 concentration in medaka (*Oryzias latipes*). Gen. Comp. Endocrinol. 285, 113272.

Kayo, D., Zempo, B., Tomihara, S., Oka, Y., Kanda, S., 2019. Gene knockout analysis reveals essentiality of estrogen receptor beta1 (Esr2a) for female reproduction in medaka. Sci. Rep. 9, 8868.

Kikuchi, Y., Hiraki-Kajiyama, T., Nakajo, M., Umatani, C., Kanda, S., Oka, Y., Matsumoto, K., Ozawa, H., Okubo, K., 2019. Sexually dimorphic neuropeptide B neurons in medaka exhibit activated cellular phenotypes dependent on estrogen. Endocrinology 160, 827–839.

Kim, D.K., Cho, E.B., Moon, M.J., Park, S., Hwang, J.I., DO Rego, J.L., Vaudry, H., Seong, J.Y., 2012. Molecular coevolution of neuropeptides gonadotropin-releasing hormone and kisspeptin with their cognate G protein-coupled receptors. Front. Neurosci. 6, 3.

Kiya, T., Oka, Y., 2003. Glutamate receptors in the terminal nerve gonadotropin-releasing hormone neurons of the dwarf gourami (teleost). Neurosci. Lett. 345, 113–116.

Knobil, E., Neill, J.D., 1994. The Physiology of Reproduction. Raven Press, New York.

Knopfel, T., Song, C., 2019. Optical voltage imaging in neurons: moving from technology development to practical tool. Nat. Rev. Neurosci. 20, 719–727.

Kobayashi, M., Aida, K., Hanyu, I., 1988. Hormone changes during the ovulatory cycle in goldfish. Gen. Comp. Endocrinol. 69, 301–307.

Kotani, M., Detheux, M., Vandenbbogaerde, A., Communi, D., Vanderwinden, J.M., le Poul, E., Brezillon, S., Tyldesley, R., Suarez-Huerta, N., Vandeput, F., Blanpain, C., Schiffmann, S.N., Vassart, G., Parmentier, M., 2001. The metastasis suppressor gene KiSS-1 encodes kisspeptins, the natural ligands of the orphan G protein-coupled receptor GPR54. J. Biol. Chem. 276, 34631–34636.

Lawrence, C., 2007. The husbandry of zebrafish (*Danio rerio*): a review. Aquaculture 269, 1–20.

Lee, Y.R., Tsunekawa, K., Moon, M.J., Um, H.N., Hwang, J.I., Osugi, T., Otaki, N., Sunakawa, Y., Kim, K., Vaudry, H., Kwon, H.B., Seong, J.Y., Tsutsui, K., 2009. Molecular evolution of multiple forms of kisspeptins and GPR54 receptors in vertebrates. Endocrinology 150, 2837–2846.

Liu, Y., Tang, H.P., Xie, R., Li, S.S., Liu, X.C., Lin, H.R., Zhang, Y., Cheng, C.H.K., 2017. Genetic evidence for multifactorial control of the reproductive Axis in zebrafish. Endocrinology 158, 604–611.

Matsuo, H., Baba, Y., Nair, R.M.G., Arimura, A., Schally, A.V., 1971. Structure of the porcine LH- and FSH-releasing hormone. I. The proposed amino acid sequence. Biochem. Biophys. Res. Commun. 43, 1334–1339.

Mezache, L., Mikhail, M., Fassan, M., Chilosi, M., Nuovo, G.J., 2015. Optimization of in situ hybridization and co-expression analysis in the central nervous system. In: Hauptmann, G. (Ed.), In Situ Hybridization Methods. Neuromethods. Humana Press, New York.

Miesenbock, G., DE Angelis, D.A., Rothman, J.E., 1998. Visualizing secretion and synaptic transmission with pH-sensitive green fluorescent proteins. Nature 394, 192–195.

Minta, A., Kao, J.P., Tsien, R.Y., 1989. Fluorescent indicators for cytosolic calcium based on rhodamine and fluorescein chromophores. J. Biol. Chem. 264, 8171–8178.

Miranda, L.A., Strussmann, C.A., Somoza, G.M., 2001. Immunocytochemical identification of GtH1 and GtH2 cells during the temperature-sensitive period for sex determination in pejerrey, *Odontesthes bonariensis*. Gen. Comp. Endocrinol. 124, 45–52.

Nagai, T., Sawano, A., Park, E.S., Miyawaki, A., 2001. Circularly permuted green fluorescent proteins engineered to sense Ca^{2+}. Proc. Natl. Acad. Sci. U. S. A. 98, 3197–3202.

Nagata, Y., 1936. Transplantation of testis into ovariotomized *Oryzias latipes*. Zool. Mag. 48, 102–108.

Nakajo, M., Kanda, S., Karigo, T., Takahashi, A., Akazome, Y., Uenoyama, Y., Kobayashi, M., Oka, Y., 2018. Evolutionally conserved function of kisspeptin neuronal system is nonreproductive regulation as revealed by nonmammalian study. Endocrinology 159, 163–183.

Nakane, R., Oka, Y., 2010. Excitatory action of GABA in the terminal nerve gonadotropin-releasing hormone neurons. J. Neurophysiol. 103, 1375–1384.

Nozaki, M., Naito, N., Swanson, P., Miyata, K., Nakai, Y., Oota, Y., Suzuki, K., Kawauchi, H., 1990. Salmonid pituitary gonadotrophs. I. Distinct cellular distributions of two gonadotropins, GTH I and GTH II. Gen. Comp. Endocrinol. 77, 348–357.

Odaka, H., Arai, S., Inoue, T., Kitaguchi, T., 2014. Genetically-encoded yellow fluorescent cAMP indicator with an expanded dynamic range for dual-color imaging. PLoS One 9, e100252.

Ohkura, S., Takase, K., Matsuyama, S., Mogi, K., Ichimaru, T., Wakabayashi, Y., Uenoyama, Y., Mori, Y., Steiner, R.A., Tsukamura, H., Maeda, K.I., Okamura, H., 2009. Gonadotrophin-releasing hormone pulse generator activity in the hypothalamus of the goat. J. Neuroendocrinol. 21, 813–821.

Ohtaki, T., Shintani, Y., Honda, S., Matsumoto, H., Hori, A., Kanehashi, K., Terao, Y., Kumano, S., Takatsu, Y., Masuda, Y., Ishibashi, Y., Watanabe, T., Asada, M., Yamada, T., Suenaga, M., Kitada, C., Usuki, S., Kurokawa, T., Onda, H., Nishimura, O., Fujino, M., 2001. Metastasis suppressor gene KiSS-1 encodes peptide ligand of a G-protein-coupled receptor. Nature 411, 613–617.

Oka, Y., Matsushima, T., 1993. Gonadotropin-releasing hormone (GnRH)-immunoreactive terminal nerve cells have intrinsic rhythmicity and project widely in the brain. J. Neurosci. 13, 2161–2176.

Oka, Y., Satou, M., Ueda, K., 1986. Ascending pathways from the spinal cord in the hime salmon (landlocked red salmon, *Oncorhynchus nerka*). J. Comp. Neurol. 254, 104–112.

Okamura, H., Yamamura, T., Wakabayashi, Y., 2013. Kisspeptin as a master player in the central control of reproduction in mammals: an overview of kisspeptin research in domestic animals. Anim. Sci. J. 84, 369–381.

Ono, Y., Uematsu, T., 1957. Mating ethogram in *oryzias latipes*. J. Fac. Sci. Hokkaido Univ. Ser. VI Zool. 13, 197–202.

Pineda, R., Aguilar, E., Pinilla, L., Tena-Sempere, M., 2010. Physiological roles of the kisspeptin/GPR54 system in the neuroendocrine control of reproduction. Prog. Brain Res. 181, 55–77.

Purves, D., 2008. Neuroscience. Sinauer, Sunderland, MA.

Ramakrishnan, S., Lee, W., Navarre, S., Kozlowski, D.J., Wayne, N.L., 2010. Acquisition of spontaneous electrical activity during embryonic development of gonadotropin-releasing hormone-3 neurons located in the terminal nerve of transgenic zebrafish (*Danio rerio*). Gen. Comp. Endocrinol. 168, 401–407.

Ramakrishnan, S., Wayne, N.L., 2009. Social cues from conspecifics alter electrical activity of gonadotropin-releasing hormone neurons in the terminal nerve via visual signals. Am. J. Physiol. Regul. Integr. Comp. Physiol. 297, R135–R141.

Royan, M.R., Kanda, S., Kayo, D., Song, W., Ge, W., Weltzien, F.-A., Fontaine, R., 2020. Gonadectomy and Blood Sampling Procedures in the Small Size Teleost Model Japanese Medaka (*Oryzias latipes*). J Vis. Exp. 166, e62006.

Saito, T.H., Nakane, R., Akazome, Y., Abe, H., Oka, Y., 2010. Electrophysiological analysis of the inhibitory effects of FMRFamide-like peptides on the pacemaker activity of gonadotropin-releasing hormone neurons. J. Neurophysiol. 104, 3518–3529.

Schally, A.V., Arimura, A., Kastin, A.J., Matsuo, H., Baba, Y., Redding, T.W., Nair, R.M.G., Debeljuk, L., White, W.F., 1971. Gonadotropin-releasing hormone - one polypeptide regulates secretion of luteinizing and follicle-stimulating hormones. Science 173, 1036–1038.

Seminara, S.B., Messager, S., Chatzidaki, E.E., Thresher, R.R., Acierno, J.S., Shagoury, J.K., BO-Abbas, Y., Kuohung, W., Schwinof, K.M., Hendrick, A.G., Zahn, D., Dixon, J., Kaiser, U.B., Slaugenhaupt, S.A., Gusella, J.F., O'rahilly, S., Carlton, M.B.L., Crowley, W.F., Aparicio, S.A.J.R., Colledge, W.H., 2003. The GPR54 gene as a regulator of puberty. N. Engl. J. Med. 349, 1614–1627.

Servili, A., le Page, Y., Leprince, J., Caraty, A., Escobar, S., Parhar, I.S., Seong, J.Y., Vaudry, H., Kah, O., 2011. Organization of two independent kisspeptin systems derived from evolutionary-ancient kiss genes in the brain of zebrafish. Endocrinology 152, 1527–1540.

Shiga, T., Oka, Y., Satou, M., Okumoto, N., Ueda, K., 1989. Retinal projections in the hime salmon (landlocked red salmon, *Oncorhynchus-nerka*). Zool. Sci. 6, 885–898.

Shimizu, A., Aida, K., Hanyu, I., 1985. Endocrine profiles during the short reproductive cycle of an autumn-spawning bitterling, *Acheilognathus rhombea*. Gen. Comp. Endocrinol. 60, 361–371.

Smith, J.T., Popa, S.M., Clifton, D.K., Hoffman, G.E., Steiner, R.A., 2006. Kiss1 neurons in the forebrain as central processors for generating the preovulatory luteinizing hormone surge. J. Neurosci. 26, 6687–6694.

Strandabø, R.A., Hodne, K., Ager-Wick, E., Sand, O., Weltzien, F.A., Haug, T.M., 2013. Signal transduction involved in GnRH2-stimulation of identified LH-producing gonadotropes from lhb-GFP transgenic medaka (*Oryzias latipes*). Mol. Cell. Endocrinol. 372, 128–139.

Tacon, P., Baroiller, J.F., LE Bail, P.Y., Prunet, P., Jalabert, B., 2000. Effect of egg deprivation on sex steroids, gonadotropin, prolactin, and growth hormone profiles during the reproductive cycle of the mouthbrooding cichlid fish *Oreochromis niloticus*. Gen. Comp. Endocrinol. 117, 54–65.

Takahashi, A., Islam, M.S., Abe, H., Okubo, K., Akazome, Y., Kaneko, T., Hioki, H., Oka, Y., 2015. Morphological analysis of the early development of telencephalic and diencephalic gonadotropin-releasing hormone neuronal systems in enhanced green fluorescent protein-expressing transgenic medaka lines. J. Comp. Neurol. 524, 896–913.

Takahashi, A., Kanda, S., Abe, T., Oka, Y., 2016. Evolution of the hypothalamic-pituitary-gonadal Axis regulation in vertebrates revealed by knockout medaka. Endocrinology 157, 3994–4002.

Tang, H.P., Liu, Y., Luo, D.J., Ogawa, S., Yin, Y.K., Li, S.S., Zhang, Y., Hu, W., Parhar, I.S., Lin, H.R., Liu, X.C., Cheng, C.H.K., 2015. The kiss/kissr systems are dispensable for zebrafish reproduction: evidence from gene knockout studies. Endocrinology 156, 589–599.

Tena-Sempere, M., Felip, A., Gomez, A., Zanuy, S., Carrillo, M., 2012. Comparative insights of the kisspeptin/kisspeptin receptor system: lessons from non-mammalian vertebrates. Gen. Comp. Endocrinol. 175, 234–243.

Tian, L., Hires, S.A., Mao, T., Huber, D., Chiappe, M.E., Chalasani, S.H., Petreanu, L., Akerboom, J., Mckinney, S.A., Schreiter, E.R., Bargmann, C.I., Jayaraman, V., Svoboda, K., Looger, L.L., 2009. Imaging neural activity in worms, flies and mice with improved GCaMP calcium indicators. Nat. Methods 6, 875–881.

Tsien, R.Y., 1989. Fluorescent probes of cell signaling. Annu. Rev. Neurosci. 12, 227–253.

Tsien, R.Y., Rink, T.J., 1980. Neutral carrier ion-selective microelectrodes for measurement of intracellular free calcium. Biochim. Biophys. Acta 599, 623–638.

Um, H.N., Han, J.M., Hwang, J.I., Hong, S.I., Vaudry, H., Seong, J.Y., 2010. Molecular coevolution of kisspeptins and their receptors from fish to mammals. Ann. N. Y. Acad. Sci. 1200, 67–74.

Umatani, C., Abe, H., Oka, Y., 2013. Neuropeptide RFRP inhibits the pacemaker activity of terminal nerve GnRH neurons. J. Neurophysiol. 109, 2354–2363.

Umatani, C., Oka, Y., 2018. Juvenile-specific burst firing of terminal nerve GnRH3 neurons suggests novel functions in addition to neuromodulation. Endocrinology 159, 1678–1689.

Umatani, C., Oka, Y., 2019. Multiple functions of non-hypophysiotropic gonadotropin releasing hormone neurons in vertebrates. Zool. Lett. 5, 23.

Uusi-Heikkila, S., Kuparinen, A., Wolter, C., Meinelt, T., Arlinghaus, R., 2012. Paternal body size affects reproductive success in laboratory-held zebrafish (*Danio rerio*). Environ. Biol. Fish. 93, 461–474.

van den Pol, A.N., 2012. Neuropeptide transmission in brain circuits. Neuron 76, 98–115.

Walter, R.O., Hamilton, J.B., 1970. Head-up movements as an indicator of sexual unreceptivity in female medaka, *Oryzias latipes*. Anim. Behav. 18, 125–127.

Watts, A.G., 2015. 60 years of neuroendocrinology: the structure of the neuroendocrine hypothalamus: the neuroanatomical legacy of Geoffrey Harris. J. Endocrinol. 226, T25–T39.

Wayne, N.L., Kuwahara, K., 2007. Beta-endorphin alters electrical activity of gonadotropin releasing hormone neurons located in the terminal nerve of the teleost medaka (*Oryzias latipes*). Gen. Comp. Endocrinol. 150, 41–47.

Wee, C.L., Song, E.Y., Johnson, R.E., Ailani, D., Randlett, O., Kim, J.Y., Nikitchenko, M., Bahl, A., Yang, C.T., Ahrens, M.B., Kawakami, K., Engert, F., Kunes, S., 2019. A bidirectional network for appetite control in larval zebrafish. eLife 8, e43775.

Yaron, Z., Gur, G., Melamed, P., Rosenfeld, H., Elizur, A., Levavi-Sivan, B., 2003. Regulation of fish gonadotropins. Int. Rev. Cytol. 225, 131–185.

Yaron, Z., Gur, G., Melamed, P., Rosenfeld, H., Levavi-Sivan, B., Elizur, A., 2001. Regulation of gonadotropin subunit genes in tilapia. Comp. Biochem. Physiol. B Biochem. Mol. Biol. 129, 489–502.

Zohar, Y., Breton, B., Fostier, A., 1986. Short-term profiles of plasma gonadotropin and estradiol-17-beta levels in the female rainbow-trout, from early ovarian recrudescence and throughout vitellogenesis. Gen. Comp. Endocrinol. 64, 172–188.

Chapter 11

The African turquoise killifish (*Nothobranchius furzeri*): biology and research applications

Gwendoline Astre*, Eitan Moses*, Itamar Harel
Department of Genetics, The Silberman Institute, The Hebrew University of Jerusalem, Jerusalem, Israel

Introduction

There are more than 1200 killifish species, which populate almost every continent aside from Australia and Antarctica. Most killifish are small fish with an average size of 2.5–7 cm, and can be roughly divided into short-lived annual species and longer-lived nonannuals. Killifish are members of the Cyprinodontiform order, also known as "toothcarps," an order of ray-finned fish that includes several families (such as Aplocheilidae, Nothobranchiidae, Cyprinodontidae, Fundulidae, and others) (Nelson et al., 2016; Scheel, 1990). In recent years, the annual African turquoise killifish (*Nothobranchius furzeri*) has emerged as an exciting model system for experimental vertebrate aging due to its naturally compressed lifespan, rapid sexual maturity, embryonic diapause (a mechanism for survival during the dry season), and short generation time (Cellerino et al., 2015; Harel and Brunet, 2015; Kim et al., 2016; Platzer and Englert, 2016) (Fig. 11.1). These life history traits are probably an adaptation to its seasonal habitat, ephemeral water pans in southeast Africa, primarily Zimbabwe and Mozambique, where water is available only during the brief rainy season (Cellerino et al., 2015; Harel and Brunet, 2015; Kim et al., 2016; Platzer and Englert, 2016) (Fig. 11.2A). The turquoise killifish was first collected in 1968 in the Gonarezhou National Park (Zimbabwe). The inbred GRZ strain, named after the name of the park **G**ona**R**e**Z**hou (which means "Place of Elephants" in the local Shona dialect) is presently the most

* Equal contribution.

Laboratory Fish in Biomedical Research. https://doi.org/10.1016/B978-0-12-821099-4.00011-0
Copyright © 2022 Elsevier Inc. All rights reserved.

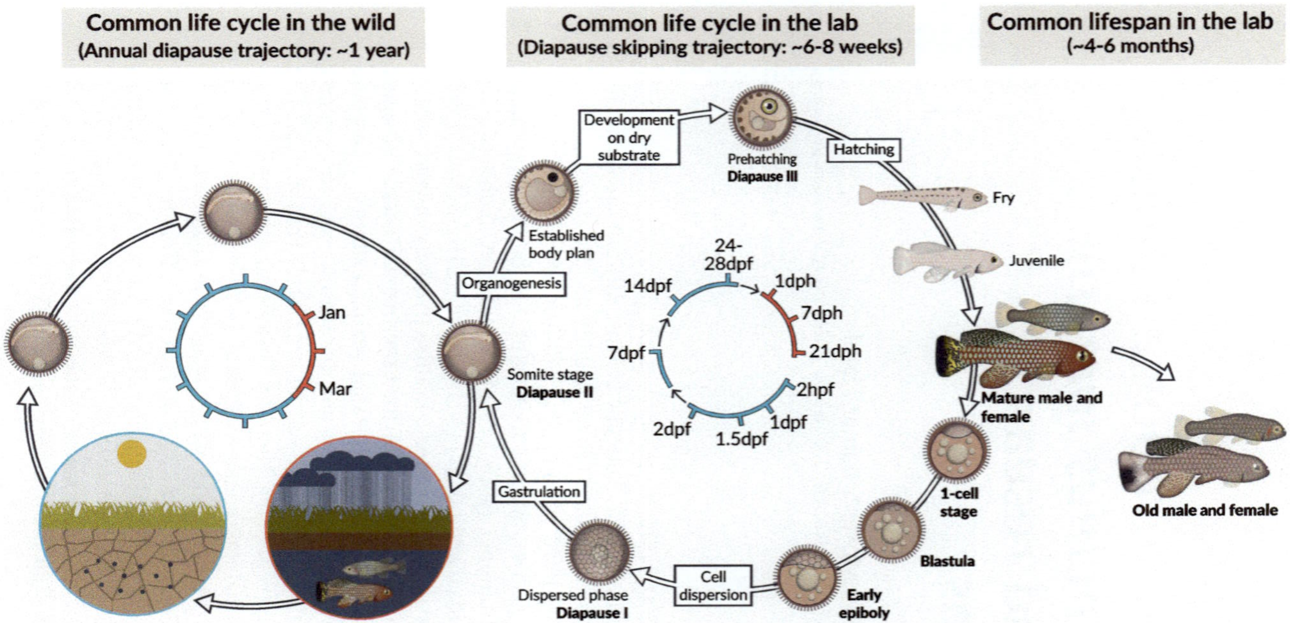

FIGURE 11.1 The life cycle of the turquoise killifish *Nothobranchius furzeri*, including the two common developmental trajectories. *Left*: In the wild, during the brief rainy season, the eggs from the previous year hatch and the fry rapidly reach sexual maturity, mate, and lay a new batch of eggs. These eggs will then enter a developmental arrest state called diapause, which commonly lasts for several months and protects the embryos during the dry season. Less common developmental trajectories, such as "diapause skipping" in the wild, or diapause that lasts for several years, are also possible and described elsewhere (Cellerino et al., 2015). *Center*: In the lab, the "diapause skipping" trajectory is largely preferred for practical reasons, and its primary stages and accompanying timeline are described. Although diapause can occur in three separate stages (diapause I, II, and III), the most common state is diapause II. A bias toward the "diapause skipping" trajectory can be achieved in the lab by controlling incubation temperatures (see the "Breeding strategies for genome engineering, including egg collection and incubation" section). Note that diapause I (dispersed phase) uncouples epiboly from gastrulation. *Right*: After approximately 3 weeks, fish reach sexual maturity. Male fish are larger and more colorful than the females. After a few months, old fish can display various signs of aging, including loss of muscle mass (i.e., sarcopenia) accompanied by bending of the spine (i.e., kyphosis), a global loss of pigmentation (or aberrant pigmentation patterns, as seen here in the tail), delayed wound healing (as seen here in the gills of the female), opaque lens, etc. *dpf*, days postfertilization.

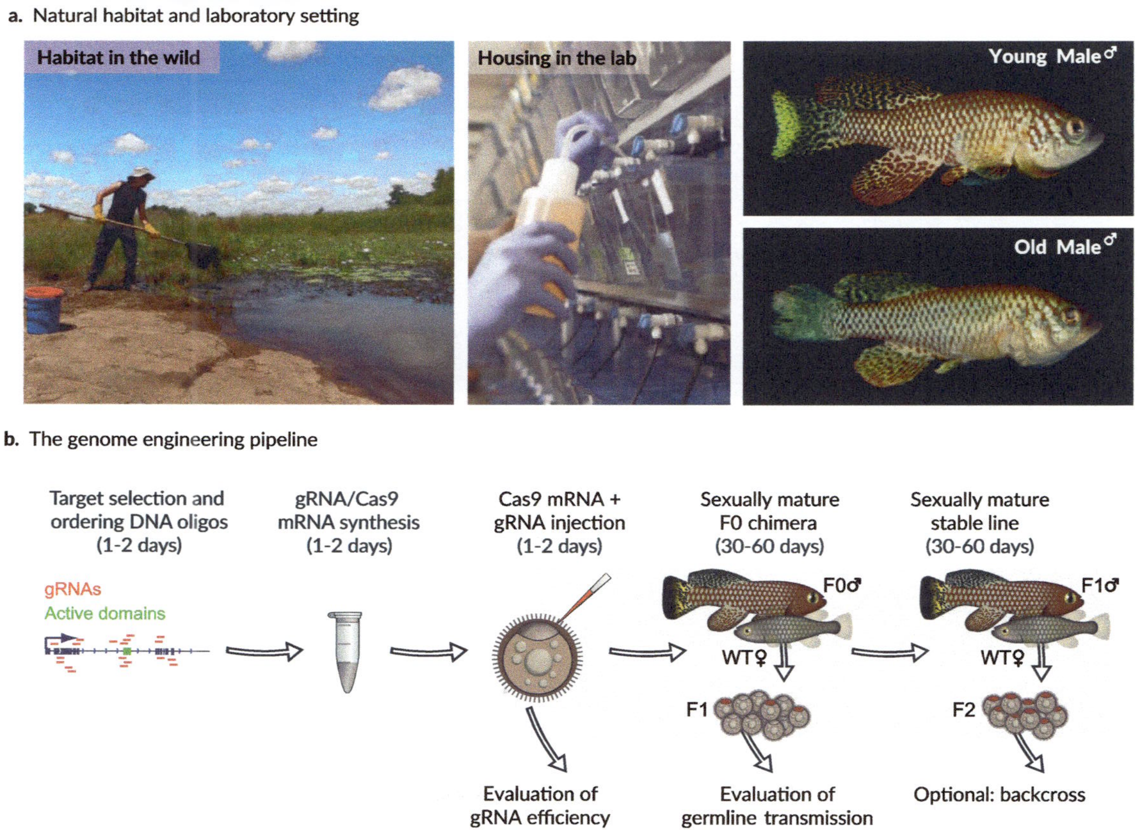

FIGURE 11.2 (A) The natural habitat of the turquoise killifish are ephemeral pans in Zimbabwe and Mozambique (*left*). In the lab, fish are housed in water recirculating systems (*middle*). An example of a young and an old male turquoise killifish (*right*). Although old fish can display various signs of aging, fish usually display only a subset. Here, loss of muscle mass (i.e., sarcopenia), a global loss of pigmentation, and an opaque lens are visible. (B) An efficient and rapid genome engineering pipeline for generating stable mutant fish lines in the turquoise killifish, including a detailed timeline. *gRNA*, guide RNA; *WT*, wild type. *Photo credits: Itamar Harel. Panel (B) is adapted with permission from Harel, I., Valenzano, D.R., Brunet, A., 2016. Efficient genome engineering approaches for the short-lived African turquoise killifish. Nat. Protoc.*

widely used laboratory strain. In this chapter we describe up-to-date husbandry guidelines to establish a large-scale killifish colony. We further provide detailed protocols for efficient genome engineering approaches (i.e. CRISPR/Cas9 and Tol2 strategies), which include reagent selection and synthesis, microinjection, genotyping, and troubleshooting.

As the most diverse group of vertebrates, fish species display extreme differences in the rate of life history traits, including lifespan. Interestingly, maximal lifespan between species is manifested as a continuum on a logarithmic scale, with more than a 1000-fold difference between extremes. On the one hand, the rockfish and the Greenland shark can live for several centuries (Nielsen et al., 2016; Tacutu et al., 2018), and on the other hand, the turquoise killifish and pygmy goby live for only several months (Cellerino et al., 2015; Depczynski and Bellwood, 2005; Harel and Brunet, 2015; Kim et al., 2016; Platzer and Englert, 2016). Aging has been studied in several fish species, either in comparative or experimental models, including salmon (Carney Almroth et al., 2012; Maldonado et al., 2000, 2002), zebrafish (Almaida-Pagan et al., 2014; Anchelin et al., 2013; Cardona-Costa et al., 2009; Gilbert et al., 2014b; Henriques et al., 2013; Kishi et al., 2003; Li et al., 2020; Novoa et al., 2019; Wang et al., 2019), medaka (Ding et al., 2010; Gopalakrishnan et al., 2013), as well as other annual killifish species (Markofsky and Milstoc, 1979; Walford and Liu, 1965; Wang et al., 2014). However, the lack of genetic and genomic tools for annual killifish, and the relatively long lifespan of other fish models, has limited large-scale exploration of aging and age-related diseases in fish.

Aging represents the primary risk factor for many human pathologies, including cardiovascular and neurodegenerative diseases, cancer, and diabetes (Niccoli and Partridge, 2012). Therefore an exciting approach for treating or postponing the onset of these pathologies, and thus extending human healthspan, is slowing down the aging processes (Lopez-Otin et al., 2013). Traditionally, the study of aging has been primarily conducted in nonvertebrate model organisms, particularly yeast (*Saccharomyces cerevisiae*), worm (*Caenorhabditis elegans*), and fly (*Drosophila melanogaster*), which are naturally short-lived and are magnificent genetic models (Lopez-Otin et al., 2013). Using these models, seminal aging studies have identified evolutionary conserved aging-related pathways, such as the insulin/insulin-like growth factor (IGF) and target of rapamycin pathways (Kenyon, 2010). However, as nonvertebrates, these models lack vertebrate-specific organs and systems (e.g., blood, bones, and an adaptive immune system), as well as many key genes, which are involved in human aging and age-related pathologies (e.g., IL8 and APOE) (Harel et al., 2015). Although vertebrate model systems, particularly the mouse (*Mus musculus*) and zebrafish (*Danio rerio*), have been utilized for genetic studies of aging and related disease, their relatively long lifespan has significantly limited their use (maximal lifespan of mice and zebrafish is 3–4 and 5 years, respectively (Tacutu et al., 2018)). Thus a new experimental vertebrate model was required for exploring and understanding the principles of vertebrate aging.

As is apparent from comparative aging studies, vertebrates age in a similar manner, just at a different pace (Austad, 2010; Harel and Brunet, 2015; Ram and Conn, 2018; Singh et al., 2019). Similarly, old turquoise killifish exhibit a classical range of aging phenotypes that are seen in humans and other vertebrates (Fig. 11.2A) (Fillit et al., 2010; Ram and Conn, 2018), including decline in fertility (Di Cicco et al., 2011; Harel et al., 2015), in wound healing and regeneration (Wendler et al., 2015), and in mitochondrial (Hartmann et al., 2011) and cognitive functions (Valenzano et al., 2006a). Additional histopathological markers include accumulation of lipofuscin in the liver (Terzibasi et al., 2008), upregulation of glial fibrillary acidic protein (GFAP) in the brain (Tozzini et al., 2012) (a marker of neuroinflammation), and increased incidences of neoplastic lesions (Genade et al., 2005; Terzibasi et al., 2008).

Importantly, the lifespan of the turquoise killifish can be manipulated by conserved interventions such as dietary restriction, temperature, and drug treatments (Baumgart et al., 2016; Terzibasi et al., 2009; Valenzano et al., 2006a, 2006b), as well as by the gut microbiota (Smith et al., 2017). In addition to exploring aging, the turquoise killifish possesses many experimental advantages, including an XY-based sexual determination system (Reichwald et al., 2015; Valenzano et al., 2009), embryonic diapause (Hu et al., 2020; Reichwald et al., 2015), which is common in other annual killifish (Podrabsky and Hand, 1999; Romney et al., 2018), and active migration of blastomeres prior to gastrulation (Dolfi et al., 2019; Naumann and Englert, 2018; Wourms, 1972). Finally, the availability of distinct wild populations and wild-derived laboratory strains with different characteristics (Terzibasi et al., 2008) has allowed for the genetic mapping and evolutionary studies of traits, (including color, sex, and lifespan (Cui et al., 2019; Kirschner et al., 2012; Reichwald et al., 2009, 2015; Valenzano et al., 2009, 2015)), as well as extensive ecological studies (Bartakova et al., 2020; Garcia and Reichard, 2020; Polacik et al., 2017; Vrtilek et al., 2018).

Recent advances in genomics and genome engineering approaches have provided efficient and reliable ways to generate precise edits to the genome of a wide range of organisms. This powerful ability, which has been traditionally reserved for a handful of classical model organisms, is now widely accessible for basic and translational research. Initial steps in developing genome engineering for the turquoise killifish were established by applying Tol2-based transgenesis, which allows for random integration and expression of exogenous genes in a temporal and tissue-specific manner (Allard et al., 2013; Harel et al., 2016; Hartmann and Englert, 2012; Valenzano et al., 2011). More recently, two transformative projects were completed in the killifish model organism: a comprehensive genome engineering platform for high-throughput genome editing using CRISPR (clustered regularly interspaced short palindromic repeats) (Harel et al., 2015, 2016) (Fig. 11.2b), and the completion of the back-to-back turquoise killifish genome projects, which provided important information for experimental and evolutionary studies (Reichwald et al., 2015; Valenzano et al., 2015). Additionally, the highly inbred GRZ strain (Harel and Brunet, 2015; Kim et al., 2016; Valdesalici and Cellerino, 2003) facilitates reference genome assembly, genome engineering, and provides a point of reference among different laboratories using the turquoise killifish.

The ability to efficiently and precisely edit endogenous genes is essential for any experimental genetic model organism. The CRISPR/Cas9 genome engineering tool has revolutionized the ability to edit the genomes of multiple model and non-model organisms (Cong et al., 2013; Jinek et al., 2012; Mali et al., 2013, and reviewed in Hsu et al., 2014). In brief, the Cas9 nuclease is guided by a short guide RNA (gRNA) molecule to a specific genomic locus. There, this ribonucleoprotein complex will generate a double strand break (DSB), which will then be repaired by the error-prone nonhomologous end joining (NHEJ), or by the more accurate homology-directed repair (HDR) (reviewed in Ran et al., 2013). As NHEJ is imprecise, it frequently introduces base-pair addition or deletions, commonly referred to as indels. These indels facilitate the generation of loss-of-function or "knockout" alleles by causing a frameshift mutation, or by deleting large genomic regions when using gRNAs in tandem. On the other hand, HDR repairs the DSB by using a homologous DNA template, and thus can be used to introduce novel sequences into the genome, also known as a "knock-in" (Doudna and Sontheimer, 2014; Heidenreich and Zhang, 2016).

As already mentioned, this system was recently developed for the killifish model, which allowed for the high-throughput generation of knockout and knock-in alleles in multiple aging and disease genes (Harel et al., 2015, 2016; Harel and Brunet, 2015). Together with the availability of a fully sequenced genome, this platform has transformed killifish into a powerful genetic model system for experimental aging research in the lab (Harel et al., 2015, 2016; Harel and Brunet, 2015). For example, knocking out the protein subunit of telomerase has allowed us to generate the fastest vertebrate model for a human age-associated telomere syndrome (e.g., dyskeratosis congenita). Overall, taking advantage of the turquoise killifish's short generation time allows for the generation of stable lines in as quickly as 2–3 months (Harel et al., 2015, 2016; Harel and Brunet, 2015). Since then, this approach was successfully used for generating additional genetic models, such as for exploring diapause (Hu et al., 2020), regeneration (Wang et al., 2020a), and Parkinson's disease (Matsui et al., 2019). In this chapter we describe up-to-date guidelines to establish a large-scale killifish facility, including housing and water parameters, food, breeding strategies, hatching and rearing of fry, and long-term storage of embryos. We describe in detail updated protocols for efficient genome engineering approaches (including CRISPR/Cas9 and Tol2 strategies), which include reagent selection and synthesis, microinjection, genotyping, and troubleshooting.

Large-scale husbandry (including water systems and water parameters)

In contrast to other laboratory fish species, the African turquoise killifish (similar to other annual killifish) undergoes a compressed lifecycle, including an "explosive" growth period, rapid sexual maturity (Dodzian et al., 2018; Harel et al., 2016; Vrtilek et al., 2018), and high fecundity, which requires specific husbandry adaptations. Several protocols have been developed for killifish housing (Dodzian et al., 2018; Harel et al., 2016; Polacik et al., 2016), and here we provide detailed step-by-step adaptations with the focus on genome engineering, lifespan, and aging studies. Additional husbandry considerations are extensively discussed in Chapter 12. Furthermore, as a vertebrate model organism, care and experimental use of turquoise killifish must be approved by and adhere to relevant institutional ethics guidelines.

Housing and water parameters

For reproducible lifespan experiments, constant housing parameters are very important. The turquoise killifish can be housed in commercially available water recirculating systems with a 12 h light/dark cycle (Harel et al., 2016; Smith et al., 2017). Optimal water temperatures range between 26 and 28°C, neutral pH (7–7.5), and water changes of approximately 10% per day are important to maintain water quality (Dodzian et al., 2018; Harel et al., 2016). It is advised that water will be reconstituted from reverse-osmosis (RO) water, added with commercial marine salt, and sodium bicarbonate for adjusting pH. As commonly used in other aquatic species, the water filtration system should include (1) filtration of large solid particles; (2) a biofilter where ammonia is converted to nitrites and nitrates by bacteria; (3) fine mechanical filtration; and (4) UV sterilization. Ammonia, nitrite, and nitrate levels (see "Materials" section) should be monitored frequently.

As the natural habitat of turquoise killifish is mostly ephemeral pans, this fish is relatively hardy and can tolerate a wide range of water parameters (including a range of temperatures, oxygen levels, pH, and salinity) that naturally occur in its seasonal habitat. Similar to its frequently murky and stagnant habitat, the killifish is comfortable in dimmer lighting (such as subdued ceiling lighting), relatively slow water flow (approximately 5–10 L/h), and higher salinity when compared to freshwater fish (commonly 700–2000 μS). Importantly, dimmer lighting reduces algae growth in water tanks, and higher salinity can inhibit velvet disease caused by a parasitic dinoflagellate *Oodinium*.

Male turquoise killifish are largely territorial; thus it is advised that for optimal growth rate, each tank should house a single male with 1–3 females. Although males that are cohoused can demonstrate aggression (Garcia and Reichard, 2020), if grown together from hatching, aggressive behavior is repressed in some cases. Fish may be housed in 1, 3 or 10 L tanks, with 1–3 L tanks to be used for individual fish housing in lifespan experiments. Killifish are excellent jumpers, therefore to decrease the possibility of strain contamination, lidded tanks and proper labeling are strongly advised. For detailed information regarding the diagnosis and treatment of common fish killifish diseases see Chapter 12, as well as "Diseases of Zebrafish in Research Facilities" from the Zebrafish International Resource Center: https://zebrafish.org/wiki/health/disease_manual/start.

Food

Feeding strategies are discussed in Chapter 12, and here we will briefly mention recommended approaches for optimal fecundity (which is essential for egg collection and genome engineering) and aging studies. Fecundity is highly dependent on food quality and availability, and according to our protocol, peak fertility is reached at 7–9 weeks of age. To optimally support the accelerated growth and rapid sexual maturity (approximately 3–4 weeks posthatching), food consumption is largely higher when compared to other laboratory fish. Therefore at least two feedings per day with high-quality food should be provided. Recently, several commercially available dry food pellets were successfully tested in killifish (Harel et al., 2016; Žák et al., 2020), paving the way to simpler and more standardized housing. Dry food should be supplemented with live/frozen food such as brine shrimp (*Artemia salina*) or blood worms (Blazek et al., 2013) (note that live and frozen blood worms can harbor potential pathogens, and a trusted source is recommended). Fry subsist on brine shrimp until the age of 2 weeks. A small amount of dry food may be given to 1-week-old fry as training but not as a primary food source.

Breeding strategies for genome engineering, including egg collection and incubation

In optimal conditions, the life cycle (from "egg to egg") of the turquoise killifish is 7–9 weeks. Embryonic development in the turquoise killifish is largely nonsynchronous and requires both aquatic and nonaquatic conditions, thus a more "hands-on" approach is required for breeding when compared to other fish models. Normal development from fertilized egg to hatching usually spans 3–4 weeks; however, developmental dormancy through the alternative diapause trajectory can also take place (such as in lower incubation temperatures or in older females) and can largely be avoided by keeping the eggs at 28°C and by breeding young mating pairs. To stimulate embryos to exit diapause a heat-shock treatment could be used, see "Reagent setup and additional protocols" section. There are two main breeding strategies: breeding for strain maintenance and breeding for microinjection and genetic manipulation (which is time sensitive).

Male killifish could be aggressive; therefore it is advised to select a male that is slightly younger and a similar size to the female. When splitting the fish prior to mating, it is advised to keep them in eye contact, such as by using a divider in a 10 L tank or by placing them in two adjacent regular tanks. For injections, plan for a yield of 100–300 eggs according to the observed fecundity in the colony. Crossing 10–20 females is usually sufficient. Cell division in the fertilized eggs is fairly slow, and two-cell stage is observed after 3–4 h. Therefore mating could be allowed for a longer period of time (4–5 h). Fish can be set up for mating weekly, and as frequently as every 3–4 days (Blazek et al., 2013; Podrabsky, 1999). The first few crosses would be mostly for training and might yield a lower number of eggs. Mating using sand trays is more natural for the turquoise killifish, although breeding tanks could be used after fish have been trained to use them. It is important to wash and autoclave sand before using it for the first time and before it is recycled for subsequent uses.

Breeding and embryo incubation for microinjection and genetic manipulation

1. Split males and females in separate tanks at least 2 days before mating. Preferably, keep them separated until mating.
2. On the morning of injection, set up several breeding groups (one male with two to three females) of sexually mature (~2-month-old) fish using 3 or 10 L tanks.
3. Place plastic trays (~10 cm in diameter, 4–6 cm in depth) filled with 2 cm of sand for the fish to lay eggs in for 2–5 h.
4. To separate the eggs from the sand, empty the sand tray into a strainer by partially submerging a strainer in a secondary water container using circular motions to let the sand fall through. Killifish eggs are ~1.2 mm, therefore use a strainer with 0.6 mm hole size.
5. Collect fertilized eggs (or embryos) using a standard plastic Pasteur pipette with the tip cut off. Keep embryos in a 6 cm Petri dish, which can be kept in Yamamoto's embryo solution (Harel et al., 2016) or autoclaved system water (Dodzian et al., 2018). In our protocol we have used Ringer's solution (RS) supplemented with methylene blue (called methylene blue solution) as the basis for all embryo solution (see "Reagent setup and additional protocols" section). Injections should start within 1–2 h for embryos to be at the single-cell or two-cell stage. See the "Microinjection" section below.
6. For disinfecting embryos postinjection, continue to the "Reagent setup and additional protocols". If high mortality is observed, try disinfecting embryos within 24 h after injection to increase survival (see the "Disinfecting embryos by iodine" and "Reagent setup and additional protocols" sections).

7. After injection, keep embryos in 6 cm Petri dishes with 5 mL of methylene blue solution (see the "Reagent setup and additional protocols" section) at a density of up to 50 embryos per dish, and monitor daily for a week. Remove damaged eggs or unfertilized (stained by methylene blue).
8. Incubate embryos at 28°C to increase synchronous embryonic development in 6 cm plates. Replace weekly with 5 mL of fresh methylene blue solution. Remove dead or unfertilized eggs.
9. Keep embryos in solution for 1–2 weeks until black eyes are visible, and proceed to transferring the embryos to a moist solid substrate.
10. Add moist coconut fiber (washed and autoclaved, see the "Reagent setup and additional protocols" section) to the bottom of the 6 cm Petri dishes, and compress it to a flat surface (approximately 0.5 cm deep) using a clean paper towel. Alternatively, moist filter paper can replace the coconut fiber (Dodzian et al., 2018).
11. Using a Pasteur pipette, transfer developed embryos (with black eyes) from the methylene blue solution to the Petri dish containing coconut fiber. Minimize transfer of liquid to keep the plate humid, but not wet. Close the lid of the Petri dish, seal with parafilm to prevent drying, and keep at 28°C. Retain undeveloped embryos in methylene blue solution and monitor their development for 1–2 weeks until they are ready to be transferred. Otherwise, proceed to heat-shock, see the "Reagent setup and additional protocols" section
12. Keep embryos spread apart, with up to ~50 embryos per 6 cm plate.

Breeding for maintenance

Similar to their natural habitat, killifish embryos require incubation of ~2 weeks in aqueous solution and ~2 weeks on a humid substrate. This period could be extended if fish enter diapause. In breeding for line maintenance, the approaches are largely similar, yet less time sensitive, and thus will be discussed in brief. There have been several protocols for line maintenance (Dodzian et al., 2018; Harel et al., 2016; Polacik et al., 2016), and their advantages are discussed in Chapter 12.

Hatching of injected eggs and rearing of fry

The growth rate of fry is highly dependent on the density in which they are placed. Here, we present the maximal density that allows normal growth of the fish. At lower densities, growth is faster (Dodzian et al., 2018). Different approaches for hatching are discussed in Chapter 12.

1. After 2 weeks, transfer all the developed embryos (with golden eyes) from the coconut fiber, using a cut plastic Pasteur pipette, into cold (4°C) humic acid (Dodzian et al., 2018; Harel et al., 2016); for preparation see the "Reagent setup and additional protocols" section. This could be done in a 0.8 L fish tank filled to approximately 1 cm in depth, and placed on the

bench. Alternatively, hatching can be done in a 28°C incubator (Dodzian et al., 2018).

2. To improve oxygen supply, provide aeration with a standard aquarium air pump connected to a tube with an air stone.
3. A day after, fill the tank with an addition of 1:1 system water (or autoclaved water with equivalent salinity) on top of the humic acid, and feed fry daily with freshly hatched brine shrimp. Repeat for 3–4 days. Be sure to remove the dead artemia that the fry have not eaten. The buildup of ammonia from dead artemia may damage the fry.
4. When embryos are hatched prematurely, the majority could end up as "belly-sliders" (i.e., embryos that have not successfully inflated their gas bladder), and many will die. Therefore retain unhatched embryos in solid substrate, and try to hatch them a week later. An alternative is to put the nonhatched embryo in a new cold humic acid solution after 24 h.
5. At 7 days posthatching, separate fry in groups of up to 10 in 1.8 L tanks using a 400 μm fry screen. Place tanks in the main water system, using a slow drip of water flow for several hours to slowly acclimate the fry.
6. Feed twice per day with brine shrimp until 2 weeks of age.
7. At 2 weeks of age, replace with 850 μm fry screen and split fish up to five in a 3 L tank.
8. Once sexually mature (3–4 weeks posthatching), separate into breeding families. Alternatively, for lifespan assays, individually house fish in a 1.8 L tank.
9. Feed adult fish 2–3 times a day with dry food supplemented with live/frozen food (such as brine shrimp or blood worms).
10. See the following table for a brief summary:

Housing					
Age	0–7 days	7–14 days	14 days-maturity (3–4 weeks)	Mature (>3–4 weeks) breeding	Mature (>3–4 weeks) lifespan
Housing density	<30 in a 0.8 L tank (out of the system)	<10 in a 1.8 L tank	<5 in a 3 L tank	<3 in a 3 L tank	1 in a 1.8 L tank
Tank screen	None	400 μm	850 μm	None	None
Feeding					
Brine shrimp	Exclusively	Mainly	Yes	Supplemental only	Yes
Dry food	None	Small amount for training only	Yes	Primarily	Yes
Feeding frequency	Once a day	Twice a day	2–3 times a day	2–3 times a day	2–3 times a day

Reagent setup and additional protocols

Embryo solution

Yamamoto's embryo solution (Harel et al., 2016) or Ringer's solution could be used to maintain embryos. Here, we describe the procedure for Ringer's solution. Dissolve two RINGER tablets/L (Millipore) in double distilled water (DDW). Methylene blue (100 μL/L) should be added to the Ringer's solution (2.3% wt/vol stock solution) to limit parasitic infection. The solution could be kept at room temperature for 2 weeks protected from light. Alternatively, autoclaved system water with methylene blue could be used.

Coconut fiber preparation

Presoak coconut fiber with distilled water and autoclave. Once cooled, squeeze out water (keeping it moist, but not dripping), and keep in closed containers at room temperature, or at 4°C for later use (use within a few weeks). Alternatively, moist filter paper can replace the coconut fiber (Dodzian et al., 2018).

Humic acid preparation

Dissolve 1 g/L humic acid in RO water added with 1 g/L of marine salt. Alternatively, you may use system water (Dodzian et al., 2018). Mix well and autoclave. Refrigerate at 4°C. This will be good for several weeks if kept refrigerated.

Disinfecting embryos by iodine

There are several commonly used approaches for disinfecting fish eggs, including H_2O_2 (Dodzian et al., 2018), iodine, and NaOCl (Harel et al., 2016). Here, we will describe the povidone-iodine (PVP-I) approach (Chang et al., 2015).

1. Prepare a 10% stock solution of PVP-I in DDW, stir carefully. As PVP-I is light sensitive, keep in a dark place (use within 6 months).
2. Freshly prepare 25 ppm iodine working solution: three drops of the stock solution in 40 mL Ringer's solution. Check once a month with an iodine strip tester (Chang et al., 2015).
3. Transfer embryos into a new 6 cm Petri dish, and wash them with 5–7 mL methylene blue solution (to remove debris and dead eggs).
4. Incubate embryos in 5–7 mL of PVP-I solution for 10 min in the same plate.
5. Wash twice in 5–7 mL methylene blue solution for 5 min in the same plate.
6. Fill the plate with methylene blue solution and incubate at 28°C.

Long-term storage of embryos and cryopreservation of sperm

For long-term storage of embryos (tested for up to a year), fertilized and disinfected eggs can be stored at 17°C to induce diapause II, which occurs after the completion of somitogenesis and prior to development of visible black eyes. Embryos could be stored either in liquid or solid conditions, though using filter papers allows for better monitoring. Eggs that have developed passed this stage should not be kept long term. Recently, a protocol for cryopreservation and in-vitro fertilization of African turquoise killifish was optimized, providing great promise for long-term storage (Antebi et al., 2020).

Receiving or shipping fertilized eggs

For initiating a colony or for distribution of strains, embryos are usually shipped in 6 or 10 cm Petri dishes with moist coconut fiber. This could be done a few days after collection, as well as when eyes are visible. Petri dishes are then sealed with parafilm and, when weather conditions are not below freezing, can be shipped via regular mail. Upon arrival, it is important to disinfect the embryos (see the "Reagent setup and additional protocols" section), and care for them as described earlier. Hatched fish should be kept separately from the colony (preferably in a quarantined area), and their health should be monitored. If fish are healthy, their disinfected progeny can eventually be moved into the main recirculating system.

Heat-shock protocol

Diapause, an alternative developmental trajectory during killifish embryogenesis, is age and temperature dependent. Accordingly, the proportion of diapaused eggs increases when females are at an older age, or egg incubation is at lower temperatures. In the laboratory, as this can slow down routine maintenance, diapaused eggs could resume normal development using a simple heat-shock protocol.

1. Freshly prepare methylene blue solution as described.
2. Add one to three eggs per 0.2 mL PCR tube, and fill with methylene blue solution. Close tubes properly to avoid desiccation.
3. Use the following program on a standard thermocycler during 2–3 days: 6 h at 28°C, 6 h at 37°C.
4. Transfer eggs to a 6 cm Petri dish filled with 5–7 mL of methylene blue solution and keep the plate at 28°C in a humidified incubator. Monitor eggs daily for a week and remove dead eggs. Continue as described in the "Hatching of injected eggs and rearing of fry" section.

Efficient genome engineering approaches

In this section, we will focus on specific adaptations from other fish protocols, tailored specifically for the biology and characteristics of the African turquoise killifish. As commonly used in other fish models, genome engineering approaches have been facilitated by microinjection protocols

(Rembold et al., 2006), transposase systems (Davidson et al., 2003; Grabher and Wittbrodt, 2007; Kawakami et al., 2000) (such as Tol1, Tol2, and Sleeping Beauty), and more recently, genome editing approaches, such as zinc finger nucleases and transcription activator-like effector nucleases (Bedell et al., 2012), and CRISPR/Cas9 (Auer et al., 2014; Chang et al., 2013; Hwang et al., 2013; Jao et al., 2013). Recently, many of these approaches were successfully developed for the turquoise killifish, including CRISPR/Cas9-based genome editing (Harel et al., 2015, 2016), transgenesis (Allard et al., 2013; Harel et al., 2016; Hartmann and Englert, 2012; Valenzano et al., 2011), and microinjection (Harel et al., 2016; Hartmann and Englert, 2012; Valenzano et al., 2011).

In the "Target selection and synthesis of reagents for CRISPR/Cas9 genome editing" section we describe a detailed protocol for CRISPR/Cas9-based genome editing in the turquoise killifish (Harel et al., 2015, 2016), which was recently demonstrated for additional disease models (Matsui et al., 2019), and for exploring diapause (Hu et al., 2020) and regeneration (Wang et al., 2020a). Knock-in alleles via HDR are described elsewhere (Harel et al., 2015, 2016). We include simple guidelines for gRNA target design, microinjection, detection of germline transmission, and minimizing off-target effects. Additionally, in the "Tol2-based transgenesis" section we describe the required adaptations for transgenesis. This step-by-step protocol offers efficient genome engineering for the turquoise killifish, and can be used in small and large killifish colonies alike. The protocol includes three main stages: target selection and synthesis of reagents, injections, and germline transmission.

Target selection and synthesis of reagents for CRISPR/Cas9 genome editing

Target selection can be achieved via CHOPCHOP v2 (Labun et al., 2016), which contains the genome sequence and gene models for the turquoise killifish (Harel et al., 2015; Valenzano et al., 2015). This platform provides detailed parameters and information, predicted efficiency and off-targets, possible restriction sites for confirming successful editing, and oligonucleotide candidates for PCR amplification of the edited region. As demonstrated for other Cas9-based target predictions (Cong et al., 2013; Doudna and Sontheimer, 2014; Jinek et al., 2012; Mali et al., 2013), gRNAs are selected according to the following structure: 5′-(N)20-NGG-3′. The core target sequence is represented by the first 20 base-pairs, while the 3′ "NGG" represents the protospacer adjacent motif site in the genome (Cong et al., 2013; Doudna and Sontheimer, 2014; Jinek et al., 2012; Mali et al., 2013).

Several promoters could be used for the in-vitro transcription (IVT) of gRNA, which might affect the selection of gRNAs. For example, when using the T7 promoter, the first two 5′ nucleotides should be limited to GG (as described in this protocol), or to GN when using the U6 promoter. It is advised to select two to five different gRNAs for each desired gene of interest, and use the most efficient one for generating the mutation. Next, Cas9 mRNA and the gRNAs are produced by IVT. gRNAs can be synthesized by T7-mediated transcription without the need of cloning. Instead, two annealed oligonucleotides are used as templates (Gagnon et al., 2014; Harel et al., 2015; Hwang et al., 2013), and one of them already contains the T7 promoter sequence. For synthesis of the Cas9 mRNA, SP6-mediated transcription of a codon-optimized nuclear localized version (Jao et al., 2013) (nCas9n) template is available. We have successfully used CRISPR ribonucleoprotein delivery (Harel, unpublished), though we found no added benefit.

Target selection

1. The genome and complete annotation of the turquoise killifish is available through https://www.ncbi.nlm.nih.gov/genome/?term=furzeri. Select a gene (or genes) of interest, and identify conserved regions between vertebrate orthologs using http://genome.ucsc.edu/. These regions are less likely to be alternatively spliced or have an alternative start codon.
2. Identify gRNA target sequences using CHOPCHOP v2 (http://chopchop.cbu.uib.no) (Labun et al., 2016). This platform accepts a range of inputs, including gene identifiers, genomic regions, or pasted sequences. CHOPCHOP ranks the proposed gRNA according to predicted efficiency and number of potential off-targets. Selecting gRNA targets in close proximity to a unique restriction site can facilitate identification of successful editing. Please note that as we are using T7 polymerase for gRNA synthesis, the first two 5′ nucleotides should be limited to Gs, as follows: 5′-GG-(N)18-NGG-3′.
3. Using either CHOPCHOP v2, Primer3 (http://bioinfo.ut.ee/primer3-0.4.0/primer3/), or PrimerBLAST (https://www.ncbi.nlm.nih.gov/tools/primer-blast/index.cgi), select oligonucleotides that are flanking the gRNA sequence, and verify the sequence by Sanger sequencing of the purified PCR product. We recommend 400–700 bp amplicon. Sanger sequencing is advised to verify that naturally occurring genetic variation would not disrupt the selected core gRNA sequence.

gRNA template preparation

1. To generate the template for the gRNA IVT, order standard oligonucleotides (e.g. 200 μM stock concentration) shown in the following table. Additional examples and a comprehensive diagram can be found in Figs. 6 and S4 and Table S1 in Harel et al. (2015) and in Fig. 2A in Harel et al. (2016).

Oligonucleotide	Sequence (5′ to 3′)
Universal reverse oligonucleotide	AAAAGCACCGACTCGGTGCCACTTTTTCAAGTTGATAACGG ACTAGCCTTATTTTAACTTGCTATTTCTAGCTCTAAAAC (used for generation of all gRNA templates)
Variable oligonucleotide	TAATACGACTCACTATA[**GG-(N)18**]GTTTTAGAGCTAGAAATA GCAAG (core target sequence in bold is unique for each gRNA)

Note that as the universal reverse oligonucleotide will be used in all the gRNA template generations, is it advised to aliquot and freeze stocks at −20°C to avoid freeze/thaw cycles.

2. Prepare the following annealing reaction in a PCR tube and mix by pipetting:

Components	Amount (μL)	Final concentration
Universal reverse oligonucleotide (200 μM)	1	20 μM
Variable oligonucleotide (200 μM)	1	20 μM
NEB 2.1 buffer 10X	1	1X
Nuclease-free water	7	
Total	10	

3. Use the following program on a standard thermocycler to anneal the oligonucleotides: 95°C 30 s, 72°C 2 min, 37°C 2 min, 25°C 2 min, 12°C 2 min, 4°C forever. After annealing, 5′ ssDNA overhangs will remain on both sides of the template. Use T4 DNA polymerase to catalyze the gap filling by adding the following to the reaction mix:

Component	Amount (μL)	Final concentration
dNTP mix (25 mM)	0.5	1.25 mM
T4 DNA polymerase (3000 units/mL)	0.5	0.15 units/μL
Nuclease-free water	9	
Total	10	

4. Using a standard thermocycler, incubate the reaction at 12°C for 20 min.
5. Clean the annealed and gap-filled template using a standard PCR purification kit, and elute in 30 μL of Elution Buffer. Expect a concentration of ~0.5 μg/mL. The template can be stored at −20°C.

Synthesis of gRNAs (1–2 days)

1. For gRNA synthesis, use 1 μL of the eluted template in a 20 μL TranscriptAid T7 High Yield Transcription Kit, according to the manufacturer's instructions (note that scaling down the volume to one-quarter reaction [5 μL] is enough).

2. Following the TURBO DNase treatment (provided in the TranscriptAid T7 kit), precipitate the RNA by adding the following components in the order listed, and mix well before adding the ethanol:

Component	Amount (μL)	Final concentration
Reaction mix	20	
Nuclease-free water	115	
3M Na acetate	15	0.1 mM
GlycoBlue coprecipitant	1	
Ethanol 100% (molecular biology grade)	300	
Total	451	

3. Mix well (or vortex briefly), and chill at −20°C for 15 min to overnight.
4. Pellet the gRNAs for 15 min in a chilled microcentrifuge (4°C) at top speed. Remove the supernatant, briefly respin the tube, and remove residual fluid with a fine-tipped pipet. Note: Do not let the pellet completely dry, otherwise it will not dissolve. Sometimes, due to a high salt content in the kit's buffer, a white ring of pellet will remain after dissolving the RNA; this does not affect the gRNA's effectiveness.
5. Resuspend the pellet in 30 μL dH_2O or 1X TE buffer by letting the tube sit on ice for 30 min. Measure concentration using NanoDrop and adjust for a final concentration of 0.5−1 μg/μL.
6. Run 2 μL of the reaction using a formaldehyde loading dye (provided in the TranscriptAid T7 kit). A discrete band should be seen. Aliquot in 2−3 μL aliquots and store at −80°C.

Synthesis of Cas9 mRNA (1−2 days)

1. Purify the nCas9n expression plasmid DNA (Addgene, #47929 Jao et al., 2013) using a standard Maxiprep Kit.
2. Linearize 5 μg of the nCas9n expression plasmid DNA with 1−2 μL of NotI at 37°C for 2 h. The remaining undigested plasmid significantly reduces mRNA synthesis. Therefore verify complete digestion by gel electrophoresis, and increase incubation time or amount of enzyme if necessary.
3. Purify the digested nCas9n expression plasmid using a PCR purification kit, and elute in 50 μL of 1X Elution Buffer.
4. Transcribe capped and polyadenylated nCas9n mRNA using the mMessage mMachine SP6 kit according to the manufacturer's instructions.
5. Following the TURBO DNase treatment, purify the resulting mRNA using lithium chloride precipitation (provided in the mMessage mMachine SP6 kit). Do not let the pellet fully dry, otherwise it will not dissolve.

6. Resuspend the nCas9n mRNA pellet in dH_2O or TE buffer and incubate at 65°C for 10 min to completely dissolve the mRNA. Adjust to a final concentration of 1–3 μg/μL.
7. Run 2 μL of the reaction using a formaldehyde loading dye (provided in the mMessage mMachine SP6 kit) to test for potential degradation. Prepare 2–3 μL aliquots and store at −80°C.

Tol2-based transgenesis

DNA transposons have been widely used as efficient tools for generating transgenic animals (Clark et al., 2011). The medaka Tol2 element encodes a functional transposase that can facilitate transposition of a DNA construct flanked by specific Tol2 DNA sequences (Kawakami, 2007). This way, a gene cassette can be randomly integrated into the host genome. This system has been demonstrated to be highly efficient in many vertebrate model systems and cell lines, including zebrafish, *Xenopus*, chicken, mouse, and human cells (Kawakami, 2007; Kwan et al., 2007).

Importantly, many steps described in the protocol for the "Target selection and synthesis of reagents for CRISPR/Cas9 genome editing" and "Evaluation of editing efficiency (1–2 days)" sections are not required for Tol2-based transgenesis as a result of the following differences: (1) there is no need for gRNA target selection and synthesis; (2) instead of nCas9n mRNA, Tol2 transposase mRNA is synthesized (the overall protocol for Tol2 mRNA synthesis resembles that of nCasn9 synthesis, with specific differences highlighted below); (3) as we are inserting a foreign sequence into the host genome, a plasmid containing a gene cassette flanked by two Tol2 sequences is required and coinjected; and finally (4) the detection of germline transmission is different because integration is random, and in many cases could be facilitated visually by the use of a fluorescent reporter.

1. Purify the PCS2FA plasmid encoding for the Tol2 transposase (Kwan et al., 2007) using a Maxiprep Kit. The PCS2FA plasmid is available from Dr. Kawakami (kokawaka@lab.nig.ac.jp). Additional details, tools, and alternatives can be found on the Tol2Kit website: http://tol2kit.genetics.utah.edu/index.php/Main_Page.
2. Linearize the PCS2FA plasmid, synthesize, and purify the Tol2 transposase mRNA by following the steps described earlier for nCas9n. (NotI restriction enzyme is also used to linearize PCS2FA.)
3. Prepare your desired construct flanked by specific Tol2 DNA sequences (Kwan et al., 2007).

4. Prepare an injection mixture (35 μL recommended volume) as shown in the following table:

Component	Amount (μL)	Final concentration
Tol2 transposase mRNA (300 ng/μL)	1–2	30 ng/μL
DNA plasmid (400 ng/μL) (containing a transgene of choice flanked by Tol2 sites)	1–2	40 ng/μL
Phenol-red (0.5% wt/vol)	2	0.1% wt/vol
Nuclease-free water	Up to 10	
Total	10	

5. Perform injections as described in step 2, and assess germline transmission visually using a fluorescent stereoscope in the F1 (when fluorescent reporters are used), or use a set of oligonucleotides unique for the transgene for genotyping.

When trying a new promoter, it is recommended to use a small surrogate marker on the same plasmid, such as a fluorescent protein under the control of the *cardiac myosin light chain* (*cmlc*) promoter (Harel et al., 2016; Huang et al., 2003) or the *gamma-crystalin* (*γ-cryst*) promoter (Davidson et al., 2003), which is specifically expressed in the lens. Note that the transgene can randomly integrate into several places in the genome, and outcrosses should be conducted until the construct is inherited in a Mendelian manner. Additionally, new transgenic constructs could be generated using the Tol2Kit (Kwan et al., 2007) with the turquoise killifish bacterial artificial chromosome library (Reichwald et al., 2015).

Microinjection

Overall, the components required for the microinjection are similar to protocols developed for other fish (Rembold et al., 2006), but require specific adaptations.

Equipment setup

- **Preparation of injection needles:** As an adaptation to their environment, the turquoise killifish eggs have a thick chorion. Thus needles have a shorter and less flexible tip than the needles used for zebrafish injections. As each needle puller or filament type will have a different setting, it is advised to compare your results with the examples presented in Fig. 5 in Harel et al. (2016). We present the parameters for borosilicate glass needle designs that have been optimized for filamented microcapillaries on a Sutter Instrument P-87 (with a trough filament). Overall, the tip is best cracked/broken by gently rubbing the tip on a Kimwipe, or breaking it on the chorion before injection (using forceps usually results in an opening that is too big). Pulled needles can be stored (for years) in a closed Petri dish with two strips of modeling clay to hold them. Importantly, for other pullers and filaments, parameters can change. Use these parameters as a starting point, and optimize based on personal experience and similarity to needle shape.

Pressure	Heat	Pull	Velocity	Time
450	43°C	40	40	250

- **Microinjection procedure (1 day)**
  1. For making the injection mold, boil 1.5%–2% agarose in 50 mL of methylene blue solution and pour 15 mL into a 10 cm Petri dish. Plates could be made in advance, sealed with parafilm and stored at 4°C.
  2. While the agarose is still hot, lay a plastic injection mold on the surface (it will float). Once the mold is removed, this will produce six troughs that will hold eggs in position. A 3D-printing design is available in Harel et al. (2016); make sure to use heat-resistant and nontoxic materials.
  3. Collect eggs for microinjection as mentioned in the "Breeding and embryo incubation for microinjection and genetic manipulation" section.
  4. Prepare the injection mixture as shown in the following table, and keep it on ice.
  5. Immediately before injections, backfill the needle with 5 μL injection mixture using fine pipette tips, as routinely used for gel loading, or alternatively, leaving them upside down in the mixture tube, and wait for the mixture to migrate to the top of the needle.
  6. Note that as a result of the thicker chorion and thicker needles, injections might require some practice. To facilitate piercing through the chorion, embryos can be stored at 4°C for 10 min before injection.

Component	Amount (μL)	Final concentration
Purified gRNA (150–300 ng/μL)	2–4	30 ng/μL
nCas9n mRNA (1–3 μg/μL)	2–6	200–300 ng/μL
Phenol-red (0.5% wt/vol)	4	0.1% wt/vol
Nuclease-free water	Up to 20	
Total	20	

1. Using a stereoscope, place the fertilized eggs into the agar troughs, and orient the single cell (or the two cells) upward toward the injection needle.
2. Using the injection apparatus, inject into the cytoplasm of the single- or two-cell stage embryos using the following parameters: 30 psi with 75 ms pulses—approximately 2.5–10 pL of injection mixture, which corresponds to about 10% of the cell's volume.
3. After injection, disinfect the eggs with PVP-I (see the "Reagent setup and additional protocols" section), monitor embryos daily, and remove dead

embryos (opaque and blue in color). Expect about 30%–50% survival of injected embryos, and about 50% chimeric embryos when using efficient gRNAs.

Note: Adjust back pressure so the needle in the liquid medium looks like a "smoking gun," with slight leaking of the red injection solution from the needle tip. Make sure that a small red diffused dot (phenol-red) is seen in the cell after injection. Injection parameters may vary slightly between different needles and injection apparatuses. To adjust the parameters to achieve a reproducible injection volume, inject into mineral oil placed on top of a micrometer slide (see the "Materials" section) and measure the diameter of a droplet. Injection volume can be calculated according to the following formula: $\frac{4}{3}\pi r^3$. Note the difference between the injection volume in zebrafish (usually between 0.5 and 2 nL).

Evaluation of editing efficiency (1–2 days)

In CRISPR injections, genome editing efficiency of individual (or pooled) gRNAs could be estimated using any of the recently developed approaches, such as restriction enzymes (in which a unique restriction site is disrupted upon successful editing, or a novel one is created Harel et al., 2015; Harel et al., 2016), high-resolution melting (Thomas et al., 2014), CRISPR-STAT (Carrington et al., 2015), and others. In cases where the indel is very small (i.e., 1–2 base-pairs), direct sequencing of the amplified genomic region could be used (Harel et al., 2016). Once an optimal gRNA is selected, additional injections should be done to generate F0 chimeric founders. Although the protocol is focused primarily on CRISPR, it could be used for Tol2-based transgenesis with standard fluorescent markers as readout for success.

1. DNA isolation protocol (Samarut et al., 2016) is performed as follows: prepare the extraction buffer, 50 mM NaOH. Prepare the neutralization buffer, 10 mM Tris HCl pH 8. Store solutions at room temperature for up to 1 year.
2. 48–72 h after injection, isolate genomic DNA from individual embryos by adding 20 μL of extraction buffer to a 1.5 mL tube. Note: Remove remaining methylene blue solution with a fine tip.
3. Crush embryos with a cleaned/autoclaved pestle, spin down, and incubate for 10 min at 95°C.
4. Vortex and spin down, and add 2.2 μL of neutralization solution (1:10 vol/vol).
5. Vortex and spin down at max speed for 1 min. Use 1–2 μL for PCR reaction and proceed to identify successful editing using the methodology of your choice.

6. When successful editing is confirmed on the "test set" of embryos, hatch the remaining injected embryos. Efficiency can range between 0% and 100%, and it is advised to continue if the efficiency is above 10%.

Germline transmission and outcrosses

Set mating pairs with sexually mature F0 chimeric founders. By setting them up with mating pairs with wild-type fish, germline transmission can be evaluated using their progeny (fertilized eggs) by either of the methods described earlier. As their germline could be chimeric, multiple types of indels could be present in an individual fish. Once germline transmission has been confirmed for an individual fish, adult F1 progeny from that successful pair should be genotyped by fin clips or from scales. Heterozygous fish at this stage are considered stable lines, and can be outcrossed to wild-type fish for two to three generations to eliminate potential off-target effects generated by the Cas9 nuclease. For efficient gRNA targets, 2–10 founders are usually sufficient to successfully generate a stable CRISPR line. For the less efficient HDR-based knock-in approach (Harel et al., 2015, 2016), 10–30 founders are usually sufficient.

Genotyping of adult fish

1. Prepare a tricaine stock solution (10 mg/mL): add 1 g sodium bicarbonate and 1 g tricaine to 100 mL molecular biology-grade water. Stir thoroughly using a magnetic stirrer. Keep at 4°C. The solution remains good for several weeks if kept refrigerated.

   Note: Apply breathing mask and safety goggles when handling the powdered tricaine.
2. For genotyping, anesthetize adult fish by adding 4–5 mL of tricaine stock solution in 200 mL of system water. After a few minutes, fish will stop moving.
3. Using a plastic spoon, gently place fish on a sponge (moistened with system water), and quickly trim 2–3 mm from the tail fin with a sterile razor blade. Alternatively, remove two to three scales with forceps or a cotton swab (Dodzian et al., 2018), which is less invasive.
4. Place the tissue in 0.2 mL tubes and process as described in the "Evaluation of editing efficiency (1–2 days)" section.
5. Gently put the fish in 1 L of fresh system water at room temperature to recover for 10 min before returning to the system.

Troubleshooting

Problem	Possible reason	Possible solution
Poor nCas9n or Tol2 mRNA synthesis or degraded product	RNA synthesis kit	Consider aliquoting the synthesis buffer, as it is sensitive to repeated freeze/thaw cycles
	Incomplete digestion of the DNA template	As mentioned earlier, remaining undigested plasmid DNA can lead to very long RNA molecules that can significantly reduce the overall efficiency. Alternatively, gel purification after digestion could be attempted.
	RNase contamination	Use RNase-free reagents and RNase decontamination solutions to clean working surface
Low survival or abnormal development of injected embryos	Nucleic acid toxicity	High concentrations of nucleic acids, primarily DNA, are toxic to the cell. According to our experience, 100 ng/μL of DNA and 330 ng/μL of RNA are safe.
	High injection volume	To prevent physical damage to the injected cell, the injection solution must be lower than 1/10th of the cell's volume (approximately 2.5 −10 pL). Possible solutions include: lowering the injection volume; decreasing the diameter of the needle opening; lowering the air pressure; or decreasing the injection pulse of the injector.
Little or no genome editing detected in injected embryos	Inefficient delivery of injection mixture	Centrifuge the injection mixture at 11,000 g in a chilled centrifuge for 5 min to prevent clogging of the needle
	mRNA degradation	nCas9n and Tol2 mRNA may be degraded. Check mRNA integrity by electrophoresis using agarose gel before making the injection solution.
	Accidental injection into the yolk	Although ssRNA can still be delivered into the cell when injected into the yolk, injections into the cell are more efficient. dsDNA cannot be shuttled this way, and has to be injected directly into the cell. If the single cell is not visible, incubate fertilized eggs at 28°C for 10–30 min to help visualize the single cell. Eggs without a visible single cell at this stage are probably unfertilized.

Fish dissection

Here, we describe a protocol for dissecting and isolating adult killifish organs, which can be used for histology, in situ hybridization, immunohistochemistry, RNA extraction, protein analysis, and other downstream molecular techniques (Fig. 11.3). Although similar protocols, such as for zebrafish (Gupta and Mullins, 2010) exist, aside from characterization of specific organs (D'Angelo, 2013), or anatomy of other killifish species (such as *Aphanius* Motamedi et al., 2019), a detailed description of *N. furzeri* has been lacking.

Experimental setup

1. Prepare two ice buckets, one for dissection and one to keep organs and solutions.
2. Prechill 1X phosphate buffered saline (PBS), and prepare two fine-tipped forceps, two curved forceps, one pair of small scissors, a razor, and one holder for scalpel with affixed scalpel blade.
3. Prepare your euthanasia solution by adding 5 mL of tricaine stock solution (see the "Genotyping of adult fish" section, step 1) to 100 mL of system water.

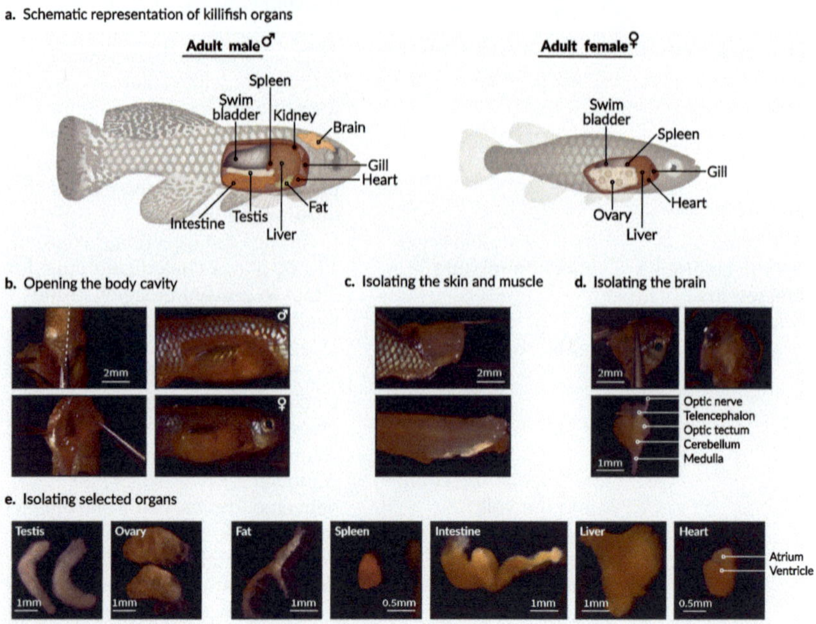

FIGURE 11.3 (A) Adult male and female turquoise killifish *Nothobranchius furzeri* are shown with the anatomical structures labeled. (B—E) A visual protocol for isolating key organs, including opening the body cavity. (B) Organs can be identified according to the anatomical diagram in (A). Isolating the skin and muscle (C), isolating the brain (D), and isolating selected organs, such as the gonads, fat tissue, spleen, intestine, liver, and heart (E).

4. Place fish in the euthanasia solution for 5 min until the operculum stops moving.
5. Using a plastic spoon, place the fish on the operating table.

Organ dissection

At this stage it is possible to bleed the fish by cutting the base of the tail while the heart is still pumping. This will allow a cleaner dissection of the organs with reduced bleeding.

Body cavity dissection

1. Use the scissors to make a cut across the ventral midline of the fish, from the gills to the cloaca. Make shallow and continuous cuts to avoid damaging the internal organs.
2. It is recommended to remove the internal organs from the body cavity to separate the organs more easily. Detach the intestine from the cloaca using scissors or fine-tipped forceps. Hold the detached end of the intestine and pull it outside of the body cavity. Most of the organs will follow.
3. Apply a few drops of 1X PBS on the operation plate and suspend the content of the body cavity. This will allow a better view of the organs, and will be very helpful in younger fish. All the organs removed from the body cavity can be easily separated from one another using scissors or a pair of fine-tipped forceps.
4. The **gastric tract** is readily recognized; it is a long yellowish tube that dominates the body cavity (the color may be affected by the contents of the intestine).
5. The **liver** is the second largest organ within the body cavity; it can be recognized by its size, distinctive color, and shape, which is similar to the mammalian liver.
6. The **gonads** are twin organs located parallel to the intestine. The **testes** are white and elongated, and are comprised of many tubules. The **ovaries** are much larger sacks filled with round eggs at different stages of maturity. The gonads will vary greatly with size as the fish matures.
7. **Adipose tissue can be seen** adjacent to the gonads, and is more visible in old fish. Although it resembles the ovary, it is more transparent and reflective.
8. The **spleen** is a small kidney-shaped dark red-brown organ, and is located near the liver in a loop of the intestine.

Trunk dissection

Detach any remaining organs from the body cavity using scissors and fine-tipped forceps.

1. The **kidneys** are two thin strips of brown tissue that run parallel to the spine ("trunk kidney"), and grow into larger bulbus structures near the head ("head kidney"). Dissect them by locating their posterior tip, and gently use a fine-tipped forceps and scissors to cut across the length of the entire kidney.
2. Detach the **skin** from the **muscle** using forceps. The muscle can be cut using a scalpel, cutting close to the spine to maximize muscle collection.

Dissecting the head

1. The **heart** is located posteriorly and ventrally to the gills. Cut the midline gently through the operculum, and the heart will be at the anterior end of the cut. It is a small, bright red, teardrop shaped, and will most likely still pulse.
2. For isolating the **brain**, use a razor blade to decapitate the head posterior to the gills. Gently remove skin and muscle until you reach the skull; clear the top of the skull.
3. Hold the head from the mouth and lift the skull from the decapitated side using forceps. The skull should come off easily, and the brain will be exposed underneath. Make sure you can see the entirety of the brain, including the **brain stem**, **olfactory bulb**, and **optic nerve**.
4. Cut the optic nerves using scissors and pull the eyes out of their sockets using fine-tipped forceps.
5. Insert a closed fine-tipped forceps underneath the brain and lift it gently. Use either a gentle lifting motion or scissors to detach the brain from the skull. Start at the posterior end, and finish with the olfactory bulb. The brain should come out of the skull in one piece.

Note: Clean working area and surgery equipment between fish with ethanol 70%.

The African turquoise killifish: available toolbox

The African turquoise killifish has been utilized to explore various biological questions, including genomics and genetics, ecology, nutrition and toxicology, embryonic development, regeneration, aging, and more. Here, we will highlight selected studies focusing mostly on "omics" datasets, aging-related phenotypes, recently developed genetic models, and embryonic development.

"Omics" datasets

In recent years, a wide range of studies have generated comprehensive high-throughput datasets exploring killifish biology, as reviewed elsewhere (Cellerino et al., 2015; Harel and Brunet, 2015; Kim et al., 2016; Muck et al.,

2018; Platzer and Englert, 2016). Here, we will highlight a partial list, focusing primarily on aging studies comparing young and old killifish. RNA-seq analyses have been performed by multiple groups to characterize the molecular signature of an aging vertebrate organ, including the skin (Petzold et al., 2013), brain (Baumgart et al., 2014b; Kelmer Sacramento et al., 2020; Petzold et al., 2013), muscle (Cencioni et al., 2019), fin (Baumgart et al., 2016), and a few organs simultaneously (Aramillo Irizar et al., 2018; Baumgart et al., 2017). Other approaches include proteomic analysis of the aging brain using mass spectrometry (Kelmer Sacramento et al., 2020), chromatin landscape of the brain (Harel et al., 2015) and the aging muscle (Cencioni et al., 2019), and sirtuin expression levels in a wide variety of organs (Kabiljo et al., 2019). Among these datasets, it is worth highlighting the killifish genome project (Reichwald et al., 2015; Valenzano et al., 2015), QTL mapping of age-associated traits (Cui et al., 2019; Kirschner et al., 2012; Ng'oma et al., 2014), as well as recent studies exploring vertebrate regeneration (Wang et al., 2020a) and diapause (Hu et al., 2020; Reichwald et al., 2015).

Available genetic models

Transgenesis using Tol2-based random integration was the first step in developing genome engineering approaches in the turquoise killifish (Allard et al., 2013; Harel et al., 2016; Hartmann and Englert, 2012; Valenzano et al., 2011). For example, this technique has allowed tracking of the cell cycle using a FUCCI reporter (Dolfi et al., 2019), a regeneration reporter using the *inhba* enhancer (Wang et al., 2020a), and organ-specific markers, such as the heart (Harel et al., 2016). The ability to manipulate endogenous genes is an essential tool for a genetic model. Indeed, the recent development of CRISPR/Cas9-based efficient genome engineering approaches for the turquoise killifish, including "knockout" and "knock-in" (Harel et al., 2015, 2016), has transformed this organism into a state-of-the-art genetic model. We have characterized a loss-of-function mutation in the gene encoding the protein component of telomerase (TERT) and show that telomerase-deficient turquoise killifish recapitulate characteristics of human pathologies much faster than any other vertebrate model.

Additionally, as a proof of principle for the versatility of this platform, we have recently targeted 13 genes encompassing the hallmarks of aging (generating six stable lines), including cellular senescence and stem cell exhaustion (*p15INK4B*), loss of proteostasis (*ATG5*), deregulated nutrient sensing (*IGF1R, RAPTOR, RPS6KB1*, and *FOXO3*), mitochondrial dysfunction (*POLG*), epigenetic alterations (*ASH2L*), genomic instability (SIRT6), and intercellular communication (*IL8* and *APOE*). We specifically selected genes whose deficiency is expected to slow down (*IGF1R, RAPTOR*, and *RPS6KB1*) or accelerate (*TERT* and *POLG*) the aging process (Lopez-Otin et al., 2013), or have been involved in age-associated diseases, such as *APOE* (Alzheimer's disease Rhinn et al., 2013), *TERT* (dyskeratosis congenita Armanios, 2009), and *p15INK4B* (cancer Okamoto et al., 1995). Since then, several studies have generated mutant lines, including a loss-of-function mutation in *α-synuclein* to study Parkinson's disease (Matsui et al., 2019), a *CBX7* knockout allele to study the mechanism of diapause (Hu et al., 2020), and the role of a specific enhancer (K-IEN) in regeneration (Wang et al., 2020a).

Neurodegeneration and brain functions

Reflecting its importance in vertebrate aging research, including humans, many killifish studies have explored the age-related decline in behavioral and cognitive functions. These studies identified and characterized a comprehensive range of phenotypes, including a decline in spontaneous exploratory behavior (Smith et al., 2017; Terzibasi et al., 2008; Valenzano et al., 2006a, 2006b; Wei et al., 2020; Yu and Li, 2012), and in learning and memory (tested by both positive and negative conditioning) (Terzibasi et al., 2007, 2008, 2009; Valenzano et al., 2006a, 2006b; Wei et al., 2020; Yu and Li, 2012).

At the cellular level, an increase in age-related neurodegeneration was observed by histological sections and staining with Fluoro Jade B, a pan marker for degenerating neurons (Terzibasi et al., 2007, 2008, 2009; Valenzano et al., 2006b; Wei et al., 2020; Yu and Li, 2012), degeneration of specific types of neurons (Matsui et al., 2019), and inflammatory-associated expression of GFAP (a key component of glia cells) (Matsui et al., 2019; Terzibasi et al., 2009; Tozzini et al., 2012; Wei et al., 2020). Additionally, an age-associated decrease in adult neurogenesis was characterized (Tozzini et al., 2012). Molecular studies have further highlighted the involvement of micro-RNAs (Ripa et al., 2017; Tozzini et al., 2014), CCNB1 (Baumgart et al., 2014a), histone deacetylases (Zupkovitz et al., 2018), NPY and NUCB2B (Montesano et al., 2019a, 2019b), α-synuclein (Matsui et al., 2019), and NGF receptors (de Girolamo et al., 2020).

Cellular aging

Aging is accompanied by changes in cellular composition, as well as a gradual loss of the ability to maintain cellular and physiological homeostasis. In the killifish, several age-dependent signatures, both molecular and histological, have been characterized. For example, lipofuscin, an autofluorescent pigment that accumulates with age in mammals, has been shown to accumulate in killifish tissues with age (Ahuja et al., 2019; Harel et al., 2015, 2016; Liu et al., 2015, 2017b; Terzibasi et al., 2007, 2008, 2009; Tozzini et al., 2013; Valenzano et al., 2006a; Wei et al., 2020; Yu and Li, 2012). Other age-dependent changes include altered cellular composition and accumulation of senescent cells, as detected by senescent-associated beta-galactosidase staining (Genade et al., 2005; Liu et al., 2017b, 2018; Terzibasi et al., 2007; Valenzano et al.,

2006a; Wang et al., 2020b; Wei et al., 2020; Yu and Li, 2012) and an analysis of their unique secretory profile, the senescence-associated secretory phenotype (Liu et al., 2018).

Killifish also display molecular signatures that suggest a decreased ability to maintain cellular homeostasis, as demonstrated by the decline of antioxidant enzymes, the increase in reactive oxidative species, and accumulation of oxidative damage (Blazek et al., 2017; Heid et al., 2017; Liu et al., 2015). Along these same lines, a major cellular component that deteriorates with age in killifish is the mitochondria, with a decrease in mitochondrial numbers, a decrease in the expression of mitogenesis- and OXPHOS-related genes, changes in mitochondrial membrane composition (Almaida-Pagan et al., 2019), as well as a reduction in respiratory chain activity (Baumgart et al., 2016; Hartmann et al., 2011). A decline in proteostasis was recently described, focusing on reduced proteasome activity and overall loss of stoichiometry in protein complexes (Kelmer Sacramento et al., 2020). Finally, telomere attrition is observed, and telomeres shorten with age in the longer-lived MZM lines. Interestingly, as a potential response to telomere attrition, TERT expression increases with age (Hartmann et al., 2009), which may contribute to the observed increased cancer incidences.

Physiological aging

A primary characteristic of the aging process is increased frailty and the decline in organ and physiological functions. Killifish display loss of color with age, reduced muscle mass, known as sarcopenia (Cencioni et al., 2019), and bending of the spine, or kyphosis (Harel and Brunet, 2015; Kim et al., 2016). Metabolic functions are also altered with age, specifically lipid metabolism, including an increase in lipid peroxidation, accumulation of sphingolipids in the heart, and a decrease in cholesterol and triglycerides (Ahuja et al., 2019; Milinkovitch et al., 2018; Wei et al., 2020). Other physiological changes include a decline in fertility and in the quality of eggs leading to slower development of the offspring (before and after hatching) (Api et al., 2018b; Blazek et al., 2013, 2017; Lee et al., 2018; Liu et al., 2017a). Interestingly, even later in life, the offspring of older parents are smaller and less fertile (Api et al., 2018a; Liu et al., 2017a).

Cancer incidence is much higher in older killifish, specifically in the liver and kidneys (Baumgart et al., 2014a; Blazek et al., 2017; Di Cicco et al., 2011; Liu et al., 2017b). Other characteristics include inflammatory signatures (Benayoun et al., 2019; Liu et al., 2017a), and a decline in regenerative capacity of the kidneys, tail, and aberrant wound healing (Benayoun et al., 2019; Hoppe et al., 2015; Hu and Brunet, 2018; Liu et al., 2017a; Wang et al., 2020a; Wendler et al., 2015). Additionally, the diversity of the gut microbiome is reduced with age (Smith et al., 2017). Finally, adaptive immunity, represented by the antibody repertoire, is highly affected with age, demonstrating how the killifish can become a powerful model for studying adaptative immunosenescence (Bradshaw and Valenzano, 2020).

Diapause

Diapause is an ability to temporarily arrest embryonic development to avoid harsh conditions. There are three developmental stages in which killifish can enter diapause (termed diapause I, II, and III), and the main stage in which diapause primarily happens is diapause II, which occurs during somitogenesis (Api et al., 2018a; Dolfi et al., 2014; Naumann and Englert, 2018). Many of the studies of killifish diapause have been performed on the American species *Austrofundulus limnaeus*, discovering the involvement of protein synthesis, cell cycle arrest, mitochondria, IGF-I, and vitamin D in diapause (Podrabsky and Hand, 1999, 2000; Romney et al., 2018). Recently, several fascinating studies in the turquoise killifish have characterized gene expression signatures during diapause (Hu et al., 2020; Reichwald et al., 2015), observed cell-cycle dynamics of diapause entry and exit, including synchronization of cell division across the embryo body (Dolfi et al., 2014), and involvement of the polycomb complex in long-term preservation of diapause (Hu et al., 2020).

Summary and future perspectives

In recent years, the African turquoise killifish has emerged as a promising experimental model for exploring aging and age-related disease, developmental processes, ecology, and evolution. This expanding interest is supported by a vibrant and growing community worldwide, which meets biennially at the International Nothobranchius Symposium. Following a symposium in Pisa, Jena, and Cologne, the fourth meeting was held in Brno (Czech Republic) as an online meeting in 2021. As an emerging model system, further standardization for husbandry is ongoing, which possibly contributes to the variances in lifespan measurements reported by different labs (Terzibasi et al., 2008; Valdesalici and Cellerino, 2003; Valenzano et al., 2015). Alternatively, it is likely that these differences also reflect natural variability in aging studies, as observed in classical genetic models of aging. Lifespan is a complex multigenic trait, further affected by intricate gene–environment interactions throughout organismal lifespan. As a consequence, lifespan protocols are continuously developed and improved, also for the well-established mouse model (Bellantuono et al., 2020; Bogue et al., 2020). Thus mouse lifespan studies performed in different labs are facing similar challenges as seen in the National Institute on Aging Interventions Testing Program (Nadon et al., 2017).

In this chapter we presented optimized protocols for husbandry and genome engineering, focusing on Tol2-based transgenesis (Allard et al., 2013; Harel et al., 2016; Hartmann and Englert, 2012; Valenzano et al., 2011) and CRISPR/Cas9-based genome editing (Harel et al., 2015, 2016). This platform should be further expanded to include classical genome editing approaches that allow for conditional gene regulation in-vivo such as Gal4/UAS (Scheer and Campos-Ortega, 1999), Cre/loxP (Dong and Stuart, 2004; Langenau et al., 2005; Pan et al., 2005), and Tet-ON (Huang et al., 2005), as well as the more recent CRISPR interference (Larson et al., 2013; Qi et al., 2013) and CRISPR activation (Dominguez et al., 2016; Gilbert et al., 2014a) (which allow reversible repression or activation of endogenous genes). Genome engineering, particularly CRISPR/Cas9, is a continuously expanding field (Heidenreich and Zhang, 2016; Moon et al., 2019; Smargon et al., 2020; Yeh et al., 2019), which undoubtedly improves existing tools and opens new possibilities. Finally, this protocol could be adapted for genome engineering approaches in other annual and nonannual killifish species, significantly expanding the current state of the art.

Materials

Reagents

- African turquoise killifish (*N. furzeri*) GRZ strain. Available from the authors upon request. CRITICAL: A breeding turquoise killifish colony must be established before attempting this protocol (see the "Reagent setup and additional protocols" section). CAUTION: As a vertebrate model organism, turquoise killifish care and use has to be approved by and must adhere to relevant institutional ethics guidelines. All turquoise killifish care and use in our lab was approved by the Stanford Subcommittee on Research Animal Care.
- Sea salt (Instant Ocean, #SS15-10).
- nCas9n expression plasmid (Jao et al., 2013) (Addgene, #47929).
- Optional: Cas9 protein with NLS (1 mg/mL, PNA Bio). The Cas9 protein can be used instead of nCas9n mRNA.
- Custom DNA oligonucleotides (once annealed will serve as the gRNA template for the in-vitro transcription), 200 μM stock concentration and standard desalted (IDT, Integrated DNA Technologies).
- mMessage mMachine SP6 Ultra Kit (Life Technologies, #AM1340).
- mMessage mMachine T7 Ultra Kit (Life Technologies, #AM1345).
- TranscriptAid T7 High Yield Transcription Kit (Thermo Scientific, #K0441).
- DNase/RNase-free distilled water (Invitrogen, #10977-015).
- 100 bp DNA ladder (Promega, #G2101).
- 1 kb DNA ladder (Promega, #G5711).
- Ethidium bromide (Thermo Scientific, #15585011).
- Agarose (Life Technologies, #16500500).

- UltraPure phenol:chloroform:isoamyl alcohol (25:24:1; Life Technologies, #15593).
- Proteinase K solution (20 mg/mL) (Ambion, #AM2546).
- High Fidelity PCR SuperMix (Invitrogen, #12532-016).
- GoTaq Green Master Mix (Promega, #M712).
- PCR lysis buffer (Viagenbiotech, 102-T).
- Phenol-red solution 0.5% wt/vol (Sigma, #P0290).
- Humic acid (Sigma, #53680). CRITICAL: Humic acid solution needs to be prepared in advance, at least a day before injections. Refrigerate at 4°C. Good for several weeks if kept refrigerated (see the "Reagent setup and additional protocols" section).
- Coconut fiber (ZooMed Item # EE-8, eight dry quarts, loose coconut fiber substrate). CRITICAL: Coconut fiber needs to be prepared in advance, at least a day before it is required (see the "Reagent setup and additional protocols" and "Hatching of injected eggs and rearing of fry" sections).
- Oxygen tablets (Pemble-Halversion, Inc., http://otabs.com/).
- RINGER tablets (Millipore, #1155250001).
- Methylene blue (2.3% wt/vol stock solution, Kordon, #37344).
- Ammonia test kit (API).
- Nitrite test kit (API).
- Nitrate test kit (API).
- NaOH pellets (Sigma, 221465).
- Hydrochloric acid, 36.5%−38.0% (JT Baker, 9535-01).
- Polyvinylpyrrolidone-iodine complex (Acros Organics, 229471000).
- Iodine test strip (Amazon, 2948-BJ).
- TOPO TA Cloning Kit (Life Technologies, #K4600).
- Zero Blunt TOPO PCR Cloning Kit (Life Technologies, #K2800-20).
- One Shot TOP10 Competent *E. Coli* (Life Technologies, #C4040).
- Agarose (Life Technologies, #16500-500).
- EndoFree Maxiprep Kit (Qiagen, #12362).
- NEB buffer 2.1 (NEB #B7202S)T4 DNA polymerase (NEB, #M0203S).
- Deoxynucleotide (dNTP) Solution Set, 100 mM (NEB, #N0446S).
- PCR Purification Kit (Qiagen, #28104).
- GlycoBlue Coprecipitant, 15 mg/mL (Invitrogen, #AM9516).
- Ethylene diamine tetra acetic acid (0.5 M), pH 8.0 (Ambion, #AM9260G).
- Sodium acetate (3 M), pH 5.5 (Ambion, #AM9740).
- QIAquick Nucleotide Removal Kit (Qiagen, #28304).
- Miniprep Kit (Qiagen, #27104).
- NotI (NEB, #R0189S).
- Tricaine (Sigma, #A5040). CAUTION: Tricaine is a hazardous chemical. When handling tricaine powder (or while handling fish that have been exposed to tricaine), protective gloves should be used. Please refer to the Material Safety Data Sheets for more details.
- Potassium chloride (KCl), 2 M (Affymetrix, #75896).

- Otohime Fish Diet (Reed Mariculture).
- Premium Grade Brine Shrimp Eggs (Brine Shrimp Direct).
- Ethanol 100% (200 proof) (Rossville Gold Shield Chemical Co.).
- Isopropanol, ACS plus (Fisher Scientific, #A416-500).
- TE buffer (1X) pH8 (Thermo Fisher, #AM9849).
- Dulbecco's Phosphate Buffered Saline DPBS (Sigma, D8537).

Equipment

- Dumont tweezers #5 (WPI, #500342).
- Gilson PIPETMAN (Fisher Scientific, #F123600G, #F144801G, #F123601G, #F123602G).
- Injection mold (well depth: 1.1 mm; well width: 0.95 mm, see Supplementary Method 1 for 3D-printing design, in STL or DWG file formats, ready to print). CRITICAL: The 3D-printed injection mold should be ordered in advance, depending on the shipping time of locally available providers.
- Phase Lock Gel Tubes Light (Eppendorf 5 Prime, #2302820).
- LB agar plates with 100 μg/mL ampicillin (Teknova, #L1004).
- Borosilicate microcapillaries with filament (Sutter, #BF100-58-10).
- Kimwipe (Kimtech Science).
- Barrier Pipet Tips, Low Binding (Genesee Scientific, #24-401, 24-404, 24-412, 24-430).
- 1.7 mL tubes (Genesee Scientific, #24-282).
- 0.22 μm filter (Genesee Scientific, #25-227).
- Pestle for 1.5 mL tube, RNase and DNase free (Argos, #P7339-901).
- Commercially available water recirculating systems with 2.8 L tanks (Aquaneering, Inc.). CRITICAL: Water recirculating systems are an efficient way to house a breeding turquoise killifish colony, and must be established before attempting this protocol (see the "Reagent setup and additional protocols" section).
- Thermal Cycler (MJ Research PTC-200).
- 60 mm × 15 mm Petri dishes (Fisher, #0875713a).
- Sutter Instrument P-87, Novato, CA.
- Nikon C-PS stereoscope and Zeiss KL 1500 LCD (Stuttgart, Germany) optic fibers.
- Injection apparatus (Applied Scientific Instrumentation, Eugene, OR), which includes an MHC model magnetic stand, an MMPI model pressure injector, a foot switch to pulse the injected solution into the embryos, an MM3 model micromanipulator, and an M-PIP model micropipette holder (Applied Scientific Instrumentation), assisted by a backpressure unit (Warner Instrument, Hamden, CT). CRITICAL: It is recommended that the injection apparatus be assembled according to the manufacturer's guidelines before attempting this protocol.

- 8-Inch Fine Mesh Strainer (OXO Good Grips, #38991).
- Stage micrometer, 1 mm divided into 0.01 mm units (Meiji Techno America, #MA285).
- Compact incubator for embryos (ThermoScientific, #50125590).
- 30 Mesh Sand (Homedepot, #200000278).
- Incubator, 28°C.

Acknowledgements

We thank the Harel lab for stimulating discussion and feedback on the manuscript. We thank Ella Yanay and Ashayma Abu-tair for help with killifish maintenance. Supported by the Zuckerman Program (I.H.), Abisch-Frenkel Foundation 19/HU04 (I.H.), NIH 1R21AG063739 (I.H.), ISF 2178/19 (I.H.), Israel Ministry of Science 3-17631 (I.H.), 3-16872 (I.H.), 3-16591 (E.M), the Moore Foundation GBMF9341 (I.H.), BSF-NSF 2020611 (I.H.), the Israel Ministry of Agriculture 12-16-0010 (I.H), and the Lady Davis Postdoctoral Fellowship (G.A).

References

Ahuja, G., Bartsch, D., Yao, W., Geissen, S., Frank, S., Aguirre, A., Russ, N., Messling, J.E., Dodzian, J., Lagerborg, K.A., et al., 2019. Loss of genomic integrity induced by lysosphingolipid imbalance drives ageing in the heart. EMBO Rep. 20 (4), e47407.

Allard, J.B., Kamei, H., Duan, C., 2013. Inducible transgenic expression in the short-lived fish *Nothobranchius furzeri*. J. Fish. Biol. 82, 1733–1738.

Almaida-Pagan, P.F., Lucas-Sanchez, A., Tocher, D.R., 2014. Changes in mitochondrial membrane composition and oxidative status during rapid growth, maturation and aging in zebrafish, *Danio rerio*. Biochim. Biophys. Acta 1841, 1003–1011.

Almaida-Pagan, P.F., Ortega-Sabater, C., Lucas-Sanchez, A., Gonzalez-Silvera, D., Martinez-Nicolas, A., Rol de Lama, M.A., Mendiola, P., de Costa, J., 2019. Age-related changes in mitochondrial membrane composition of *Nothobranchius furzeri*.: comparison with a longer-living *Nothobranchius* species. Biogerontology 20 (1), 83–92.

Anchelin, M., Alcaraz-Perez, F., Martinez, C.M., Bernabe-Garcia, M., Mulero, V., Cayuela, M.L., 2013. Premature aging in telomerase-deficient zebrafish. Dis. Models Mech. 6, 1101–1112.

Antebi, A., Dolfi, L., Suen, T.K., Ripa, R., 2020. Optimization of cryopreservation and in vitro fertilization techniques for the African turquoise killifish *Nothobranchius furzeri*. bioRxiv. https://doi.org/10.1101/2020.04.14.040824.

Api, M., Biondi, P., Olivotto, I., Terzibasi, E., Cellerino, A., Carnevali, O., 2018a. Effects of parental aging during embryo development and adult life: the case of *Nothobranchius furzeri*. Zebrafish 15, 112–123.

Api, M., Notarstefano, V., Olivotto, I., Cellerino, A., Carnevali, O., 2018b. Breeders age affects reproductive success in *Nothobranchius furzeri*. Zebrafish 15, 546–557.

Aramillo Irizar, P., Schäuble, S., Esser, D., Groth, M., Frahm, C., Priebe, S., Baumgart, M., Hartmann, N., Marthandan, S., Menzel, U., et al., 2018. Transcriptomic alterations during ageing reflect the shift from cancer to degenerative diseases in the elderly. Nat. Commun. 9, 327.

Armanios, M., 2009. Syndromes of telomere shortening. Annu. Rev. Genom. Hum. Genet. 10, 45–61.

Auer, T.O., Duroure, K., De Cian, A., Concordet, J.P., Del Bene, F., 2014. Highly efficient CRISPR/Cas9-mediated knock-in in zebrafish by homology-independent DNA repair. Genome Res. 24, 142–153.

Austad, S.N., 2010. Cats, "rats," and bats: the comparative biology of aging in the 21st century. Integr. Comp. Biol. 50, 783–792.

Bartakova, V., Nagy, B., Polacik, M., Blazek, R., Lamtane, H., Reichard, M., 2020. Genetic diversity of a widespread annual killifish from coastal Tanzania. BMC Evol. Biol. 20, 1.

Baumgart, M., Barth, E., Savino, A., Groth, M., Koch, P., Petzold, A., Arisi, I., Platzer, M., Marz, M., Cellerino, A., 2017. A miRNA catalogue and ncRNA annotation of the short-living fish *Nothobranchius furzeri*. BMC Genomics 18, 693.

Baumgart, M., Di Cicco, E., Rossi, G., Cellerino, A., Tozzini, E.T., 2014a. Comparison of captive lifespan, age-associated liver neoplasias and age-dependent gene expression between two annual fish species: *Nothobranchius furzeri* and *Nothobranchius korthause*. Biogerontology 16 (1), 63–69.

Baumgart, M., Groth, M., Priebe, S., Savino, A., Testa, G., Dix, A., Ripa, R., Spallotta, F., Gaetano, C., Ori, M., et al., 2014b. RNA-seq of the aging brain in the short-lived fish *N. furzeri* - conserved pathways and novel genes associated with neurogenesis. Aging Cell 13 (6), 965–974.

Baumgart, M., Priebe, S., Groth, M., Hartmann, N., Menzel, U., Pandolfini, L., Koch, P., Felder, M., Ristow, M., Englert, C., et al., 2016. Longitudinal RNA-seq analysis of vertebrate aging identifies mitochondrial complex I as a small-molecule-sensitive modifier of lifespan. Cell Syst. 2, 122–132.

Bedell, V.M., Wang, Y., Campbell, J.M., Poshusta, T.L., Starker, C.G., Krug 2nd, R.G., Tan, W., Penheiter, S.G., Ma, A.C., Leung, A.Y., et al., 2012. In vivo genome editing using a high-efficiency TALEN system. Nature 491, 114–118.

Bellantuono, I., de Cabo, R., Ehninger, D., Di Germanio, C., Lawrie, A., Miller, J., Mitchell, S., Navas-Enamorado, I., Potter, P., Tchkonia, T., 2020. A toolbox for the longitudinal assessment of healthspan in aging mice. Nat. Protoc. 15, 540–574.

Benayoun, B.A., Pollina, E.A., Singh, P.P., Mahmoudi, S., Harel, I., Casey, K.M., Dulken, B.W., Kundaje, A., Brunet, A., 2019. Remodeling of epigenome and transcriptome landscapes with aging in mice reveals widespread induction of inflammatory responses. Genome Res. 29, 697–709.

Blazek, R., Polacik, M., Kacer, P., Cellerino, A., Rezucha, R., Methling, C., Tomasek, O., Syslova, K., Terzibasi Tozzini, E., Albrecht, T., et al., 2017. Repeated intraspecific divergence in life span and aging of African annual fishes along an aridity gradient. Evolution 71, 386–402.

Blazek, R., Polacik, M., Reichard, M., 2013. Rapid growth, early maturation and short generation time in African annual fishes. EvoDevo 4, 24.

Bogue, M.A., Philip, V.M., Walton, D.O., Grubb, S.C., Dunn, M.H., Kolishovski, G., Emerson, J., Mukherjee, G., Stearns, T., He, H., 2020. Mouse Phenome Database: a data repository and analysis suite for curated primary mouse phenotype data. Nucleic Acids Res. 48, D716–D723.

Bradshaw, W.J., Valenzano, D.R., 2020. Extreme genomic volatility characterizes the evolution of the immunoglobulin heavy chain locus in cyprinodontiform fishes. Proc. Biol. Sci. 287, 20200489.

Cardona-Costa, J., Perez-Camps, M., Garcia-Ximenez, F., Espinos, F.J., 2009. Effect of gametes aging on their activation and fertilizability in zebrafish (*Danio rerio*). Zebrafish 6, 93–95.

Carney Almroth, B., Johnsson, J.I., Devlin, R., Sturve, J., 2012. Oxidative stress in growth hormone transgenic coho salmon with compressed lifespan–a model for addressing aging. Free Radic. Res. 46, 1183–1189.

Carrington, B., Varshney, G.K., Burgess, S.M., Sood, R., 2015. CRISPR-STAT: an easy and reliable PCR-based method to evaluate target-specific sgRNA activity. Nucleic Acids Res. 43, e157.

Cellerino, A., Valenzano, D.R., Reichard, M., 2015. From the bush to the bench: the annual *Nothobranchius* fishes as a new model system in biology. Biol. Rev. Camb. Phil. Soc. 91 (2), 511–533.

Cencioni, C., Heid, J., Krepelova, A., Rasa, S.M.M., Kuenne, C., Guenther, S., Baumgart, M., Cellerino, A., Neri, F., Spallotta, F., et al., 2019. Aging triggers H3K27 trimethylation hoarding in the chromatin of *Nothobranchius furzeri* skeletal muscle. Cells 8 (10), 1169.

Chang, C.T., Colicino, E.G., DiPaola, E.J., Al-Hasnawi, H.J., Whipps, C.M., 2015. Evaluating the effectiveness of common disinfectants at preventing the propagation of *Mycobacterium* spp. isolated from zebrafish. Comp. Biochem. Physiol. C Toxicol. Pharmacol. 178, 45–50.

Chang, N., Sun, C., Gao, L., Zhu, D., Xu, X., Zhu, X., Xiong, J.W., Xi, J.J., 2013. Genome editing with RNA-guided Cas9 nuclease in zebrafish embryos. Cell Res. 23, 465–472.

Clark, K.J., Urban, M.D., Skuster, K.J., Ekker, S.C., 2011. Transgenic zebrafish using transposable elements. Methods Cell Biol. 104, 137–149.

Cong, L., Ran, F.A., Cox, D., Lin, S., Barretto, R., Habib, N., Hsu, P.D., Wu, X., Jiang, W., Marraffini, L.A., et al., 2013. Multiplex genome engineering using CRISPR/Cas systems. Science 339, 819–823.

Cui, R., Medeiros, T., Willemsen, D., Iasi, L.N.M., Collier, G.E., Graef, M., Reichard, M., Valenzano, D.R., 2019. Relaxed selection limits lifespan by increasing mutation load. Cell 178, 385–399.e320.

D'Angelo, L., 2013. Brain atlas of an emerging teleostean model: *Nothobranchius furzeri*. Anat. Rec. 296, 681–691.

Davidson, A.E., Balciunas, D., Mohn, D., Shaffer, J., Hermanson, S., Sivasubbu, S., Cliff, M.P., Hackett, P.B., Ekker, S.C., 2003. Efficient gene delivery and gene expression in zebrafish using the sleeping beauty transposon. Dev. Biol. 263, 191–202.

de Girolamo, P., Leggieri, A., Palladino, A., Lucini, C., Attanasio, C., D'Angelo, L., 2020. Cholinergic system and NGF receptors: insights from the brain of the short-lived fish *Nothobranchius furzeri*. Brain Sci. 10 (6), 394.

Depczynski, M., Bellwood, D.R., 2005. Shortest recorded vertebrate lifespan found in a coral reef fish. Curr. Biol. 15, R288–R289.

Di Cicco, E., Tozzini, E.T., Rossi, G., Cellerino, A., 2011. The short-lived annual fish *Nothobranchius furzeri* shows a typical teleost aging process reinforced by high incidence of age-dependent neoplasias. Exp. Gerontol. 46, 249–256.

Ding, L., Kuhne, W.W., Hinton, D.E., Song, J., Dynan, W.S., 2010. Quantifiable biomarkers of normal aging in the Japanese medaka fish (*Oryzias latipes*). PLoS One 5, e13287.

Dodzian, J., Kean, S., Seidel, J., Valenzano, D.R., 2018. A protocol for laboratory housing of turquoise killifish (*Nothobranchius furzeri*). J. Vis. Exp. (134), 57073.

Dolfi, L., Ripa, R., Antebi, A., Valenzano, D.R., Cellerino, A., 2019. Cell cycle dynamics during diapause entry and exit in an annual killifish revealed by FUCCI technology. EvoDevo 10, 29.

Dolfi, L., Ripa, R., Cellerino, A., 2014. Transition to annual life history coincides with reduction in cell cycle speed during early cleavage in three independent clades of annual killifish. EvoDevo 5, 32.

Dominguez, A.A., Lim, W.A., Qi, L.S., 2016. Beyond editing: repurposing CRISPR-Cas9 for precision genome regulation and interrogation. Nat. Rev. Mol. Cell Biol. 17, 5–15.

Dong, J., Stuart, G.W., 2004. Transgene manipulation in zebrafish by using recombinases. Methods Cell Biol. 77, 363–379.

Doudna, J.A., Sontheimer, E.J., 2014. The use of CRISPR/Cas9, ZFNs, and TALENs in generating site-specific genome alterations. Preface. Methods Enzymol. 546, xix–xx.

Fillit, H.M., Rockwood, K., Woodhouse, K., 2010. Brocklehurst's Textbook of Geriatric Medicine and Gerontology. Elsevier Health Sciences.

Gagnon, J.A., Valen, E., Thyme, S.B., Huang, P., Ahkmetova, L., Pauli, A., Montague, T.G., Zimmerman, S., Richter, C., Schier, A.F., 2014. Efficient mutagenesis by Cas9 protein-mediated oligonucleotide insertion and large-scale assessment of single-guide RNAs. PLoS One 9, e98186.

Garcia, D., Reichard, M., 2020. The effect of population density on growth and survival of a Neotropical annual killifish. J. Fish. Biol. 97, 298–301.

Genade, T., Benedetti, M., Terzibasi, E., Roncaglia, P., Valenzano, D.R., Cattaneo, A., Cellerino, A., 2005. Annual fishes of the genus *Nothobranchius* as a model system for aging research. Aging Cell 4, 223–233.

Gilbert, L.A., Horlbeck, M.A., Adamson, B., Villalta, J.E., Chen, Y., Whitehead, E.H., Guimaraes, C., Panning, B., Ploegh, H.L., Bassik, M.C., et al., 2014a. Genome-scale CRISPR-mediated control of gene repression and activation. Cell 159, 647–661.

Gilbert, M.J., Zerulla, T.C., Tierney, K.B., 2014b. Zebrafish (*Danio rerio*) as a model for the study of aging and exercise: physical ability and trainability decrease with age. Exp. Gerontol. 50, 106–113.

Gopalakrishnan, S., Cheung, N.K., Yip, B.W., Au, D.W., 2013. Medaka fish exhibits longevity gender gap, a natural drop in estrogen and telomere shortening during aging: a unique model for studying sex-dependent longevity. Front. Zool. 10, 78.

Grabher, C., Wittbrodt, J., 2007. Meganuclease and transposon mediated transgenesis in medaka. Genome Biol. 8 (Suppl. 1), S10.

Gupta, T., Mullins, M.C., 2010. Dissection of organs from the adult zebrafish. J. Vis. Exp. (37), 1717.

Harel, I., Benayoun, B.A., Machado, B., Singh, P.P., Hu, C.K., Pech, M.F., Valenzano, D.R., Zhang, E., Sharp, S.C., Artandi, S.E., et al., 2015. A platform for rapid exploration of aging and diseases in a naturally short-lived vertebrate. Cell 160 (5), 1013–1026.

Harel, I., Brunet, A., 2015. The African turquoise killifish: a model for exploring vertebrate aging and diseases in the fast lane. Cold Spring Harbor Symp. Quant. Biol. 80, 275–279.

Harel, I., Valenzano, D.R., Brunet, A., 2016. Efficient genome engineering approaches for the short-lived African turquoise killifish. Nat. Protoc. 11 (10), 2010–2028.

Hartmann, N., Englert, C., 2012. A microinjection protocol for the generation of transgenic killifish (Species: *Nothobranchius furzeri*). Dev. Dyn. 241, 1133–1141.

Hartmann, N., Reichwald, K., Lechel, A., Graf, M., Kirschner, J., Dorn, A., Terzibasi, E., Wellner, J., Platzer, M., Rudolph, K.L., et al., 2009. Telomeres shorten while Tert expression increases during ageing of the short-lived fish *Nothobranchius furzeri*. Mech. Ageing Dev. 130 (5), 290–296.

Hartmann, N., Reichwald, K., Wittig, I., Drose, S., Schmeisser, S., Luck, C., Hahn, C., Graf, M., Gausmann, U., Terzibasi, E., et al., 2011. Mitochondrial DNA copy number and function decrease with age in the short-lived fish *Nothobranchius furzeri*. Aging Cell 10, 824–831.

Heid, J., Cencioni, C., Ripa, R., Baumgart, M., Atlante, S., Milano, G., Scopece, A., Kuenne, C., Guenther, S., Azzimato, V., et al., 2017. Age-dependent increase of oxidative stress regulates microRNA-29 family preserving cardiac health. Sci. Rep. 7, 16839.

Heidenreich, M., Zhang, F., 2016. Applications of CRISPR-Cas systems in neuroscience. Nat. Rev. Neurosci. 17, 36–44.

Henriques, C.M., Carneiro, M.C., Tenente, I.M., Jacinto, A., Ferreira, M.G., 2013. Telomerase is required for zebrafish lifespan. PLoS Genet. 9, e1003214.

Hoppe, B., Pietsch, S., Franke, M., Engel, S., Groth, M., Platzer, M., Englert, C., 2015. MiR-21 is required for efficient kidney regeneration in fish. BMC Dev. Biol. 15, 43.

Hsu, P.D., Lander, E.S., Zhang, F., 2014. Development and applications of CRISPR-Cas9 for genome engineering. Cell 157, 1262–1278.

Hu, C.K., Brunet, A., 2018. The African turquoise killifish: a research organism to study vertebrate aging and diapause. Aging Cell 17, e12757.

Hu, C.K., Wang, W., Brind'Amour, J., Singh, P.P., Reeves, G.A., Lorincz, M.C., Alvarado, A.S., Brunet, A., 2020. Vertebrate diapause preserves organisms long term through Polycomb complex members. Science 367, 870–874.

Huang, C.J., Jou, T.S., Ho, Y.L., Lee, W.H., Jeng, Y.T., Hsieh, F.J., Tsai, H.J., 2005. Conditional expression of a myocardium-specific transgene in zebrafish transgenic lines. Dev. Dyn. 233, 1294–1303.

Huang, C.J., Tu, C.T., Hsiao, C.D., Hsieh, F.J., Tsai, H.J., 2003. Germ-line transmission of a myocardium-specific GFP transgene reveals critical regulatory elements in the cardiac myosin light chain 2 promoter of zebrafish. Dev. Dyn. 228, 30–40.

Hwang, W.Y., Fu, Y., Reyon, D., Maeder, M.L., Tsai, S.Q., Sander, J.D., Peterson, R.T., Yeh, J.R., Joung, J.K., 2013. Efficient genome editing in zebrafish using a CRISPR-Cas system. Nat. Biotechnol. 31, 227–229.

Jao, L.E., Wente, S.R., Chen, W., 2013. Efficient multiplex biallelic zebrafish genome editing using a CRISPR nuclease system. Proc. Natl. Acad. Sci. U.S.A. 110, 13904–13909.

Jinek, M., Chylinski, K., Fonfara, I., Hauer, M., Doudna, J.A., Charpentier, E., 2012. A programmable dual-RNA-guided DNA endonuclease in adaptive bacterial immunity. Science 337, 816–821.

Kabiljo, J., Murko, C., Pusch, O., Zupkovitz, G., 2019. Spatio-temporal expression profile of sirtuins during aging of the annual fish *Nothobranchius furzeri*. Gene Expr. Patterns 33, 11–19.

Kawakami, K., 2007. Tol2: a versatile gene transfer vector in vertebrates. Genome Biol. 8 (Suppl. 1), S7.

Kawakami, K., Shima, A., Kawakami, N., 2000. Identification of a functional transposase of the Tol2 element, an Ac-like element from the Japanese medaka fish, and its transposition in the zebrafish germ lineage. Proc. Natl. Acad. Sci. U.S.A. 97, 11403–11408.

Kelmer Sacramento, E., Kirkpatrick, J.M., Mazzetto, M., Baumgart, M., Bartolome, A., Di Sanzo, S., Caterino, C., Sanguanini, M., Papaevgeniou, N., Lefaki, M., et al., 2020. Reduced proteasome activity in the aging brain results in ribosome stoichiometry loss and aggregation. Mol. Syst. Biol. 16 (6), e9596.

Kenyon, C., 2010. The genetics of ageing. Nature 464, 504–512.

Kim, Y., Nam, H.G., Valenzano, D.R., 2016. The short-lived African turquoise killifish: an emerging experimental model for ageing. Dis. Models Mech. 9, 115–129.

Kirschner, J., Weber, D., Neuschl, C., Franke, A., Bottger, M., Zielke, L., Powalsky, E., Groth, M., Shagin, D., Petzold, A., et al., 2012. Mapping of quantitative trait loci controlling lifespan in the short-lived fish *Nothobranchius furzeri*–a new vertebrate model for age research. Aging Cell 11, 252–261.

Kishi, S., Uchiyama, J., Baughman, A.M., Goto, T., Lin, M.C., Tsai, S.B., 2003. The zebrafish as a vertebrate model of functional aging and very gradual senescence. Exp. Gerontol. 38, 777–786.

Kwan, K.M., Fujimoto, E., Grabher, C., Mangum, B.D., Hardy, M.E., Campbell, D.S., Parant, J.M., Yost, H.J., Kanki, J.P., Chien, C.B., 2007. The Tol2kit: a multisite gateway-based construction kit for Tol2 transposon transgenesis constructs. Dev. Dyn. 236, 3088–3099.

Labun, K., Montague, T.G., Gagnon, J.A., Thyme, S.B., Valen, E., 2016. CHOPCHOP v2: a web tool for the next generation of CRISPR genome engineering. Nucleic Acids Res. 44, W272–W276.

Langenau, D.M., Feng, H., Berghmans, S., Kanki, J.P., Kutok, J.L., Look, A.T., 2005. Cre/lox-regulated transgenic zebrafish model with conditional myc-induced T cell acute lymphoblastic leukemia. Proc. Natl. Acad. Sci. U.S.A. 102, 6068–6073.

Larson, M.H., Gilbert, L.A., Wang, X., Lim, W.A., Weissman, J.S., Qi, L.S., 2013. CRISPR interference (CRISPRi) for sequence-specific control of gene expression. Nat. Protoc. 8, 2180–2196.

Lee, Y., Drake, A.C., Thomas, N.O., Ferguson, L.G., Chappell, P.E., Shay, K.P., 2018. Dietary resveratrol increases mid-life fecundity of female *Nothobranchius guentheri*. Comp. Biochem. Physiol. C Toxicol. Pharmacol. 208, 71–76.

Li, C., Barton, C., Henke, K., Daane, J., Treaster, S., Caetano-Lopes, J., Tanguay, R.L., Harris, M.P., 2020. celsr1a is essential for tissue homeostasis and onset of aging phenotypes in the zebrafish. eLife 9, e50523.

Liu, S., Zheng, Z., Ji, S., Liu, T., Hou, Y., Li, S., Li, G., 2018. Resveratrol reduces senescence-associated secretory phenotype by SIRT1/NF-κB pathway in gut of the annual fish *Nothobranchius guentheri*. Fish Shellfish Immunol. 80, 473–479.

Liu, T., Liu, S., Ma, L., Li, F., Zheng, Z., Chai, R., Hou, Y., Xie, Y., Li, G., 2017a. Oogenesis, vitellogenin-mediated ovarian degeneration and immune response in the annual fish *Nothobranchius guentheri*. Fish Shellfish Immunol. 66, 86–92.

Liu, T., Ma, L., Zheng, Z., Li, F., Liu, S., Xie, Y., Li, G., 2017b. Resveratrol inhibits age-dependent spontaneous tumorigenesis by SIRT1-mediated post-translational modulations in the annual fish *Nothobranchius guentheri*. Oncotarget 8 (33), 55422–55434.

Liu, T., Qi, H., Ma, L., Liu, Z., Fu, H., Zhu, W., Song, T., Yang, B., Li, G., 2015. Resveratrol attenuates oxidative stress and extends life span in the annual fish *Nothobranchius guentheri*. Rejuvenation Res. 18 (3), 225–233.

Lopez-Otin, C., Blasco, M.A., Partridge, L., Serrano, M., Kroemer, G., 2013. The hallmarks of aging. Cell 153, 1194–1217.

Maldonado, T.A., Jones, R.E., Norris, D.O., 2000. Distribution of beta-amyloid and amyloid precursor protein in the brain of spawning (senescent) salmon: a natural, brain-aging model. Brain Res. 858, 237–251.

Maldonado, T.A., Jones, R.E., Norris, D.O., 2002. Intraneuronal amyloid precursor protein (APP) and appearance of extracellular beta-amyloid peptide (abeta) in the brain of aging kokanee salmon. J. Neurobiol. 53, 11–20.

Mali, P., Yang, L., Esvelt, K.M., Aach, J., Guell, M., DiCarlo, J.E., Norville, J.E., Church, G.M., 2013. RNA-guided human genome engineering via Cas9. Science 339, 823–826.

Markofsky, J., Milstoc, M., 1979. Aging changes in the liver of the male annual cyprinodont fish, *Nothobranchius guentheri*. Exp. Gerontol. 14, 11–20.

Matsui, H., Kenmochi, N., Namikawa, K., 2019. Age- and alpha-synuclein-dependent degeneration of dopamine and noradrenaline neurons in the annual killifish *Nothobranchius furzeri*. Cell Rep. 26, 1727–1733.e1726.

Milinkovitch, T., Lefrancois, C., Durollet, M., Thomas-Guyon, H., 2018. Influence of temperature on age-related lipid peroxidation in a short-lived vertebrate (*Nothobranchius furzeri*). Fish Physiol. Biochem. 44, 343–347.

Montesano, A., Baumgart, M., Avallone, L., Castaldo, L., Lucini, C., Tozzini, E.T., Cellerino, A., D'Angelo, L., de Girolamo, P., 2019a. Age-related central regulation of orexin and NPY in the short-lived African killifish *Nothobranchius furzeri*. J. Comp. Neurol. 527 (9), 1508–1526.

Montesano, A., De Felice, E., Leggieri, A., Palladino, A., Lucini, C., Scocco, P., de Girolamo, P., Baumgart, M., D'Angelo, L., 2019b. Ontogenetic pattern changes of nucleobindin-2/nesfatin-1 in the brain and intestinal bulb of the short lived African turquoise killifish. J. Clin. Med. 9 (1), 103.

Moon, S.B., Ko, J.-H., Kim, Y.-S., 2019. Recent advances in the CRISPR genome editing tool set. Exp. Mol. Med. 51, 1–11.

Motamedi, M., Shamsaldini, F., Teimori, A., Hesni, M.A., 2019. Histomicroscopy and normal anatomy of the adult killifish *Aphanius hormuzensis* (Teleostei; Aphaniidae) from the Persian Gulf coastal environment. Microsc. Res. Tech. 82, 466–480.

Muck, J., Kean, S., Valenzano, D.R., 2018. Chapter 28 - the short-lived African turquoise killifish (*Nothobranchius furzeri*): a new model system for research on aging. In: Ram, J.L., Conn, P.M. (Eds.), Conn's Handbook of Models for Human Aging, second ed. Academic Press, pp. 377–386.

Nadon, N.L., Strong, R., Miller, R.A., Harrison, D.E., 2017. NIA interventions testing program: investigating putative aging intervention agents in a genetically heterogeneous mouse model. EBioMedicine 21, 3–4.

Naumann, B., Englert, C., 2018. Dispersion/reaggregation in early development of annual killifishes: phylogenetic distribution and evolutionary significance of a unique feature. Dev. Biol. 442, 69–79.

Nelson, J.S., Grande, T., Wilson, M.V.H., 2016. Fishes of the World. John Wiley & Sons, Hoboken, New Jersey, p. 1 (online resource).

Ng'oma, E., Reichwald, K., Dorn, A., Wittig, M., Balschun, T., Franke, A., Platzer, M., Cellerino, A., 2014. The age related markers lipofuscin and apoptosis show different genetic architecture by QTL mapping in short-lived *Nothobranchius* fish. Aging (6), 468–480.

Niccoli, T., Partridge, L., 2012. Ageing as a risk factor for disease. Curr. Biol. 22, R741–R752.

Nielsen, J., Hedeholm, R.B., Heinemeier, J., Bushnell, P.G., Christiansen, J.S., Olsen, J., Ramsey, C.B., Brill, R.W., Simon, M., Steffensen, K.F., et al., 2016. Eye lens radiocarbon reveals centuries of longevity in the Greenland shark (*Somniosus microcephalus*). Science 353, 702–704.

Novoa, B., Pereiro, P., Lopez-Munoz, A., Varela, M., Forn-Cuni, G., Anchelin, M., Dios, S., Romero, A., Martinez-Lopez, A., Medina-Gali, R.M., et al., 2019. Rag1 immunodeficiency-induced early aging and senescence in zebrafish are dependent on chronic inflammation and oxidative stress. Aging Cell 18, e13020.

Okamoto, A., Hussain, S.P., Hagiwara, K., Spillare, E.A., Rusin, M.R., Demetrick, D.J., Serrano, M., Hannon, G.J., Shiseki, M., Zariwala, M., et al., 1995. Mutations in the p16INK4/MTS1/CDKN2, p15INK4B/MTS2, and p18 genes in primary and metastatic lung cancer. Cancer Res. 55, 1448–1451.

Pan, X., Wan, H., Chia, W., Tong, Y., Gong, Z., 2005. Demonstration of site-directed recombination in transgenic zebrafish using the Cre/loxP system. Transgenic Res. 14, 217–223.

Petzold, A., Reichwald, K., Groth, M., Taudien, S., Hartmann, N., Priebe, S., Shagin, D., Englert, C., Platzer, M., 2013. The transcript catalogue of the short-lived fish *Nothobranchius furzeri* provides insights into age-dependent changes of mRNA levels. BMC Genomics 14, 185.

Platzer, M., Englert, C., 2016. *Nothobranchius furzeri*: a model for aging research and more. Trends Genet. 32, 543–552.

Podrabsky, J.E., 1999. Husbandry of the annual killifish Austrofundulus limnaeus with special emphasis on the collection and rearing of embryos. Environ. Biol. Fish. 54, 421–431.

Podrabsky, J.E., Hand, S.C., 1999. The bioenergetics of embryonic diapause in an annual killifish, austrofundulus limnaeus. J. Exp. Biol. 202 (Pt 19), 2567–2580.

Podrabsky, J.E., Hand, S.C., 2000. Depression of protein synthesis during diapause in embryos of the annual killifish Austrofundulus limnaeus. Physiol. Biochem. Zool. 73 (6), 799–808.

Polacik, M., Blazek, R., Reichard, M., 2016. Laboratory breeding of the short-lived annual killifish *Nothobranchius furzeri*. Nat. Protoc. 11, 1396–1413.

Polacik, M., Smith, C., Reichard, M., 2017. Maternal source of variability in the embryo development of an annual killifish. J. Evol. Biol. 30, 738–749.

Qi, L.S., Larson, M.H., Gilbert, L.A., Doudna, J.A., Weissman, J.S., Arkin, A.P., Lim, W.A., 2013. Repurposing CRISPR as an RNA-guided platform for sequence-specific control of gene expression. Cell 152, 1173–1183.

Ram, J.L., Conn, P.M., 2018. Conn's Handbook of Models for Human Aging. Academic Press.

Ran, F.A., Hsu, P.D., Wright, J., Agarwala, V., Scott, D.A., Zhang, F., 2013. Genome engineering using the CRISPR-Cas9 system. Nat. Protoc. 8, 2281–2308.

Reichwald, K., Lauber, C., Nanda, I., Kirschner, J., Hartmann, N., Schories, S., Gausmann, U., Taudien, S., Schilhabel, M.B., Szafranski, K., et al., 2009. High tandem repeat content in the genome of the short-lived annual fish *Nothobranchius furzeri*: a new vertebrate model for aging research. Genome Biol. 10, R16.

Reichwald, K., Petzold, A., Koch, P., Downie, B.R., Hartmann, N., Pietsch, S., Baumgart, M., Chalopin, D., Felder, M., Bens, M., et al., 2015. Insights into sex chromosome evolution and aging from the genome of a short-lived fish. Cell 163 (6), 1527–1538.

Rembold, M., Lahiri, K., Foulkes, N.S., Wittbrodt, J., 2006. Transgenesis in fish: efficient selection of transgenic fish by co-injection with a fluorescent reporter construct. Nat. Protoc. 1, 1133–1139.

Rhinn, H., Fujita, R., Qiang, L., Cheng, R., Lee, J.H., Abeliovich, A., 2013. Integrative genomics identifies APOE epsilon4 effectors in Alzheimer's disease. Nature 500, 45–50.

Ripa, R., Dolfi, L., Terrigno, M., Pandolfini, L., Savino, A., Arcucci, V., Groth, M., Terzibasi Tozzini, E., Baumgart, M., Cellerino, A., 2017. MicroRNA miR-29 controls a compensatory response to limit neuronal iron accumulation during adult life and aging. BMC Biol. 15, 9.

Romney, A.L.T., Davis, E.M., Corona, M.M., Wagner, J.T., Podrabsky, J.E., 2018. Temperature-dependent vitamin D signaling regulates developmental trajectory associated with diapause in an annual killifish. Proc. Natl. Acad. Sci. U.S.A. 115, 12763–12768.

Samarut, E., Lissouba, A., Drapeau, P., 2016. A simplified method for identifying early CRISPR-induced indels in zebrafish embryos using High Resolution Melting analysis. BMC Genomics 17, 547.

Scheel, J.J., 1990. Atlas of Killifishes of the Old World (Neptune City, N.J. (1 T.F.H. Plaza, 3rd and Union Aves., Neptune City 07753): T.F.H. Publications.

Scheer, N., Campos-Ortega, J.A., 1999. Use of the Gal4-UAS technique for targeted gene expression in the zebrafish. Mech. Dev. 80, 153–158.

Singh, P.P., Demmitt, B.A., Nath, R.D., Brunet, A., 2019. The genetics of aging: a vertebrate perspective. Cell 177, 200–220.

Smargon, A.A., Shi, Y.J., Yeo, G.W., 2020. RNA-targeting CRISPR systems from metagenomic discovery to transcriptomic engineering. Nat. Cell Biol. 1–8.

Smith, P., Willemsen, D., Popkes, M., Metge, F., Gandiwa, E., Reichard, M., Valenzano, D.R., 2017. Regulation of life span by the gut microbiota in the short-lived African turquoise killifish. eLife 6, e27014.

Tacutu, R., Thornton, D., Johnson, E., Budovsky, A., Barardo, D., Craig, T., Diana, E., Lehmann, G., Toren, D., Wang, J., et al., 2018. Human ageing genomic resources: new and updated databases. Nucleic Acids Res. 46, D1083–D1090.

Terzibasi, E., Lefrançois, C., Domenici, P., Hartmann, N., Graf, M., Cellerino, A., 2009. Effects of dietary restriction on mortality and age-related phenotypes in the short-lived fish *Nothobranchius furzeri*. Aging Cell 8 (2), 88–99.

Terzibasi, E., Valenzano, D.R., Benedetti, M., Roncaglia, P., Cattaneo, A., Domenici, L., Cellerino, A., 2008. Large differences in aging phenotype between strains of the short-lived annual fish *Nothobranchius furzeri*. PLoS One 3, e3866.

Terzibasi, E., Valenzano, D.R., Cellerino, A., 2007. The short-lived fish *Nothobranchius furzeri* as a new model system for aging studies. Exp. Gerontol. 42 (1–2), 81–89.

Thomas, H.R., Percival, S.M., Yoder, B.K., Parant, J.M., 2014. High-throughput genome editing and phenotyping facilitated by high resolution melting curve analysis. PLoS One 9, e114632.

Tozzini, E.T., Baumgart, M., Battistoni, G., Cellerino, A., 2012. Adult neurogenesis in the short-lived teleost *Nothobranchius furzeri*: localization of neurogenic niches, molecular characterization and effects of aging. Aging Cell 11, 241–251.

Tozzini, E.T., Dorn, A., Ng'oma, E., Polacik, M., Blazek, R., Reichwald, K., Petzold, A., Watters, B., Reichard, M., Cellerino, A., 2013. Parallel evolution of senescence in annual fishes in response to extrinsic mortality. BMC Evol. Biol. 13, 77.

Tozzini, E.T., Savino, A., Ripa, R., Battistoni, G., Baumgart, M., Cellerino, A., 2014. Regulation of microRNA expression in the neuronal stem cell niches during aging of the short-lived annual fish *Nothobranchius furzeri*. Front. Cell. Neurosci. 8, 51.

Valdesalici, S., Cellerino, A., 2003. Extremely short lifespan in the annual fish *Nothobranchius furzeri*. Proc. Biol. Sci. 270 (Suppl. 2), S189–S191.

Valenzano, D.R., Benayoun, B.A., Singh, P.P., Zhang, E., Etter, P.D., Hu, C.K., Clément-Ziza, M., Willemsen, D., Cui, R., Harel, I., et al., 2015. The African turquoise killifish genome provides insights into evolution and genetic architecture of lifespan. Cell 163 (6), 1539–1554.

Valenzano, D.R., Kirschner, J., Kamber, R.A., Zhang, E., Weber, D., Cellerino, A., Englert, C., Platzer, M., Reichwald, K., Brunet, A., 2009. Mapping loci associated with tail color and sex determination in the short-lived fish *Nothobranchius furzeri*. Genetics 183, 1385–1395.

Valenzano, D.R., Sharp, S., Brunet, A., 2011. Transposon-mediated transgenesis in the short-lived African killifish *Nothobranchius furzeri*, a vertebrate model for aging. G3 1, 531–538.

Valenzano, D.R., Terzibasi, E., Cattaneo, A., Domenici, L., Cellerino, A., 2006a. Temperature affects longevity and age-related locomotor and cognitive decay in the short-lived fish *Nothobranchius furzeri*. Aging Cell 5, 275–278.

Valenzano, D.R., Terzibasi, E., Genade, T., Cattaneo, A., Domenici, L., Cellerino, A., 2006b. Resveratrol prolongs lifespan and retards the onset of age-related markers in a short-lived vertebrate. Curr. Biol. 16, 296–300.

Vrtilek, M., Zak, J., Polacik, M., Blazek, R., Reichard, M., 2018. Longitudinal demographic study of wild populations of African annual killifish. Sci. Rep. 8, 4774.

Walford, R.L., Liu, R.K., 1965. Husbandry, life span, and growth rate of the annual fish, *Cynolebias adloffi* E. Ahl. Exp. Gerontol. 1, 161–168.

Wang, N., Luo, Z., Jin, M., Sheng, W., Wang, H.T., Long, X., Wu, Y., Hu, P., Xu, H., Zhang, X., 2019. Exploration of age-related mitochondrial dysfunction and the anti-aging effects of resveratrol in zebrafish retina. Aging (Albany NY) 11, 3117–3137.

Wang, W., Hu, C.K., Zeng, A., Alegre, D., Hu, D., Gotting, K., Ortega Granillo, A., Wang, Y., Robb, S., Schnittker, R., et al., 2020a. Changes in regeneration-responsive enhancers shape regenerative capacities in vertebrates. Science (New York, NY) 369 (6508), eaaz3090.

Wang, X., Ren, Y., Du, X., Song, L., Chen, F., Su, F., 2020b. Effects of late-onset dietary intake of salidroside on insulin/insulin-like growth factor-1 (IGF-1) signaling pathway of the annual fish *Nothobranchius guentheri*. Arch. Gerontol. Geriatr. 91, 104233.

Wang, X., Shang, X., Luan, J., Zhang, S., 2014. Identification, expression and function of apolipoprotein E in annual fish *Nothobranchius guentheri*: implication for an aging marker. Biogerontology 15, 233–243.

Wei, J., Qi, H., Liu, K., Zhao, C., Bian, Y., Li, G., 2020. Effects of metformin on life span, cognitive ability, and inflammatory response in a short-lived fish. J Gerontol A Biol Sci Med Sci 75 (11), 2042–2050.

Wendler, S., Hartmann, N., Hoppe, B., Englert, C., 2015. Age-dependent decline in fin regenerative capacity in the short-lived fish *Nothobranchius furzeri*. Aging Cell 14, 857–866.

Wourms, J.P., 1972. The developmental biology of annual fishes. 3. Pre-embryonic and embryonic diapause of variable duration in the eggs of annual fishes. J. Exp. Zool. 182, 389–414.

Yeh, C.D., Richardson, C.D., Corn, J.E., 2019. Advances in genome editing through control of DNA repair pathways. Nat. Cell Biol. 1–11.

Yu, X., Li, G., 2012. Effects of resveratrol on longevity, cognitive ability and aging-related histological markers in the annual fish *Nothobranchius guentheri*. Exp. Gerontol. 47, 940–949.

Žák, J., Dyková, I., Reichard, M., 2020. Good performance of turquoise killifish (*Nothobranchius furzeri*) on pelleted diet as a step towards husbandry standardization. Sci. Rep. 10, 8986.

Zupkovitz, G., Lagger, S., Martin, D., Steiner, M., Hagelkruys, A., Seiser, C., Schöfer, C., Pusch, O., 2018. Histone deacetylase 1 expression is inversely correlated with age in the short-lived fish *Nothobranchius furzeri*. Histochem. Cell Biol. 150 (3), 255–269.

Chapter 12

Challenges in keeping annual killifish

Martin Reichard[1,2,3], Radim Blažek[1,3], Iva Dyková[3], Jakub Žák[1,4], Matej Polačik[1]

[1]*Institute of Vertebrate Biology, Czech Academy of Sciences, Brno, Czech Republic;* [2]*Department of Ecology and Vertebrate Zoology, Faculty of Biology and Environmental Protection, University of Łodz, Łodz, Poland;* [3]*Department of Botany and Zoology, Faculty of Science, Masaryk University, Brno, Czech Republic;* [4]*Department of Zoology, Faculty of Science, Charles University, Prague, Czech Republic*

Introduction

African annual killifish live in shallow temporary pools in tropical and subtropical Africa. Conditions in their natural habitat have direct implications on requirements for their successful captive maintenance. Most challenges (and advantages) of their captive breeding are directly linked to their natural history, which involves unusual embryo development, rapid growth, strong competition for reproductive opportunities among males, and short lifespan. These challenges are associated with the standardization of embryo development, provision of appropriate food, amelioration of male aggression, and susceptibility to diseases.

Nothobranchius is a species-rich genus from East Africa, one of the annual killifish lineages, which consist of more than 90 formally described species. Most *Nothobranchius* species live in habitats that are inundated once a year and complete a single generation every year. Species from equatorial regions may experience two rainy seasons each year, giving rise to two generations in a year. In addition, some species (such as *Nothobranchius guentheri* and *Nothobranchius korthausae*) live in habitats that may not desiccate annually, and others (e.g., *Nothobranchius furzeri*) live in regions where seasonal rains may be too weak in some years, leading to "dry" years, with no pool inundation (Reichard, 2015). The life history of these species has been shaped by local environmental conditions and its consequences may be observed in captivity. This includes extremely rapid maturation (2–3 weeks) and short lifespan (3–8 months, typically 6 months) of *N. furzeri* from semiarid regions

Laboratory Fish in Biomedical Research. https://doi.org/10.1016/B978-0-12-821099-4.00001-8
Copyright © 2022 Elsevier Inc. All rights reserved.

and relatively long lifespan (2 years) and the ability to complete embryo development in aqueous media in *N. guentheri* from semipermanent habitats.

In this chapter, we focus on describing the challenges for captive breeding of *Nothobranchius* species, and of *N. furzeri* in particular, arising from their life history. We also discuss current approaches to mitigating some of those challenges. Specifically, we (1) summarize availability of different strains of *N. furzeri* (and of other *Nothobranchius* species) for experimental work, (2) provide updates on the existing protocols for laboratory maintenance and breeding of *N. furzeri*, (3) describe the course and potential pitfalls of *N. furzeri* embryo development, (4) report current progress toward diet standardization, and (5) define major health issues associated with captive *Nothobranchius*.

Available strains and species

There are several *Nothobranchius* species available in captivity for experimental work. The most commonly used species, *N. furzeri*, is available in the form of several strains with variable inbreeding history and variation in life history traits, including expected lifespan (Cellerino et al., 2016; Reichard et al., 2020). *N. furzeri* inhabits southern Mozambique and a small adjacent part of Zimbabwe. It consists of two phylogenetic clades: northern Chefu clade from a more arid region and southern Limpopo clade. These clades can be subdivided at a finer scale (Bartáková et al., 2013, 2015). Chefu clade fish (including the most common strains GRZ, MZM 04-10, and MZCS 222) have shorter lifespans (Terzibasi-Tozzini et al., 2013; Blažek et al., 2017). *N. furzeri* displays two discrete male color morphs: red (red caudal fin, with or without black marginal band) and yellow (yellow caudal fin, always with black marginal band) (Fig. 12.1A and B), but they have no direct link to phylogeny. The two morphs largely coexist in nature and populations vary in the proportion of red and yellow males, without any apparent association with other phenotypic traits. In general, the yellow morph is more common in the driest part of the *N. furzeri* range, while the red morph is most common in the Limpopo clade. Color morphs cannot be distinguished in *N. furzeri* females (Fig. 12.1C).

Some *N. furzeri* strains are easier to maintain and breed than others. In particular, MZCS 222 (producing both red and yellow males) appears the least problematic, with rapid growth, high fecundity, and low egg mortality (Polačik et al., 2017; Žák et al., 2020; Žák and Reichard, 2021). This strain originates from the import of 20 males and 40 females in 2011. The core stock is bred at the Institute of Vertebrate Biology in Brno, Czech Republic, with a focus on maintaining it relatively outbred. The GRZ strain is the most inbred and its uniform genetic background makes it superior for many experimental approaches (Harel et al., 2015). The GRZ strain (exclusively yellow males) originates from the most arid part of the *N. furzeri* range (Gona Re Zhou,

FIGURE 12.1 **Turquoise killifish and related species.** Yellow (A) and red (B) male forms and female (C) *Nothobranchius furzeri*. *Nothobranchius kadleci* (D) replaces *N. furzeri* north of the Save river, while *Nothobranchius orthonotus* (E) and *Nothobranchius pienaari* (F) largely coexist with both species across their range.

Zimbabwe) and has been maintained in captivity since 1969 (Foersch, 1975). Its egg production is low at young age and there is a shorter time window for optimal fecundity (Žák and Reichard, 2021). It also has a shorter lifespan. This set of life history traits may be a combination of inbreeding, inadvertent artificial selection over 50 years in captivity, and natural genetic background (Willemsen et al., 2020; Žák and Reichard, 2021). Given occasional problems with the maintenance of GRZ, other strains have started to be used for research, such as MZM 04-10 (Baumgart et al., 2017) and ZMZ1002 (Hu and Brunet, 2018; Hu et al., 2020). These strains were collected from sites relatively close to the GRZ. At the IVB, several other *N. furzeri* strains derived from wild ancestors in 2011–15, including strains from other clades, are maintained (Cellerino et al., 2016), and their lifespan and telomere length have been well characterized (Reichard et al., 2020).

In Central Mozambique (north of the Save river), *N. furzeri* is replaced by its sister species *Nothobranchius kadleci* (Dorn et al., 2011; Reichard et al., 2017), also available for experimental work (Cellerino et al., 2016). This species differs in coloration (Fig. 12.1D), is less aggressive, and has a longer lifespan, but is genetically similar and produces fertile hybrids with *N. furzeri* in captivity (Ng'oma et al., 2014). Species coexisting locally with *N. furzeri* are also available for experimental work. *Nothobranchius orthonotus* (Fig. 12.1E) is closely related but larger and tends to be more aggressive than *N. furzeri*. It also has relatively short has a lifespan and rapid growth. *Nothobranchius pienaari* (Fig. 12.1F) is smaller, less aggressive, and lives longer. Populations of *N. orthonotus* from wet regions are sometimes referred to as *Nothobranchius kuhntae* (Terzibasi-Tozzini et al., 2013), but its taxonomic status is unclear (Vrtílek and Reichard, 2016). *N. guentheri* from Zanzibar is another species commonly used in laboratory work (e.g., Liu et al., 2018). It lives approximately 2 years and is easier to maintain in captivity. *N. guentheri* strains used in experimental work are not well characterized, but appear to arise from strains held in captivity over a long period and are readily available among hobby fish breeders. Other *Nothobranchius* species are sometimes used for comparative purposes, such as *N. korthausae* (Baumgart et al., 2015) and *Nothobranchius rachovii* (Hsu and Chiu, 2009).

Housing and breeding

While this section specifically discusses the maintenance and breeding of *N. furzeri*, it also applies to other *Nothobranchius* species, which are generally easier to keep but live at a slower pace. Detailed housing and breeding protocols for *N. furzeri* are available, summarizing killifish culture from different labs and highlighting particular issues, central for particular research agenda (Polačik et al., 2016; Dodzian et al., 2018; Muck et al., 2018). We briefly outline key points for successful maintenance of *N. furzeri*, but suggest referring to those protocols for an in-depth guide to the culture of *N. furzeri*.

Behavior and its association to welfare

Compared to other laboratory fish models (zebrafish, medaka, guppy), *N. furzeri* are naturally more aggressive and so require relatively more space. Males are more aggressive than females and larger individuals tend to chase smaller ones, potentially leading to injuries or mortality in social tanks. To avoid this, more males should be kept in social tanks to spread possible aggression of the dominant fish among more individuals. We recommend four or more similarly sized males, and a female-biased sex ratio to minimize aggression in social groups. Longer coexistence creates stable hierarchies and decreases the level of aggression. Different *N. furzeri* strains vary in their aggression levels. In our experience, MZM 0410 exhibits very little aggression and GRZ displays aggression only occasionally, while some outbred strains (e.g., MZCS 222) are commonly more aggressive (R. Blažek, personal observations). Although killifish do not require any kind of shelter, socially housed fish may use structures of internal filters as a refuge from undesirable social interactions or when disturbed.

Annual killifish do not live in shoals and keeping them in isolation does not interfere behaviorally with their welfare. The minimal suggested volume of water per single adult fish is 2 L (Harel et al., 2015; Polačik et al., 2016), which applies to both single fish housing and social tanks. *N. furzeri* are bold and not seriously stressed by handling. Given the dynamic conditions in their natural habitats, fluctuations in environmental conditions (for example, during water exchange) are not a concern for *N. furzeri*.

Types of killifish housing

Juvenile killifish can be maintained at higher densities, but as fish grow their density should be reduced. We recommend a maximum of 200 individuals per 20 L tank for fish 10 mm long, 50 individuals for fish 15 mm long, and 20 individuals for fish 20 mm long. Within each group of juveniles, there are always some faster growing individuals. These tend to be aggressive toward smaller fish, displacing them from feeding spots and constraining their growth. The largest individuals (which are mostly males) should be removed from the communal tank and either kept separately or with others of comparable size. After removal of the larger fish, smaller individuals remaining in the communal tank grow and mature faster. After cessation of major growth period (at age of 4–5 weeks, depending on strain and conditions), the larger fish can be moved back into the communal tanks.

Adult fish can be housed in individual tanks, trios, or social tanks, depending on the focus of the particular experiment. Individual housing enables most standardization (no aggression or socially suppressed feeding rates, comparable growth rates among fish). Social housing maximizes the size of a particular cohort at a given time and is least laborious for larger sets of fish. Trio housing is intermediate, enabling sufficient standardization, minimizing aggression, and allowing regular egg collection. In all three types of housing, a barren bottom (transparent or opaque) is recommended as it facilitates the removal of debris.

Social tanks can be 20–80 L large and should be equipped with air-driven filters that should be cleaned every 2 weeks (or more often if required). Siphoning debris from the tank bottom and partial water exchange (30%) are recommended twice a week (Polačik et al., 2016) or 10% daily (Harel et al., 2016). If the water temperature decreases after water exchange, it is advisable to wait at least an hour before feeding. Dominant fish may control the feeding spot so food should be spatially distributed during feeding to enable sufficient consumption by all fish in the tank. Stable social groups display considerably less aggression, but transfer of individuals among tanks is not problematic. We commonly house approximately 12 adult fish in 24 L tanks and 15–20 adult fish in 40 L tanks, but larger tanks with proportionately more fish can also be used.

Individually housed fish require 2 L or larger tanks. Commercial zebrafish racks meet the basic requirements and enable automated control of water parameters. Custom-made racks are also used, with higher design flexibility. Water recirculation through the system should enable sufficient water exchange in individual tanks but avoid excessive water flow within the tanks, as *Nothobranchius* are poor swimmers and do not thrive in flowing water. Depending on rack design, frequent siphoning of uneaten food and debris from tank bottoms may be necessary and cannot be satisfactorily controlled by means of a sloping tank and automated siphoning, given the high waste production of fish fed on bloodworms. The main advantages of individual housing are standardization of environmental and social conditions and ease of repeated measurements on individual fish without the need for marking. Individual housing also yields more comparable growth rates among individuals. The disadvantage is the need to move fish out of their home tanks for reproduction. Adult females need to oviposit at least once a week (ideally twice a week). Failure to do so may obstruct the proper function of their ovaries. While some females manage to lay their eggs without male presence, others suffer malfunction of the ovaries after a prolonged period of reproductive deprivation, potentially leading to their mortality. Note that individually held fish tend to be more aggressive when encountering another individual. This is particularly important for reproduction. Some individually housed males could occasionally harm females during spawning. To mitigate this, the spawning pair should be inspected for signs of male aggression. Apart from visual inspection, the first sign of increased aggression is the presence of scales in spawning containers. To minimize the risk of female injuries, two females can be placed with a single male in a spawning tank. Alternatively, the spawning period can be initially reduced to 30 min. If there are no signs of male aggression after the initial pairing, the next spawning can take longer. Dodzian et al. (2018) recommend the selection of smaller males for breeding to mitigate spawning aggression, but note that this represents a mode of artificial selection on a particular phenotype, which may interfere with some research questions.

Trio housing combines the advantages of automated water quality maintenance without the need to transfer fish between tanks for reproduction. A female-biased sex ratio decreases potential aggression from the male, though housing in pairs (one male, one female) is also possible. Individual tanks should match the recommended space requirements (2 L per adult fish) and larger tanks are available for standard commercial zebrafish racks. Trio housing is the preferred method for stock maintenance (breeding outside experimental design) if social tanks are not available in the fish-breeding facility and maintenance of genetic variability is not required.

Water quality and sanitation

To sustain their rapid growth and high reproductive output, *N. furzeri* must consume a large amount of food (Vrtílek and Reichard, 2015). Digestion is rapid and leads to high levels of waste production. This comes with demands on the maintenance of water quality. Deterioration in water quality may lead to increased incidence of diseases and increased mortality (Ramsay et al., 2009).

For social tanks, separate hand nets should ideally be used for each tank. After removal of any dead individual, the net must be disinfected using bleach or a saturated solution of kitchen salt. Before entering the fish room, hands should be washed with antibacterial soap and thoroughly rinsed, as many antibacterial agents may be harmful for fish and other aquatic organisms. We recommend regular disinfection of all equipment used for fish feeding and breeding. Moreover, all tanks should be disinfected prior to the introduction of a new fish cohort. Bleach (5% sodium hypochlorite solution) or peracetic acid can be used as a disinfectant. Care must be taken that all equipment treated with disinfectant (especially glass beads or fine sand as the spawning substrate) is disinfectant free when reused. This is achieved, for example, by using commercial sodium thiosulfate-based bleach neutralizers.

Reproduction

The fecundity of *N. furzeri* is directly linked to their condition and matches available resources (Vrtílek and Reichard, 2015). High quality and quantity of food results in the production of an average 40–70 eggs per 2 h of spawning, when eggs are collected twice a week during separate spawning events in special containers. A much lower yield per female is achieved from social tanks or tanks with trios (e.g., Grégoir et al., 2018; Hu et al., 2020) due to egg cannibalism. Eggs are produced daily, with a new batch ovulated each morning and released during the day (Žák et al., 2019).

For spawning in separate tanks, females should be isolated for 24–48 h to maximize egg yield. The bottom of spawning tanks can be covered with a layer of fine sand (grain size < 1 mm) or glass beads (diameter 0.25–0.50 mm) that serve as a spawning substrate. Alternatively, a false mesh bottom (as described in Polačik et al., 2016) can be used. Fish must become accustomed to this type of setting before standard yield of eggs is produced, and this typically takes two to three spawnings. The main advantage of this method is that eggs can be collected directly from the bare bottom of the container using a Pasteur pipette. After the initiation of spawning, females lay all their eggs within 2–2.5 h. An increase in water temperature of approximately 2–4°C (up to 30°C) helps to initiate spawning. Unless eggs are needed immediately (e.g., for microinjection), we recommend postponing egg collection for 5–6 h after the fish are removed. At this stage, the chorion is hard and the eggs are more resilient to damage during handling. Additionally, fertilization success can be estimated under a binocular microscope at this stage. Fertilized eggs develop a perivitelline space (visually clear as a double-layered structure), while unfertilized eggs have no perivitelline space (Fig. 12.2A and B). Collected eggs can be incubated in a

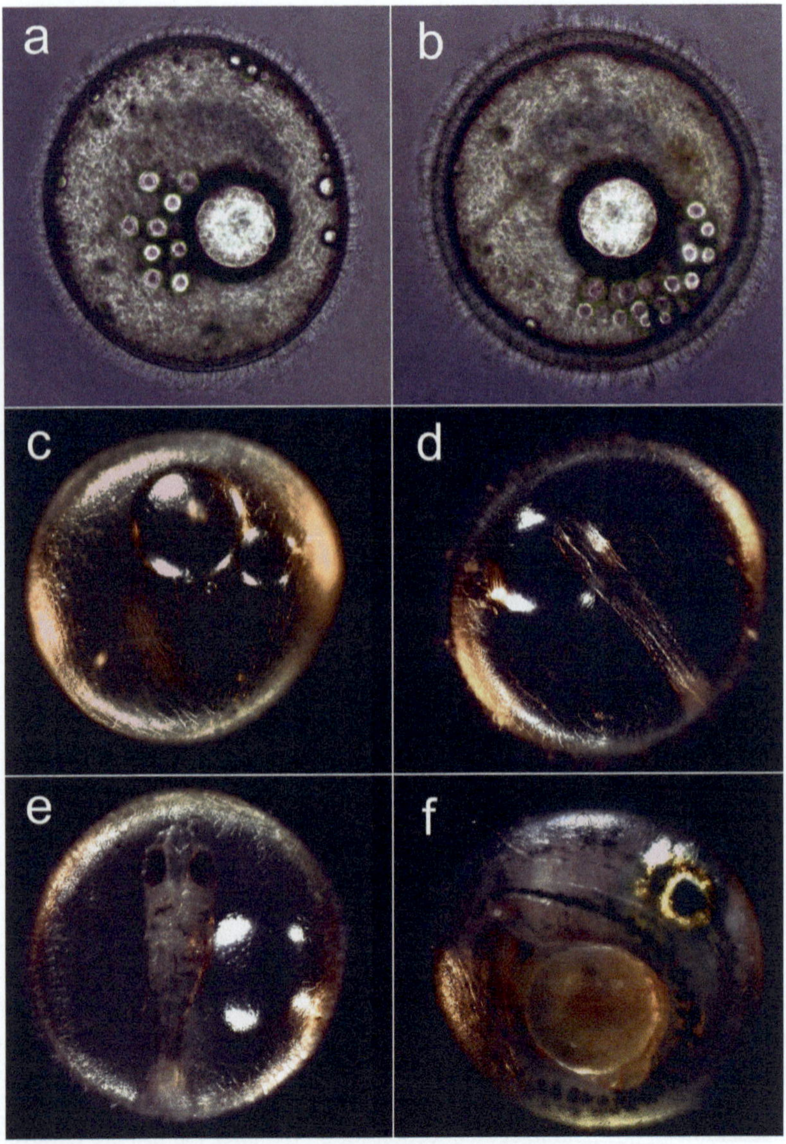

FIGURE 12.2 **Killifish embryos.** Unfertilized (A) and fertilized (B) killifish embryos 24 h postspawning and diapausing embryos at DI (C) and DII (D) stages, development between DII and DIII (E), and DIII stage embryo (F).

sealed Petri dish with a moist filter paper (see Harel et al., 2016) or moist peat substrate, or inside a larger volume of spawning substrate in a sealed plastic (Ziploc) bag (see later). More detail is provided in the next chapter.

In social tanks, a spawning substrate can be provided in the form of peat (or coconut fiber) inserted in a container. The substrate must be autoclaved before the initial use and if reused repeatedly. Acidic peat moss should be avoided; we recommend as high a pH as possible ($pH > 6$) for the peat substrate. Placing the peat in a glass (rather than plastic) container ensures stability on the tank bottom. We recommend a container at least 10 cm deep, with a 2–4 cm layer of peat. When the container is shallower, peat may spill out during spawning as the sediment is disturbed. The spawning substrate should be replaced at least every 2 weeks. Upon removal, the peat should be gently squeezed to get rid of excessive moisture and wrapped in a filter paper. After 2 days of air drying the substrate should be unwrapped, placed in a Ziploc plastic bag, and stored in a temperature-controlled incubator. This incubation method enables space-saving storage of a large number of eggs, but lacks precision in estimates of egg quantity and survival during incubation. For regular checks of their developmental rate, the bag needs to be opened for egg inspection.

Embryo development

The main difference between *N. furzeri* (and *Nothobranchius* species in general) and other laboratory fishes is the relatively long period of embryo development, and the requirement of at least a short dry phase to sustain high embryo survival and hatching success. This brings many advantages, including the possibility of long-term embryo storage and easy postal shipment of the eggs, but comes with a set of challenges that require the modification of standard fish facility equipment. Under natural conditions, the embryonic period typically lasts longer than the posthatching lifespan. Most of this time is spent in the form of developmental dormancy, evolved to survive through the long dry period. Dormancy enables the embryo to resist harsh environmental conditions and conserves the limited energy reserves available to the embryo throughout its development. In the lab, dormancy stages may be manipulated and minimized, if rapid turnover of generations is required.

Developmental diapauses

Annual killifish embryos can, uniquely among vertebrates (and perhaps even invertebrates), pause their development at three well-defined developmental stages. These are historically termed Diapause I (DI), Diapause II (DII), and Diapause III (DIII) (Wourms, 1972). All three diapauses are facultative and any of them can be entered, skipped, and have variable duration. This leads to a multitude of possible developmental trajectories (Wourms, 1972).

Typical DI (Fig. 12.2C) occurs at the dispersed cell stage, which is specific to annual fish. During epiboly, ameboid cells that traveled over the yolk surface become round in shape and remain relatively evenly dispersed (Wourms, 1972; Podrabsky et al., 2017). Alternatively, the embryo may pause its development at a slightly advanced stage when the cells partly reaggregate, as described for the South American annual killifish *Austrolebias* spp. (Arezo et al., 2017). Under standard laboratory conditions (aqueous medium, 25°C), *N. furzeri* enter DI spontaneously but rarely, with a usual duration of 7–15 days (Polačik et al., 2016). However, some embryos may only "slow down" their development compared to the other embryos from the same clutch for a period of 1–3 days in this phase. This introduces variability in the developmental pace, which levels up with the entry into DII, but whether these embryos entered a very brief DI is unclear.

The embryo may enter DII (Fig. 12.2D) around the middle of embryonic development. At 25°C, this stage is reached 4–6 days after the exit from DI. At this stage, the head with otic vesicles, optic cups, and olfactory placodes is clearly distinguishable, and the embryo has 32 or more somite pairs (Wourms, 1972; Polačik et al., 2016; Podrabsky et al., 2017). The heart beats with a frequency of approximately 10 heart beats per minute. As in DI, there is some variability in the exact developmental point when DII is entered. Some *N. furzeri* embryos pause their development at the stage with scattered melanocytes at the hindbrain region and partly pigmented optic cups and eyes. Unusual cases of DII-like dormancy when the embryo is partly pigmented, with a heartbeat and circulating blood cells visible in the vasculature, but with no apparent developmental advance over weeks, have also been observed. These "half diapausing" embryos might represent a link between the normal and escape (direct) developmental trajectory (see later; Podrabsky et al., 2010; Furness et al., 2015).

DIII (Fig. 12.2F) may occur in fully developed, ready to hatch embryos. It is reached after 5–7 days of development past DII at 25°C. It is characterized by a markedly depressed metabolic rate (Levels et al., 1986; Podrabsky and Hand, 1999) and is difficult to distinguish morphologically from the closely preceding developmental stages. In *Nothobranchius* spp., DIII is reliably broken when developed embryos are placed in water.

Erratic development with high intrapopulation developmental variability is not present in natural conditions, where embryos develop more synchronously. They are deposited in a clay-like incubation substrate and experience gradual changes in moisture levels, and seasonal and diurnal temperature fluctuations. Specific environmental cues govern embryo development in line with seasonal changes in the habitat. Inundated habitats with their hypoxic substrate maintain the embryos in DI. When the pool desiccates, DI is broken and the entire embryo pool proceeds to DII. Variability only occurs at the end of the dry season when embryo banks consist of a mixture of DII and DIII embryos (Polačik et al., 2021).

Developmental variability

The laboratory environment differs from the natural conditions in several important aspects, related to practical issues but giving rise to challenges in laboratory embryo husbandry. Specifically, the incubation medium is typically aqueous, normoxic, and the incubation temperature is held stable. If dry incubation is used, the incubation substrate is different from that of natural pools (Polačik et al., 2016; Dodzian et al., 2018). This substitute environment does not control development and the intrinsic programming, most likely epigenetic maternal effects (Romney and Podrabsky, 2017), become important in governing the course of embryo development and entrance into particular diapauses. This is especially obvious when the common laboratory incubation temperature of 25°C is used (Furness et al., 2015). Consequently, embryos under laboratory conditions develop erratically, with a proportion of embryos developing directly without entering any diapause within about 3 weeks and others taking up to several months to reach the prehatching stage (e.g., Wourms, 1972; Polačik et al., 2017).

This developmental variability has consequences for experimental work using *N. furzeri* embryos. In contrast to nondiapausing fish models, elapsed time since fertilization (or degree-days) does not reliably signal actual developmental stage in an embryo and regular/continuous embryo staging is needed if a particular stage is required for experimental purposes. The need for the continuous staging requires embryo incubation in a transparent aqueous medium or on top of a substrate when a dry incubation method is used. Embryos should be disinfected and cleaned prior to incubation to avoid fungal and microbial infections (see especially the "Mycobacteriosis" section) and enable a clear view. Disinfection (detailed bleaching protocol described in Polačik et al., 2016; Dodzian et al., 2018) can be achieved using 5% sodium hypochloride solution or, alternatively, 0.008‰ peracetic acid, iodine, or H_2O_2. Disinfected eggs are moved into moderately moist autoclaved or boiled peat and incubated further. Notwithstanding, embryo mortality of 30%–50% is common, especially in inbred strains such as the GRZ.

Manipulation of embryo development in the laboratory

The embryonic development of *N. furzeri* can be manipulated under laboratory conditions. Developmental pace correlates positively with incubation temperature and with the likelihood of the escape developmental trajectory. For example, the median time from fertilization to the emergence of the neural keel (i.e., pre-DII stage, DI skipped) is around 23 days at 18.5°C; 16 days at 20°C; and 11 days at 23°C. More important than its influence on developmental pace is the direct development-forcing effect of increased temperature. At 18°C, most *N. furzeri* embryos enter DI for a week or two, and the vast majority enter DII for at least a couple of months. DII can subsequently be broken by exposing the embryos to 28°C for a week, and the eggs are ready to hatch within the next 2 weeks (Polačik et al., 2016). Depending on storage conditions, embryo development can be completed over a period of 3–4 weeks, because exposure of freshly spawned eggs directly to temperatures of 28–30°C will induce direct development in the majority of eggs. On the other hand, embryo development can also be held in DII for up to 2–3 years. For detailed protocols for each option, we refer readers elsewhere (Polačik et al., 2016; Dodzian et al., 2018). Exit from diapause and continuing development until hatching can be induced by removal from cold temperature and exposure to temperatures of 28–30°C. This triggers development at a pace to reach the prehatching stage within 2–3 weeks. We recommend collection of the required number of eggs (allowing 50% extra to cope with failure to hatch or early juvenile mortality) at the DII stage from the incubation substrate using fine forceps before placing at DII-breaking high temperature.

For shipment of eggs, the external temperature in both sender's and recipient's regions should not fall below 5°C. The most tolerant stage for shipment is DII. Embryos are sent either in incubation substrate enclosed in a Ziploc bag or in Petri dishes, depending on the preferred type of incubation.

The embryonic period is developmentally very sensitive, with even minor alterations having significant impacts. Potential risks of manipulating *N. furzeri* embryo development include the risk that its effects may extend beyond the embryonic phase and influence the posthatch performance of the fish. Notably, development, including DII compared to direct (escape) development without DII, represents an alternative developmental strategy, with different morphological, physiological, and molecular processes (Podrabsky et al., 2010). Thus entry into DII (DI and DIII have not been investigated in this respect) may not be simply a pause in development but rather a switch between two alternative life histories. Developmentally related phenotypic plasticity has been documented in wild-derived *N. furzeri*. Nondiapausing embryos hatched at a smaller size but with a larger yolk sac. They also grew faster, matured earlier, reached a smaller maximum body size, and had a shorter lifespan compared to those that entered DII (Polačik et al., 2014). On the other hand, no such trade-offs of DII entry on individual posthatching performance were detected in another study, where the diapausing and escape fish of inbred GRZ strain had comparable growth, reproductive output, and lifespan (Hu et al., 2020).

Diet

In the wild, *N. furzeri* is an opportunistic predator from the entire water column and consumes crustaceans, aquatic insect larvae, and floating terrestrial insects (Polačik and Reichard, 2010). Feeding activity of wild fish peaks in the morning but the digestive system contains some food throughout the 24 h cycle (Žák et al., 2019). The natural diet of *N. furzeri* juveniles consists of invertebrates that are small enough to be gulped by small fish. Growth of *N. furzeri* is plastic and directly related to food availability and fish density (Vrtílek and Reichard, 2015; Vrtílek et al., 2019). This should be considered when fish are fed in captivity.

The standardization of diet and feeding procedure in the laboratory has a profound effect on the success of *N. furzeri* as a biological model. At present, the most common food for adult *N. furzeri* in captivity is chironomid larvae (commonly referred to as bloodworms), which are delivered live or frozen (Polačik et al., 2016; Dodzian et al., 2018). Other, less frequently used food includes aquatic larvae of *Chaoborus* sp. midges, *Daphnia* spp., and sludge worms (*Tubifex tubifex*) (Genade et al., 2005). The use of live food risks introducing infections (Nenoff and Uhlemann, 2006) and even freezing the food at −80°C does not eliminate viruses and bacterial spores (Kim et al., 2007). Bloodworms and sludge worms live in polluted habitats and their origin is typically unknown, posing an even greater threat to the killifish colony (Armitage, 1995; Papa et al., 2019). While cultivation of food in captivity is expensive but possible, it does not solve the problem of diet standardization. Only the development of standardized dry food specifically designed for *N. furzeri* can allow the standardization of dietary nutrient contents across experiments and laboratories and enable precise experimental manipulation of nutrients and chemical compounds.

Habituation of *N. furzeri* to commercially available dry food has received interest (Cellerino et al., 2016; Harel et al., 2016; Polačik et al., 2016; Powell et al., 2016; Hu et al., 2020), but the current consensus is that dry food may be used only as a supplementary dietary resource (Polačik et al., 2016). Recently, we have shown that adult *N. furzeri* readily accept commercial BioMar pellets as the exclusive food, with condition and performance comparable to that of the bloodworm diet (Žák et al., 2020). However, no suitable dry food is currently available for juvenile killifish, with *Artemia* nauplii remaining the only food source for juvenile *N. furzeri* for at least the first week posthatching. This is comparable to the situation in zebrafish, where feeding juvenile fish with dry food is possible but suboptimal for growth and survival (Goolish et al., 1999). The optimal onset of conditioning *N. furzeri* to dry food is at the subadult age, corresponding to the first appearance of nuptial coloration in males. Conditioning to dry food is easier and faster in communal tanks than for individually housed fish. The first dose of dry food is more readily accepted following 1 day of fasting. Weaning (cofeeding dry and live food) lasts 2−7 days, after which fish readily accept dry food. In communal tanks, sorting fish by body size during the rapid growth phase (typically the first 2 months posthatching) is important to prevent wide variability in body size within the age cohort. Žák et al. (2020) fed a combination of dry and presoaked pellets, but presoaking is only necessary for pellet sizes above 1 mm. The optimal size of dry pellets for *N. furzeri* is 75%−100% size of the fish eye pupil.

The current protocol for the use of commercial dry food provides direction for further development of a standardized diet. It should be noted that BioMar pellets are well accepted but this diet leads to serious adiposity in *N. furzeri*, which is a sign of dietary nutrient imbalance (Žák et al., 2020). This is not surprising given the natural diet of *N. furzeri* (Polačik and Reichard, 2010) contains only about half the crude lipid content of BioMar pellets. A *N. furzeri*-specific formula should be derived from the macronutrient content of their natural diet to provide the correct balance of proteins and lipids. Feeding trials using formulated diets (with known ingredient composition) and commercially available diets with suitable macronutrient compositions are currently being performed. Recent advances in the development of a killifish-specific formula promise the availability of an optimized *N. furzeri* diet in the near future.

Confirmation that *N. furzeri* accepts dry food should not lead to the use of a high diversity of commercially available diets. This scenario has occurred in zebrafish husbandry and current use of different diets has led to an unfortunate variability in zebrafish performance among research labs (Tye and Masino, 2019). When dietary manipulation is not part of the study, no more than 10 dry food types/formulae/brands should ideally be used in research laboratory fish (Tye and Masino, 2019). It should be emphasized that the exact composition of commercial food is not known and the macronutrient content of the food is often met through a mix of various ingredients, which may even differ among particular batches of the same brand or by country of production (Jenkins et al., 2014; Sprague et al., 2016). This variability may be overcome through the development of an open formula diet.

Open formula diets have known proportions of ingredients of natural origin. The transparency of open formula diet composition significantly improves research reproducibility (Barnard et al., 2009). The disadvantage of open formula diets is that natural ingredients are slightly variable among batches and cannot be chemically defined, although this also applies to commercially available diets. Another option for controlled laboratory experiments is the development of a purified diet made of chemically defined ingredients (Barnard et al., 2009). Fish fed purified (and open formula) diets have often inferior performance and survival (Watts et al., 2012), but it should be emphasized that the primary purpose of these diets is the reproducibility of results rather than the best performance. The development of standardized dry food would allow more exact nutritional research on *N. furzeri* and enable more scientifically robust aging interventions via nutrient manipulation and pharmaceutical additives.

Diseases

Even the best-equipped and well-cared laboratory facilities are endangered by the introduction of various diseases via feeding live or frozen food. Most disease outbreaks are caused by bacteria but also by eukaryotic agents with direct life cycles and low host specificity. To minimize the risk of disease introduction, the protocol for maintenance and breeding of *Nothobranchius* fishes under laboratory conditions (Polačik et al., 2016) should be followed, with maximum emphasis on the quality control of food.

Velvet disease

The disease manifests macroscopically as minute yellowish spots, petechial hemorrhages and erosions on skin, or edema and hyperplasia of gill filaments. Heavily infected fish exhibit a yellowish velvety coat over the whole body surface (Fig. 12.3A). The causative agent, the trophic stage of the parasitic dinoflagellate *Piscinoodinium pillulare* firmly attached to host cell, can be diagnosed microscopically in wet mounts or histological sections (Fig. 12.3B and C). A detailed description of the life cycle that includes trophonts as well

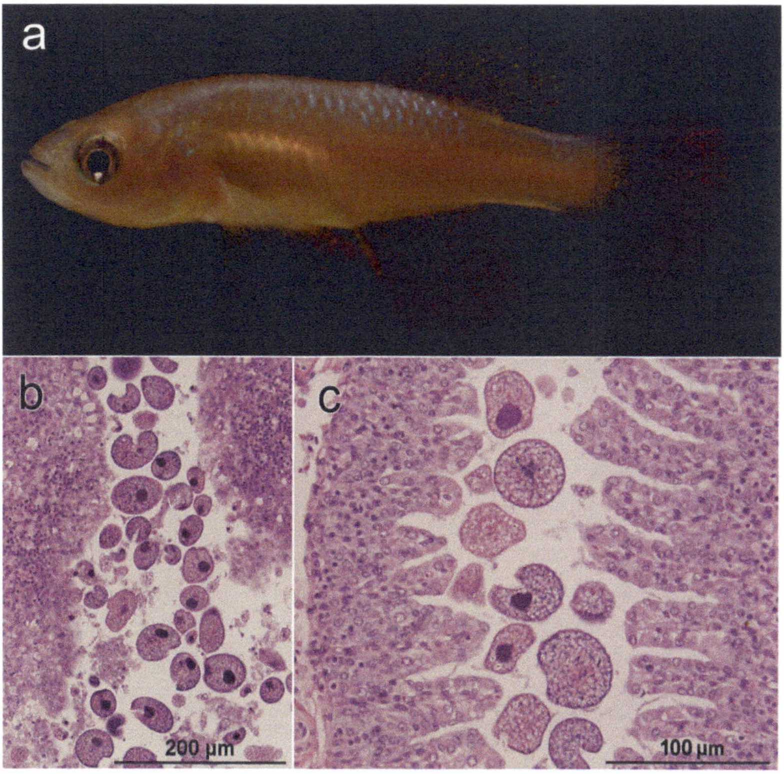

FIGURE 12.3 **Velvet disease.** *Nothobranchius furzeri* with velvet disease manifested mainly on fins. Agent trophonts seen attached to fin epithelium (A). Trophonts of variable size detached from gill surface (B, C). Histological section stained with hematoxylin and eosin.

as free-living stages (tomonts dividing into tomites and motile gymnospores) is given in Lom and Dyková (1992) and Kent and Fournie (2007). Velvet disease can be treated using a short-term bath in 10% solution of NaCl (for 10–30 s) (Polačik et al., 2016). Long-term housing of killifish at 4000 μS cm^{-1} successfully limits velvet disease and does not interfere with *N. furzeri* condition.

Glugea sp. infections

Infections caused by *Glugea* sp. with light microscopical and ultrastructural characteristics of *Glugea anomala* (microsporidia) were described in *Nothobranchius eggersi* and *N. korthausae* by Lom et al. (1995). The agent develops intracellularly in a wide range of hosts, transforming the infected host cells into specific structures called xenomas (Lom and Dyková, 2005; Dyková and Lom, 2007). In advanced stages of infection, xenomas are seen as giant macroscopically visible formations (Fig. 12.4A) localized in almost all organs. Fish-to-fish transmission occurs by ingestion of spores liberated from mature xenomas, localized most frequently in the subepithelial layer of the intestinal wall (Fig. 12.4B). Since connective tissue target cells are present in all organs,

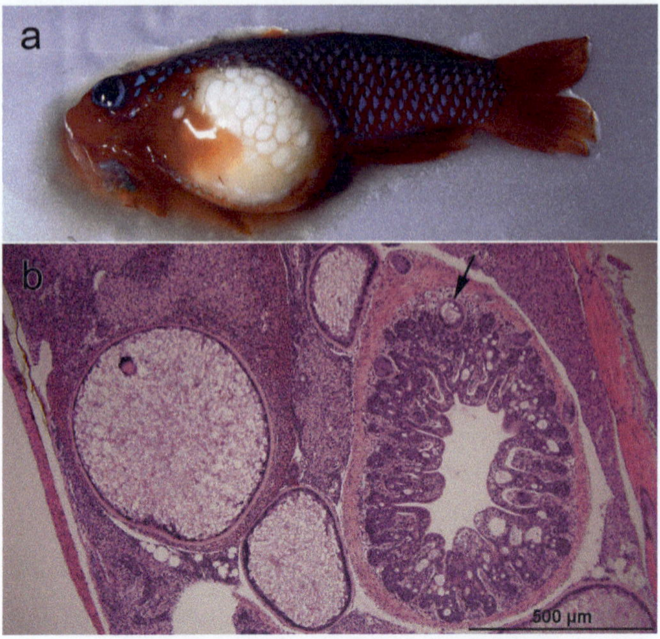

FIGURE 12.4 ***Glugea* sp. infection.** *Nothobranchius cardinalis* with advanced stage of *Glugea* sp. infection exhibiting giant xenoma formations in body cavity (A). Hematoxylin and eosin-stained histological section through body cavity organs of *Nothobranchius furzeri* showing large mature and much smaller developing intestinal xenomas (*arrow*) (B).

spores are also liberated from mature xenomas in the ovaries, although the oocytes themselves are not infected, i.e., they cannot spread infection if adequately treated. It is impossible to control the infection.

Mycobacterioses

Mycobacterial infections, diagnosed in numerous species of wild, cultured, and aquarium fish, represent a considerable threat to *Nothobranchius* research breeding units. The ubiquitous distribution of several *Mycobacterium* spp. infectious to fish (Davidovich et al., 2020; Gauthier and Rhodes, 2009) facilitates their introduction to laboratory colonies via live and frozen food. Severe spontaneous mycobacterial infections of juveniles are manifested as a loss of neutral buoyancy due to swimbladder lesions as early as 7 days post-hatch (Dyková et al., 2021). In adult fish, mycobacterioses may become systemic. Nonspecific gross symptoms such as faded coloration, exophthalmos, skin erosions, and abdominal distention are associated with enlargement and severe alterations in the kidney, liver, spleen, and swimbladder (Fig. 12.5A). The mycobacterial etiology of lesions dominated by massive proliferation of macrophages in the interstitial tissue of the kidney can only be identified using Ziehl–Neelsen staining, developed for the identification of acid-fast bacteria (Fig. 12.5B). It is impossible to detect mycobacteria in hematoxylin and eosin-

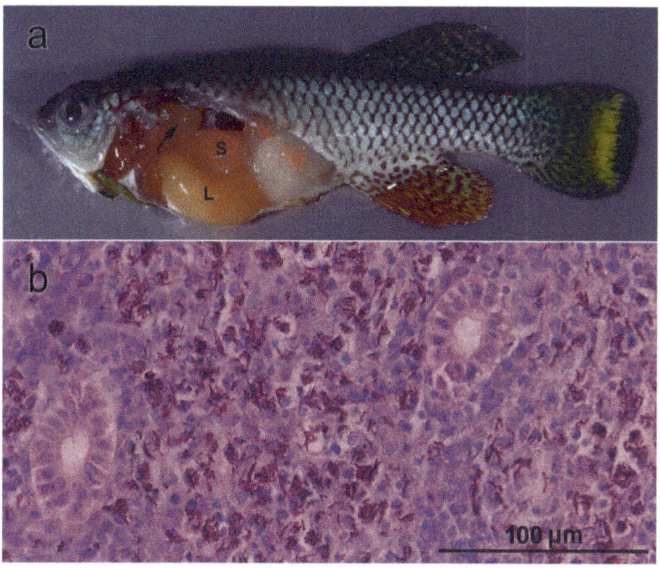

FIGURE 12.5 **Mycobacteriosis.** *Nothobranchius furzeri* with pronounced enlargement of kidneys (*arrow*) associated with mycobacterial infection (L, liver, S, spleen) (A). Mycobacterial infection visualized in kidney tissue section with Ziehl–Neelsen staining specific for acid-fast bacteria (B).

stained histological sections, even though mycobacteria often almost fill up macrophages or accumulate extracellularly. Regular examination of moribund juveniles (using imprints from altered swimbladder or histological examination of the whole body) can keep the potential mycobacterial load of breeding tanks under control.

Summary of the challenges

In this chapter, we aimed to characterize the most challenging issues associated with the maintenance, reproduction, and laboratory use of *N. furzeri*. One unknown effect is the lack of circadian fluctuation in water temperature in captivity (Žák and Reichard, 2020). Another important challenge is the development of a standardized diet, which is currently the subject of intensive research. We also need to establish whether the developmental trajectory of the embryo has any significant impact on posthatching life history, as current evidence is equivocal. An associated issue is the need to standardize embryo incubation conditions across laboratories to increase repeatability of experimental results. Finally, the source of commonly reported proliferative changes to multiple organs needs to be clarified, especially resolving whether these are of spontaneous origin or induced by immunological responses to pathogens. We conclude that despite these ongoing challenges, *N. furzeri* culture is well established and *N. furzeri* constitutes an excellent resource for biomedical, evolutionary, and toxicological research.

Acknowledgments

Contributions: ID: diseases; MP: embryo development; RB: housing and breeding; JŽ: diet; MR: general concept, other chapters. Financial support came from Czech Science foundation projects 19-01781S (to MR), 19-20873S (to RB), and 18-26284S (to MP).

References

Arezo, M.J., Papa, N.G., Berois, N., Clivio, G., Montagne, J., De la Piedra, S., 2017. Annual killifish adaptations to ephemeral environments: diapause i in two austrolebias species. Dev. Dyn. 246, 848–857.

Armitage, P.D., 1995. Chironomidae as food. In: Armitage, P.D., Cranston, P.S., V Pinder, L.C. (Eds.), The Chironomidae. Springer, Dordrecht, pp. 423–435.

Barnard, D.E., Lewis, S.M., Teter, B.B., Thigpen, J.E., 2009. Open- and closed-formula laboratory animal diets and their importance to research. J. Am. Assoc. Lab. Anim. Sci. 48, 709–713.

Bartáková, V., Reichard, M., Blažek, R., Polačik, M., Bryja, J., 2015. Terrestrial fishes: rivers are barriers to gene flow in annual fishes from the African savanna. J. Biogeogr. 42, 1832–1844.

Bartáková, V., Reichard, M., Janko, K., Polačík, M., Blažek, R., Reichwald, K., Cellerino, A., Bryja, J., 2013. Strong population genetic structuring in an annual fish, *Nothobranchius furzeri*, suggests multiple savannah refugia in southern Mozambique. BMC Evol. Biol. 13, 196.

Baumgart, M., Barth, E., Savino, A., Groth, M., Koch, P., Petzold, A., Arisi, I., Platzer, M., Marz, M., Cellerino, A., 2017. A miRNA catalogue and ncRNA annotation of the short-living fish *Nothobranchius furzeri*. BMC Genomics 18, 1–12.

Baumgart, M., Di Cicco, E., Rossi, G., Cellerino, A., 2015. Comparison of captive lifespan, age-associated liver neoplasias and age-dependent gene expression between two annual fish species: *Nothobranchius furzeri* and *Nothobranchius korthause*. Biogerontology 16, 63–69.

Blažek, R., Polačik, M., Kačer, P., Cellerino, A., Řežucha, R., Methling, C., Tomášek, O., Syslová, K., Terzibasi Tozzini, E., Albrecht, T., Vrtílek, M., Reichard, M., 2017. Repeated intraspecific divergence in life span and aging of African annual fishes along an aridity gradient. Evolution 71, 386–402.

Cellerino, A., Valenzano, D.R., Reichard, M., 2016. From the bush to the bench: the annual *Nothobranchius* fishes as a new model system in biology. Biol. Rev. 91, 511–533.

Davidovich, N., Morick, D., Carella, F., 2020. Mycobacteriosis in aquatic invertebrates: a review of its emergence. Microorganisms 8, 1–20.

Dodzian, J., Kean, S., Seidel, J., Valenzano, D.R., 2018. A protocol for laboratory housing of turquoise killifish (*Nothobranchius furzeri*). J. Vis. Exp. e57073.

Dorn, A., Ng'oma, E., Janko, K., Reichwald, K., Polačik, M., Platzer, M., Cellerino, A., Reichard, M., 2011. Phylogeny, genetic variability and colour polymorphism of an emerging animal model: the short-lived annual *Nothobranchius* fishes from southern Mozambique. Mol. Phylogenet. Evol. 61, 739–749.

Dyková, I., Lom, J., 2007. Histopathology of Protistan and Myxozoan Infections in Fishes. An Atlas. Academia Praha, 219 pp.

Dyková, I., Žák, J., Reichard, M., Součková, K., Slabý, O., Blažek, R., 2021. Swim bladder as a primary site of mycobacterial infection in Nothobranchius 'belly sliders'. Dis. Aquat. Org. 145, 111–117.

Foersch, W., 1975. *Nothobranchius furzeri* (Jubb 1971). J. Am. Killifish Assoc. 8, 315–321.

Furness, A.I., Reznick, D.N., Springer, M.S., Meredith, R.W., 2015. Convergent evolution of alternative developmental trajectories associated with diapause in African and South American killifish. Proc. R. Soc. B Biol. Sci. 282, 20142189.

Gauthier, D.T., Rhodes, M.W., 2009. Mycobacteriosis in fishes: a review. Vet. J. 180, 33–47.

Genade, T., Benedetti, M., Terzibasi Tozzini, E., Roncaglia, P., Valenzano, D.R., Cattaneo, A., Cellerino, A., 2005. Annual fishes of the genus *Nothobranchius* as a model system for aging research. Aging Cell 4, 223–233.

Goolish, E.M., Okutake, K., Lesure, S., 1999. Growth and survivorship of larval zebrafish *Danio rerio* on processed diets. N. Am. J. Aquacult. 61, 189–198.

Grégoir, A.F., Thoré, E.S.J., Philippe, C., Pinceel, T., Brendonck, L., Vanschoenwinkel, B., 2018. Squeezing out the last egg—annual fish increase reproductive efforts in response to a predation threat. Ecol. Evol. 8, 6390–6398.

Harel, I., Benayoun, B.A., Machado, B., Singh, P.P., Hu, C.K., Pech, M.F., Valenzano, D.R., Zhang, E., Sharp, S.C., Artandi, S.E., Brunet, A., 2015. A platform for rapid exploration of aging and diseases in a naturally short-lived vertebrate. Cell 160, 1013–1026.

Harel, I., Valenzano, D.R., Brunet, A., 2016. Efficient genome engineering approaches for the short-lived African turquoise killifish. Nat. Protoc. 11, 2010–2028.

Hsu, C.Y., Chiu, Y.C., 2009. Ambient temperature influences aging in an annual fish (*Nothobranchius rachovii*). Aging Cell 8, 726–737.

Hu, C.K., Brunet, A., 2018. The African turquoise killifish: a research organism to study vertebrate aging and diapause. Aging Cell 17, e12757.

Hu, C., Wang, W., Amour, J.B., Singh, P.P., Reeves, G.A., 2020. Vertebrate diapause preserves organisms long term through Polycomb complex members. Science 367, 870–874.

Jenkins, J.A., Chair, H.L., Bart, J.D., Bowker, P.R., Bowser, P.R., MacMilan, J.R., Nickum, J.G., Rose, J.D., Sorensen, P.W., Whitledge, G.W., Rachlin, J.W., Warkentine, B.E., Bart, H.L., 2014. Guidelines for the Use of Fishes in Research. American Fisheries Society, Bethesda, Maryland.

Kent, M.L., Fournie, J.W., 2007. Parasites of fishes (chapter 7). In: Baker, D.G. (Ed.), Flynn's Parasites of Laboratory Animals. American College of Laboratory Animal Medicine, Blackwell Publishing, pp. 69–116.

Kim, J.H., Gomez, D.K., Choresca, C.H., Park, S.C., 2007. Detection of major bacterial and viral pathogens in trash fish used to feed cultured flounder in Korea. Aquaculture 272, 105–110.

Levels, P.J., Gubbels, R.E., Denuce, J.M., 1986. Oxygen consumption during embryonic development of the annual fish *Nothobranchius korthausae* with special reference to diapause. Comp. Biochem. Physiol. Part A Mol. Integr. Physiol 84, 767–770.

Liu, S., Zheng, Z., Ji, S., Liu, T., Hou, Y., Li, S., Li, G., 2018. Resveratrol reduces senescence-associated secretory phenotype by SIRT1/NF-κB pathway in gut of the annual fish *Nothobranchius guentheri*. Fish Shellfish Immunol. 80, 473–479.

Lom, J., Dyková, I., 1992. Protozoan parasites of fishes. In: Developments in Aquaculture and Fisheries Science, vol. 26. Elsevier Science Publishers, Amsterdam-London-New York-Tokyo, 315 pp.

Lom, J., Dyková, I., 2005. Microsporidian xenomas in fish seen in wider perspective. Folia Parasitica 52, 69–81.

Lom, J., Noga, E.J., Dyková, I., 1995. Occurrencre of a microsporean with characteristics of *Glugea anomala* in ornamental fish of the family Cyprinodontidae. Dis. Aquat. Org. 21, 239–242.

Muck, et al., 2018. The short-lived African turquoise killifish (*Nothobranchius furzeri*): a new model system for research on aging. In: Ram, J.L., Conn, P.M. (Eds.), Conns Handbook of Models for Human Aging.

Nenoff, P., Uhlemann, R., 2006. First report of an atypical mycobacterial infection of the mangrove killifish *Rivulus magdalenae* due to living fish food (*Tubifex tubifex*) infected by *Mycobacterium marinum*. Dtsch. Tierärztliche Wochenschr. 230–232.

Ng'oma, E., Groth, M., Ripa, R., Platzer, M., Cellerino, A., 2014. Transcriptome profiling of natural dichromatism in the annual fishes *Nothobranchius furzeri* and *Nothobranchius kadleci*. BMC Genomics 15, 754.

Papa, N.G., Chalar, C., Berois, N., Arezo, M.J., 2019. Annual killifish: an approach to the choriogenins of *Austrolebias charrua* egg envelope. Environ. Biol. Fish. 102, 829–844.

Podrabsky, J.E., Garrett, I.D.F., Kohl, Z.F., 2010. Alternative developmental pathways associated with diapause regulated by temperature and maternal influences in embryos of the annual killifish *Austrofundulus limnaeus*. J. Exp. Biol. 213, 3280–3288.

Podrabsky, J.E., Hand, S.C., 1999. The bioenergetics of embryonic diapause in an annual killifish, *Austrofundulus limnaeus*. J. Exp. Biol. 202, 2567–2580.

Podrabsky, J.E., Riggs, C.L., Romney, A.L., Woll, S.C., Wagner, J.T., Culpepper, K.M., Cleaver, T.G., 2017. Embryonic development of the annual killifish *Austrofundulus limnaeus*: an emerging model for ecological and evolutionary developmental biology research and instruction. Dev. Dyn. 246, 779–801.

Polačik, M., Blažek, R., Reichard, M., 2016. Laboratory breeding of the short-lived annual killifish *Nothobranchius furzeri*. Nat. Protoc. 11, 1396–1413.

Polačik, M., Blažek, R., Řežucha, R., Vrtílek, M., Terzibasi Tozzini, E., Reichard, M., 2014. Alternative intrapopulation life-history strategies and their trade-offs in an African annual fish. J. Evol. Biol. 27, 854–865.

Polačik, M., Reichard, M., 2010. Diet overlap among three sympatric African annual killifish species *Nothobranchius* spp. from Mozambique. J. Fish. Biol. 77, 754–768.

Polačik, M., Smith, C., Reichard, M., 2017. Maternal source of variability in the embryo development of an annual killifish. J. Evol. Biol. 30, 738–749.

Polačik, M., Vrtílek, M., Reichard, M., Žák, J., Blažek, R., Podrabsky, J., 2021. Embryo ecology: developmental synchrony and asynchrony in the embryonic development of wild annual fish populations. Ecol. Evol. 11 (9), 4945–4956.

Powell, M.L., George, B., Allison, D.B., Watts, S.A., Austad, S.N., 2016. The impact of restricting a formulated diet on growth, reproduction, and survival in Nothobranchius *f*urzeri. In: 2nd Notho Symposium 2016 T5.

Ramsay, J., Watral, V., Schreck, C., Kent, M., 2009. Husbandry stress exacerbates mycobacterial infections in adult zebrafish, *Danio rerio* (Hamilton). J. Fish. Dis. 32, 931–941.

Reichard, M., 2015. The evolutionary ecology of African annual fishes. In: Annual Fishes: Life History Strategy, Diversity, and Evolution, pp. 133–158.

Reichard, M., Janáč, M., Polačik, M., Blažek, R., Vrtílek, M., 2017. Community assembly in *Nothobranchius* annual fishes: nested patterns, environmental niche and biogeographic history. Ecol. Evol. 7, 2294–2306.

Reichard, M., Giannetti, K., Ferreira, T., Vrtilek, M., Polacik, M., Blazek, R., Ferreira, M.G., 2020. Environment and sex control lifespan and telomere length in wild-derived African killifish. bioRXiv. https://doi.org/10.1101/2020.09.03.280792.

Romney, A.L., Podrabsky, J.E., 2017. Transcriptomic analysis of maternally provisioned cues for phenotypic plasticity in the annual killifish, *Austrofundulus limnaeus*. EvoDevo 8, 6.

Sprague, M., Dick, J.R., Tocher, D.R., 2016. Impact of sustainable feeds on omega-3 long-chain fatty acid levels in farmed Atlantic salmon, 2006 – 2015. Sci. Rep. 6, 21892.

Terzibasi Tozzini, E., Dorn, A., Ng'Oma, E., Polačik, M., Blažek, R., Reichwald, K., Petzold, A., Watters, B., Reichard, M., Cellerino, A., 2013. Parallel evolution of senescence in annual fishes in response to extrinsic mortality. BMC Evol. Biol. 13, 77.

Tye, M., Masino, M.A., 2019. Dietary contaminants and their effects on zebrafish embryos. Toxics 7, 46.

Vrtílek, M., Reichard, M., 2015. Highly plastic resource allocation to growth and reproduction in females of an African annual fish. Ecol. Freshwater Fish 24, 616–628.

Vrtílek, M., Reichard, M., 2016. Patterns of morphological variation among populations of the widespread annual killifish *Nothobranchius orthonotus* are independent of genetic divergence and biogeography. J. Zool. Syst. Evol. Res. 54, 289–298.

Vrtílek, M., Žák, J., Polačik, M., Blažek, R., Reichard, M., 2019. Rapid growth and large body size in annual fish populations are compromised by density-dependent regulation. J. Fish. Biol. 95, 673–679.

Watts, S.A., Powell, M., D'Abramo, L.R., 2012. Fundamental approaches to the study of zebrafish nutrition. ILAR J. 53, 144–160.

Willemsen, D., Cui, R., Reichard, M., Valenzano, D.R., 2020. Intra-species differences in population size shape life history and genome evolution. eLife 9, e55794.

Wourms, J.P., 1972. The developmental biology of annual fishes III. Pre-embryonic and embryonic diapause of variable duration in the eggs of annual fishes. J. Exp. Zool. 182, 389–414.

Žák, J., Reichard, M., 2020. Fluctuating temperatures extend median lifespan, improve reproduction and reduce growth in turquoise killifish. Exp. Gerontol. 140, 111073.

Žák, J., Dyková, I., Reichard, M., 2020. Good performance of turquoise killifish (*Nothobranchius furzeri*), on pelleted diet as a step towards husbandry standardization. Sci. Rep. 10, 8986.

Žák, J., Reichard, M., 2021. Reproductive senescence in a short-lived fish. J. Anim. Ecol. 90, 492–502.

Žák, J., Vrtílek, M., Reichard, M., 2019. Diel schedules of locomotor, reproductive and feeding activity in wild populations of African annual killifish. Biol. J. Linn. Soc. 182, 435–450.

Chapter 13

Mexican tetra (*Astyanax mexicanus*): biology, husbandry, and experimental protocols

Diana P. Baumann, Andrew Ingalls
Reptile & Aquatics, Stowers Institute for Medical Research, Kansas City, MO, United States

Biology and natural history

Astyanax mexicanus, also known as the Mexican tetra, is a teleost species originating in South America. Cataloged in 1936 by Hubbs and Innes as the first blind fish in the Characidae family, it became available in the US aquarium trade in the 1940s, and early investigations were carried out by Breder (1942). While its classification was initially in question, it is now understood that two distinct forms of the species can be found in the wild: the surface morphology, a sighted river-dwelling fish found in the North and Central Americas, and the troglobitic morphology, a blind cave-dwelling fish indigenous to the Sierra de El Abra, Sierra de Guatemala, and Sierra de Colmena regions of Mexico (Fig. 13.1).

A. mexicanus has been recorded in 29 distinct cave localities across the regions, having diverged from their surface counterparts between 20,000 and 1 million years ago (Fumey et al., 2018; Elipot et al., 2014). This broad range in time can be attributed to the complex regional geomorphology and, more pertinently, a lack of detailed understanding regarding the number and timeline of colonizations for this species. The long-standing hypothesis for the age of divergence in these populations involves two separate surface lineages: "old" and "new" (Gross, 2012). The "old" lineage is comprised of an ancient, extinct variety of the surface morph, while the "new" lineage descends from the surface morph currently inhabiting the rivers of Mexico (Gross, 2012). This hypothesis is being revisited by studies such as Fumey et al. (2018), suggesting that the basis for these two groupings is an inaccurate model for evolution and that all cave populations more likely diverged as recently as approximately

Laboratory Fish in Biomedical Research. https://doi.org/10.1016/B978-0-12-821099-4.00003-1
Copyright © 2022 Elsevier Inc. All rights reserved.

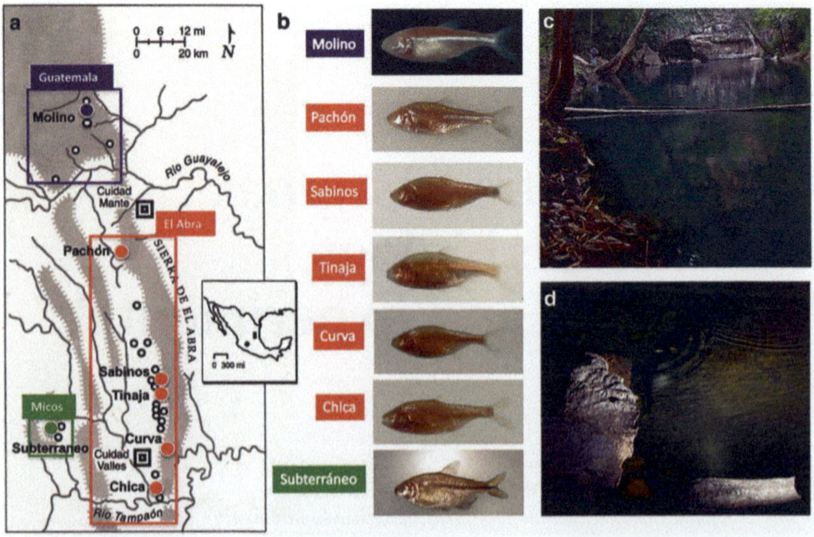

FIGURE 13.1 Astyanax caves, cavefish populations, and habitats. (A) Map showing the distribution of caves in the El Abra region of Tamaulipas and San Luis Potosí, Mexico. Boxes outlined in blue, red, and green show locations of Astyanax caves (black outlined and colored spheres) in the Sierra de Guatemala, Sierra de El Abra, and Micos regions, respectively. *Inset*: Mexico map showing the locations of *Astyanax mexicanus* cavefish in the El Abra epicenter (*right shaded rectangle*) and *Astyanax aeneus* cavefish in Guerrero (*left shaded sphere*). (B) Most frequently studied cavefish populations in the Guatemala (blue label), El Abra (red labels), and Micos (green label) regions. (C) A surface fish habitat at El Nacimiento del Río Choy. (D) A cavefish habitat in El Sótano de Las Piedras. *Figure is taken from Jeffery, W.R., 2020.* Astyanax *surface and cave fish morphs. EvoDevo 11, 14. https://doi.org/10.1186/s13227-020-00159-6.*

20,000 years ago. This furthers the intrigue surrounding this model organism, which clearly shows a strong capacity for phenotypical adaptations in a short period of time.

One of the key reasons why these fish have emerged so prominently as a new model organism is that all cave morphs studied so far have proven to be interfertile, both with one another and with the surface morph. This has opened a window to study the evolution of the cave morphs as they have adapted to survival in perpetually dark and nutrient-deprived environments. Trait diversity, both phenotypical and behavioral, can be investigated to determine the genetic basis for these changes. Much of the work currently being conducted revolves around trait evolution and comparative approaches. However, as the field continues to grow, an increasing number of studies are beginning to explore *A. mexicanus* as a model for human disease, in addition to their differing biological characteristics (McGaugh et al., 2020b; Rohner, 2018a, 2018b).

Two cave morph traits that have received the most attention are the loss of pigmentation (Jeffery, 2009; McCauley et al., 2004; Stahl and Gross, 2015) and eye degeneration, leading to the complete absence of eyes in multiple cave morphs (Jeffery, 2009; Krishnan and Rohner, 2017). Other physical changes include an increase in the number of taste buds and teeth as well as structural changes to the jaw, leading to altered feeding behavior (Schemmel, 1980). More recent work looks at craniofacial asymmetry and changes to visual dendrite morphology (Gross et al., 2014; Tanvir et al., 2020). Jeffery (2001) provided a thorough overview of both constructive and regressive changes in *A. mexicanus*, while Yoshizawa (2015) delivered a wide-ranging summary of behavioral evolution. Evolutionary adaptations include changes in circadian rhythm (Beale et al., 2013) and sleep patterning (Duboué et al., 2011; Jaggard et al., 2018; McGaugh et al., 2020a; Yoshizawa et al., 2015). Brain atlas studies have subsequently revealed neuroanatomical changes in brain structures associated with behavior (Jaggard et al., 2020; Loomis et al., 2019).

Genetic manipulation has been achieved through a variety of mechanisms, including transcription activator-like effector nucleases (TALEN), Tol2 transposase, and clustered regularly interspaced short palindromic repeats (CRISPR) (Klaassen et al., 2018; Ma et al., 2015; Stahl et al., 2019a, 2019b).

Disease modeling studies include heart regeneration (Stockdale et al., 2018). Work on metabolism is providing insights into diabetes and obesity (Aspiras et al., 2015; Salin et al., 2010; Volkof, 2016; Xiong et al., 2018). Novel acoustic studies have recently increased the understanding of communication (Hyacinthe et al., 2019), as well as the acoustic startle response (Paz et al., 2020).

For further information, Keene et al. (2016) examined the evolution and biology of the Mexican tetra in a comprehensive tome; Kowalko (2020) provided a review of the genetic foundations of evolved behavioral traits; and a summary of methods, resources, meetings, websites, and conferences is covered by Jeffery (2020).

Husbandry

To maintain high-quality, reproducible research, the husbandry of any model organism is paramount. As the prevailing aquatic model organism, zebrafish have steered the advancement of husbandry for many species, including *A. mexicanus*. However, as *A. mexicanus* grows in popularity, defined differences have begun to emerge. The following overview details husbandry elements the authors find critical for *A. mexicanus* to thrive in a research environment.

Housing

Depending on the needs of the facility and the research, tanks can vary in shape, volume, and material (Fig. 13.2).

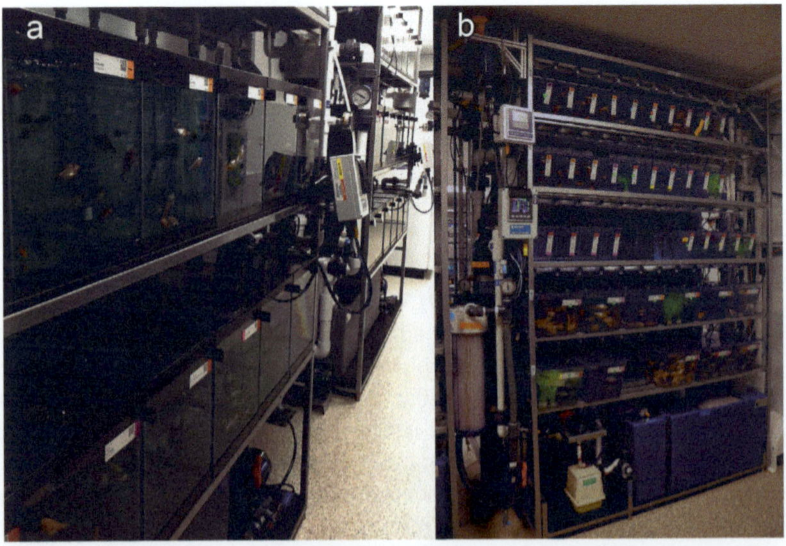

FIGURE 13.2 Common recirculating aquaculture system tank setups for *Astyanax mexicanus*. (A) Modular rack system housing 38 L tanks. (B) Modular rack system housing 3 and 10 L tanks. *Figure is taken from Stowers Institute Cavefish Team.*

According to Harper and Lawrence (2011), "the density at which fish are kept in captivity exerts profound effects on their health, productivity, and welfare." Therefore research facilities must balance a desire to maximize the census with the welfare of the animals. This balance is not always simple, as some teleosts, such as the surface morph of *A. mexicanus*, exhibit problematic behaviors at higher densities (territoriality), while also displaying the behaviors of a social species (schooling), indicating a need for companionship. Thus the regulation of both low and high stocking densities is critical for this morph of *A. mexicanus* (Burchards et al., 1985). Requirements for density regulation among the individual cave localities are currently undetermined, and their housing densities remain similar to that of the surface morphs.

Independent of the number of fish within a tank, many teleosts respond to the volume of their environment. This can be seen in the surface morph of *A. mexicanus*, which displays increasing territoriality as the tank size decreases (Burchards et al., 1985). Additionally, Gallo and Jeffery (2012) found that surface *A. mexicanus* will regulate their growth rate according to their environmental space, a trait known as space-dependent growth (SDG). Some cavefish localities, such as Pachón, also possess a reduced version of the SDG trait, while others, such as Tinaja, do not. As such, many facilities have adopted a "sliding scale" (Harper and Lawrence, 2011) of densities as the fish age, increase in size, and change in behavior and physiology. While no standards exist for *A. mexicanus*, a conservative guideline of tank sizes and densities determined by age and purpose is provided in Table 13.1.

TABLE 13.1 Densities and tank volume by purpose and age.

Purpose	Age category	Age range (months postfertilization)	Density (fish per liter)	Typical volume (liters)	Maximum capacity (fish)
Nursery	Larval to early juvenile	0–3	8–10	3	Up to 35
Grow-out	Early juvenile	3–6	6–8	3	Up to 35
Grow-out	Late juvenile	6–12	1.5–3	10–16	Up to 60
Maintenance	Adult	12+	0.5–1	16–38	Up to 60
Breeding	Adult	12+	0.5–1	38	Up to 60

Arguably the most crucial element of housing is the medium in which the animals live: the water. The American Veterinary Medical Association (AVMA) states that animals should have "an environment appropriate to their care and use, with thoughtful consideration for their species-typical biology and behavior" (AVMA, 2020a). For *A. mexicanus*, this means understanding both river and cave environments. Due to the inherent difficulties of maintaining multiple environments in a laboratory, the authors suggest a unified approach. Table 13.2 indicates average water chemistry from several tests taken in March 2017 and July 2019 from the Pachón cave, Rio Subterraneo pool, and Nacimiento Rio Choy.

The authors have found that *A. mexicanus* can thrive in a wide range of multiple water-quality parameters, and Table 13.3 provides values utilized at Stowers Institute for Medical Research. Based on the field data presented in Table 13.2, a higher range for alkalinity and general hardness may be more appropriate.

Aggression behavior

Important adaptations to feeding and social behavior have led to stark differences in cave and surface fish. Studies have begun to explore the "fight or feed" behaviors in *A. mexicanus*. These behaviors are unique to the surface and cave morphs, respectively, yet linked by common genetic components (Elipot et al., 2013; Rétaux and Elipot, 2013). This section will focus on the surface morph, which exhibits an "erratic viciousness" (Breder, 1943), unlike the cave morph, which is generally nonaggressive.

In the wild, surface morphs have a wide range of territory, compelling them to school for protection against larger predators. During the dry season, their

TABLE 13.2 Location water chemistry.

Parameter	Value
Specific conductance (μS/cm)	742 ± 296
Temperature (°C)	25.6 ± 1.3
pH	7.6 ± 0.4
Ammonia (ppm)	0.05 ± 0.02
Nitrite (ppm)	0.003 ± 0.002
General hardness (ppm)	104 ± 20[a]
Alkalinity (ppm)	280 ± 92
Dissolved oxygen (ppm)	7.01 ± 0.25[b]

[a]*Four values were over-range (>120 ppm).*
[b]*Four values recorded in percent saturation (not included).*

TABLE 13.3 Stowers Institute water quality parameters.

Parameter	Set point	Acceptable range
Specific conductance (μS/cm)	800	700–1100
Temperature (°C)	23.0	16.0–28.0
pH	7.65	7.3–8.0
Ammonia (ppm)	0	<0.5
Nitrite (ppm)	0	<0.5
Nitrate (ppm)	0	<60
General hardness (ppm as $CaCO_3$)	In range	20–120
Alkalinity (ppm as $CaCO_3$)	In range	20–120
Dissolved oxygen (% saturation)	>90	80–103

environments can diminish into small, separated pools. This decrease in volume evokes a behavioral shift toward territoriality and dominance, a trait that has also been documented within the laboratory environment (Burchards et al., 1985). Schooling behavior is not completely diminished, however. Even at smaller tank volumes, a dominant female will lead the social structure. Kowalko et al. (2013a) noted the presence of schooling and shoaling for surface morphs in a laboratory environment; behaviors not witnessed in the cave morphs. This social drive conflicts with the territorial behavior and can lead to increased aggression as subordinates attempt to establish their place within the hierarchy.

Feeding conditions can also play a role in the aggressive behavior of *A. mexicanus*. When food is in short supply, resource guarding and increased territoriality can be observed. Conversely, when food is in overabundance, an excess of energy may lead to increased agonistic behaviors. Burchards et al. (1985) showed a significant increase in attacks by the Rio Teapao surface morph prior to satiation, revealing a relationship between aggressive behavior and food supply.

While aggressive tendencies continue throughout the life of the surface morph, Ingalls et al. (2020a) began to explore a trend in aggression-related deaths when compared to age. This study showed a significant increase in these incidences from 3 to 6 months, peaking at 5 months. It also indicated a strong negative correlation between aggression and Fulton's condition factor, suggesting an inverse relationship between the welfare of the fish and its aggressive behavior. This could be due, in part, to the start of territorial and dominant behaviors to establish the hierarchy. Aggression can be difficult to spot before a physical incident or death, though there are certain social behaviors associated with this trait. Among the most defining agonistic behaviors are "fin spreading," "snake swimming," "tail beating," "ramming," and "biting" (Burchards et al., 1985; Espinasa et al., 2005), with "ramming" and "biting" being the most common. If a submissive individual subjected to these behaviors has no way to escape or hide, the persistent attacks will result in its eventual mortality.

There are many potential prevention techniques when dealing with aggression. It is important to note that although no single technique will eliminate aggression from the species, the inclusion of any of these methods has the potential to vastly improve welfare. For subordinate fish to escape aggression from the dominant female, hiding locations must be available for retreat. The authors have found success using several items, such as artificial plants, honeycomb alcoves, multitiered stands, and PVC pipes as depicted in Fig. 13.3. Blocking visual contact between tanks can also reduce cross-tank aggression.

Environmental factors also play a significant role in the behavior and aggression of *A. mexicanus*. For example, with a change in tank size and/or density, it is possible to further reduce aggression. Burchards et al. (1985) showed that ramming movements are dependent on the available space and reduce as tank volume increases. Palstra and Planas (2011) stated that an increase in water flow, leading to sustained exercise, may result in reduced aggression. As noted, food denial can increase the potential for aggression. Rétaux and Elipot (2013) discussed that when denied food, even cave morphs are 75% more aggressive than their "normal-fed" or "overfed" counterparts.

In some cases, the severity of this behavior requires more drastic intervention, and isolation of the aggressor may be required. If a hyperaggressive

FIGURE 13.3 Enrichment items used to reduce aggression. (A) Honeycomb alcoves. (B) PVC pipes and plastic plants. (C) Multitiered stands. (D) Barriers to block visual contact between tanks. *Figure is taken from Stowers Institute Cavefish Team.*

fish does not respond to these methods, it may be necessary to cull the individual. This will lead to the reestablishment of the pyramid hierarchy and a temporary increase in tank-wide aggression.

Nutrition and feeding

Due to the novel status of *A. mexicanus*, many nutritional requirements remain unknown. This section will use the requirements of other warm water, omnivorous Characids, such as the neon tetra (*Paracheirodon innesi*) and the *Hyphessobrycon* genus to propose a nutritional baseline for *A. mexicanus*.

Regarding macronutrients, Sealey et al. (2009) recommended a diet containing at least 45% crude protein for optimal mass gain. Darve et al. (2016) stated that neon tetras perform well at lipid levels ranging from 10% to 15%, with a peak at 12.3%. There is no dietary requirement for carbohydrates in teleosts. However, it is commonly used as an alternative source of energy, saving protein and lipids from glucose conversion (Harper and Lawrence, 2011). This nutrient profile is preferred from the start of exogenous feeding until the adult life stage. Due to the unknown species-specific demands for vitamins and minerals, formulated diets carry an overabundance. Texts, such as *Nutrition and Feeding of Fish and Crustaceans* and *The Vitamins: Fundamental Aspects in Nutrition and Health*, contain more information on micronutrients.

The two morphologies, surface and cave, exhibit different feeding adaptations. For instance, the cave morph has double the number of teeth per maxillary bone and a protruding lower maxilla. The cave morph also has a several-fold increase in taste buds present in the mouth, pharynx, lips, and external surface of the head. These adaptations may have led to a posture that promotes bottom feeding, as the surface morph is cited as feeding at a much steeper angle than the cave morph (Jeffery, 2001; Schemmel, 1980; Yamamoto et al., 2009). However, using quantitative trait locus on F2 fish, Kowalko et al. (2013b) concluded that, in the case of feeding posture, they could not find a strong correlation with morphological traits such as number of taste buds, craniofacial morphology, or body depth. Cave and surface morph evolution also diverge in several areas of the gut, including plasticity, morphology, and homeostasis (Riddle et al., 2018, 2019).

There are also behavioral differences that lend themselves to the cave morph's foraging capability. For example, Yoshizawa et al. (2010) showed that cavefish utilize detection in their lateral line system, allowing them to sense the specific swimming frequencies of crustaceans such as copepods. This behavior, known as vibration attraction behavior (VAB), is absent from most surface fish, although a few have been noted to demonstrate it at low levels (Yoshizawa et al., 2015). Espinasa et al. (2020) tested the presence of VAB in the wild among several cave localities (Pachón, Tinaja, and Molino), noting differential expression of the trait between distinct pools within the same cave. This suggests VAB may express a level of plasticity based on the biological load of each microenvironment. Due to this trait, the authors have found it beneficial to halt the water flow before feeding. This can act as a feeding signal for the fish, as well as a potential reduction of vibration interference.

At 5.5 days postfertilization (dpf) (Hinaux et al., 2011), a rudimentary digestive tract has developed, and the yolk sac has regressed. This is when exogenous feeding will begin, although full functionality will not be complete until metamorphosis occurs. While this period remains unknown for *A. mexicanus*, other warm water Characids such as neon tetra and black tetra cross this milestone at approximately 20 dpf (Lipscomb et al., 2020).

The growth and energy requirements during the larval stage are at the highest. Carbohydrates should be at a minimum ($<5\%$) to increase digestibility (Harper and Lawrence, 2011). The feeding regimen should include a significant live feed component, such as rotifers or *Artemia*, due to their balanced nutritional content and high digestibility (Lipscomb et al., 2020).

During this life stage, larvae have not developed a strong swimming ability; as such, it is best if food remains suspended in the water column. Continuous feeding such as polyculture or multiple feedings per day will lead to the highest survival rate. Toward the end of this life stage, a weaning protocol can transition the organism into microparticulate diets. The authors have used Skretting Gemma Micro diets for this life stage; however, many suitable feeds exist.

Once metamorphosis has occurred, *A. mexicanus* enters a period of rapid growth and maturation (McMenamin and Parichy, 2013) known as the juvenile stage. While protein and lipids remain high to allow for maximal growth and sexual maturation, carbohydrates are now more acceptable due to the developed digestive system. The juvenile cave morph will begin to exhibit a foraging behavior, eating primarily at the bottom of the tank. The surface morph will prefer to eat at the surface, then in the water column, and finally at the bottom of the tank, so feeds should be stable for an extended period with the ability to both float and sink.

Due to the rapid growth rate of juvenile *A. mexicanus*, smaller live feeds, such as *Artemia* and rotifers, no longer provide a sustainable nutritional profile. The feeding regimen should constitute microparticulate feeds that range in size as the fish's gape size increases. Three or more feedings per day will promote high growth rates. The authors have used Skretting Gemma Micro and Gemma Diamond diets for this life stage. However, there are many suitable alternative feeds with similar properties.

Once the fish have reached adult maturation, at 1-year postfertilization, the goal is no longer high growth rates. Maintenance of the stock allows for a slight reduction in protein and lipids. While not in high demand, these macronutrients are still important for energy production and continued sexual maturation. Carbohydrates are now used in feeds at higher levels for energy supplementation. Diets can vary during this stage, depending on the fish's gape size. There are many feeds available that meet these nutrient guidelines, of which Piscine Pellets, Skretting Gemma Diamond, and Skretting Clean Assist are examples.

To shift from maintenance to reproduction, adults require a high concentration of protein (>45%). Lipids (10%–15%) are also critical, as certain fatty acids (n3 and n6 polyunsaturated fatty acids and highly unsaturated fatty acids) are important for reproductive functions. Carbohydrates become a low priority to maximize lipid and protein potential. Vitamins and minerals are also more important during breeding. For example, elevated vitamins C and A can have positive effects on gamete production and embryo quality (Harper and Lawrence, 2011). Due to the energy demand, a higher feeding frequency may be beneficial to maintain the energy lost through reproduction. Additionally, it may be helpful to supplement the diet with high-lipid feeds such as small black soldier fly larvae or Mysis shrimp.

Colony management

A. mexicanus remains a promising model organism in evolutionary and developmental studies due to the interfertility of the cave and surface morphs. Housing multiple genetic lines, maintaining population diversity, and cultivating F_1 and F_2 hybrids are all factors in the success of a facility using *A. mexicanus*. Due to this species' significant maturation time, proactive colony management, breeding, and larval care are of critical importance to maintaining healthy stock numbers.

Sexing

The first stage in a successful breeding program requires a healthy and robust broodstock that regularly produces embryos. This necessitates the ability to reliably identify sexes of the fish available to establish suitable breeding ratios. There are several possible methods of sexual identification for *A. mexicanus*: visual observation, cotton swabbing, and gamete collection. Each of these techniques induces a varying degree of stress on the fish and must be considered according to the certainty required. While gamete collection is the most reliable, other techniques, either individually or in tandem, may provide sufficient confidence to forgo the procedure. For *A. mexicanus*, the authors have noted sexual maturation as early as 9 months, while other facilities have indicated generations as short as 6 months (Borowsky, 2008a). However, to provide the highest success rate, the authors suggest that sexing occur at approximately 1-year postfertilization.

All *A. mexicanus* share dimorphic traits that allow for the identification of sex; however, each morph may also exhibit additional traits specific to only that locality. It is important to note that while these methods may help identify sexes, they are not a guaranteed method for success. These dimorphic traits can be subtle or even absent from a sexually mature adult and should only be used as a guide for further study. *A. mexicanus* females frequently have a distinct, deeper body shape than males, which can appear more concave between the pelvic and anal fins. This deep body shape extends to the transverse view of a female, where one can see a wider abdominal region. Additionally, females tend to have a pointed cranial region, compared to the males' shortened and rounded upper and lower maxilla. Finally, the authors note that one of the best methods for visual identification of any morph of *A. mexicanus* is the shape of the anal fin: a concave posterior margin in females and an approximately linear margin in males (Fig. 13.4).

The surface morph has a distinct coloration that appears on the anal fin of females. The first two anterior rays will sometimes exhibit a white coloration and thickening that makes them opaque. Additionally, the females may have a patch of red or orange coloration throughout the anal fin. The Pachón locality

FIGURE 13.4 *Astyanax. mexicanus* female (*upper*) and male (*lower*) showing anatomical differences in overall body shape, craniofacial shape, and shape and color of the anal fin. *Figure is taken from Stowers Institute Cavefish Team.*

has a distinct body coloration difference, with the males exhibiting a golden hue versus the females' pale yellow or slightly pink hue. This extends to the anal fins, where males will have opaque and yellow coloration toward the anterior half of the fin. Like the Pachón, fish from the Tinaja locality have sexually dimorphic anal fins. The males' fins are opaque toward the anterior half, and a yellow hue is typically present on all fins, while females may have a pale pink tint to their fins. The bodies of the Molino locality do not exhibit any apparent dimorphism in color. However, like the other cave morphs, the males do have distinct opaque coloration on the anterior half of the anal fin. Molino females also have a small protrusion anterior to their anal fins, related to the urogenital pore, which has not been noted with any other cavefish locality. Due to the nature of hybrids, sexual identification is varied and can encompass traits from both lineages. For this reason, second-generation hybrids can be difficult to sex. First-generation hybrids will typically follow the same visual identification as a surface morph, assuming it is part of the cross (Fig. 13.5).

While visual identification has proven to be an adequate, noninvasive method for sexual identification, testing the anal fin for the presence of hook-like structures known as denticles using a cotton swab will provide more consistent results on sexually mature males. To perform this technique, gently brush the anterior portion of the anal fin in a caudal to cranial direction along the side of the fin. If the fish in question is a sexually mature male, the cotton will catch on the denticles. If the swab does not catch, the fish is either a

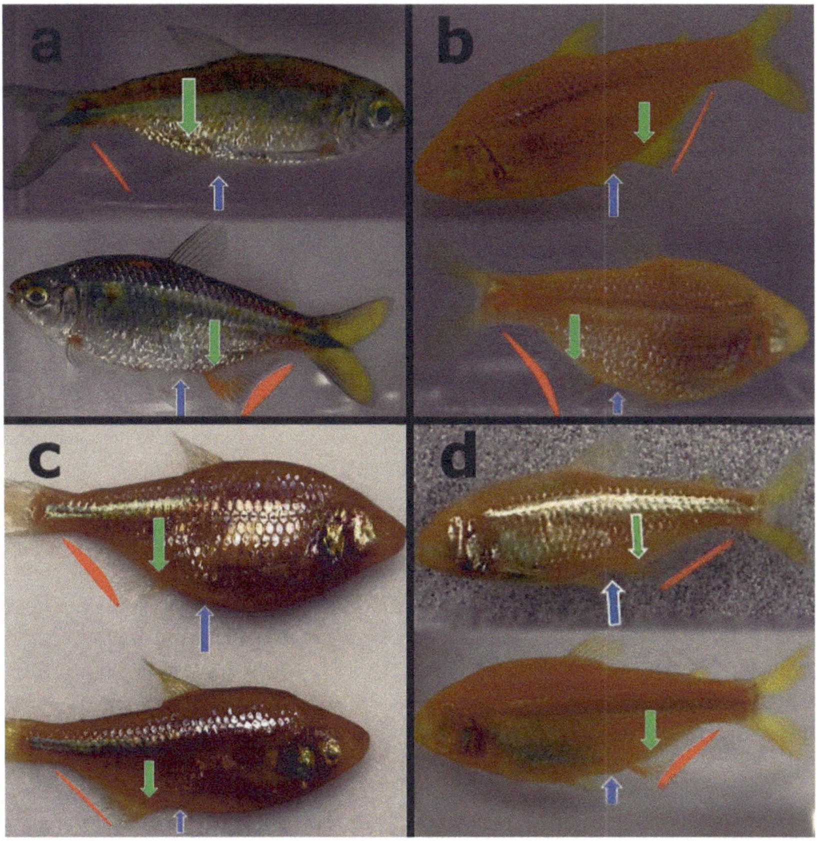

FIGURE 13.5 Sexual dimorphism observed in *Astyanax mexicanus* from the major localities. Red line, indicates anal fin margin curvature, blue arrow pointing up indicates curvature of abdomen, and green arrow pointing down indicates anal fin coloration. Images are not to scale. (A) Surface morph: upper image is male, lower image is female. (B) Pachón locality: upper image is male, lower image is female. (C) Tinaja locality: upper image is female, lower image is male. (D) Molino locality: upper image is male, lower image is female. *Figure is taken from Stowers Institute Cavefish Team.*

female or an immature male, and other methods should be used to confirm the sex. For further verification, fish can be observed under a microscope to validate the presence of denticles (Fig. 13.6).

Finally, gamete collection is the most successful method of sexual identification and can be used when other methods prove unreliable.

Breeding and embryo handling

When creating a breeding program for any organism, it is important to understand and replicate its reproductive cycle as closely as possible to reduce

FIGURE 13.6 Micrograph of male anal fin. *Inset*: Hook-like structures known as denticles. *Figure is taken from Stowers Institute Cavefish Team.*

stress, maximize viable embryos, and prolong breeding capabilities. The breeding cycle of *A. mexicanus* has not been studied extensively in the wild; however, insights from ecological and laboratory settings are available. Elipot et al. (2014) tracked three years of spawning data, indicating there was no correlation between spawning efficiency and either the seasonal or lunar cycle. Additionally, they show that the quantity and viability of embryos were consistent throughout the year, indicating that both surface and cave morphs are capable of continual production. Furthermore, Pachón fry have been sighted in both the rainy and dry seasons in the wild, suggesting there is no natural seasonality of the cave morph (Espinasa et al., 2017). The authors concur with several studies observing that most spawning events occur for both cave and surface morphs during the dark hours (Elipot et al., 2014; Simon et al., 2019). It should also be noted that surface morphs produce more embryos per breeding event than cave morphs (Simon et al., 2019). Females may spawn multiple times during a breeding week, which has been validated in other facilities (Elipot et al., 2014; Borowsky, 2008a). The authors recommend resting a brood for a period of 1–4 weeks after each breeding week for optimal production.

Cave and surface morphs have identical courtship behaviors prior to spawning. This similarity suggests breeding behavior may not have changed during the divergent evolution (Simon et al., 2019). The cause for the activation of breeding events is unclear, though hormone production from the female may play a large role in the stimulation process (Simon et al., 2019). Additionally, photoperiod and temperature/water changes are important signals for many teleosts, including *A. mexicanus*. The authors cycle the temperature of a breeding rack from 22 to 26°C over the course of the week to inducing spawning behavior.

The courtship ritual of *A. mexicanus* occurs in five quick phases that are repeated many times over the course of 1 h and can involve many males

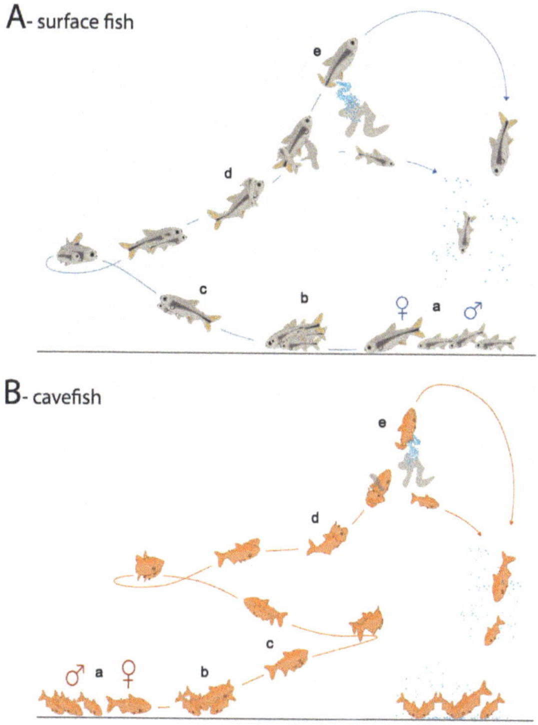

FIGURE 13.7 Description of spawning behavior in *Astyanax mexicanus*. (A) (surface fish) and (B) (cavefish) show realistic drawings capturing the distinct phases of *A. mexicanus* behavior in the dark, after careful observation of many spawning events. The swimming movements and body postures of the male and females are identical in surface fish and cavefish. (a) The female will begin to swim along the bottom. (b) The males will chase the female. (c) A male will swim next to the female in a rapid synchronized event known as "quiver swimming." (d) The pair will swim upward with the male wrapping around the back of the female. (e) The female will release a sudden burst of energy and contraction, causing her to twitch, culminating in the deposition of eggs as she propels in an upward direction (Simon et al., 2019). *Figure is taken from Simon, V., Hyacinthe, C., Rétaux, S., 2019. Breeding behavior in the blind Mexican cavefish and its river-dwelling conspecific. PLoS One 14(2), e0212591. https://doi.org/10.1371/journal.pone.0212591.*

signaling a single female. It is interesting to note that the hierarchical pyramid does not impact breeding behaviors as the fish indicate no preference, with all females breeding at least once during the spawning event (Simon et al., 2019). However, Plath et al. (2006) suggested that surface morph females will prefer larger males when they spawn in the light. The five stages of courtship will only last 2–3 s in total but are repeated as many times as required to elicit the intended response (Simon et al., 2019) (Fig. 13.7).

Like other teleosts, *A. mexicanus* exhibits egg cannibalism. For this reason, the authors suggest the use of a spawning trap to catch the deposited eggs before they can be consumed. This will allow for collection of the embryos

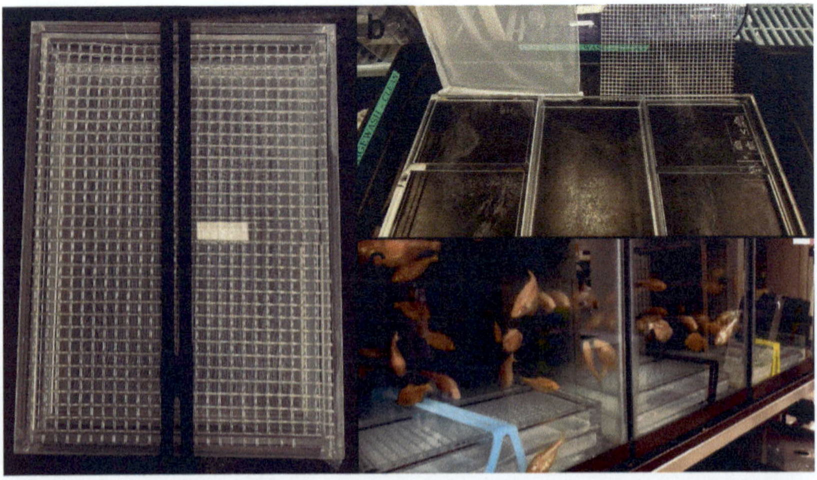

FIGURE 13.8 Spawning traps used to collect embryos. (A) Assembled. (B) Dismantled to show individual components. (C) Placed in tanks housing breeder fish. *Figure is taken from Stowers Institute Cavefish Team.*

with minimal stress to the fish and the highest retention of embryos for the researcher (Fig. 13.8).

Once the traps have been removed from the tank, the embryos must be separated from the organic detritus that has accumulated. Embryos should be collected in a sieve and gently washed with fish water or embryo media to remove the unwanted organic matter. When separating embryos into Petri dishes, the authors recommend limiting to 50 embryos per dish to reduce issues associated with water quality, fungal infestation, and stunted developmental growth. The embryos should be submerged in embryo media or fish water. Fungal growth can be inhibited by dosing the media with methylene blue, an antifungal agent.

After collection, the embryo stage should be determined (hours post-fertilization) using a developmental staging table (Hinaux et al., 2011) and a microscope. The recommended incubation temperature for optimal embryo survivability and development is 23°C. Faster development was noted at 28°C, although this produced a significantly decreased survival rate (Hinaux et al., 2011).

The embryo itself is surrounded by a soft chorion, similar to zebrafish. Unlike zebrafish, however, the chorion has an adhesive surface, which can lead to egg clusters. *A. mexicanus* have transparent embryos, making developmental tracking simple. Hinaux et al. (2011) also indicated that both Pachón and surface morphs have similar developmental milestone timepoints and hatch at similar times, which is approximately 24 h postfertilization.

Once the embryos hatch, the authors recommend transferring all surviving larvae into new Petri dishes to quickly remove the shed chorion debris. This is a critical step for the survival of cave morphs, as organic waste build-up can lead to infections and diminished development. Finally, Petri dishes should be cleaned daily with removal of dead larvae and embryos to ensure optimal survival.

Larviculture

Proper larviculture is crucial as it is the most difficult and critical stage in the lifecycle of any teleost. *A. mexicanus* are born as altricial larvae roughly 24 h postfertilization. During the first 30 min of hatching they untwist from their embryonic state. While the developmental timeline of surface and cave morphs is similar in the embryonic stages, clear differences in pigmentation and eye development begin to emerge during larval growth. After the first several days, cave morph eyes begin to degenerate, never exceeding 0.3 mm in diameter (Hinaux et al., 2011) (Fig. 13.9).

The most pivotal time in larval development occurs at 5.5 dpf, when exogenous feeding, bladder inflation, and other major milestones occur (Hinaux et al., 2011). This development can be delayed by several environmental factors, including temperature, medium, density, and photoperiod. Currently, little is known of the environmental robustness of larval *A. mexicanus*. The authors suggest rearing larvae (0–6 dpf) in a controlled temperature incubator accompanied by daily media changes. The larval stage continues until the beginning of metamorphosis.

Once the larvae reach 6 dpf, they should be placed on system in a low-flow environment. While their swim bladders have inflated, they are not yet able to resist strong currents and swimming exhaustion could lead to mortality. Due to the low turnover rate in these tanks, frequent cleanings are recommended to remove biological build-up from feed or waste. The authors suggest a pipette for detritus removal to lessen the chance of overstressing the larvae. Finally, if rearing in a low-flow environment, ensure the baffle screens are small enough

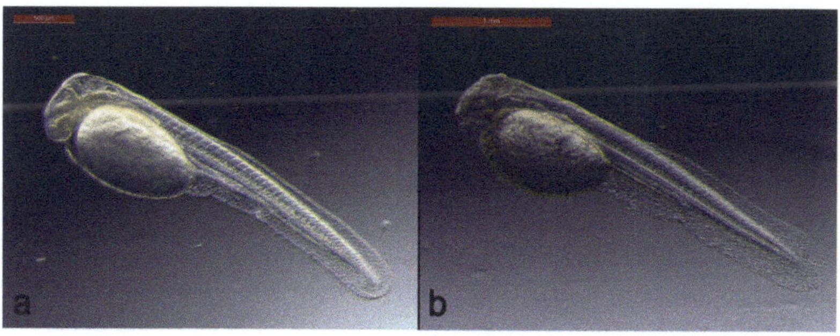

FIGURE 13.9 Larvae of cave and surface morph *Astyanax mexicanus*. (A) Pachón locality larvae at approximately 30 h postfertilization. (B) Surface morph larvae at approximately 40 h postfertilization. *Figure is taken from Stowers Institute Cavefish Team.*

(400 μm) to prevent the larvae from escaping. Alternatively, rearing in a static tank, while posing significant challenges, allows for polyculture using invertebrates such as rotifers to continuously feed the fish. This can be advantageous for survival and growth when monitored closely. If choosing to rear in static tanks, periodic water changes are recommended.

Animal welfare and health management

Welfare indicators

It is now well established that fish are sentient beings deserving of compassion and humane treatment (Broom, 2007; Brown, 2015; Saraiva and Arechavala-Lopez, 2019). This extends to their welfare and protection from suffering. Physical welfare indicators can include growth rate, fertility, fecundity, disease incidence, and mortality (Relić et al., 2010), while behavioral ones include foraging and exploratory behavior, food-anticipatory behavior, feed intake, swimming activity, and aggression (Martins et al., 2012). Even if the growth rate is healthy, this does not confirm the absence of welfare issues and instead a range of welfare indicators should be considered (Broom, 1986). It is also important to examine the welfare of each fish in the tank, not simply as a group since dominance can factor into an individual's wellbeing within the environment. Traditionally, fish care involved providing adequate and appropriate physical and chemical environmental properties. Today there is growing interest in maintaining good mental health in addition to physical health, using a positive welfare approach to develop standards and monitoring strategies (Fife-Cook and Franks, 2019). To eliminate or reduce unpleasant emotional states, environments should be provided that remove boredom and fear, in addition to pain and hunger (Johansen et al., 2006).

Directly related to animal welfare, stress can be defined as "the nonspecific response of the body to any demand made upon it" (Selye, 1974). It is not inherently negative, simply the body's way of adapting to maintain homeostasis. However, if a stressor becomes too intense or chronic, the response mechanisms may not be able to maintain balance and the fish's health and wellbeing will begin to decline as a result. In a laboratory environment, these stressors may include elements such as water quality, low oxygen, handling, capture, tank density, transport, presence of a predator, startle response stimuli such as observation, and technical procedures. These lead to the release of cortisol, a principal corticosteroid, as a primary response. Cortisol levels have been used as a welfare indicator to rate the severity of various acute and chronic stressors in many species of teleosts, including *A. mexicanus* (Félix et al., 2013; Gallo and Jeffery, 2012; Martins et al., 2012; Ramsay et al., 2009; Ruane and Komen, 2003; Scott and Sorensen, 1994; Scott et al., 2008). The prominent methods for collection are whole-body and blood draw, both of which are stressful, invasive, and terminal in smaller fish. To date, the most well-studied noninvasive method in teleosts is waterborne cortisol. However, despite the consistency in the procedure itself, this method requires validation when applied to any new species (Félix et al., 2013).

Recently, Ingalls et al. (2020b) correlated whole-body and waterborne cortisol levels for the surface morph of *A. mexicanus*, indicating a significant and strong positive Pearson correlation between the rate of release into the water and the level of whole-body cortisol over the course of a 1 h experiment. This suggests that waterborne cortisol measurements can be used as an indicator of the welfare of *A. mexicanus*. However, before this becomes a common welfare tool, more work needs to be done to create a scale of stress levels as they relate to absolute cortisol release rates.

Health screening

Daily observation and health monitoring of all animals are essential components for colony maintenance. Flashlights can prove particularly useful when viewing tanks on lower shelves farther away from the light source, as well as observing fish toward the back of larger tanks or hidden among enrichment items. A fish that has isolated deserves additional attention as that can provide the first indicator of illness, injury, or aggression avoidance. Additionally, care should be taken with fish housed on shifted light cycles, since observations may have to be conducted at unusual times, or with red filtered lights during the dark cycle.

Even if fish appear healthy, subclinical infections can be present and health status might vary throughout the colony. Sampling animals for histopathology or disease screening methods such as PCR evaluation for commonly identified pathogens offer valuable information. The authors recommend sampling fish a minimum of twice a year with the goal being able to understand the biological baseline. Knowing which pathogens are present will assist with mitigation and control decisions, as well as contribute to the interpretation of research results. Underlying disease can affect fish morphology and physiology causing off-protocol variation and negatively impacting studies (CCAC, 2005; Harper and Lawrence, 2011; Johansen et al., 2006). Few facilities housing *A. mexicanus* currently have a sentinel program. This may pose additional pathogenic risks for the facility when accepting outside shipments. The authors recommend having the originating facility send fish for sampling to provide prior knowledge of potential issues. Alternatively, screening the shipment postarrival can be an acceptable substitution. For fish arriving with external parasites, these may be reduced with the use of a salt dip (Francis-Floyd et al., 2019). The authors used a 3% salt solution for up to 10 min of exposure to address the confirmed presence of monogenean parasites and have not detected any subsequent to arrival (Francis-Floyd, 1995).

When fish are found dead, a necropsy may provide valuable insight into the cause of death. Repeated necropsies will lead to a better understanding of the expected layout, volume, shape, and color of organs, which may assist with future diagnoses.

Common health issues

The most common cause of health issues in captive fish arises from poor water quality. Whenever disease is suspected or observed, the water chemistry should be evaluated for out-of-range parameters (Casebolt et al., 1998). Fish suffering from a bacterial or fungal infection should be isolated or euthanized to avoid spreading the infection. Dead fish should be rapidly removed from the housing system as they will be readily consumed by tankmates.

As discussed earlier, aggression is of significant concern with the surface morphs and several commonly seen health issues, such as wounds and frayed fins, result from this characteristic behavior. These may arise due to direct attack or be self-inflicted during escape maneuvers. Injured fish should be isolated to facilitate healing and, after flushing, cyanoacrylate glue can be used to seal noninfected wounds (Nemetz and Macmillan, 1988). Trauma may also be observed in the form of rostral abrasions, particularly in submissive fish, affecting the front part of the nose and mouth.

Absence of natural breeding stimuli, combined with space-restricted housing and plentiful food, can lead to egg binding or dystocia. This is caused by a build-up of eggs in the ovaries coupled with an inability to release them, leading to a markedly distended abdomen beyond that of a normally gravid female. Due to the compression exerted by the enlarged ovaries, other organs in the coelomic cavity may have their functions compromised (Khoo, 2019). The authors have found that manually expressing ova within the housing tank by gentle massage of the abdomen can relieve symptoms and allow females to resume a normal breeding cycle.

Exophthalmia, which can affect one or both eyes, occurs occasionally and may indicate a growth or fluid build-up within or behind the eye, or an underlying infection. Fish exhibiting exophthalmia can also suffer subsequent trauma to the eye. If the cause is a systemic infection, other symptoms may be observed in addition to the swollen, protruding eye. Hargis (1991) noted that although eye disease is rarely fatal it can affect growth or lead to other diseases and cautions that eye conditions are aggravated by housing stressors, including overcrowding and aggression.

Other health issues the authors have regularly observed in the colony are abnormal growths, bloating and bristling in Surface Pachón F_2 hybrids, and spinning or corkscrew spiral swimming behaviors. Congenital defects such as missing opercula, deformed mouths, and scoliosis are best dealt with by screening embryos and removing those affected to prevent the introduction of abnormalities to the breeding colony. Spinal deformities are commonly observed with aging (Fig. 13.10).

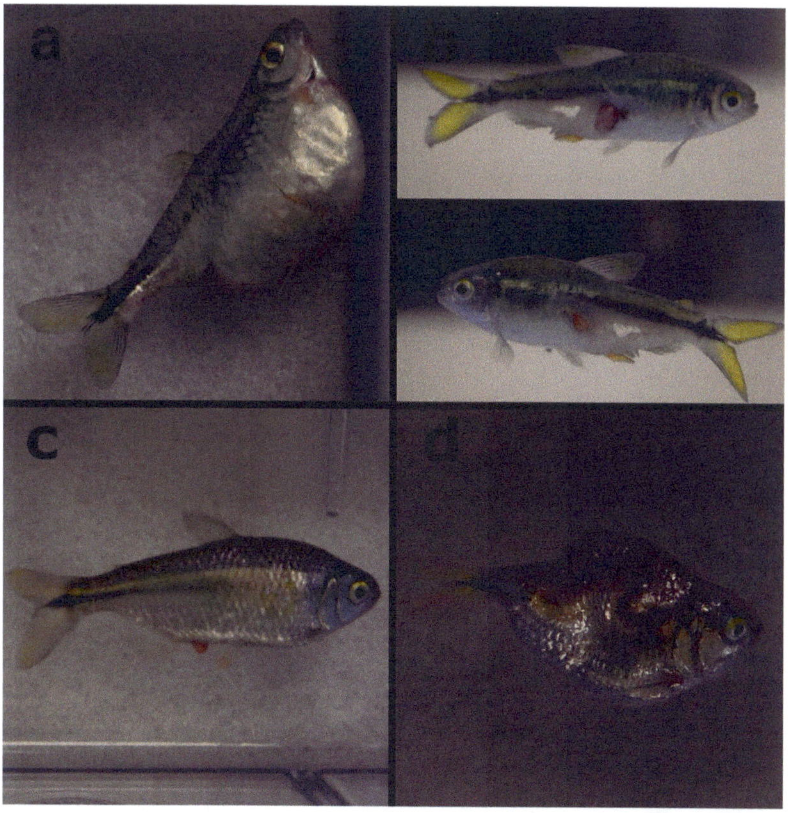

FIGURE 13.10 Observed health issues. (A) Distended abdomen. (B) Wounds and frayed fins. (C) Prolapse. (D) Skeletal deformity. *Figure is taken from Stowers Institute Cavefish Team.*

Anesthesia

With practice and skill, it is possible to manually restrain fish for brief actions such as transfer between tanks or swabbing. However, anesthesia should be used for any procedure that causes pain or goes beyond momentary distress, as well as when movement by a conscious fish could negatively impact the procedure. The level of anesthesia required will be determined by the duration and potential for pain of the planned procedure. Stoskopf (1985) produced a table describing stages of anesthesia in fish, which was later adapted for inclusion in his 1993 manual on fish medicine (Brown, 1993). Carter et al. (2011) provided a discussion of the appropriateness of the stage of anesthesia to the severity of the procedure.

Fish should be fasted for up to 24 h preanesthesia and removal of stressors will aid in a smooth induction. A recovery tank should be prepared containing fish water prior to initiation of anesthesia. Careful handling of anesthetized fish is paramount to avoid damage to the protective mucous layer (Brown, 1993; Stoskopf and Posner, 2008).

MS-222, also known as tricaine methanesulfonate, tricaine, ethyl 3-aminobenzoate methanesulfonate, TMS, Tricaine-S, and Finquel is the most used chemical immersion method. Since MS-222 is acidic, it should be buffered to the approximate pH of fish water to avoid exposing fish to acid stress (Harper and Lawrence, 2011). This is readily accomplished using sodium bicarbonate. A dose of MS-222 between 25 and 200 mg/L is recommended, with doses at the lower end of the range suitable for sedation. Neiffer and Stamper (2009) suggest 60–300 mg/L for fish in the Characidae family. Temperature and hardness of water affect solution potency, and this should be considered when determining dose (Stoskopf and Posner, 2008).

Of nonchemical methods, rapid chilling is the most common. Fish are placed in fish water at 2–4°C until loss of righting occurs and the fish is unresponsive. Care should be taken that fish do not come into contact with the ice used to chill the water and that no chlorine is present in the immersion solution. This can be accomplished by placing a beaker of chilled fish water within a secondary beaker containing the ice melt solution. Hypothermia has been documented from fisheries to be effective for short-term anesthesia; however, more work is needed to determine suitability for most species (Hovda and Linley, 2000). Sudden temperature drops can lead to osmoregulatory changes affecting body fluids, ions, and hematology (Ross and Ross, 2008; Summerfelt and Smith, 1990). Since insufficient research exists regarding the level of anesthesia or analgesia provided, cooling should not be considered for invasive procedures, but instead considered as an effective method of restraint for brief, noninvasive procedures. While working with fish immobilized through cooling, it is important to work quickly and minimize handling to avoid warming up the animal. Dipping gloved fingertips in chilled water assists with this.

Both methods will render the rapid loss of consciousness suitable for restraint and short procedures such as imaging, fin clipping, and tagging. Eames et al. (2010) performed a comparison of MS-222 against ice-cold water as anesthesia methods noting that MS-222 has been reported to increase blood glucose in teleost fish, which has a particular impact for those performing glucose-related studies in *A. mexicanus*. The method selected will need to be compatible with the research and follow the relevant local and national laws for the country in which the research is taking place.

Euthanasia

Animals may need to be euthanized or humanely killed for a variety of reasons during or at the conclusion of a study. Illness or injury may necessitate the action to end suffering for an individual fish. Some animals may be unsuitable for experimental or breeding colony purposes due to being an undesired genotype, unneeded sex, or exhibiting an unfavorable phenotype. Finally, in many research studies, the conclusion of a project requires availability of the whole animal carcass to study the tissues. In all cases, the method selected must "ensure that animals are killed with minimum pain, suffering and distress" (European Parliament, 2010) and should comply with the regulations in the relevant country. When determining a euthanasia method, in addition to being humane, the method must be compatible with the aims of the study, be appropriate for the life stage of the fish, and offer reliability, reproducibility, and irreversibility. Training of personnel performing euthanasia is essential.

Although elimination of all pain and distress may not be achievable, the goal is to minimize it to the extent that is practicably possible. Additionally, the AVMA Guidelines recommend carrying out the procedure in a quiet, low-light environment with as little stimulatory input as possible, in addition to fasting the fish prior for 12–14 h (AVMA, 2020b). The most commonly used method is immersion of the fish in a sufficient volume of euthanasia solution (Neiffer and Stamper, 2009). If multiple animals are being euthanized concurrently, an adequate volume is required for all fish to exhibit normal behavioral movements prior to loss of consciousness. For euthanasia of multiple fish, whether simultaneously or separately over a period of time, it is important to refresh the solution periodically as it will be depleted through absorption during the process.

Indicators of death include loss of opercular and ocular movement for a minimum of 30 min. The CCAC guidelines recommend a two-step method with the use of an adjunctive step after initial anesthesia (CCAC, 2005, 2010). Suitable second steps include decapitation, pithing, exsanguination, and freezing. The guidelines note that if a physical method is utilized, physical destruction of the brain should be accomplished through pithing or crushing.

MS-222 is the most frequently used chemical immersion method, causing anesthesia prior to euthanasia. Fish should remain immersed for a minimum of 30 min following cessation of opercular and ocular movement (AVMA 2020b; CCAC, 2010). In zebrafish (*Danio rerio*), rapid chilling (hypothermic shock) has been shown to have a shorter mean time for euthanasia and be less aversive, leading multiple labs to utilize this method with *A. mexicanus* (Wilson et al., 2009; Wong et al., 2014). However, careful observation is required for full-grown fish due to their larger size and consequent increased volume-to-surface area ratio. Eggs and embryos can be destroyed by placing in a dilute solution of sodium or calcium hypochlorite. This method can also be used for larval fish after anesthesia has been achieved. For studies where chemical contamination would compromise data, euthanasia can be accomplished through decapitation followed by pithing. In the field, 500 mg/L MS-222 has proven to be effective, followed by decapitation as an adjunctive method (Peuß, 2020; personal communication, September 9).

While developing the protocol for any study, it is important to consider and establish humane endpoints. New procedures or transgenes can lead to unexpected outcomes, requiring a plan to minimize pain and suffering should this situation arise. Disposal of tissues and carcasses after completion of the study should be performed in a manner consistent with country, municipal, and institutional requirements. This will prevent the unwanted build-up of biological waste that could attract vermin. Both human and environmental safety should be part of the planning process to minimize exposure and risk of disease transmission. For hazardous carcasses contaminated with chemical, biological, or radioactive materials, all applicable guidelines and regulations must be followed to prevent unwanted contact and exposures.

Procedures

Training is an important component of all animal work and those new to a procedure should train on euthanized fish before progressing to live animals whenever possible. For many procedures such as injections and oral gavage, it can be useful to first dissect a fish to gain an understanding of the body wall thickness and organ layout. Experience with a procedure results in greater efficiency performing it, leading to the benefit of decreased duration required for anesthesia. Choose appropriate restraint and anesthesia for the procedure to be performed, taking into consideration the duration of unconsciousness required. Have a recovery tank prepared prior to starting any procedure and use wet gloves when handling fish to minimize damage to the protective mucus layer. If fish do not readily recover from chemical anesthesia it can help to pipette water over the gills once the fish is in the recovery tank. Prepare the surface and equipment before working with fish; sponges cut with grooves to accommodate the size of fish can be moistened with fish water and used to hold anesthetized animals during a procedure. If a microscope will be needed, focus on the staging platform that will be used prior to placing a fish to save time. To minimize the number of handling and anesthesia sessions a fish experiences, consider whether multiple procedures can be performed at the same time such as gamete collection and Visible Implant Elastomer (VIE) tagging to mark the identified sex for future reference.

Visible Implant Elastomer tagging

A useful method to distinguish individuals while cohousing is the use of VIE tags (Northwest Marine Technology, 2020). The elastomer, a member of the silicone family, is injected subdermally which then hardens into a pliable, solid tag, providing externally visible markings. It is available in multiple colors, some of which contain fluorescent pigments. Combining colors with multiple injection locations provides the capability to individually identify large numbers of fish. Details on the preparation and injection of the elastomer for small fish can be found on the Northwest Marine Technology site.

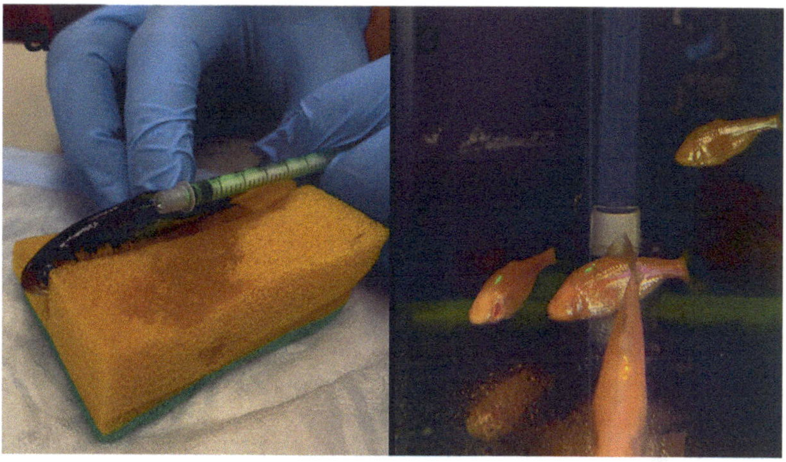

FIGURE 13.11 Visible Implant Elastomer tagging. (A) Placement of an anesthetized surface morph *Astyanax mexicanus* in a moistened sponge trough and tagging in a dorsal location. (B) Tagged cave morph *A. mexicanus*. *Figure is taken from Stowers Institute Cavefish Team.*

It is important to stop dispensing the elastomer before the bevel reaches the opening to avoid any exposed elastomer, which prevents the wound from healing over the tag end, leading to tag loss. Green and orange tags work well on pigmented surface morphs, while cave morphs more readily display a variety of colors due to the reduction of pigmentation. A VI light (deep violet 405 nm) can be helpful if fluorescent colors are utilized to increase tag visibility, particularly in darker tank locations. No infections have occurred due to tagging for the authors, and tag retention has been excellent (Fig. 13.11).

Embryo dechorionation

It may be necessary to remove embryos from their chorions for easier observation, chemical exposure, or surgical manipulation. The authors use a similar manual dechorionation procedure to that detailed in *The Zebrafish Book* (Westerfield, 1993). Additionally, *A. mexicanus* embryos can be dechorionated using Pronase, detailed in the ZIRC embryo bleaching protocol (Zebrafish.org, 2020).

Care is needed to avoid damaging developing embryos particularly prior to epiboly, and once dechorionated, gentle handling is required. The use of glass pipettes may help minimize damage. At some stages, embryos may be prone to sticking to plastic, and so glass or 1% agarose-coated plastic Petri dishes may be beneficial for holding purposes. Embryos fixed in 4% paraformaldehyde in phosphate-buffered saline can be dechorionated, dehydrated in an increasing methanol series, and stored at −20°C prior to in situ hybridization (Bilandžija et al., 2013; Ma et al., 2014).

Gamete collection

Gametes can be collected either for individual studies, such as unfertilized eggs for RNA extraction (Ma et al., 2018), or for performing in vitro fertilization (IVF) (Borowsky, 2008b; Peuß et al., 2019). Allow animals a minimum of one week, preferably two, to recover between collections. For a detailed procedure on gamete collection in *A. mexicanus*, see Peuβ et al. (2019). Use of Ovaprim injections in females has been found to increase spawning success by 10%–20% and can also lead to success with females that have not spawned for multiple years (Yoshizawa, 2020, personal communication, October 12).

When selecting a female that appears stimulated for breeding, it is common to observe rapid swimming and a pink hue on the abdomen due to the vasculature under the skin near the ovaries although this is more difficult to observe in surface fish. While expressing ova, avoid rubbing along the scales to prevent skin trauma. The number of ova produced, typically between 20 and 2000, cannot be estimated by viewing the fish due to the presence of fat pads. In addition to ova, plugs are commonly expelled, more so in cavefish than surface fish. Plugs can resemble clusters of eggs that have not correctly developed or were not expelled in the previous session. Occasionally, vasculature will be expelled along with the plugs. Ova that appear like clear bubbles indicate an immature fish, which should be allowed time for further maturation. Overdeveloped ova will turn white and cloudy, with uneven color distribution giving a patchy appearance (Fig. 13.12).

For males, selection of an animal able to produce gametes is much easier; if there is a stimulated female in the tank, the males are likely to provide samples. Males may also exhibit a color change with the whole body taking on a pinkish hue during the increased activity. Milt may vary in opacity depending on the strain of fish used. That from surface fish is commonly opaque; however, cave morph milt may vary from opaque to clear. Typical values of milt per fish will range between 2 and 20 μL, noting that this volume does not correlate with fish size (Fig. 13.13).

In vitro fertilization

IVF is a useful procedure to obtain large numbers of synchronously developing embryos, haploid embryos, or embryos for early pressure or heat-shock treatments (Westerfield, 1993). In *A. mexicanus*, IVF is also useful to create hybrids of surface and cave morphs to avoid the risk of aggression during natural breeding. Additionally, IVF can be used to obtain developing embryos during daylight hours by acclimatizing fish to shifted light cycles (Peuß et al., 2019).

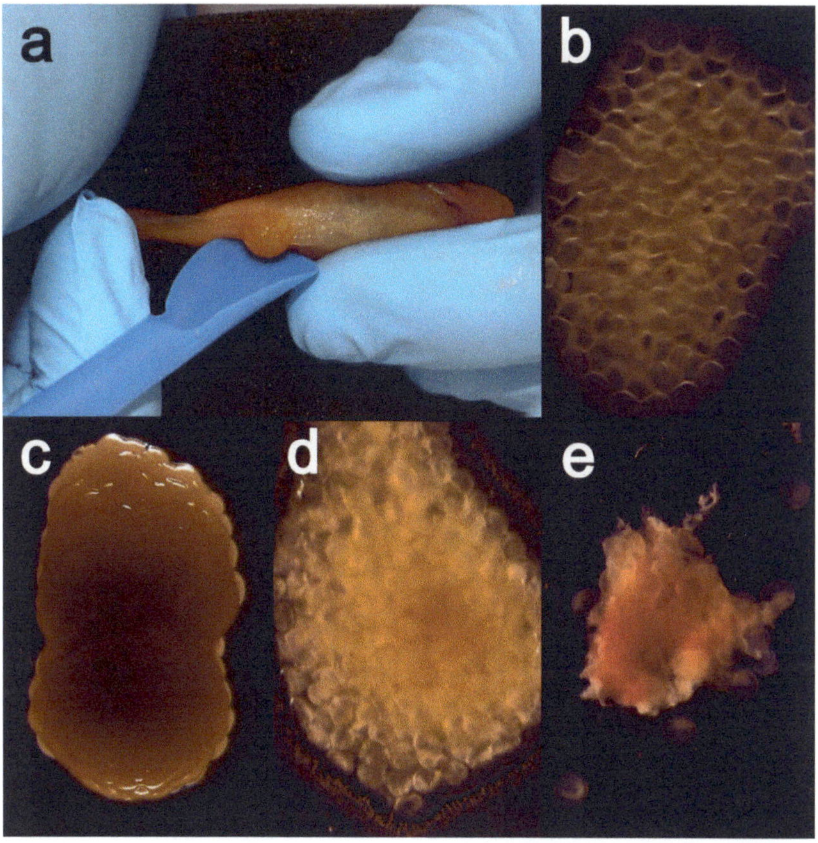

FIGURE 13.12 Female gamete collection from *Astyanax mexicanus*. (A) Manual expression of ova from an anesthetized female. (B) Micrograph of good ova. (C) Micrograph of underdeveloped ova. (D) Micrograph of overdeveloped ova. (E) Micrograph of expressed plug. *Figure is taken from Stowers Institute Cavefish Team.*

When performing IVF, check the milt quality using a microscope and add water to activate the sperm. Motility and sperm count are the best indicators of sperm quality and should be verified to avoid wasting ova with poor-quality sperm. Once fertilized, ova will become slightly more translucent before entering the one-cell stage of development (Fig. 13.14).

Yasui et al. (2015), in their work with the yellowtail tetra (*Astyanax altiparanae*), demonstrated improvements in sperm motility by treating male fish with hormonal spawning inducers, along with developing new procedures for short-term storage of both sperm and oocytes. Interestingly, they also found that housing males with ovulating female fish improved sperm quality.

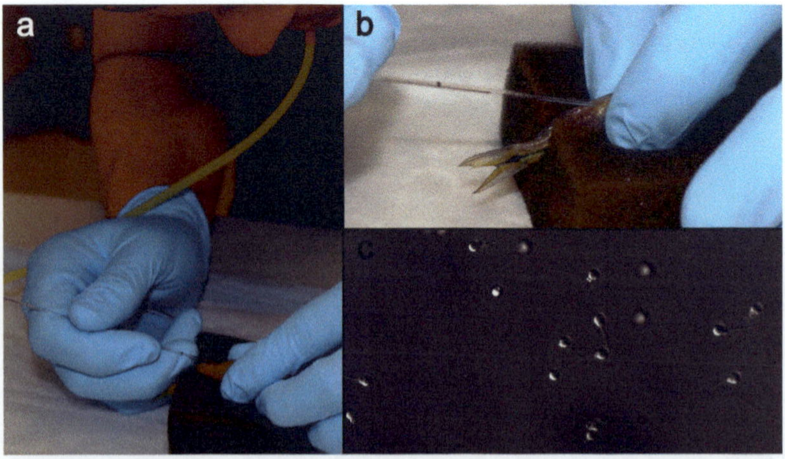

FIGURE 13.13 Male gamete collection from *Astyanax mexicanus*. (A) Milt collection from an anesthetized male using an aspirator tube. (B) Milt entering the capillary tube. (C) Micrograph of sperm. *Figure is taken from Stowers Institute Cavefish Team.*

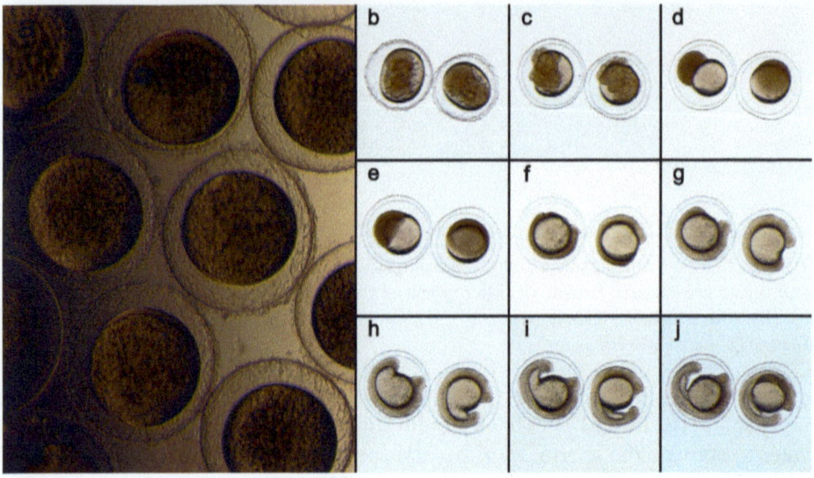

FIGURE 13.14 A) One-cell stage *Astyanax mexicanus* embryos. (B–J) Embryo development time lapse in *A. mexicanus* of the Pachón locality from <1 h postfertilization (hpf) to approximately 18 hpf. *Figure is taken from Stowers Institute Cavefish Team.*

Injections

Although immersion is a common method of providing chemical exposure to a fish, it is difficult to know the level of uptake, and more precise dosages may be required. Drug administration through injection is one method allowing exact amounts to be delivered, at a known time, to a specific body location. To calculate the dose required, fish should first be weighed, which may be performed as part of the process to reduce the need for additional anesthesia. Ensure there are no air bubbles in the syringe or needle prior to injection (Kinkel et al., 2010). Penney and Volkoff (2014) mentioned the use of performing preliminary experiments with a dye to verify correct needle placement.

Intraperitoneal injections are most common and may be used to administer a variety of substances from experimental agents such as stimulants and inhibitors, to injectable ovulating and spermiating agents for reproductive improvements. Insert the needle along the midline between the pelvic girdle and the anus, pointing cranially. Fast fish for a minimum of 12 h prior to injection; this is most easily accomplished by remembering to mark the tanks the previous night to avoid any accidental morning feedings.

Schroeder (2018) noted that intramuscular and subcutaneous injections can be a source of trauma for smaller-sized research species, making them difficult to administer successfully. For intramuscular injections, use a dorsal approach to the epaxial muscles on either side of the midline to avoid traumatizing the lateral line or inadvertently entering the coelom. Using the smallest needle possible, insert it until the bevel disappears prior to injecting. Larvae can be embedded in agarose and injected using fine glass capillary needles, such as those used for embryo microinjections (Seger et al., 2011).

Embryo microinjection

Embryo microinjection is a useful technique to manipulate the genome using editing technologies such as TALEN and CRISPR. Kowalko et al. (2016) discussed methods used to achieve this in *A. mexicanus* following collection of single-cell stage eggs, remarking that needle length is important to achieve the correct level of flexibility to perform the injections. Further modifications are provided by Stahl et al. (2019a) who noted that at room temperature, the single-cell stage lasts for approximately 40 min. They also suggest removing water from the top of the eggs in the injection plate to facilitate chorion penetration. A sample needle-pulling program is provided in Stahl et al. (2019b).

Blood collection

Survival blood collection is difficult from small research species so blood is most often collected as a terminal procedure from anesthetized fish. The simplest procedure is tail ablation and collection from the severed vessels (Whitman, 2004). Care should be taken to avoid cutting into the coelom to avoid significant contamination by other body fluids. A similar and less precise method involves decapitation with an appropriate blade or scissors and the use of a microcapillary tube to collect blood from severed vessels or perhaps even from the heart; this procedure requires more skill and knowledge of internal anatomy. With practice, it may be possible to obtain blood via cardiac puncture using a ventral approach inserting the needle in a cranial direction. As it is difficult to determine the depth of penetration into the heart and significant trauma usually results, this is usually a terminal procedure. In larger fish, it may be possible to collect from the caudal vein in the caudal peduncle using either a ventral or a lateral approach. Hemostasis can be applied following this technique so it may be a survival technique when performed by a skilled technician and with careful monitoring of the fish during recovery.

Oral gavage

Riddle et al. (2018) used a pulled microcapillary needle to deliver solution into the gastrointestinal tract through the mouth of anesthetized *A. mexicanus*, following a procedure adapted from work with zebrafish larvae by Cocchiaro and Rawls (2013). Harper and Lawrence (2011) noted in reference to zebrafish that fish should be fasted first to assure that the solution easily flows into the gastrointestinal lumen.

Cryopreservation

More than 200 species of fish have had spermatozoa cryopreserved (Torres et al., 2016) and, in the research environment, cryopreservation can form an integral component of the disaster plan. To avoid housing numerous genetic lines not currently on study, a robust cryopreservation program can provide a critical service. Although some initial studies on embryo cryopreservation have been carried out, more work is needed before this becomes an established practice (Khosla et al., 2020).

Using a modified version of Matthews et al. (2018), the authors had initial success preserving sperm from Surface, Tinaja, Pachón, and Molino fish. Each sample was normalized to 40 μL using E400 sperm extender to reduce the density, then preserved using raffinose, skim milk, methanol, and bicine buffer cryoprotective medium with 10% MeOH. Sealed samples were cooled on dry ice for 20–60 min before placing in liquid nitrogen. Prefreezing motility ranged from 70% to 100%. Short-term storage (<24 h) motility ranged from 10% to 60% when thawed at 28°C and activated using 1 μL of sample to 9 μL of fish water. Samples in long-term storage (>30 days) have been used to successfully perform IVF.

Acknowledgments

The authors wish to gratefully acknowledge input and comments from Elizabeth Evans, Franchesca Hutton-Lau, David Jewell, Molly Miller, and Zachary Zakibe. Additionally, to the Stowers Institute for Medical Research, Nicolas Rohner, and the Rohner lab members for providing us with the opportunity to work with such an exciting model organism.

References

American Veterinary Medical Association, 2020a. Animal Welfare: What Is it? Available at: https://www.avma.org/resources/animal-health-welfare/animal-welfare-what-it. (Accessed 29 September 2020).

American Veterinary Medical Association, 2020b. AVMA Guidelines for the Euthanasia of Animals: 2020 Edition. https://www.avma.org/sites/default/files/2020-01/2020-Euthanasia-Final-1-17-20.pdf.

Aspiras, A.C., Rohner, N., Martineau, B., Borowsky, R.L., Tabin, C.J., 2015. Melanocortin 4 receptor mutations contribute to the adaptation of cavefish to nutrient-poor conditions. Proc. Natl. Acad. Sci. U.S.A. 112 (31), 9668–9673. https://doi.org/10.1073/pnas.1510802112.

Beale, A., Guibal, C., Tamai, T.K., Klotz, L., Cowen, S., Peyric, E., et al., 2013. Circadian rhythms in Mexican blind cavefish *Astyanax mexicanus* in the lab and in the field. Nat. Commun. 4, 2769. https://doi.org/10.1038/ncomms3769.

Bilandžija, H., Ma, L., Parkhurst, A., Jeffery, W.R., 2013. A potential benefit of albinism in *Astyanax* cavefish: downregulation of the *oca*2 gene increases tyrosine and catecholamine levels as an alternative to melanin synthesis. PLoS One 8 (11), e80823. https://doi.org/10.1371/journal.pone.0080823.

Borowsky, R., 2008a. Astyanax mexicanus, the Blind Mexican cave fish: a model for studies in development and morphology. CSH Protoc. https://doi.org/10.1101/pdb.emo107.

Borowsky, R., 2008b. In vitro fertilization of Astyanax mexicanus. CSH Protoc. https://doi.org/10.1101/pdb.prot5092.

Breder Jr., C.M., 1942. Descriptive ecology of La Cueva Chica, with especial reference to the blind fish, Anoptichthys. Zoologica 27, 7–15.

Breder Jr., C.M., 1943. A note on erratic viciousness in *Astyanax mexicanus* (Philippi). Copeia 2, 82–84.

Broom, D.M., 1986. Indicators of poor welfare. Br. Vet. J. 142, 524–526.

Broom, D.M., 2007. Cognitive ability and sentience: which aquatic animals should be protected? Dis. Aquat. Org. 75 (2), 99–108. https://doi.org/10.3354/dao075099.

Brown, C., 2015. Fish intelligence, sentience and ethics. Anim. Cognit. 18 (1), 1–17. https://doi.org/10.1007/s10071-014-0761-0.

Brown, L.A., 1993. Anesthesia and restraint. In: Stoskopf, M.K. (Ed.), Fish Medicine. W.B. Saunders Company, pp. 79–90.

Burchards, H., Dölle, A., Parzefall, J., 1985. Aggressive behaviour of an epigean population of *Astyanax mexicanus* (Characidae, Pisces) and some observations of three subterranean populations. Behav. Process. 11 (3), 225–235. https://doi.org/10.1016/0376-6357(85)90017-8.

Canadian Council on Animal Care, 2005. Guidelines on: The Care and Use of Fish in Research, Teaching and Testing. https://www.ccac.ca/Documents/Standards/Guidelines/Fish.pdf.

Canadian Council on Animal Care, 2010. CCAC Guidelines on: Euthanasia of Animals Used in Science. https://www.ccac.ca/Documents/Standards/Guidelines/Euthanasia.pdf.

Carter, K.M., Woodley, C.M., Brown, R.S., 2011. A review of tricaine methanesulfonate for anesthesia of fish. Rev. Fish Biol. Fish. 21, 51–59. https://doi.org/10.1007/s11160-010-9188-0.

Casebolt, D.B., Speare, D.J., Horney, B.S., 1998. Care and use of fish as laboratory animals: current state of knowledge. Lab. Anim. Sci. 48 (2), 124–136.

Cocchiaro, J.L., Rawls, J.F., 2013. Microgavage of zebrafish larvae. J. Vis. Exp. (72), e4434. https://doi.org/10.3791/4434.

Darve, S.I., Wani, G.B., Hukam, S., 2016. Dietary lipid requirement for Neon tetra (*Paracheirodon innesi*): a valuable ornamental fish. SKUAST J. Res. 18 (2), 146–153.

Duboué, E.R., Keene, A.C., Borowsky, R.L., 2011. Evolutionary convergence on sleep loss in cavefish populations. Curr. Biol. 21 (8), 671–676. https://doi.org/10.1016/j.cub.2011.03.020.

Eames, S.C., Philipson, L.H., Prince, V.E., Kinkel, M.D., 2010. Blood sugar measurement in zebrafish reveals dynamics of glucose homeostasis. Zebrafish 7 (2), 205–213. https://doi.org/10.1089/zeb.2009.0640.

Elipot, Y., Hinaux, H., Callebert, J., Rétaux, S., 2013. Evolutionary shift from fighting to foraging in blind cavefish through changes in the serotonin network. Curr. Biol. 23 (1), 1–10. https://doi.org/10.1016/j.cub.2012.10.044.

Elipot, Y., Legendre, L., Père, S., Sohm, F., Rétaux, S., 2014. *Astyanax* transgenesis and husbandry: how cavefish enters the laboratory. Zebrafish 11 (4), 291–299. https://doi.org/10.1089/zeb.2014.1005.

Espinasa, L., Yamamoto, Y., Jeffery, W.R., 2005. Non-optical releasers for aggressive behavior in blind and blinded *Astyanax* (Teleostei, Characidae). Behav. Process. 70 (2), 144–148. https://doi.org/10.1016/j.beproc.2005.06.003.

Espinasa, L., Bonaroti, N., Wong, J., Pottin, K., Queinnec, E., Rétaux, S., 2017. Contrasting feeding habits of post-larval and adult *Astyanax* cavefish. Subterr. Biol. 21, 1–17. https://doi.org/10.3897/subtbiol.21.11046.

Espinasa, L., Heintz, C., Rétaux, S., Yoshizawa, M., Agnès, F., Ornelas-Garcia, P., et al., 2020. Vibration Attraction Response (VAB) is a plastic trait in Blind Mexican tetra (*Astyanax mexicanus*), variable within subpopulations inhabiting the same cave. J. Fish. Biol. https://doi.org/10.1111/jfb.14586. Advance online publication.

European Parliament and Council, 2010. Directive 2010/63/EU on the Protection of Animals Used for Scientific Purposes. https://eur-lex.europa.eu/eli/dir/2010/63/2019-06-26.

Félix, A.S., Faustino, A.I., Cabral, E.M., Oliveira, R.F., 2013. Noninvasive measurement of steroid hormones in zebrafish holding-water. Zebrafish 10 (1), 110–115. https://doi.org/10.1089/zeb.2012.0792.

Fife-Cook, I., Franks, B., 2019. Positive welfare for fishes: rationale and areas for future study. Fishes 4 (2), 31.

Francis-Floyd, R., 1995. The Use of Salt in Aquaculture. University of Florida IFAS Fact Sheet VM 86. https://ufdc.ufl.edu/IR00002495/00001.

Francis-Floyd, R., Klinger, R., Petty, D., Pouder, D., 2019. Monogenean Parasites of Fish. University of Florida IFAS Extension, Publication #FA28. https://edis.ifas.ufl.edu/fa033.

Fumey, J., Hinaux, H., Noirot, C., Thermes, C., Rétaux, S., Casane, D., 2018. Evidence for late Pleistocene origin of *Astyanax mexicanus* cavefish. BMC Evol. Biol. 18 (1), 43. https://doi.org/10.1186/s12862-018-1156-7.

Gallo, N.D., Jeffery, W.R., 2012. Evolution of space dependent growth in the teleost *Astyanax mexicanus*. PLoS One 7 (8), e41443. https://doi.org/10.1371/journal.pone.0041443.

Gross, J.B., 2012. The complex origin of *Astyanax* cavefish. BMC Evol. Biol. 12, 105. https://doi.org/10.1186/1471-2148-12-105.

Gross, J.B., Krutzler, A.J., Carlson, B.M., 2014. Complex craniofacial changes in blind cave-dwelling fish are mediated by genetically symmetric and asymmetric loci. Genetics 196 (4), 1303–1319. https://doi.org/10.1534/genetics.114.161661.

Hargis Jr., W.J., 1991. Disorders of the eye in finfish. Annu. Rev. Fish Dis. 1, 95–117.

Harper, C., Lawrence, C., 2011. The Laboratory Zebrafish. CRC Press, Boca Raton, Florida.

Hinaux, H., Pottin, K., Chalhoub, H., Père, S., Elipot, Y., Legendre, L., et al., 2011. A developmental staging table for *Astyanax mexicanus* surface fish and Pachón cavefish. Zebrafish 8 (4), 155–165. https://doi.org/10.1089/zeb.2011.0713.

Hovda, J., Linley, T.J., 2000. The potential application of hypothermia for anesthesia in adult Pacific Salmon. N. Am. J. Aquacult. 62 (1), 67–72. https://doi.org/10.1577/1548-8454(2000)062<0067:TPAOHF>2.0.CO;2.

Hubbs, C.L., Innes, W.T., 1936. The first known blind fish of the family Characidae: a new genus from Mexico. In: Occasional Papers of the Museum of Zoology University of Michigan, 342, pp. 1–7.

Hyacinthe, C., Attia, J., Rétaux, S., 2019. Evolution of acoustic communication in blind cavefish. Nat. Commun. 10, 4231. https://doi.org/10.1038/s41467-019-12078-9.

Ingalls, A., Merryman, S., Baumann, D.P., 2020a. Length-mass relationship and aggression of *Astyanax mexicanus* (Mexican cavefish) reared in a laboratory environment. In: Aquaculture America 2020, Honolulu, HI, 9-12 February.

Ingalls, A., Merryman, S., Baumann, D.P., 2020b. Non-invasive measurement of cortisol in *Astyanax mexicanus* (Mexican tetra) as it relates to environmental stressors. In: American Fisheries Society Virtual Annual Meeting, 14-25 September.

Jaggard, J.B., Stahl, B.A., Lloyd, E., Prober, D.A., Duboue, E.R., Keene, A.C., 2018. Hypocretin underlies the evolution of sleep loss in the Mexican cavefish. eLife 7, e32637. https://doi.org/10.7554/eLife.32637.

Jaggard, J.B., Lloyd, E., Yuiska, A., Patch, A., Fily, Y., Kowalko, J.E., et al., 2020. Cavefish brain atlases reveal functional and anatomical convergence across independently evolved populations. Sci. Adv. 6 (38), eaba3126. https://doi.org/10.1126/sciadv.aba3126.

Jeffery, W.R., 2001. Cavefish as a model system in evolutionary developmental biology. Dev. Biol. 231 (1), 1–12. https://doi.org/10.1006/dbio.2000.0121.

Jeffery, W.R., 2009. Chapter 8. Evolution and development in the cavefish *Astyanax*. Curr. Top. Dev. Biol. 86, 191–221. https://doi.org/10.1016/S0070-2153(09)01008-4.

Jeffery, W.R., 2020. *Astyanax* surface and cave fish morphs. EvoDevo 11, 14. https://doi.org/10.1186/s13227-020-00159-6.

Johansen, R., Needham, J.R., Colquhoun, D.J., Poppe, T.T., Smith, A.J., 2006. Guidelines for health and welfare monitoring of fish used in research. Lab. Anim. 40 (4), 323–340. https://doi.org/10.1258/002367706778476451.

Keene, A., Yoshizawa, M., McGaugh, S., 2016. Biology and Evolution of the Mexican Cavefish. Elsevier Inc., New York.

Khoo, L., 2019. Reproductive disorders. In: Smith, S.S. (Ed.), Fish Diseases and Medicine. CRC Press, Boca Raton, FL.

Khosla, K., Kangas, J., Liu, Y., Zhan, L., Daly, J., Hagedorn, M., et al., 2020. Cryopreservation and laser nanowarming of zebrafish embryos followed by hatching and spawning. Adv. Biosyst. e2000138. https://doi.org/10.1002/adbi.202000138. Advance online publication.

Kinkel, M.D., Eames, S.C., Philipson, L.H., Prince, V.E., 2010. Intraperitoneal injection into adult zebrafish. J. Vis. Exp. (42), 2126. https://doi.org/10.3791/2126.

Klaassen, H., Wang, Y., Adamski, K., Rohner, N., Kowalko, J.E., 2018. CRISPR mutagenesis confirms the role of *oca2* in melanin pigmentation in *Astyanax mexicanus*. Dev. Biol. 441 (2), 313–318. https://doi.org/10.1016/j.ydbio.2018.03.014.

Krishnan, J., Rohner, N., 2017. Cavefish and the basis for eye loss. Philos. Trans. R. Soc. Lond. Ser. B Biol. Sci. 372 (1713), 20150487. https://doi.org/10.1098/rstb.2015.0487.

Kowalko, J.E., Rohner, N., Rompani, S.B., Peterson, B.K., Linden, T.A., Yoshizawa, M., et al., 2013a. Loss of schooling behavior in cavefish through sight-dependent and sight-independent mechanisms. Curr. Biol. 23 (19), 1874–1883. https://doi.org/10.1016/j.cub.2013.07.056.

Kowalko, J.E., Rohner, N., Linden, T.A., Rompani, S.B., Warren, W.C., Borowsky, R., et al., 2013b. Convergence in feeding posture occurs through different genetic loci in independently evolved cave populations of *Astyanax mexicanus*. Proc. Natl. Acad. Sci. U.S.A. 110, 16933–16938. https://doi.org/10.1073/pnas.1317192110.

Kowalko, J.E., Ma, L., Jeffery, W.R., 2016. Genome editing in *Astyanax mexicanus* using transcription activator-like effector nucleases (TALENs). J. Vis. Exp. (112), 54113. https://doi.org/10.3791/54113.

Kowalko, J., 2020. Utilizing the blind cavefish *Astyanax mexicanus* to understand the genetic basis of behavioral evolution. J. Exp. Biol. 223 (Pt Suppl. 1), jeb208835. https://doi.org/10.1242/jeb.208835.

Lipscomb, T.N., Yanong, R.P., Ramee, S.W., DiMaggio, M.A., 2020. Histological, histochemical and biochemical characterization of larval digestive system ontogeny in black tetra *Gymnocorymbus ternetzi* to inform aquaculture weaning protocols. Aquaculture 520, 734957. https://doi.org/10.1016/j.aquaculture.2020.734957.

Loomis, C., Peuß, R., Jaggard, J.B., Wang, Y., McKinney, S.A., Raftopoulos, S.C., et al., 2019. An adult brain atlas reveals broad neuroanatomical changes in independently evolved populations of Mexican cavefish. Front. Neuroanat. 13, 88. https://doi.org/10.3389/fnana.2019.00088.

Ma, L., Parkhurst, A., Jeffery, W.R., 2014. The role of a lens survival pathway including sox2 and αA-crystallin in the evolution of cavefish eye degeneration. EvoDevo 5, 28. https://doi.org/10.1186/2041-9139-5-28.

Ma, L., Jeffery, W.R., Essner, J.J., Kowalko, J.E., 2015. Genome editing using TALENs in blind Mexican cavefish, *Astyanax mexicanus*. PLoS One 10 (3), e0119370. https://doi.org/10.1371/journal.pone.0119370.

Ma, L., Strickler, A.G., Parkhurst, A., Yoshizawa, M., Shi, J., Jeffery, W.R., 2018. Maternal genetic effects in *Astyanax* cavefish development. Dev. Biol. 441 (2), 209–220. https://doi.org/10.1016/j.ydbio.2018.07.014.

Martins, C.I.M., Galhardo, L., Noble, C., Damsgård, B., Spedicato, M.T., Zupa, W., et al., 2012. Behavioural indicators of welfare in farmed fish. Fish Physiol. Biochem. 38, 17–41. https://doi.org/10.1007/s10695-011-9518-8.

Matthews, J.L., Murphy, J.M., Carmichael, C., Yang, H., Tiersch, T., Westerfield, M., et al., 2018. Changes to extender, cryoprotective medium, and in vitro fertilization improve zebrafish sperm cryopreservation. Zebrafish 15 (3), 279–290. https://doi.org/10.1089/zeb.2017.1521.

McCauley, D.W., Hixon, E., Jeffery, W.R., 2004. Evolution of pigment cell regression in the cavefish *Astyanax*: a late step in melanogenesis. Evol. Dev. 6 (4), 209–218. https://doi.org/10.1111/j.1525-142X.2004.04026.x.

McGaugh, S.E., Passow, C.N., Jaggard, J.B., Stahl, B.A., Keene, A.C., 2020a. Unique transcriptional signatures of sleep loss across independently evolved cavefish populations. J. Exp. Zool. B Mol. Dev. Evol. https://doi.org/10.1002/jez.b.22949. Advance online publication.

McGaugh, S.E., Kowalko, J.E., Duboué, E., Lewis, P., Franz-Odendaal, T.A., Rohner, N., et al., 2020b. Dark world rises: the emergence of cavefish as a model for the study of evolution, development, behavior, and disease. J. Exp. Zool. B Mol. Dev. Evol. https://doi.org/10.1002/jez.b.22978. Advance online publication.

McMenamin, S.K., Parichy, D.M., 2013. Metamorphosis in teleosts. Curr. Top. Dev. Biol. 103, 127–165. https://doi.org/10.1016/B978-0-12-385979-2.00005-8.

Neiffer, D.L., Stamper, M.A., 2009. Fish sedation, analgesia, anesthesia, and euthanasia: considerations, methods, and types of drugs. ILAR J. 50 (4), 343–360. https://doi.org/10.1093/ilar.50.4.343.

Nemetz, T.G., Macmillan, J.R., 1988. Wound healing of incisions closed with a cyanoacrylate adhesive. Trans. Am. Fish. Soc. 117 (2), 190–195. https://doi.org/10.1577/1548-8659(1988)117<0190:WHOICW>2.3.CO;2.

Palstra, A.P., Planas, J.V., 2011. Fish under exercise. Fish Physiol. Biochem. 37 (2), 259–272. https://doi.org/10.1007/s10695-011-9505-0.

Paz, A., McDole, B., Kowalko, J.E., Duboue, E.R., Keene, A.C., 2020. Evolution of the acoustic startle response of Mexican cavefish. J. Exp. Zool. B Mol. Dev. Evol. https://doi.org/10.1002/jez.b.22988. Advance online publication.

Penney, C.C., Volkoff, H., 2014. Peripheral injections of cholecystokinin, apelin, ghrelin and orexin in cavefish (*Astyanax fasciatus mexicanus*): effects on feeding and on the brain expression levels of tyrosine hydroxylase, mechanistic target of rapamycin and appetite-related hormones. Gen. Comp. Endocrinol. 196, 34–40. https://doi.org/10.1016/j.ygcen.2013.11.015.

Peuß, R., Zakibe, Z., Krishnan, J., Merryman, M.S., Baumann, D.P., Rohner, N., 2019. Gamete collection and in vitro fertilization of *Astyanax mexicanus*. J. Vis. Exp. (147) https://doi.org/10.3791/59334.

Plath, M., Rohde, M., Schröder, T., Taebel-Hellwig, A., Schlupp, I., 2006. Female mating preferences in blind cave tetras *Astyanax fasciatus* (Characidae, Teleostei). Behaviour 143 (1), 15–32. https://doi.org/10.1163/156853906775133560.

Ramsay, J.M., Feist, G.W., Varga, Z.M., Westerfield, M., Kent, M.L., Schreck, C.B., 2009. Whole-body cortisol response of zebrafish to acute net handling stress. Aquaculture 297 (1–4), 157–162. https://doi.org/10.1016/j.aquaculture.2009.08.035.

Relić, R.R., Hristov, S.V., Vučinić, M.V., Poleksić, V.D., Zoran, Z., Marković, Z.Z., 2010. Principles of fish welfare assessment in farm rearing conditions. J. Agric. Sci. 55 (3), 273–282. https://doi.org/10.2298/JAS1003273R.

Rétaux, S., Elipot, Y., 2013. Feed or fight: a behavioral shift in blind cavefish. Commun. Integr. Biol. 6 (2), e23166. https://doi.org/10.4161/cib.23166.

Riddle, M.R., Boesmans, W., Caballero, O., Kazwiny, Y., Tabin, C.J., 2018. Morphogenesis and motility of the *Astyanax mexicanus* gastrointestinal tract. Dev. Biol. 441 (2), 285–296. https://doi.org/10.1016/j.ydbio.2018.06.004.

Riddle, M.R., Damen, F., Aspiras, A., Tabin, J.A., McGaugh, S., Tabin, C.J., 2019. Evolution of Gastrointestinal Tract Morphology and Plasticity in Cave-Adapted Mexican tetra, *Astyanax mexicanus*. bioRxiv 852814. https://doi.org/10.1101/852814.

Rohner, N., 2018a. Cavefish as an evolutionary mutant model system for human disease. Dev. Biol. 441 (2), 355–357. https://doi.org/10.1016/j.ydbio.2018.04.013.

Rohner, N., 2018b. "Out of the dark" cavefish are entering biomedical research. In: Hirata, H., Iida, A. (Eds.), Zebrafish, Medaka, and Other Small Fishes. Springer, Singapore. https://doi.org/10.1007/978-981-13-1879-5_13.

Ross, L.G., Ross, B., 2008. Chapter 11: Anaesthesia of fish: IV. Non-chemical methods. In: Ross, L.G., Ross, B. (Eds.), Anaesthetic and Sedative Techniques for Aquatic Animals, third ed. https://doi.org/10.1002/9781444302264.ch11

Ruane, N.M., Komen, H., 2003. Measuring cortisol in the water as an indicator of stress caused by increased loading density in common carp (*Cyprinus carpio*). Aquaculture 218 (1–4), 685–693. https://doi.org/10.1016/S0044-8486(02)00422-2.

Salin, K., Voituron, Y., Mourin, J., Hervant, F., 2010. Cave colonization without fasting capacities: an example with the fish *Astyanax fasciatus mexicanus*. Comp. Biochem. Physiol. A Mol. Integr. Physiol. 156 (4), 451–457. https://doi.org/10.1016/j.cbpa.2010.03.030.

Saraiva, J.L., Arechavala-Lopez, P., 2019. Welfare of fish—no longer the elephant in the room. Fishes 4 (3), 39. https://doi.org/10.3390/fishes4030039.

Schemmel, C., 1980. Studies on the genetics of feeding behaviour in the cave fish *Astyanax mexicanus f. anoptichthys*. An example of apparent monofactorial inheritance by polygenes. Z. Tierpsychol. 53 (1), 9–22. https://doi.org/10.1111/j.1439-0310.1980.tb00730.x.

Schroeder, P., 2018. Pain sensitivity in fish. CAB Rev. 13 (049) https://doi.org/10.1079/PAVSNNR201813049.

Scott, A.P., Sorensen, P.W., 1994. Time course of release of pheromonally active gonadal steroids and their conjugates by ovulatory goldfish. Gen. Comp. Endocrinol. 96 (2), 309–323. https://doi.org/10.1006/gcen.1994.1186.

Scott, A.P., Hirschenhauser, K., Bender, N., Oliveira, R., Earley, R.L., Sebire, M., et al., 2008. Non-invasive measurement of steroids in fish-holding water: important considerations when applying the procedure to behaviour studies. Behaviour 145 (10), 1307–1328. http://www.jstor.org/stable/40296047.

Sealey, W.M., Barrows, F.T., Casten, M., Hardy, R.W., 2009. Dietary protein source and level affects growth in neon tetras. N. Am. J. Aquacult. 71 (4), 320–324. https://doi.org/10.1577/A08-017.1.

Seger, C., Hargrave, M., Wang, X., Chai, R.J., Elworthy, S., Ingham, P.W., 2011. Analysis of Pax7 expressing myogenic cells in zebrafish muscle development, injury, and models of disease. Dev. Dyn. 240 (11), 2440–2451. https://doi.org/10.1002/dvdy.22745.

Selye, H., 1974. Stress without Distress. New American Library, New York.

Simon, V., Hyacinthe, C., Rétaux, S., 2019. Breeding behavior in the blind Mexican cavefish and its river-dwelling conspecific. PLoS One 14 (2), e0212591. https://doi.org/10.1371/journal.pone.0212591.

Stahl, B.A., Gross, J.B., 2015. Alterations in Mc1r gene expression are associated with regressive pigmentation in *Astyanax* cavefish. Dev. Gene. Evol. 225 (6), 367–375. https://doi.org/10.1007/s00427-015-0517-0.

Stahl, B.A., Jaggard, J.B., Chin, J., Kowalko, J.E., Keene, A.C., Duboué, E.R., 2019a. Manipulation of gene function in Mexican cavefish. J. Vis. Exp. (146) https://doi.org/10.3791/59093.

Stahl, B.A., Peuß, R., McDole, B., Kenzior, A., Jaggard, J.B., Gaudenz, K., et al., 2019b. Stable transgenesis in *Astyanax mexicanus* using the Tol2 transposase system. Dev. Dyn. 248 (8), 679–687. https://doi.org/10.1002/dvdy.32.

Stockdale, W.T., Lemieux, M.E., Killen, A.C., Zhao, J., Hu, Z., Riepsaame, J., et al., 2018. Heart regeneration in the Mexican cavefish. Cell Rep. 25 (8), 1997–2007.e7. https://doi.org/10.1016/j.celrep.2018.10.072.

Stoskopf, M., Posner, L.P., 2008. Chapter 21 - Anesthesia and restraint of laboratory fish. In: Fish, R.E., Brown, M.J., Danneman, P.J., Karas, A.Z. (Eds.), American College of Laboratory Animal Medicine, Anesthesia and Analgesia in Laboratory Animals, second ed. Academic Press, pp. 519–534. https://doi.org/10.1016/B978-012373898-1.50025-5.

Summerfelt, R.C., Smith, L.S., 1990. Anesthesia, surgery and related techniques. In: Schreck, C.B., Moyle, P.B. (Eds.), Methods for Fish Biology. American Fisheries Society, Bethesda, MD, pp. 213–272.

Tanvir, Z., Rivera, D., Severi, K.E., Haspel, G., Soares, D., 2020. Evolutionary and homeostatic changes in morphology of visual dendrites of Mauthner cells in *Astyanax* blind cavefish. J. Comp. Neurol. https://doi.org/10.1002/cne.25056. Advance online publication.

Torres, L., Hu, E., Tiersch, T.R., 2016. Cryopreservation in fish: current status and pathways to quality assurance and quality control in repository development. Reprod. Fertil. Dev. https://doi.org/10.1071/RD15388. Advance online publication.

Volkof, H., 2016. Feeding behavior, starvation response, and endocrine regulation of feeding in Mexican blind cavefsh (*Astyanax fasciatus mexicanus*). In: Keene, A.M., Yoshizawa, M., McGaugh, S. (Eds.), Biology and Evolution of the Mexican Cavefish. Elsevier, New York, pp. 269–290. https://doi.org/10.1016/B978-0-12-802148-4.00014-1.

Westerfield, M., 1993. The Zebrafish Book. A Guide for the Laboratory Use of Zebrafish (*Danio rerio*), fourth ed. University of Oregon Press, Eugene.

Whitman, K.A., 2004. Chapter 4: Aseptic bacterial examination of finfish. In: Whitman, K.A. (Ed.), Finfish and Shellfish Bacteriology Manual: Techniques and Procedures. Blackwell Publishing, Iowa State Press, Iowa, USA.

Wilson, J.M., Bunte, R.M., Carty, A.J., 2009. Evaluation of rapid cooling and tricaine methanesulfonate (MS222) as methods of euthanasia in zebrafish (*Danio rerio*). J. Am. Assoc. Lab. Anim. Sci. 48 (6), 785–789.

Wong, D., von Keyserlingk, M.A., Richards, J.G., Weary, D.M., 2014. Conditioned place avoidance of zebrafish (*Danio rerio*) to three chemicals used for euthanasia and anaesthesia. PLoS One 9 (2), e88030. https://doi.org/10.1371/journal.pone.0088030.

Xiong, S., Krishnan, J., Peuß, R., Rohner, N., 2018. Early adipogenesis contributes to excess fat accumulation in cave populations of *Astyanax mexicanus*. Dev. Biol. 441 (2), 297–304. https://doi.org/10.1016/j.ydbio.2018.06.003.

Yamamoto, Y., Byerly, M.S., Jackman, W.R., Jeffery, W.R., 2009. Pleiotropic functions of embryonic sonic hedgehog expression link jaw and taste bud amplification with eye loss during cavefish evolution. Dev. Biol. 330 (1), 200–211. https://doi.org/10.1016/j.ydbio.2009.03.003.

Yasui, G.S., Senhorini, J.A., Shimoda, E., Pereira-Santos, M., Nakaghi, L.S., Fujimoto, T., et al., 2015. Improvement of gamete quality and its short-term storage: an approach for biotechnology in laboratory fish. Animal 9 (3), 464–470. https://doi.org/10.1017/S1751731114002511.

Yoshizawa, M., Goricki, S., Soares, D., Jeffery, W.R., 2010. Evolution of a behavioral shift mediated by superficial neuromasts helps cavefish find food in darkness. Curr. Biol. 20 (18), 1631–1636. https://doi.org/10.1016/j.cub.2010.07.017.

Yoshizawa, M., Robinson, B.G., Duboué, E.R., Masek, P., Jaggard, J.B., O'Quin, K.E., et al., 2015. Distinct genetic architecture underlies the emergence of sleep loss and prey-seeking behavior in the Mexican cavefish. BMC Biol. 13, 15. https://doi.org/10.1186/s12915-015-0119-3.

Yoshizawa, M., 2015. Behaviors of cavefish offer insight into developmental evolution. Mol. Reprod. Dev. 82 (4), 268–280. https://doi.org/10.1002/mrd.22471.

Zebrafish.org, 2020. Embryo Surface Sanitation (Egg Bleaching) Protocol [ZIRC Public Wiki]. Available at: https://zebrafish.org/wiki/protocols/ess. (Accessed 30 October 2020).

Northwest Marine Technology, Inc., 2020. Visible Implant Elastomer Project Manual. http://www.nmt.us/wp-content/uploads/2017/11/VIE-Project-Manual-Nov-2017-1.pdf. (Accessed 19 September 20).

Further reading

Combs Jr., G.F., McClung, J.P., 2017. The Vitamins: Fundamental Aspects in Nutrition and Health. Elsevier.

Guillaume, J., Kaushik, S., Bergot, P., Metailler, R., 2001. Nutrition and Feeding of Fish and Crustaceans. Springer, London.

Chapter 14

The housing, care, and use of a laboratory three-spined stickleback colony

Ellen Blaker, Marion Sebire, Tim Ellis, Ioanna Katsiadaki
Centre for Environment, Fisheries and Aquaculture Science (Cefas), The Nothe, Barrack Road, Weymouth, Dorset, United Kingdom

Introduction

The three-spined stickleback (*Gasterosteous aculeatus*), hereafter referred to as 3SSB, is a small teleost fish ubiquitous to the Northern Hemisphere. The species is euryhaline and lives in fresh, brackish, and sea water. There are various other species of 3SSB, of genera *Gasterosteous*, *Pungitius*, *Apeltes*, and *Culaea*, within the order Gasterosteiformes. However, none have the popularity of *Gasterosteous aculeatus* as a study species, which is due to its environmental ubiquity and suitability as a model across several research areas. This popularity, its small size, and ease of laboratory culture have made it a frequently kept laboratory fish.

The 3SSB is actually a marine species and over time, anadromous populations evolved, migrating to fresh water to reproduce. During ice ages, many such populations became isolated in fresh water at different geographic locations and multiple time points. Freshwater populations were both repeatedly isolated from, and recolonized by, their marine ancestor (Bell and Foster, 1994). This has resulted in a remarkable adaptive radiation, with the species complex consisting of many different ecotypes with diverse traits in morphology (Colosimo et al., 2004; Shapiro et al., 2004), physiology (Ishikawa et al., 2019; Kitano et al., 2010), sex determination (Kitano et al., 2009; Peichel et al., 2004), behavior (Boughman, 2001; Dingemanse et al., 2007; Kozak and Boughman, 2009), and life history (Baker et al., 2015). The consistent associations of particular phenotypes with certain ecological conditions across different locations provide strong evidence for adaptation and repeated selection, making the 3SSB an excellent model for studying evolution (Peichel et al., 2001).

Laboratory Fish in Biomedical Research. https://doi.org/10.1016/B978-0-12-821099-4.00009-2
Copyright © 2022 Crown Copyright. Published by Elsevier Inc. All rights reserved.

The early sequencing of the 3SSB genome (Kingsley, 2003) and subsequent explosion in research activity cemented the species' "supermodel" status in studying the molecular basis of vertebrate evolution (Kingsley and Peichel, 2007). The high-quality reference genome is available via the UC-Santa Cruz and ENSEMBL genome browsers, and comparative results based on resequencing 21 other 3SSB genomes can also be explored and downloaded at "sticklebrowser" (Jones et al., 2012). The relatively small genome of the 3SSB implies small intron size and lack of second genome duplication event (as in many other fish species), which aided the speed of annotation and the functional characterization of genes of interest.

The early works of Tinbergen (1951, 1952) highlighted the fascinating reproductive behavior displayed by the male 3SSB, and contributed significantly to the study of animal behavior and ethology. Over the past 20 years, our laboratory has focused on developing tools and tests for use in toxicology, capitalizing on the unique traits of the 3SSB to detect endocrine disrupting chemicals (EDCs) (Katsiadaki et al., 2007). Briefly, these traits are relevant to the presence of a genetic sex marker (Griffiths et al., 2000; Peichel et al., 2004) and a xenoandrogen-specific endpoint. The latter is a rare trait among fishes and is based on spiggin (Jakobsson et al., 1999), the glue protein produced in the male kidney under androgen stimulation and used to build a nest for breeding. Spiggin is readily induced in female 3SSB exposed to chemicals with androgenic action (Katsiadaki et al., 2002), while its production in androgen-stimulated female 3SSB is inhibited by simultaneous exposure to antiandrogens (Katsiadaki et al., 2006). The development and validation of international test protocols aimed at detecting EDCs using 3SSB have resulted in two regulatory tests under the OECD test guideline program: the fish sexual development test (TG234) and the androgenized female stickleback screen (GD 148).

Today, there are thousands of publications, including book series, discussing 3SSB as research and experimental animals. General aspects of their ecology and biology are well documented in books by the late Robert Wootton (Wootton, 1976, 1984), while other more specialized areas are reviewed in the "Biology of the three-spined stickleback" (Östlund-Nilsson et al., 2006) and the "Evolutionary biology of the threespine stickleback" (Bell and Foster, 1994). The 3SSB research community meets every 3 years at an "International Conference on Stickleback Behavior and Evolution," where information flows freely between research groups.

Nevertheless, despite its popularity as a vertebrate model across many research disciplines, there is limited practical guidance available (e.g., Divino and Schultz, 2015) on how to care for 3SSBs in the laboratory, or how to establish a long-term colony. Here we aim to share our knowledge, gained from over 20 years of practical experience in 3SSB care, with the wider research community. Although much of this information may be common knowledge within the 3SSB fraternity, this chapter is intended as a centralized resource for both existing and new researchers.

Supply, quarantine, and disinfection

Source

New stocks (taken from the wild or from another establishment) are needed to establish a new colony and to enable outbreeding in long-established colonies. To our knowledge, there are no registered laboratory suppliers of 3SSB, at least in Europe. Fish will therefore need to be sourced from either another laboratory, an ornamental fish supplier, a freshwater fish farm (where they often occur in water supply channels), or collected from the wild. Our colony was started with wild-caught fish and has been outbred at intervals with fish supplied from trout farms, where 3SSB were also present in ponds. Any source should be confirmed as suitable with respect to national regulations on animal experimentation, environmental protection, and aquatic animal health.

We limit new introductions into our laboratory and follow a quarantine procedure due to the inherent risks of introducing pathogens. Quarantined stocks should be held separately from any established colony, ideally in a different room with tight biosecurity.

Our quarantine rules include:

- Introduced stocks are housed in a separate room until their health status has been assessed as pathogen free.
- A sample of incoming stocks is health checked: samples are taken for bacteriology, virology, and histopathology testing and depending on the results, fish may be therapeutically treated.
- As a precaution, stocks originating from fresh water are held at 7–10‰ salinity for 4 weeks to prophylactically treat white spot (*Ichthyophthirius multifiliis*) and other ectoparasite infections. (Similarly, seawater 3SSB could be "dipped" in fresh water [dechlorinated and pH balanced] for up to 5 min to eliminate marine ectoparasites; however, close observations are needed to check for adverse effects and the fish should be returned to sea water if signs are seen.)

If production of a new colony is intended, the most biosecure way is to start from fertilized eggs (unhatched embryos) from another laboratory. Fertilized embryos pose a much lower biosecurity risk (due to reduction in horizontally transmitted pathogens) and can additionally be rinsed and disinfected (e.g., sodium hypochlorite Westerfield, 2007 or methylene blue at 1 mg/L) before transfer to the main holding facility. Alternatively, sexually mature fish can be obtained in early spring, held in quarantine, and humanely killed after production of fertilized embryos (naturally or by in vitro fertilization [IVF], see the "Natural breeding" and "In vitro fertilization" sections).

Arrival

Tanks should be set up in advance to house the transported fish, with water to match the expected temperature and salinity during transport. While unpacking, consideration must be given to biosecurity, avoidance of sudden changes in water quality, and acclimation to the new environmental conditions. In our facility, the fish and transport water are carefully emptied into a suitable holding vessel. As a priority, dissolved oxygen, temperature, ammoniacal-nitrogen, and pH are recorded. If the fish have been transported in water with compressed oxygen, there may be a build-up of dissolved carbon dioxide; this will reduce the pH of the water and decrease the proportion of toxic NH_3. As carbon dioxide is released from the water, the pH and the proportion of toxic NH_3 will increase. To mitigate this risk, the transport water is diluted with tank water; the rate of addition is balanced to avoid potential "shock" from different water qualities (additions of 25%, then 33%, then 50% over a 2-h period, with water quality checks after each addition). Aeration is added to the vessel to aid removal of carbon dioxide and restore oxygen levels, ensuring that associated water movements do not affect the fish's ability to maintain position.

Despite planning and temperature control during transport, temperature changes do occur, and we have not observed adverse effects after transfer to tanks with a $\pm 2^{\circ}C$ change from transport water. The fish are netted into the receiving tank to minimize transfer of the transport water (and pathogens); however, such handling should be kept to a minimum to reduce stress and damage and it is also advisable to avoid changes in light intensity to allow the animals to settle as changes in light can cause a stress response (Mork and Gulbrandsen, 1994). If reception is completed during the morning and no adverse effects or abnormal behaviors are observed, food can be offered in the afternoon.

Basic husbandry

Housing

3SSB can be maintained in a variety of housing systems, with transparent tanks (glass or plastic) conventionally being used. We routinely use glass tanks, but opaque fiberglass tanks have proved suitable for general temporary holding. Transparent tanks are commonly used as they allow better visual checks of health and welfare, assessment of sexual dimorphism and sexual maturity, as well as associated behaviors (including male nest building, courtship, and potential aggression). Consideration should be given to screening transparent tanks, as behavior may be affected by visual cues from nearby tank populations (Fernandes Silva et al., 2019).

All our tanks have lids or covers to reduce potential cross-contamination, avoid unnecessary disturbance (if opaque), and aid containment of the fish. To further prevent escape, all inflow and drain/outflow holes have apertures that are either much smaller than the diameter of the fish's head or are screened with plastic mesh. As fish grow, larger apertures and mesh sizes are used to aid waste removal. 3SSB appear to be sensitive to acoustic noise (Purser and Radford, 2011) and our observations indicate strong reactions to sudden loud noises or new vibrations. Tanks should therefore be located to minimize noise and disturbance, and ideally any equipment causing vibration (e.g., pumps) should not be in the same room.

For tank cleaning, outflow screens are cleaned daily with a soft bristle brush (e.g., toothbrush) to prevent blockage from food and/or fecal matter and potential overflow. Detritus (uneaten food, feces) is removed from the tank floor by siphoning as required; the frequency depends upon food inputs and temperature. To restrict the siphon flow (and potential for inadvertent damage to the fish) we use a small (4 mm) bore tube with a straight plastic pipette attached at the tank end. The detritus and siphoned water are collected in a jug. Every 2–4 weeks, depending on season and stage (see the "Stocking density and water quality" section), the tank glass is cleaned with either a glass or acrylic pad to remove algae and enable a clear view of the fish. The gravel substratum is also siphoned to reduce detritus and improve water flow through the under-gravel filter.

Environmental enrichment

3SSB are highly social and interactive, showing complex behaviors in the laboratory. To allow full expression of these behaviors, we consider a barren environment should be avoided where possible. As a minimum we provide a gravel (~5–10 mm Ø) substratum of 4–10 cm depth (dependent on tank size). 3SSB do interact with the gravel, by searching interstices and moving particles around. We have, however, recorded a possible adverse effect of gravel enrichment, i.e., ingestion and retention in the stomach, but this is rare and associated with interventions such as protracted removal of fish from the tank. The hygiene of the substratum is maintained via siphon cleaning and under-gravel filters (a downwelling water flow generated by air uplifts, with the additional benefit of biological filtration). 3SSB show a strong antipredator response to shadows and movements from above (Giles, 1984), so placement of tanks on low shelving (or where technicians are likely to be above the tanks) should be avoided; if such placement is necessary, then provision of opaque lids and refuges such as artificial caves are recommended. In our experience, artificial plants are an optional enrichment in holding tanks with mixed populations as they can cause general disturbance during tank cleaning or repositioning when dislodged. We do, however, provide both artificial plants and refuges when sexually mature males are held as single sex populations: these act as visual barriers and reduce aggressive behaviors such as nipping, flashing, and chasing.

Stocking density and water quality

The stocking density at which fish are kept changes as the fish grow (see the "Establishing and maintaining a successful colony" section). We have found that keeping groups of sexually mature males at a density of >1 fish/L contributes to minimizing aggression.

3SSB have a relatively rare trait among small laboratory fish species as they can grow and reproduce at a range of salinities. In the laboratory, 3SSBs are traditionally reared in fresh water (<7‰). Although our experience is largely with fresh water, we have found stocks can be easily transferred to full strength sea water (35‰). We transfer fish gradually, increasing salinity by ≤10‰ per week. If stocks need to be transferred to a different salinity for experimental purposes, acclimation at the new salinity regime takes place ≥2 weeks prior to the experiment.

3SSB can be maintained in a variety of water qualities and associated maintenance systems, including flow-through, recirculation, and static systems. In static systems, filtration or regular water changes are used to maintain water quality, and fish biomass and feeding should be controlled to ensure ammonia does not become a problem. In our experience, flow-through systems are preferable as they reduce labor requirements and the risk of water quality issues by maintaining oxygen levels and flushing out metabolic waste products (ammonia and carbon dioxide). During the spring and summer months (see the "Annual cycle of temperature and photoperiod" section), biomass will increase rapidly due to fish growth. The associated increases in feed input and temperature during this period could potentially have a detrimental impact on water quality. Typically, for an established system, fish loading at <0.3–0.5 g/mL/min presents no water quality issues (with the additional biological filtration from the under-gravel filter). As with most small fish species, unionized ammonia (NH_3) should be kept <0.05 mg/L and nitrite <0.10 mg/L (Noga, 2010). The pH of fresh water should be approximately neutral but can range between 6.5 and 7.5.

In our laboratory, the freshwater supply is drawn from the potable water mains, which is dechlorinated (through activated charcoal filters) and UV light treated before use. Residual chlorine and chloride are regularly checked and a full suite of analyses for microcontaminants is conducted twice a year. Additional measures may be needed if the water supply comes from natural sources.

Photoperiod and temperature

In general, we keep our fish in conditions that imitate the seasonal changes in photoperiod and temperature. The photoperiod used in most laboratories is between 8:16 LD and 18:6 LD (h Light: h Dark) and is linked to water temperature to avoid unnatural combinations (i.e., cold temperature with long daylength). Although 3SSB grow at any photoperiod within this range, a controlled seasonal change is used to optimize sexual maturation. Increasing daylength (coupled with rising temperature to a lesser extent) is the main environmental cue for 3SSB to grow and mature sexually (Baggerman, 1957). The annual cycle of photoperiod and temperature conditions applied in our laboratory is summarized in the "Annual cycle of temperature and photoperiod" section.

Lighting (intensity, wavelength, diel transitions, seasonal changes) is an area that merits further investigation in the context of holding fish in general. We use a moderate intensity throughout the life cycle (about 500 lux at the water surface) although this can increase (to 1000 lux) during the reproductive season. Standard fluorescent or LED lighting systems with a lower kelvin output have proved suitable. Higher kelvin outputs are undesirable as they promote algal growth, which is not an inherent requirement for holding 3SSB. We also provide dawn and dusk periods of about 30 min, with a gradual fade-up and -down of light intensity.

Handling

Handling represents a stressful event for any fish and should be kept to a minimum. In our laboratory, handling is necessary for sample weighing (to monitor growth), transfers between tanks, and gamete stripping. It is typically thought that retention in water, avoiding emersion in air, will reduce handling stress. Brydges et al. (2009) used opercular beat rate and blood cortisol level to compare stress responses to net handling (air exposure) and scoop handling (kept in water); the former seemed to cause a greater physiological stress response. In contrast, Thompson et al. (2016) assessed diving and scototaxis tests of anxiety-like behavior, and suggested that net handling induced lower anxiety than scoop handling.

Feeding

Feeding requirements

3SSB are predominantly carnivorous, visual predators feeding on a broad range of live prey (Wootton, 1976). In captivity, they will take a variety of food and will readily feed on bloodworm (aquatic larvae of *Chironomus* midge species, named due to their red coloration) as observed in the wild (Dukowska et al., 2009). This preference may be partly due to their well-documented attraction to red (e.g., red-colored male preference in a review by Bell and Foster, 1994; red-colored algae/thread for their nest, (Östlund-Nilsson and Holmlund, 2016). Prepackaged frozen gamma-irradiated bloodworm is readily available and forms the staple diet of our 3SSB colony. Rather than feeding bloodworm ad lib, we predetermine food ration (% body weight day^{-1}), which changes through the life cycle to reflect differing requirements (e.g., juvenile and breeding fish in summer conditions will have a higher nutritional demand than adult fish over-wintering). The daily ration is fed across discrete meals at intervals through the working day depending on the size of the fish (summarized in Table 14.1). However, rations should be responsive to uneaten food, poor water quality, negative/abnormal behavior, and lower than expected growth rates.

Daily feed (g) is calculated as a percentage of the total biomass (TB) of the fish (% TB/day). TB is defined as mean fish weight (g) × total number of fish

TABLE 14.1 Summary of diet, feeds, and rations used for three-spined stickleback at Cefas Weymouth.

Stage	Age/size (average weight in g)	Environmental condition (see the "Annual cycle of temperature and photoperiod" section)	Diet	Minimum number of feeds per day	Estimated ration (% body weight/day)
Embryos	Up to 3 dph at 17°C	Summer	No diet needed until yolk sac is absorbed		
Early larvae	<4 wph (<0.1)	Summer	Newly hatched *Artemia*	3	To satiation
Late larvae	4–6 wph (≤0.1)	Summer	Chopped bloodworm and newly hatched *Artemia*	3	To satiation
Juvenile	6–12 wph (0.1–0.2)	Summer	Whole mini bloodworm	3	30
Juvenile	12–24 wph (0.2–1)	Autumn–winter	Whole bloodworms	3	30
Adult	24–36 wph (>1)	Winter	Whole bloodworms	2	10–15
Adult	36–40 wph (>1)	Spring	Whole bloodworms	2	15–20
Adult	>40 wph (>1)	Summer	Whole bloodworms and discus diet	2	20–25

bw, Body weight; *dph*, days posthatching; *wph*, weeks posthatching.

and the total is then divided by the number of meals. Although frozen bloodworm blocks can be portioned and weighed, this is often impractical, time consuming, and difficult for low rations. We have therefore developed a method whereby the total weight of bloodworm required per day is defrosted, added to dechlorinated water at a 1:3 ratio to create a suspension, and the required amounts are dispensed volumetrically using disposable graduated pipettes. This method has proved quick, accurate, repeatable, and consistent enabling appropriate amounts to be fed to all stocks by different technicians.

First feeding and weaning

When the larvae begin to swim up from the tank floor (see the "Embryo development" section) we provide newly hatched *Artemia* spp. nauplii as a first feed. *Artemia* is added to the tank three times per day in excess, i.e., some remains at the next feed. The amount of *Artemia* is increased as the larvae grow. We do not routinely switch to larger on-grown (enriched) *Artemia* metanauplii and continue with newly hatched until weaning.

At 4–6 weeks posthatch (ca. 0.1 g), we start weaning onto a bloodworm diet. Due to the small mouth gape, finely chopped bloodworms (produced with a kitchen blender) are fed in combination with *Artemia* nauplii. As juvenile 3SSB are predominantly visual feeders, food movement is thought to be a cue for feeding and consideration should be given to how to introduce nonlive feed. Our flow-through systems have an inherent current, so small pieces of bloodworm move with the flow as they sink and are readily eaten. In static systems, feeding on the bloodworm diet should be tested by dropping a few small pieces into the tank and allowing them to sink or by introducing them over an air stone. If feeding is not stimulated, finely diced bloodworm may be introduced shortly after live *Artemia*, which stimulates active feeding behavior.

Artemia feeds are continued until all fish are readily feeding on the diced bloodworm diet; when this occurs, *Artemia* feeds are reduced to one per day. As the fish grow, they show a preference for bloodworm over *Artemia*, likely because of the latter's small size. A single daily *Artemia* feed is continued to provide a nutritional supplement and ensure access to food for any smaller fish in the population.

At 6–12 weeks posthatch (average 0.15 g) we just feed whole "mini bloodworm," i.e., smaller than normal whole bloodworm, and after that they are fed exclusively on whole normal-sized bloodworm. We have tried providing a mix with other food items (e.g., thawed irradiated adult *Artemia*), but the 3SSB preferentially select bloodworm, although discus diet, for its beef heart content, is taken during breeding.

Establishing and maintaining a successful colony

The 3SSB has strong circannual rhythm and successful maintenance of a laboratory colony requires the environmental cues of photoperiod and

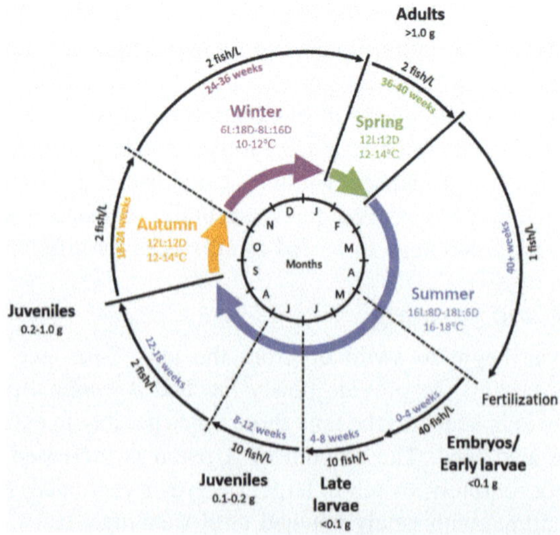

FIGURE 14.1 Annual cycle for the three-spined stickleback. Maximum stocking density expressed in fish/liter (fish/L). Conditions and duration represent our production regime, inclusive of light regime and temperature. Time periods indicate age in weeks posthatch.

temperature. The annual cycle we employ at the Cefas Weymouth laboratory is summarized in Fig. 14.1.

Annual cycle of temperature and photoperiod

We limit changes (increase/decrease) to temperature and photoperiod to 1°C per day and 2 h per day, respectively. Early larval and juvenile stages are provided with a long photoperiod, 18:6 (L:D h), and warm temperatures, 16–18°C. These conditions are regarded as "summer" and promote higher activity, feeding, and growth rates. After ≤18 weeks (depending on fertilization date), the stocks are shifted to "autumn," i.e., photoperiod is reduced to 12:12 (L:D h) and temperature is reduced to 12–14°C. Stocks are held in autumn conditions for 4–6 weeks before changing to "winter," i.e., photoperiod is further reduced to 6–8:18–16 (L:D h) and temperature can also be further reduced to 10–12°C depending on practicalities. Reduction of temperature seems to be less essential than photoperiod for successful sexual maturation the following spring.

Our males reach sexual maturity before the females and it is not uncommon to see some males develop breeding coloration even during winter conditions. In contrast, female maturation appears to depend on environmental "spring" cues. The change to "spring" (12:12 L:D h; 12–14°C) and subsequently "summer" (16–18:8–6 L:D h, 16–18°C) conditions stimulates further males to come into breeding condition. In our experience, females need more time to mature, and gravid fish are not expected until after 4 weeks in "spring" and 4 weeks in "summer" conditions. We have tried, via photoperiod and temperature manipulations, to shift spawning and produce out-of-season embryos. However, our attempts have thus far failed, confirming the strong circannual rhythm present in the species.

Genitors and embryo production

The lifespan of wild 3SSB is only 1 year in warmer environments, extending to 2–3 years at higher latitudes (Wootton, 1976, 1984), and they are seasonal breeders, starting to spawn in early spring like the majority of temperate fish species. In our laboratory, we tend to keep a cohort for 1 year and produce new stocks each year using the previous years' fish as genitors. We aim to complete breeding early in the annual cycle to allow larvae/juveniles a long summer growth period (up to 18 weeks).

In the wild, 3SSB minimize the risk of inbreeding by avoiding kin as breeding partners (Frommen and Bakker, 2006). As a rule, we cross 3SSB from different batches (no full siblings) each season (April–July) to produce the following year's populations. Additionally, to further avoid potential adverse effects from inbreeding and maintain a wild geno/phenotype in our long-established laboratory colony, every 4–5 years we have introduced new genitors (mostly wild-sourced females).

Sexual dimorphism

During the breeding season (April–August), the male displays the secondary sexual characteristics (SSCs) of red coloration on the throat and ventral side, and blue irises (Fig. 14.2A). Outside the breeding season, however, there are no strong SSCs and sex identification is more difficult. However, there is sexual dimorphism in certain cranioskeletal features, including serration of the dorsal spines in males (Kitano et al., 2007; Leinonen et al., 2011). In our laboratory, we have observed an additional subtle SSC in head shape: females have a slight indent in the posterior–superior cranium dorsal to the eyes (Fig. 14.2B). We have found this to be an accurate method for sex identification but are unaware if it is specific to our colony or universal. Genetic sex can be determined without terminal sampling (see the "Genetic sex determination" section).

Controlling sexual maturity and gamete production

Manipulating male sexual maturation via photoperiod and stocking density changes (removing dominant males to allow other fish to enter breeding) has proved easy. On the other hand, females appear to be harder to stimulate to mature sexually. Only a proportion of females will mature under standard conditions to provide eggs; our observations suggest that maturation is not

FIGURE 14.2 Key morphological characteristics of reproductively mature three-spined stickleback (3SSB) and breeding components. (A) Breeding male, showing its red nuptial coloration; (B) cranial sexual dimorphism in the 3SSB colony at Cefas Weymouth; (C) a gravid female displaying the head-up posture; (D) a completed nest with an entrance; (E) overly gravid female showing a swollen cloaca and the beginning of berry appearance on her belly region; (F) female showing a berry appearance after absorption of ovarian fluid.

related to female size. Over the years we have noticed that the proportion of females entering active vitellogenesis can be increased by:

1. Removing excess males from the tank, aiming for a ratio of one breeding male to 10 females.
2. Reducing the stocking density by ~50%.

Removal of excess males is also important for the welfare of the fish; mature dominant males are aggressive toward both rival males and gravid females. We remove the dominant males once they show their characteristic nuptial colors and house separately in all-male 10 L tanks at a density of one fish/L. We continue removal until all male fish have been removed from the tank before reintroducing the early dominant males that were held separately. We find that this population manipulation ensures that the females enter and remain in breeding condition, and the apparently fittest males are used for breeding. It is also a simple way to reduce stocking density without handling the females.

We have maintained records of phenotypic sex for 11 separate batches of fish reared over 4 years and have found that our populations of 3SSB generally have a 1:1 sex ratio (batch average 52% M:48% F).

Natural breeding

In the wild during early spring, male 3SSB with nuptial coloration (red throat and blue irises, Fig. 14.2A) establish a territory and build a tunnel-shaped nest using plant material (often filamentous algae), sand, and various debris, which he glues together with spiggin (Fig. 14.2D). Once the nest is completed, the male displays courtship behavior toward females, which signals readiness to spawn with a head-up posture (Fig. 14.2C). The male leads the female to the nest with a zigzag dance (for most populations) or rolling display (Kitano et al., 2007). The female then enters the nest, lays her eggs, and exits at the other end. The male follows the female and fertilizes the eggs. The male then chases the female away to prevent her eating the eggs and continues parental care (e.g., nest guarding, fanning water over the eggs) until larvae leave the nest.

Although it is easy to replicate natural reproductive behavior in the laboratory, we typically only use natural breeding for experimental observations. For colony maintenance, we find artificial fertilization of eggs is more efficient, and reproduction is better controlled. However, when setting up conditions for natural breeding in the laboratory, considerations include:

- Once males show their breeding coloration, they are ready to build a nest. To enable this, they should be isolated in individual breeding tanks >5 L to provide space to establish a territory and express the ritual dance.
- Gravel/sand substratum and nest material are needed for nest building. The gravel pieces should be smaller than the mouth gape so males can move them to form the pit in which the nest is built. The gravel/sand can cover the floor of the tank or be limited to an area (e.g., in a Petri dish).
- Nest material can be any water weed with long threads. In laboratories, the most commonly used materials are Canadian waterweed (*Elodea canadensis*) and synthetic or cotton threads (Barber et al., 2009; Östlund-Nilsson and Holmlund, 2016). Care should be taken if opting for nest material collected from the wild as this could pose a biosecurity issue and is difficult to decontaminate.
- Although most males build a nest in $\leq$1 day after transfer to such an environment, more reluctant nest builders can be motivated by visual presentation of a gravid female (in a neighboring tank or confined within in a transparent container) for 30 min periods.
- A nest is completed and ready for spawning once the male "creeps through" the nest, i.e., enters one end and exits the other. It is not uncommon for males to dismantle, relocate, and rebuild their nests, particularly when they can see other males in neighboring tanks.

- Once a nest is completed, a gravid female should be introduced. If nesting males are left in isolation without the opportunity to breed for more than 2 weeks, we find that they lose motivation to breed and deconstruct their nests.
- Upon presentation of the female, spawning usually takes place within 10 min. However, female choice appears to be very strong, with certain females apparently choosing not to breed with a particular male. If no spawning occurs after 20 min, we remove the female and introduce a different female.
- The fish should be observed during breeding: nesting males can become aggressive when interacting with a gravid female, causing stress and/or injury, which needs to be prevented by removing the female.
- Once the male has fertilized the eggs in the nest, the female should be removed immediately. The fertilized eggs can either be left in the nest to be cared for by the male until hatching and beyond or they can be removed for experimental use or artificial raising (see the "Embryo development" and "On-growing" sections).

In vitro fertilization

There are several advantages of IVF over natural breeding for laboratory 3SSB, including time and space saving, controlled selection of genitors, and economy of production. A typical laboratory female produces an average of 100–150 eggs per batch. A single male's sperm (used for IVF) can fertilize up to 15 egg batches and provide around 2000 half sibling embryos within a short time (e.g., 1 h). In contrast, natural breeding is not as efficient and requires more labor and time to achieve the same number of half siblings.

The IVF protocol was developed by Barber and Arnott (2000), with further adaptations in other laboratories, including our own (Sebire et al., 2015). Our IVF protocol is detailed in Annex 11 of OECD TG234 (OECD, 2011) and we routinely obtain ≥95% fertilization rate (in most cases 100%). There are three steps to the procedure:

1. Obtaining sperm (steps 1–8 in Annex 11 TG234). The procedure is terminal for the male and it is our normal practice to keep a tissue sample (e.g., tail fixed in 100% ethanol) for future genetic assignment of parental origin of offspring.
2. Obtaining eggs from the females (steps 1 and 2 of fertilization in Annex 11 TG234). Females can spawn several clutches of eggs in a season (batch spawners), which lasts for around 4 months in our colony. The length of the interspawning interval depends on food supply but is usually 3–5 days. Eggs are spherical and vary in diameter from 1.1 to 1.8 mm, and weigh ca. 2 mg (wet weight) or 0.3–0.4 mg (dry weight) (Wallace and Selman, 2006). A female ready to spawn can be identified in the tank as they display

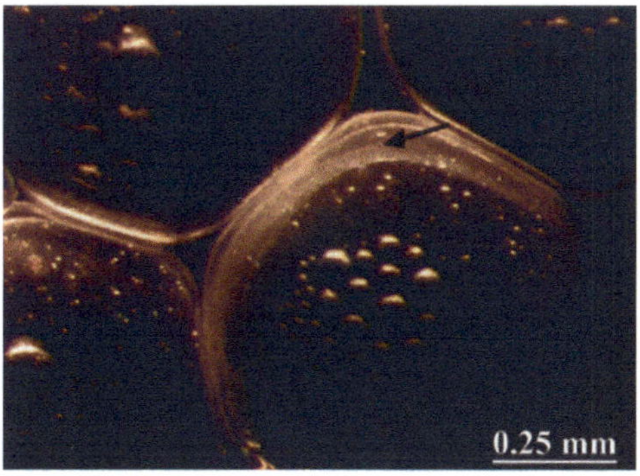

FIGURE 14.3 Fertilized eggs showing the vitelline membrane (*arrow*) separating from the egg proper, developing the perivitelline space.

the "head-up" posture (Fig. 14.2C) and a typical gravid appearance: a rotund abdomen and "swollen" cloaca with protruding eggs in some cases. Although some researchers anesthetize females prior to collecting eggs, we consider that manual stripping can be conducted by experienced workers very quickly without anesthesia. We have observed that egg color varies, ranging from bright yellow to almost white, even between full siblings kept in the same tank. Females can be overly gravid (Fig. 14.2E) if not allowed to breed; such fish should have their eggs manually stripped within 3 days of this stage. If individual eggs are visible under the skin (referred to locally as a "berry" appearance, Fig. 14.2F), this means that ovarian fluid has been largely reabsorbed and the female has become egg-bound. These females will never be able to release eggs and although in most cases they do not present a welfare concern, consideration should be given to humane killing.

3. Fertilization (steps 3–6 of fertilization in Annex 11 TG234). Fertilized egg can be identified as the vitelline membrane separates from the chorion resulting in a perivitelline space (Fig. 14.3). Once fertilization is confirmed, a picture of the batch of embryos is taken and uploaded to counting software, e.g., ImageJ with a suitable plugin enabled to aid counting and ensure accurate records.

Embryo development

Egg development rate after fertilization is temperature dependent; the lower the water temperature, the longer the development of the embryo takes. At 18°C, hatching takes place approximately 7 days postfertilization (dpf), i.e., 126°C days (°C days). Embryo mortality can be assessed visually or checked under a microscope as dead embryos usually have an opaque white appearance, which contrasts with the translucent color of a healthy embryo. Dead embryos can become infected by fungus and develop visible mycelia, which then spread to infect adjacent healthy embryos. For this reason, it is important to check and remove dead embryos several times through the working day.

We have found that reconstituted water (referred to as dilution water in Busquet et al., 2014) produces higher hatching and survival rates than our aquarium water supply during the early developmental stages (ova, first feeding). Fertilized embryos are kept in either 1 L flasks or 300 mL crystallizing dishes in either a temperature-controlled room or water bath. The flasks/dishes are filled to 100 mL with reconstituted water and refreshed every 2 days via a partial (25%) water change. Larger water changes (50%) are given on the days during and around hatching to mitigate water quality deterioration due to decomposing chorions. The tops of the flask/crystallizing dishes are covered with parafilm to reduce contamination by airborne fungal spores, with a small hole left for aeration. Air is passed through a 0.2 μm syringe filter (to reduce fungal contamination) and introduced as a steady but gentle stream of bubbles via a rigid plastic tube (pipette).

After hatching, the larvae are not transferred until they are considered as "independent feeders", meaning the larvae are able to swim up independently through the water column looking for food and the yolk sac absorption is almost complete. This larval form is the threshold between nonprotected/protected life stage as defined by European Directive 2010/63. However, this legislative criterion is variable between fish species and its interpretation varies largely across different countries. For example, for zebrafish, the independent feeding stage has been shown to be at 120 h postfertilization, i.e., 5 dpf at 26–28°C (Strähle et al., 2012). For the 3SSB we advise the larvae to become protected from 9 dpf at 18°C (equivalent to 162°C days). Furthermore the transcriptomes of five early embryonic stages comprising developmental leaps have been described in detail (Kaitetzidou et al., 2019).

On-growing

When the larvae are considered independent feeders as defined earlier, they are transferred without handling to on-growing tanks. We use 10 L tanks with a fitted standpipe to an initial depth of 5 L and a flow-through of 0.05–0.15 L/min. The tanks are barren to enable any mortalities and moribund larvae to be seen and removed or humanely killed. Aeration is added to the tank and set to a gentle stream of bubbles to avoid generating strong water currents. Larvae are transferred from incubation flasks by gentle horizontal placement of the flask within the tank (it is removed the next day after the larvae have swum out). Larval transfer from incubation dishes is simpler: we use the flow-through tank as the water bath, so the tank water depth is simply increased via the standpipe to above the level of the dish and the inflow is redirected into the dish so the larvae can swim out. The larvae (then juveniles) remain in the 10 L tank, with the water depth being increased as fish size increases.

At 4 weeks posthatch, juveniles (0.05 g) are transferred to a 20 L tank. This tank is set up with gravel (~4 cm deep) and an under-gravel filter with two air lifts (positioned facing each other to reduce turbulence) and supplied with aeration via two 2.5 cm air stones. The water in-flow rate is set at 0.2–0.4 L/min. For batches of fish <70, the 10 L tank size may be retained. Transfer is by netting and the opportunity is taken to assess the mean weight by temporary holding in a preweighed jug of tank water. The fish are counted as they swim out of the partially submerged jug into the new tank, ensuring the lip of the jug is beneath the tank water surface; the fish are never poured.

Around 12 weeks posthatch, the juvenile 3SSBs are moved to larger tanks (100–150 L) to reduce stocking density. These tanks have the same filter and gravel set-up as previously and overflow mesh size can be increased (e.g., Ø 2–5 mm). The flow rate is set at 0.3–0.5 L/min.

Monitoring growth

We routinely sample-weigh stocks to monitor fish growth and ensure the appropriate amount of feed is provided. This occurs up to twice a month in summer conditions and once a month during the winter. Where possible it is combined with moving fish between tanks to minimize handling. We consider the benefits of interim sample-weighing, to assess total biomass and monitor growth, outweigh the costs of handling. Our size at age data have been collated (Fig. 14.4); from 50 weeks posthatch, breeding males and females start to differ in weight, resulting in a larger variation of their size. This plot enables us to identify stocks that diverge from expected growth and implement corrective measures, and to update frequently food amounts based on expected growth.

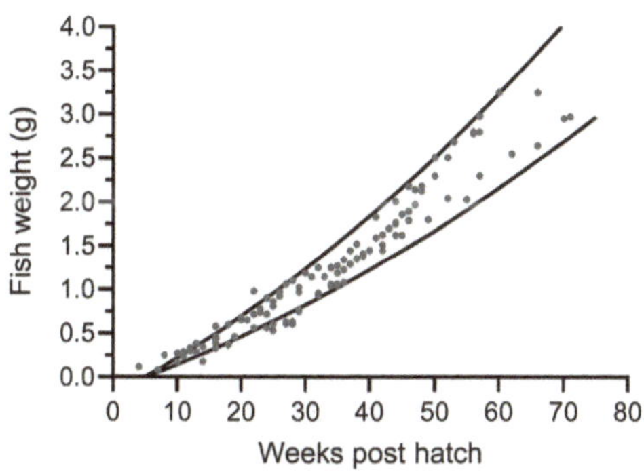

FIGURE 14.4 The expected size (g) at age of the Cefas three-spined stickleback stocks. The lines represent the ±20% of the fitted polynomial model curve and can be used as the range for expected weights.

Procedures

Blood sampling

Although the 3SSB is a small fish species, sufficient blood (5−50 μL, size and sex dependent) can be collected by terminal sampling of adults to enable individual measurement of parameters, including steroid levels (Brydges et al., 2009) and vitellogenin (Katsiadaki et al., 2012). Briefly, fish are terminally anesthetized (MS222 5 mg/L) and removed when in the surgical anesthesia stage. The caudal peduncle is cut with a sharp blade, and the blood collected by placing a heparinized microcapillary tube onto the severed blood vessel underneath the spine. Individual steroid levels can also be measured noninvasively via release into the water; we initially developed the noninvasive method for salmonids (Ellis et al, 2004, 2005; Scott and Ellis, 2007) and extended and validated its application to 3SSB, which enabled a time series of samples from the same individual (Sebire et al, 2007, 2009).

Spine clipping

While some noninvasive methods have been used (Barber and Ruxton, 2000; Ward et al., 2002), spine clipping has been commonly used in behavioral studies to individually mark fish (the three spines enable a variety of marks), and for nonterminal tissue sampling for DNA analysis where it is considered a refinement for fin or skin sampling (Baube, 1997; Bell and Stamps, 2004; Kraak et al., 1999). Briefly, individual fish are captured and confined in a net in the water until the technician is ready. The fish is gently held between two fingers, either in the air or, preferably, water. Using a sprung iris scissors tip, the selected spine is lifted and cut to remove around 70%; it should not be cut closer to the base to avoid skin damage. Spines are hard structures, so protective goggles should be worn for eye protection, particularly when handling in the air. The handling and clipping procedure should take <15 s. After spine clipping, the fish should be monitored until they have regained appropriate orientation, operculum ventilation, and swimming behavior, which can be expected to occur within 10−20 s. In the following few days, the fish should be checked for signs of fungal infection.

Genetic sex determination

For genetic sex determination we use the primers designed by Peichel et al. (2004) and a modified procedure. DNA can be collected using either a spine clip or a mucus swab (Sebire et al., 2015). Mucus swabbing has been used successfully in our laboratory for >7 years and has also been reported by Breacker et al. (2016). Briefly, the fish is captured and held in the net out of water, and mucus is collected by rubbing a sterile cotton bud on each side of the fish. The fish is then released to an individual vessel (pending confirmation of sex) and the cotton bud placed in an Eppendorf tube. We do not anesthetize the fish for this procedure as it can be concluded very fast and appears to be less stressful than the anesthesia (Tilley et al., 2020), and after swabbing, fish should be monitored as for spine clipping.

Conclusion

Our long experience with a laboratory colony of 3SSB supports the notion that 3SSBs are highly amendable to laboratory conditions. Although zebrafish present practical advantages such as high rearing temperatures, continuous breeding, and short generation time, none of these traits are typical for the vast majority of fish that live in temperate zones and have an annual reproductive cycle. The 3SSB presents all the advantages of a small teleost and is amendable to laboratory culture while offering further advantages such as ecological relevance and evolutionary perspective in a changing environment. Sticklebacks, in comparison to the salmonids and cyprinids (including zebrafish) held in our aquarium facility, are largely free from welfare issues such as deformities of the spine, jaw, and operculum, and although aggression does occur within populations, this appears to be limited to the breeding season and can be managed to avoid adverse effects. The most enjoyable aspect of working with 3SSBs is the ready (and fascinating) display of behaviors, such as their response to human presence (approaching feeding stations and following technicians), elaborate sexual rituals, and cognitive abilities (Bensky et al., 2017).

There is always a need to improve and optimize/standardize further the care of 3SSBs (as with all laboratory fish species) to refine scientific use. Future areas of work on 3SSB to facilitate this includes the production of recognized genetic lines. Although our colony is over 20 years old, we have made efforts to outbreed and ensure that they remain "wild type." Given the numerous 3SSB ecotypes, there is a strong potential to generate specific inbred genetic lines differing in particular traits (e.g., age at senescence). A significant contributor to this development is a reliable sperm cryopreservation technique (Aoki et al., 1997). Another area of future work is to further standardize the environmental conditions (i.e., optimum light spectrum) with a focus on diet, contributing to standardizing gut microflora, a potential confounding factor in experiments (Vatsos, 2016).

This chapter is an initial step toward the production of a complete guide for the care of 3SSBs in the laboratory. We have described our practices for sourcing, quarantine and biosecurity, housing, environmental conditions, husbandry, and reproduction. We have included many anecdotal and empirically derived tips that we consider key contributors to successful laboratory culture and refinement and we would welcome feedback and the experience of other researchers and animal care technicians.

Acknowledgments

We would like to thank all the Cefas aquarists, both at the Lowestoft laboratory and at Weymouth site, where the colony has been based for over 20 years for their contribution to its success. You are far too many to thank individually! We also want to acknowledge continuous funding support from Defra, UK. Any reference to products or services do not imply or constitute endorsement or recommendation.

References

Aoki, K., Okamoto, M., Tatsumi, K., Ishikawa, Y., 1997. Cryopreservation of medaka spermatozoa. Zool. Sci. 14 (4), 641–644.

Baggerman, B., 1957. An experimental study on the timing of breeding and migration in the three-spined stickleback (*Gasterosteus aculeatus* L.). Arch. Neerl. Zool. 12, 105–318.

Baker, J.A., Wund, M.A., Heins, D.C., King, R.W., Reyes, M.L., Foster, S.A., 2015. Life history plasticity in female threespine stickleback. Heredity 115, 322–334.

Barber, I., Arnott, S.A., 2000. Split-clutch IVF: a technique to examine indirect fitness consequences of mate preferences in sticklebacks. Behaviour 137, 1129–1140.

Barber, I., Nairn, D., Huntingford, F.A., 2009. Nests as ornaments: revealing construction by male sticklebacks. Behav. Ecol. 12 (4), 390–396.

Barber, L., Ruxton, G.D., 2000. The importance of stable schooling: do familiar sticklebacks stick together? Proc. Biol. Sci. R. Soc. 267, 151–155.

Baube, C., 1997. Manipulations of signalling environment affect male competitive success in three-spined sticklebacks. Anim. Behav. 53 (4), 819–833.

Bell, M.A., Foster, S.A., 1994. The Evolutionary Biology of the Threespine Stickleback. Oxford University Press, New York.

Bell, A., Stamps, J., 2004. Development of behavioral differences between individuals and populations of sticklebacks, *Gasterosteus aculeatus*. Anim. Behav. 68 (6), 1339–1348.

Bensky, M.K., Paitz, R., Pereira, L., Bell, A.M., 2017. Testing the predictions of coping styles theory in threespined sticklebacks. Behav. Process 136, 1–10.

Boughman, J.W., 2001. Divergent sexual selection enhances reproductive isolation in sticklebacks. Nature 411, 944–946.

Breacker, C., Barber, I., Norton, W.H.J., McDearmid, J.R., Tilley, C.A., 2016. A low-cost method of skin swabbing for the collection of DNA samples from small laboratory fish. Zebrafish 14 (1).

Brydges, N.M., Boulcott, P., Ellis, T., Braithwaite, V.A., 2009. Quantifying stress responses induced by different handling methods in three species of fish. Appl. Anim. Behav. Sci. 116, 295–301.

Busquet, F., Strecker, R., Rawlings, J.M., Belanger, S.E., Braunbeck, T., Carr, G.J., Cenijn, P., Fochtman, P., Gourmelon, A., Hübler, N., Kleensang, A., Knöbel, M., Kussatz, C., Legler, J., Lillicrap, A., Martínez-Jerónimo, F., Polleichtner, C., Rzodeczko, H., Salinas, E., Schneider, K.E., Scholz, S., van den Brandhof, E.J., van der Ven, L.T.M., Walter-Rohde, S., Weigt, S., Witters, H., Halder, M., 2014. OECD validation study to assess intra- and inter-laboratory reproducibility of the zebrafish embryo toxicity test for acute aquatic toxicity testing. Regul. Toxicol. Pharmacol. 69, 496–511.

Colosimo, P.F., Peichel, C.L., Nereng, K., Blackman, B.K., Shapiro, M.D., Schluter, D., Kingsley, D.M., 2004. The genetic architecture of parallel armor plate reduction in threespine sticklebacks. PLoS Biol. 2, 635–641.

Dingemanse, N.J., Wright, J., Kazem, A.J.N., Thomas, D.K., Hickling, R., Dawnay, N., 2007. Behavioural syndromes differ predictably between 12 populations of three-spined stickleback. J. Anim. Ecol. 76, 1128–1138.

Divino, J., Schultz, E., 2015. Juvenile Threespine Stickleback Husbandry: Standard Operating Procedures of the Schultz Lab.

Dukowska, M., Grzybkowska, M., Marszał, L., Zieba, G., 2009. The food preferences of three-spined stickleback, *Gasterosteus aculeatus* L., downstream from a dam reservoir. Oceanol. Hydrobiol. Stud. 38, 39–50.

Ellis, T., James, J., Scott, A., 2005. Branchial release of free cortisol and melatonin by rainbow trout. J. Fish. Biol. 67 (2), 535–540.

Ellis, T., James, J., Stewart, C., 2004. A non-invasive stress assay based upon measurement of free cortisol released into the water by rainbow trout. J. Fish. Biol. 65 (5), 1233–1252.

Fernandes Silva, P., de Leaniz, C.G., Luchiari, A.C., 2019. Corrigendum to 'Fear contagion in zebrafish: a behaviour affected by familiarity'. Anim. Behav. 153, 95–103.

Frommen, J.G., Bakker, T.C.M., 2006. Inbreeding avoidance through non-random mating in sticklebacks. Biol. Lett. 2 (2), 232–235.

Giles, N., 1984. Development of the overhead fright response in wild and predator-naive three-spined sticklebacks, *Gasterosteus aculeatus* L. Anim. Behav. 32, 276–279.

Griffiths, R., Orr, K.J., Adam, A., Barber, I., 2000. DNA sex identification in the three-spined stickleback. J. Fish Biol. 57, 1331–1334.

Ishikawa, A., Kabeya, N., Ikeya, K., Kakioka, R., Cech, J.N., Osada, N., Leal, M.C., Inoue, J., Kume, M., Toyoda, A., Tezuka, A., Nagano, A.J., Yamasaki, Y.Y., Suzuki, Y., Kokita, T., Takahashi, H., Lucek, K., Marques, D., Takehana, Y., Naruse, K., Mori, S., Monroig, O., Ladd, N., Schubert, C.J., Matthews, B., Peichel, C.L., Seehausen, O., Yoshizaki, G., Kitano, J., 2019. A key metabolic gene for recurrent freshwater colonization and radiation in fishes. Science 364 (6), 886–889.

Jakobsson, S., Borg, B., Haux, C., Hyllner, S.J., 1999. An 11-ketotestosterone induced kidney-secreted protein : the nest building glue from male three-spined stickleback, *Gasterosteus aculeatus*. Fish Physiol. Biochem. 20, 79–85.

Jones, F.C., Grabherr, M.G., Chan, Y.F., Russell, P., Mauceli, E., Johnson, J., Swofford, R., Pirun, M., Zody, M.C., White, S., Birney, E., Searle, S., Schmutz, J., Grimwood, J., Dickson, M.C., Myers, R.M., Miller, C.T., Summers, B.R., Knecht, A.K., Brady, S.D., Zhang, H., Pollen, A.a., Howes, T., Amemiya, C., Baldwin, J., Bloom, T., Jaffe, D.B., Nicol, R., Wilkinson, J., Lander, E.S., Di Palma, F., Lindblad-Toh, K., Kingsley, D.M., 2012. The genomic basis of adaptive evolution in threespine sticklebacks. Nature 484, 55–61.

Kaitetzidou, E., Katsiadaki, I., Lagnel, J., Antonopoulou, E., Sarrapoulou, E., 2019. Unravelling paralogous gene expression dynamics during three-spined stickleback embryogenesis. Sci. Rep. 9, 3752.

Katsiadaki, I., Morris, S., Squires, C., Hurst, M.R., James, J.D., Scott, A.P., 2006. Use of the three-spined stickleback (*Gasterosteus aculeatus*) as a sensitive *in vivo* test for detection of environmental antiandrogens. Environ. Health Perspect. 114, 115–121.

Katsiadaki, I., Sanders, M., Sebire, M., Nagae, M., Soyano, K., Scott, A.P., 2007. Three-spined stickleback: an emerging model in environmental endocrine disruption. Environ. Sci. 14, 263–283.

Katsiadaki, I., Sanders, M.B., Henrys, P.A., Scott, A.P., Matthiessen, P., Pottinger, T.G., 2012. Field surveys reveal the presence of anti-androgens in an effluent-receiving river using stickleback-specific biomarkers. Aquat. Toxicol. 122–123, 75–85.

Katsiadaki, I., Scott, A.P., Hurst, M.R., Matthiessen, P., Mayer, I., 2002. Detection of environmental androgens: a novel method based on enzyme-linked immunosorbent assay of spiggin, the stickleback (*Gasterosteus aculeatus*) glue protein. Environ. Toxicol. Chem. 21, 1946–1954.

Kingsley, D., 2003. Sequencing the Genome of Threespine Sticklebacks (*Gasterosteus aculeatus*). National Human Genome Research Institute (NHGRI), Bethesda, MD.

Kingsley, D.M., Peichel, C.L., 2007. The molecular genetics of evolutionary change in sticklebacks. In: Ostlund-Nilsson, S., Mayer, I., Huntingford, F.A. (Eds.), Biology of the Three-Spine Stickleback. CRC Press, pp. 41–81.

Kitano, J., Lema, S.C., Luckenbach, J.A., Mori, S., Kawagishi, Y., Kusakabe, M., Swanson, P., Peichel, C.L., 2010. Adaptive divergence in the thyroid hormone signaling pathway in the stickleback radiation. Curr. Biol. 20 (23), 2124–2130.

Kitano, J., Mori, S., Peichel, C., 2007. Phenotypic divergence and reproductive isolation between sympatric forms of Japanese threespine sticklebacks. Biol. J. Linn. Soc. (91), 671–685.

Kitano, J., Mori, S., Peichel, C.L., KItano, J.U.N., Mori, S., Peichel, C.L., 2007. Sexual dimorphism in the external morphology of the threespine stickleback (*Gasterosteus aculeatus*). Copeia 2007, 336–349.

Kitano, J., Ross, J.A., Mori, S., Kume, M., Jones, F.C., Chan, Y.F., Absher, D.M., Grimwood, J., Schmutz, J., Myers, R.M., Kingsley, D.M., Peichel, C.L., 2009. A role for a neo-sex chromosome in stickleback speciation. Nature 461, 1079–1083.

Kozak, G.M., Boughman, J.W., 2009. Learned conspecific mate preference in a species pair of sticklebacks. Behav. Ecol. 20, 1282–1288.

Kraak, S., Bakker, T., Mundwiler, B., 1999. Sexual selection in sticklebacks in the field: correlates of reproductive, mating, and paternal success. Behav. Ecol. 10 (6), 696–706.

Leinonen, T., Cano, J.M., Merilä, J., 2011. Genetic basis of sexual dimorphism in the threespine stickleback *Gasterosteus aculeatus*. Heredity 106, 218–227.

Mork, O.I., Gulbrandsen, J., 1994. Vertical activity of four salmonid species in response to changes between darkness and two intensities of light. Aquaculture 127, 317–328.

Noga, E., 2010. Fish Diseases: Diagnosis and Treatment. Wiley-Blackwell, Ames, IA.

OECD, 2011. Fish Sexual Development Test, OECD Test Guideline 234, Adopted 28th July 2011.

Östlund-Nilsson, S., Holmlund, M., 2016. The artistic three-spined stickleback (*Gasterosteous aculeatus*). Behav. Ecol. Sociobiol. 53, 214–220.

Östlund-Nilsson, S., Mayer, I., Huntingford, F.A., 2006. Biology of the Three-Spined Stickleback. CRC Press, Boca Raton.

Peichel, C.L., Nereng, K.S., Ohgi, K.A., Cole, B.L.E., Colosimo, P.F., Buerkle, C.A., Schluter, D., Kingsley, D.M., 2001. The genetic architecture of divergence between threespine stickleback species. Nature 414, 901–905.

Peichel, C.L., Ross, J.A., Matson, C.K., Dickson, M., Grimwood, J., Schmutz, J., Myers, R.M., Mori, S., Schluter, D., Kingsley, D.M., 2004. The master sex-determination locus in threespine sticklebacks is on a nascent Y chromosome. Curr. Biol. 14, 1416–1424.

Purser, J., Radford, A.N., 2011. Acoustic noise induces attention shifts and reduces foraging performance in three-spined sticklebacks (*Gasterosteus aculeatus*). PLoS One 6 (2), e17478.

Scott, A., Ellis, T., 2007. Measurement of fish steroids in water - a review. Gen. Comp. Endocrinol. 153 (1–3), 392–400.

Sebire, M., Elphinstone Davis, J., Hatfield, R., Winberg, S., Katsiadaki, I., 2015. Prozac affects stickleback nest quality without altering androgen, spiggin or aggression levels during a 21-day breeding test. Aquat. Toxicol. 168, 78–89.

Sebire, M., Katsiadaki, I., Scott, A., 2009. Further refinement of the non-invasive procedure for measuring steroid production in the male three-spined stickleback *Gasterosteus aculeatus*. J. Fish. Biol. 75 (8), 2082–2094.

Sebire, M., Katsiadaki, I., Scott, A., 2007. Non-invasive measurement of 11-ketotestosterone, cortisol and androstenedione in male three-spined stickleback (*Gasterosteus aculeatus*). Gen. Comp. Endocrinol. 152 (1), 30–38.

Shapiro, M.D., Marks, M.E., Peichel, C.L., Blackman, B.K., Nereng, K.S., Jónsson, B., Schluter, D., Kingsley, D.M., 2004. The genetic and developmental basis of evolutionary pelvic reduction in threespine sticklebacks. Nature 248, 717–723.

Strähle, U., Scholz, S., Geisler, R., Greiner, P., Hollert, H., Rastegar, S., Schumacher, A., Selderslaghs, I., Weiss, C., Witters, H., Braunbeck, T., 2012. Zebrafish embryos as an alternative to animal experiments-a commentary on the definition of the onset of protected life stages in animal welfare regulations. Reprod. Toxicol. 33, 128–132.

Thompson, R.R.J., Paul, E.S., Radford, A.N., Purser, J., Mendl, M., 2016. Routine handling methods affect behaviour of three-spined sticklebacks in a novel test of anxiety. Behav. Brain Res. 306, 26–35.

Tinbergen, N., 1952. The curious behavior of the stickleback. Sci. Am. 187, 22–26.

Tilley, C.A., Carreño Gutierrez, H.C., Sebire, M., Obasaju, O., Reichmann, F., Katsiadaki, I., Barber, I., Norton, W.H.J., 2020. Skin swabbing is a refined technique to collect DNA from model fish species. Sci. Rep. 10 (1), 18212.

Tinbergen, N., 1951. The Study of Instinct. Oxford University Press, London.

Vatsos, I.N., 2016. Standardizing the microbiota of fish used in research. Lab. Anim. 51 (4), 353–364.

Wallace, R.A., Selman, K., 2006. Physiological aspects of oogenesis in two species of sticklebacks, *Gasterosteus aculeatus* L. and *Apeltes quadracus* (Mitchill). J. Fish Biol. 14, 551–564.

Ward, A.J.W., Botham, M.S., Hoare, D.J., James, R., Broom, M., Godin, J.-G.J., Krause, J., 2002. Association patterns and shoal fidelity in the three-spined stickleback. Proc. Biol. Sci. R. Soc. 269, 2451–2455.

Westerfield, M., 2007. The Zebrafish Book. A Guide for the Laboratory Use of Zebrafish (*Danio rerio*). University of Oregon Press, Eugene, OR.

Wootton, R.J., 1984. A Functional Biology of Sticklebacks, Functional Biology Series. Croom Helm Ltd, Beckenham.

Wootton, R.J., 1976. The Biology of the Sticklebacks. Academic Press, London.

Chapter 15

Goldfish (*Carassius auratus*): biology, husbandry, and research applications

Ayelén M. Blanco[1,2], Suraj Unniappan[2]
[1]*Department of Bioloxía funcional e ciencias da saúde, University of Vigo, Vigo, Pontevedra, Spain;* [2]*Department of Veterinary Biomedical Science, Western College of Veterinary Medicine, University of Saskatchewan, Saskatoon, SK, Canada*

The goldfish, *Carassius auratus*, is a species native to China that was introduced to the West around the 17th century. For centuries, presumably since the Song dynasty (960–1279), the goldfish has been domesticated and bred in captivity for ornamental purposes. The process of selective crossings over time has led to the development of more than 300 morphologically divergent varieties of the species (Chen, 1956; Smartt, 2001). While the goldfish has been historically considered an ornamental species, its ease of handling, convenient size, and easy adaptation to a range of environments (including laboratory conditions), among other important advantages, have made this species widely used in research (Blanco et al., 2017b). The use of goldfish in scientific work dates back to 1901, when William L. Underwood described its role as a devourer of mosquito larvae (Underwood, 1901). Since then, use of this species in research has increased considerably over the years (Blanco et al., 2017b). This chapter will provide an overview of the current knowledge on the husbandry, care, and research applications of goldfish.

Taxonomy and phylogeny

The goldfish was described as *C. auratus* in 1758 when the Swedish botanist, zoologist, and physician Carl Linnaeus published his book *Systema Naturae*, 10th Edition (Linnaeus, 1758). *C. auratus* is one of six species within the genus *Carassius* and one of more than 4000 species within the order of Cypriniformes (Stout et al., 2016). The species is comprised of several subspecies (e.g., auratus, buergeri, grandoculis, red var., etc.). A complete taxonomic classification of the goldfish is shown in Table 15.1.

Laboratory Fish in Biomedical Research. https://doi.org/10.1016/B978-0-12-821099-4.00012-2
Copyright © 2022 Elsevier Inc. All rights reserved.

TABLE 15.1 Taxonomic classification of the goldfish (*Carassius auratus*).

Kingdom	Metazoa (Animalia)
Superphylum	Deuterostomia
Phylum	Chordata
Subphylum	Vertebrata (Craniata)
Infraphylum	Gnathostomata
Superclass	Osteichthyes (Euteleostomi)
Class	Actinopterygii
Subclass	Neopterygii
Infraclass	Teleostei
Supercohort	Clupeocephala
Cohort	Otomorpha
Subcohort	Ostariophysi
Superorder	Cypriniphysae
Order	Cypriniformes
Suborder	Cyprinoidei
Family	Cyprinidae
Subfamily	Cyprininae
Genus	*Carassius*
Species	*auratus*

Fig. 15.1 shows the phylogenetic position of goldfish within the vertebrate lineage. Goldfish belongs to the ray-finned fishes (Actinopterygii), whose phylogeny is further detailed in Fig. 15.2. Within the ray-finned fishes, the goldfish belongs to the Infraclass Teleostei, which includes more than 25,000 species, representing almost 50% of all living vertebrates (Helfman et al., 2009). Apart from two rounds of whole-genome duplications (1R and 2R) that likely occurred at the base of the vertebrate lineage, the teleost lineage has experienced an additional whole-genome duplication event (3R) (Glasauer and Neuhauss, 2014). Following such event, the teleost genome was originally tetraploid, but a process of rediploidization secondarily returned it to the diploid state (Glasauer and Neuhauss, 2014). Within the teleosts, the Cypriniforms represent the largest group of freshwater fishes, and include species important for aquaculture (such as common carp) and research applications

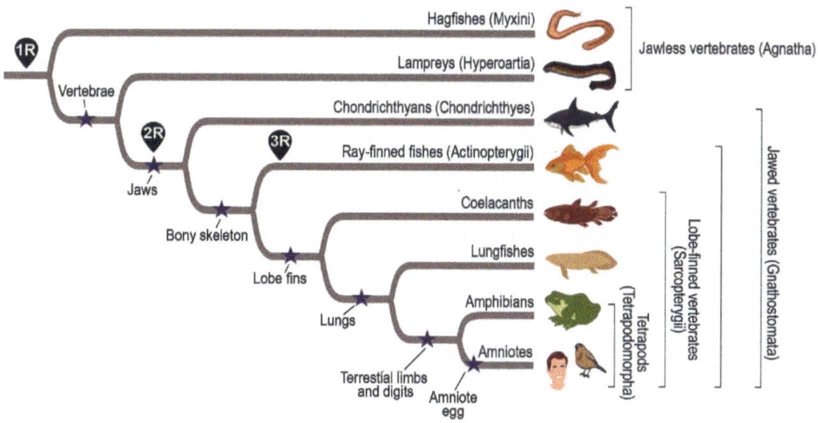

FIGURE 15.1 Simplified phylogeny of the vertebrate lineage. Diagram shows the main vertebrate groups, and indicates key evolutionary events considered for taxonomic classification. Black pin symbols indicate the whole-genome duplication events that occurred during evolution (1R, first round of duplication; 2R, second round; 3R, third round).

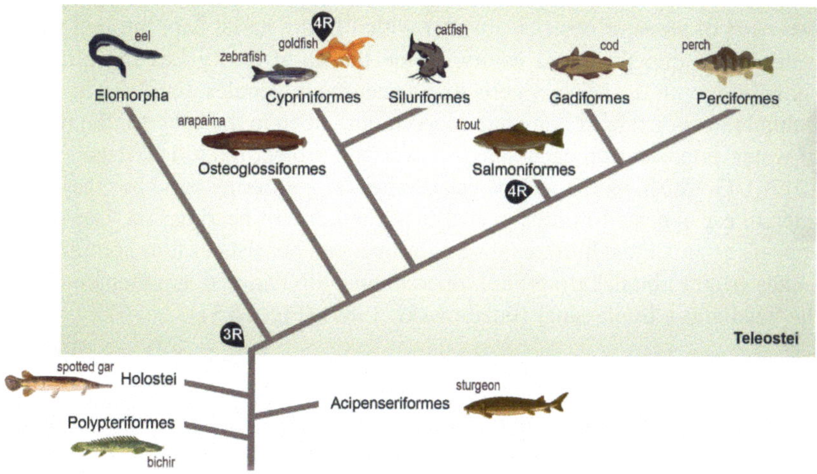

FIGURE 15.2 Simplified phylogeny of the ray-finned fishes (Class Actinopterygii). Diagram shows the main ray-finned fish groups. Black pin symbols indicate the whole-genome duplication events that occurred during the teleost evolution (3R, third round of duplication, common to all teleosts; 4R, fourth round of duplication, lineage-specific).

(such as zebrafish and goldfish). Importantly, the goldfish and carps (Cyprininae subfamily), as well as other teleostean groups (e.g., the Salmonidae Family), underwent a fourth round of whole-genome duplication (4R) later in evolution, and thus possess a tetraploid genome (Lv et al., 2017; Wang et al., 2012).

Morphology/anatomy

External anatomy

The goldfish is one of the fish species with more interspecific variations in morphological traits, existing as more than 300 varieties, each with different external morphological features (Chen, 1956; Ota and Abe, 2016; Smartt, 2001). Our focus here will be to describe the morphology of the common goldfish. The common goldfish has an elongated and laterally compressed body, typically orange or golden, with five sets of fins following the same pattern as most of the Cyprinidae family: dorsal (single), caudal (single, with the dorsal and ventral lobes equal in length), anal (single), ventral or pelvic (pair), and pectoral (pair), with a hard serrate spine at the origin of the dorsal and anal fins. The average size of an adult fish is 10–20 cm (standard length), but can reach up to 45 cm. The body is entirely covered by large, uniform scales, overlapping in a regular pattern. The number of scales along the length of the lateral line is 25–30, and from the base of the dorsal spine to the lateral line, 5–6 (Smartt, 2001). Unlike the common carp, the goldfish lacks barbels. The eyes are situated on both sides of the head and have relatively little movement. Very close to the eyes, on the dorsal surface of the snout, there are two pairs of nares or nostrils, covered with a fleshy narial flap, through which water flows into the nasal cavity. Inside the nasal cavity lies the olfactory epithelium with the sensory cells that detect the molecules dissolved in water. Behind the eyes, a hard bony flap (operculum), open in the back for the release of water, is present on each side to cover and protect the gills (Blanco et al., 2017b). Goldfish, as the rest of fish, do not have external ears, but they have internal ear bones or otoliths in charge of the sense of hearing. The lateral line appears as an almost horizontal straight line, and consists of four commissural canals (supraorbital, infraorbital, operculomandibular, and supratemporal) on the head and a trunk canal (Puzdrowski, 1989) (Fig. 15.3).

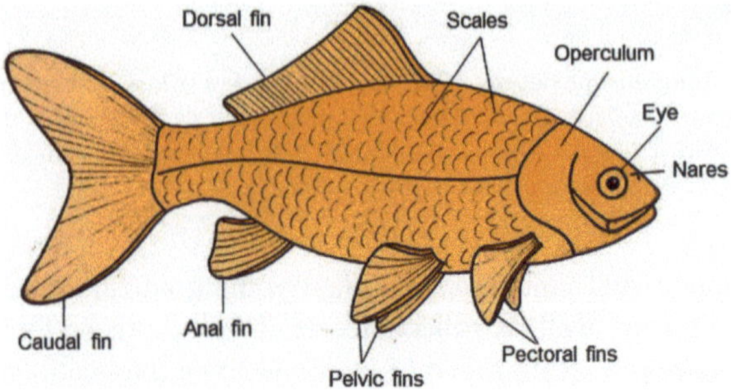

FIGURE 15.3 External anatomy of goldfish. Image was originally drawn for use in this book chapter.

Internal anatomy

Internal goldfish anatomy is made up of all the usual organs of highly developed animals (Fig. 15.4). We will limit this discussion to a brief description of the morphology of the main organs. The brain is a highly complex and organized system divided into discrete areas. Among them, the most anterior region is the prominent olfactory bulbs, which are the primary olfactory structures. The secondary olfactory projections originating in the olfactory bulbs are part of the lateral and medial olfactory tracts, which form relatively long stalks (Rupp et al., 1996). The rest of the telencephalon, the optic tectum, the hypothalamus, and the cerebellum show no great difference to other related fish, including zebrafish. However, the goldfish and zebrafish brain differs mainly in the morphology of the paired vagal lobes: while these lobes in goldfish form a large, separate brain part caudal to the cerebellum, in the zebrafish they consist only of slight protrusions of the dorsal rhombencephalon (Rupp et al., 1996).

Below the brain lies the pituitary gland or hypophysis. Morphologically, the pituitary gland is typically divided into two regions: the adenohypophysis and the neurohypophysis. The adenohypophysis occupies most of the hypophysis and is comprised of three conserved lobes: (1) the proximal pars distalis, containing gonadotrophs and somatotrophs, (2) the rostral pars distalis, containing mainly lactotrophs, and a small number of thyrotrophs and corticotrophs, and (3) the neurointermediate lobe, containing the pars intermedia whose main function is the secretion of melanocyte-stimulating hormone that regulates pigmentation (Baker, 1981; Kaul and Vollrath, 1974). The neurohypophysis is located in the floor of the diencephalon and functions as a neurosecretory structure innervating the adenohypophysis, unlike in the tetrapodean pituitaries (Kleeman and Cutler, 1963). Apart from this, two other important differences between the teleostean and mammalian pituitary glands are worth mentioning: (1) the distribution of endocrine cells within the

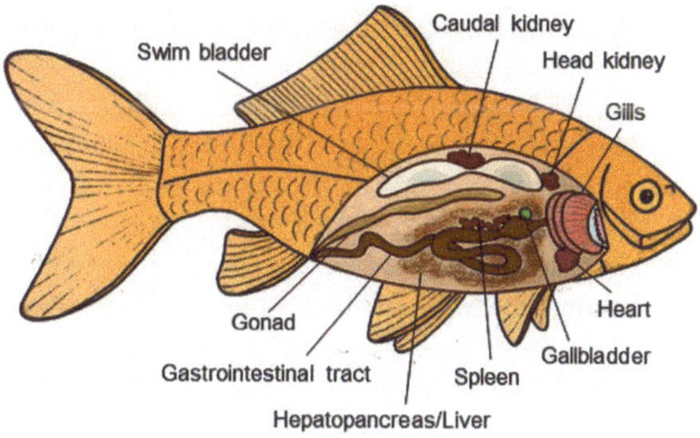

FIGURE 15.4 **Internal anatomy of goldfish.** Image was originally drawn for in this book chapter.

pituitary is highly regionalized in teleosts, while a dispersed organization of endocrine cells is found in mammals; (2) the teleostean pituitary lacks a functional hypothalamo-hypophyseal blood portal system and instead its endocrine cells are directly innervated (Kaul and Vollrath, 1974).

The goldfish heart has the classical four chambers: (1) sinus venous, (2) atrium, (3) ventricle, and (4) bulbus arteriosus. In addition, an atrioventricular region and a muscularized conus arteriosus are present to provide support for the atrioventricular and ventricular-bulbar valves, respectively (Garofalo et al., 2012). The sinus venous of the goldfish heart is mostly made up of connective tissue. The atrium is a single, large chamber, with a complex network of thin trabeculae that divide the atrial cavity into lacunae, progressively smaller moving from the lumen toward the epicardium. The ventricle is sac shaped (rounded and with a blunt apex) and, among the classification of fish ventricle based on the myocardial arrangement, it responds to a type II ventricle, which is characterized for presenting both an external vascularized compacta layer and an inner spongiosa layer. Finally, the bulbus arteriosus is a highly muscularized structure, in continuity with the ventral aorta (Garofalo et al., 2012).

The abdominal cavity of goldfish is mostly occupied by the gastrointestinal tract, an elongated, relatively undifferentiated long tube. The most notable feature of this tract is the lack of a stomach comparable to that of other vertebrates. Instead, a thickened straight tube capable of great expansion, known as the intestinal bulb, replaces the stomach. The intestinal bulb, which extends into the posterior end of the abdominal cavity, is larger and straight at its anterior end, and narrows gradually until it merges into the intestine proper, which occupies the ventral portion of the abdominal cavity (Mcvay and Kaan, 1940). The goldfish digestive tract could thus be divided into mouth, buccal cavity, pharynx, esophagus, intestinal bulb, intestine, and rectum. These regions are, in general, not very well demarcated, but can be distinguished by changes in the characteristics of the epithelial lining, or by valve-like structures at the junctions between the adjoining regions (Sarbahi, 1951). The folds of the intestinal mucous membrane present a zig-zag pattern, characterized by straight lines and angles in the anterior portion, and a convoluted lining in the posterior region (Sarbahi, 1951).

The goldfish gastrointestinal tract is in tight relationship with the liver, which in this species is diffuse and consists of an elongated mass formed by narrow hepatic lobes, right and left, extending on each side of the intestinal bulb. The two lobes join at their anterior and posterior ends to form the anterior median and posterior median hepatic lobes, respectively. Between the right lobe of the liver and the intestinal bulb is the gall bladder (Sarbahi, 1951). Different to the mammalian model, the liver cells of goldfish are not arranged into plates or cords, but they are closely packed between the irregularly distributed capillaries (sinusoids). In addition, no intercellular bile canaliculi are formed in goldfish liver cells; instead, each cell has a single intracellular bile canaliculus, which consists of channels inside the liver cells that collects bile secreted by hepatocytes (Yamamoto, 1965). Embedded in the substance of the goldfish liver is the rudimentary, also diffused, pancreas, thus forming a structure that some authors have referred to as a hepatopancreas. Both bile and pancreatic ducts open on the roof of the intestinal bulb, behind the valve-like structure that separates the bulb from the esophagus (Sarbahi, 1951).

The spleen of goldfish is normally an elongated organ, surrounded by a connective tissue capsule, lying inside the serosal lining of the intestine. The splenic parenchyma contains the same elements as in other vertebrates (blood vessels, ellipsoids, red pulp, lymphoid tissue or white pulp, and macrophages), although the lymphoid tissue is scattered and appears as diffuse layers surrounding arteries and melanomacrophage centers (Fänge and Nilsson, 1985). Goldfish lymphocytes of the spleen (and also of the thymus) possess surface immunoglobulins (Ig) (Warr et al., 1976). Interestingly, goldfish contains only one type of serum Ig, which is composed of polypeptide chains similar to those of IgM from mammals, but exists as a tetramer in the serum of Cyprinidae (Shelton and Smith, 1970). Unlike mammals, the fish splenic tissue does not contain germinal centers (Fänge and Nilsson, 1985).

The goldfish kidney, like in many other fish, is a distinctly bilobed organ consisting of a cranial and a caudal kidney (Ogawa, 1962). The caudal kidney is the actual analog of the mammalian kidney, containing nephrons surrounded by hematopoietic and lymphoid tissue dispersed throughout the organ. Nephrons typically have a well-vascularized glomerulus, a ciliated neck segment, two proximal segments, a ciliated intermediate segment, a distal segment, and a collecting duct system (Reimschuessel, 2001). Instead, the cranial or head kidney, located around the posterior cardinal vein, is an endocrine and immune organ containing endocrine and hematopoietic tissue (Reimschuessel, 2001). The endocrine tissue contains interrenal cells (interrenal tissue; cortical), which produce cortisol, and chromaffin cells (medullary) producing catecholamines, making it the functional analog of the mammalian adrenal gland (Sampour, 2008). The hematopoietic tissue produces lymphoid (B and T lymphocytes) and myeloid (phagocytic) cells, which form an integral analog of the mammalian bone marrow (van Muiswinkel, 1995). In addition, the goldfish head kidney contains thyroid follicles, apart from the typical pharyngeal thyroid (Chavin, 1956). Such follicles respond to physiological alterations similar to those of the normally located thyroid. Indeed, the goldfish head kidney thyroid accumulates about twice as much radioiodine as does the pharyngeal thyroid (Chavin and Bouwman, 1965).

Like other teleosts, the gonads of goldfish lack medullary tissue, and contain only the equivalent to the cortex of other vertebrates. Morphologically, the female gonads (ovaries) consist of a pair of hollow sac-like organs attached to the dorsal body wall. Each ovary is formed of oogonia, oocytes and their surrounding follicle cells, supporting tissue or stroma, and vascular and nervous tissue. Based on the pattern of oocyte development, the goldfish ovary is classified as asynchronous, meaning that it contains oocytes at all stages of development. The teleost ovary has an ovarian cavity continuous with the oviduct, where mature oocytes are discharged before moving to the oviduct. In the goldfish, such cavity is formed by the fusion of the distal edges of the ovary with the dorsolateral coelomic wall (Nagahama, 1983). Male gonads (testes) are elongated paired organs, also attached to the dorsal body wall. Based on the differentiation of the germinal tissue, goldfish testicles, as in most teleosts, are classified as the lobular type, thus being composed of numerous lobules, separated from each other by a thin layer of connective tissue. Within the lobules, primary spermatogonia undergo numerous mitotic divisions to produce cysts containing several spermatogonial cells. As spermatogenesis proceeds, cysts then expand and rupture, and sperm is liberated into the lobular lumen, which is continuous with the sperm duct (Nagahama, 1983). The sperm duct in the testis of goldfish is not a simple duct but a cephalocaudal succession of complicatedly anastomosing cavities (Takahashi and Takano, 1972).

The goldfish is not characterized by large adipose tissue depots, but low levels of fat depots can be found all around the body of the fish (subcutaneous fat) or in the abdominal cavity around the digestive tract (visceral or mesenteric fat) (Weil et al., 2013). Visceral fat is highly dependent on the fish nutritional status. In addition, visceral lipid stores are mobilized for ovarian development in female goldfish (Delahunty and de Vlaming, 1980). Interestingly, goldfish adipocytes have been demonstrated to be glucose-responsive insulin-producing cells, thus having properties similar to that of pancreatic β cells (Blanco et al., 2020a).

Geography and natural habitat

The goldfish is native to East Asia. It was first selectively bred in ancient China more than 1000 years ago, and taken to neighboring countries, including Japan, Korea, and Taiwan, at the very beginning of the 16th century. The first introduction of the goldfish to Europe was to Portugal in the 17th century, and from there it was imported to other European countries. During the course of the 18th century the goldfish had been distributed generally over Europe, and by 1791 it was clearly established in Russia. The late 1850s seems to be the accepted date of introduction of the goldfish in the United States, where it is now established or reported in all states except Alaska (Smartt, 2001). Currently, the goldfish is spread worldwide, with a few exceptions such as Greenland and Antarctica (Brown et al., 2018).

Natural habitats of the species include shallow (up to 20 m deep) slow-moving rivers, lakes, and ponds, with sandy or muddy bottoms with dense submerged vegetation (Robison and Buchanan, 2020). Goldfish also frequently occupy stagnant backwaters that do not experience currents. Its native climate is subtropical to tropical, and it lives preferentially in warm or cold waters (10–21°C) with a pH close to neutrality (7–7.5) (Kottelat and Freyhof, 2007), although it can tolerate more extreme conditions, such as high levels of turbidity (Wallen, 1951), temperature fluctuations (Spotila et al., 1979), low levels of dissolved oxygen (Walker and Johansen, 1977), high salinity (Schwartz, 1964), and even certain types of water pollution (Robison and Buchanan, 2020).

Behavior

Social behavior

Goldfish are gregarious, and typically display shoaling and schooling behavior (Leem et al., 2012). Such behavior of the goldfish has been subject to several studies, which demonstrated that goldfish in larger shoals show less "timid" behavior (for example, lurking in the weeds or rapid darting and turning) and spend more time foraging than those in smaller shoals, probably because of a reduced need for individual vigilance (Magurran and Pitcher, 1983; Pitcher and Magurran, 1983). The goldfish do not form social hierarchies, and can be described as a "friendly" or docile species toward other fish, rarely harming another goldfish or an individual from another species, and exhibiting very little male–male aggression.

Feeding behavior

The goldfish is a benthic species that is characterized by constant swimming, exploring, and/or chasing its food, i.e., it displays a foraging behavior. It preferentially forages in sandy substrates rather than gravel, pebbles, or cobbles (Smith and Gray, 2011). In the wild, goldfish is omnivorous and its diet consists of planktonic organisms, small crustaceans, mollusks, worms, insect larvae, and various plant matter (Muus and Dahlström, 1981; Navarrete Salgado et al., 2009). Laboratory studies on feeding behavior have demonstrated that the goldfish, like many other fish species and mammals, shows increased activity in anticipation of a forthcoming meal, known as feeding anticipatory activity, within a few days after the establishment of a regular feeding cycle (Spieler, 1992). Food sources in nature may vary predictably in both place and time, and so it could be plausible that the anticipation of daily food availability occurs in the wild as well for optimal foraging.

Reproductive behavior

The goldfish has a prolonged breeding season that starts in the late spring and continues throughout the summer as long as water temperatures remain above 15.6°C (Robison and Buchanan, 2020). Reproduction takes place in shallow waters with dense vegetation. During the mating ritual, males chase the females around, bumping and nudging them to prompt them to release her eggs.

A female may be accompanied by several males who fertilize the eggs as they are released. Depending on its size, female goldfish can lay between 4000 and 400,000 eggs. Goldfish eggs are adhesive and attach to aquatic vegetation, typically dense plants such as *Cabomba* or *Elodea*, or roots. Hatching occurs within 48–72 h, and under optimal conditions, fry grow rapidly (Brown et al., 2018). Sexual maturity occurs at 1–3 years of age, and the normal lifespan is 6 or 7 years, although a maximum lifespan of 30 years has been reported (Robison and Buchanan, 2020).

Husbandry

Aquatic facility and housing

In a facility dedicated to goldfish housing, as with all other fish species, the primary housing enclosures are the tanks. Tanks will vary in material, shape, and size depending on the research scope. All these are factors that should be given careful consideration when designing the experimental projects, as they affect the welfare of the fish and the functionality of the life-support system. The most common materials that goldfish holding tanks are made from include glass, thermoplastic polymers (polycarbonate), and acrylic. Glass tanks are the most frequently implemented housing element in small-scale projects, as they are easily available at most pet shops. Glass possesses important advantages as material for tank construction: (1) it is inert, therefore there is no risk of chemical leaching to fish, (2) it is relatively inexpensive, and (3) it does not scratch easily. However, it is heavy and fragile, entailing health and safety challenges to personnel who maintain them. As an alternative to glass, most goldfish tanks in modern facilities are made from thermoplastic polymers, mainly polycarbonate, although others such as polysulfone are also used. Polycarbonate tanks are also relatively inexpensive, and have the significant advantage of being lightweight, so are easily movable within the facility. Nevertheless, their main disadvantage is that they contain bisphenol-A, which has been demonstrated to cause serious reproductive problems in vertebrates, including fish (Canesi and Fabbri, 2015). Howdeshell and coworkers (Howdeshell et al., 2003) demonstrated that polycarbonate tanks leach bisphenol-A into the water; however, levels are minimal in new tanks and significantly higher in old cages, thus suggesting that regular tank replacement may be a reasonable strategy for managing the drawback of this type of tank. Finally, acrylic tanks have become popular for goldfish and other fish research species as they combine the main advantages of both glass and polycarbonate tanks. Thus acrylic is lightweight and, although technically also a thermoplastic polymer, it does not contain bisphenol-A, meaning that no leaching side effect can be encountered. However, certain disadvantages can also be found for acrylic: it is significantly prone to scratching, can be damaged by organic solvents such as ethanol, and has low heat tolerance.

Goldfish tanks are typically rectangular, although, considering the wide tolerance range of this species, other tank designs can be used. Size of tanks is also variable and can be adjusted to the particular needs of the research project. The stocking density is dependent on fish size and the size of tanks available. For instance, for fry (2.20 ± 0.005 cm, 0.33 ± 0.006 g), it has been reported that a stocking density of $300/m^3$ provides the best fish growth (Shete et al., 2013). As a general rule, overcrowding must be avoided (Ababouch, 2014; M'balaka et al., 2012). If required by the experimental design, goldfish can also be housed individually. Priestley and coworkers demonstrated that goldfish kept individually or in groups of 4, 8, or 16 fish/50-L tank show similar growth rate and body condition (Priestley et al., 2006a).

Goldfish tanks can be kept independently, but most aquatic facilities have recirculating aquaculture systems (RAS) that immensely facilitate routine cleaning and disinfection procedures (Badiola et al., 2018). The distinguishing feature of these systems is that the water used to house the fish is cleaned and reused, minimizing water consumption and waste. These systems are equipped with different filters (mechanical, biological, chemical), aeration mechanisms, and an ultraviolet light for disinfection, which ensure an optimal water treatment. RAS aquaria are available at different scales, from "stand-alone" systems (typically in the form of a rack filled with tanks on the top section, and with a compact filtration assembly condensed on the bottom section) to multilinked racks with a common filtration assembly and sump system. If not using a RAS, each goldfish tank should be provided with a filter system, which could be either external or internal. The filter system should be equipped with at least mechanical and biological filters, and should filter an absolute minimum flow rate of five times the volume of the tank per hour. With some RAS, or in individual tanks with some external filters, the waterfall generated during filtration may be able to aerate the water sufficiently. Alternatively, and certainly for individual tanks with internal filters, adding an air pump is highly recommended (Badiola et al., 2018).

In addition to water filtration, two other aspects should be considered in a goldfish housing facility: temperature control and illumination. The temperature in a fish housing system must be controlled and maintained within a stable and optimum range. This can be accomplished through a combination of adjusting the air temperature of the room and the use of specialized water heaters and chillers that regulate water temperature via thermostats. Adequate illumination is facilitated by the built-in lights in the room. If additional lighting is required, fluorescent lamps may be set above individual aquaria. If lights are added to the aquaria, special attention should be paid to the type of light chosen, as some wavelengths may affect study parameters. For instance, it was demonstrated that goldfish show a better growth performance under white light than red light (Imanpoor et al., 2011). A similar observation was reported by Ruchin (2004), who not only demonstrated that goldfish kept under red light show a significantly lower growth rate than those kept under luminescent, white light, but also that keeping goldfish under green light can result in a higher growth rate (Ruchin, 2004). Red light appears to also negatively affect goldfish sex maturation, as suggested by a study of Yun and coworkers, showing that this type of light inhibits the expression of important genes involved in reproduction, such as the kisspeptin system (Yun et al., 2015). Apart from the type of light, another key factor that should be considered when talking about illumination is the photoperiod, i.e., the recurring cycle of uninterrupted light and dark period fish are exposed to. Photoperiod has been reported to affect several physiological and behavioral aspects of goldfish, including feeding, growth, metabolism, oxidative stress, reproduction, thermal resistance, and locomotor activity (López-Olmeda et al., 2006; Sarkar and Upadhyay, 2011; Spieler and Noeske, 1984; Wei et al., 2019). Therefore an appropriate photoperiod should be established according to the scope of the research.

Water quality

Proper water quality is essential to maintain healthy fish colonies. Recommended water quality parameters for goldfish are: temperature 10–30°C (optimal range = 20–24), pH 6.0–8.4 (optimal 7.0–8.0), hardness ($CaCO_3$) 100–250 mg/L, conductivity 180–480 mS/cm, ammonia (total) <0.02 mg/L, nitrite <0.2 mg/L, nitrate <40 mg/L, oxygen 5.0–8.0 mg/L, and chlorine 0.002 mg/L (Jepson, 2016). Among these parameters, probably the most critical factor in maintaining fish health is inadequate oxygen levels. Water contains much less oxygen than air, particularly at high water temperatures. To prevent stress, water oxygen levels should be higher than 5 ppm. Next to oxygen, the most limiting factor in fish culture is toxic nitrogen compounds. Nitrogen compounds are introduced into the water mainly via fish waste (excretion across the gills) or excessive uneaten food. Such compounds can be present in two forms, as the relatively nontoxic ionized ammonium (NH_4) or as toxic nonionized ammonia (NH_3), depending on the water temperature and pH (NH_4 will predominate in water with low temperature and pH, while NH_3 increases with elevated temperature and pH). As ammonia is formed, bacteria called *Nitrosomas* will oxidize this ammonia to nitrite (NO_2^-). Nitrite is also toxic to fish, although to a lesser extent than ammonia; fish can withstand roughly twice the amount of nitrite in their water when compared to ammonia. Nitrite is then converted into nitrate (NO_3^-) by bacteria called *Nitrobacter*. Nitrate is not as harmful to fish as ammonia or nitrite, but it is still harmful in large amounts. Under anaerobic conditions, nitrate can be turned into harmless nitrogen gas. However, the parameters needed to create this condition are not commonly present in most aquaria, and hence, periodic partial water changes are the easiest and best way to keep low levels of nitrate. Levels of nitrogenous compounds, as well as of dissolved oxygen and other parameters, such as pH, alkalinity, hardness, and chlorine, should be checked frequently using strips, test kits, and/or electronic probes. In an established system, weekly monitoring is adequate.

Export and transportation

Goldfish are usually obtained from commercial suppliers that breed fish at large scale to export them to both pet shops for ornamental purposes or aquatic facilities for research applications. Appropriate transportation of fish is vital to ensure that fish arrive healthy and in good condition to the destination. Fish transport is a lengthy process that consists of a series of sequential actions, including harvesting, packing, actual transport, unpacking, and stocking. There are basically two major types of transporting live fish: in tanks (the so-called *open system*) or packed in polyethylene bags (*closed system*) (Lim et al., 2007).

The most common system for transporting fish to research aquatic facilities is the closed system, in which all factors necessary to meet fish survival are self-sustained (Lim et al, 2003, 2007). It consists of packing the fish in sealed polyethylene bags filled with overoxygenated water that is usually pretreated with chemicals or drugs. Bags used for this system have a bottom with either a rectangular base or a seam. In the latter case, the two bottom corners are usually tied together with rubber bands or heat sealed to avoid fish being trapped or squashed in the corners. Goldfish are packed at a relatively high density for transportation. However, the number of fish per bag is highly dependent on the volume of transport water, size of the fish, and transport time. For instance, for consignments from Singapore to Europe with a 30-h transit time, the average loading density of goldfish (average body weight 13.8 g) is 272 g/L (Lim et al., 2003). Since goldfish can tolerate relatively low temperatures (Ford and Beitinger, 2005), they are usually packed at 15−18°C, so that their metabolic rate is low. To improve the survival of fish during transport, fish are usually starved prior to transport to prevent regurgitation of partially digested food, minimize water pollution, and reduce stress during transport (Nemato, 1957). The optimal duration of starvation for goldfish is 2 days (Lim et al, 2003, 2007). Once fish are placed in the bags, an adequate supply of dissolved oxygen is added, which also depends on the number and size of fish and the duration of the transport. For instance, an oxygen-to-water volume ratio ranging from 4:1 to 6:1, or an oxygen-to-fish ratio of 14 mL oxygen/g fish, is commonly used for air transport of goldfish (Lim et al, 2003, 2007). Then, the opening of the bag is closed, either by heat sealing or using rubber (elastic) bands or a metal clip. After packing, the bags with fish are placed in a styrofoam (polystyrene) box (usually four to eight bags per box), which serves to provide thermal insulation to prevent sudden changes in the water temperature. The styrofoam box is usually placed inside a cardboard box of similar size for additional safety.

Once fish arrive at their destination, acclimation to the new tanks and recovery from the transport are two important operations. Acclimation includes floating the sealed bags in the new housing tanks until the temperature of the transport water equals that of the housing water. Then, fish should be hand netted and introduced in the housing tanks, either directly or after being first transferred to a basin of clean water. Water quality is vital in the post-shipment recovery, and all quality parameters should be as close as possible to the optimal value for the fish. Adding salt to the water has been proven to be an effective way to improve post-transport survival of fish (Lim et al, 2003, 2007). Indeed, goldfish has been reported to show good growth, food consumption rate, and no signs of stress when kept in waters with a salinity of up to 6% for long periods (Luz et al., 2008). The first day after arrival, it is normal that fish are unwilling to eat, but appetite should be recovered by the second or third day.

Handling

Proper handling of fish is essential, both to prevent injury and to reduce stress. Even minimal handling causes a profound stress response in fish, which, apart from likely influencing the response of fish to experimental manipulations, may weaken the endocrine and immune systems, predisposing fish to infection, ultimately causing disease or even death (Wendelaar Bonga, 1997). In goldfish, it was demonstrated that experimental handling stress affects the gills and skin integrity, facilitating ulcerative diseases (Dror et al., 2006). To minimize stress, all fish handling procedures must be quick and carried out cautiously. Conscious fish can be captured from their tanks using a net, plastic bag, or jar, avoiding chasing as much as possible. After capture, fish must be held firmly, but gently, without squeezing and avoiding touching the eyes and gills, which are very sensitive. Fish can be held with bare hands or wearing gloves, previously moistened in both cases. Abrasive materials such as paper towels should be avoided. The time that fish is out of the water must be minimal; if prolonged, to moisten fish with water is highly recommended.

Feeding/nutrition

In captivity, goldfish are commonly fed dried flake or pellet food, typically obtained commercially. An example of the complete composition of a typical goldfish custom regular diet (Blanco et al., 2016) is shown in Table 15.2. A nutritional study on the utilization of a commercial food by goldfish demonstrated that this species consumes food at 3.6% of body weight, and that the satiation and gastric evacuation times last for 93 min and 22 h, respectively (Al-Dubakel et al., 1999). While dried flakes or pellets constitute the most common diet for goldfish, it has been demonstrated that goldfish fed *Artemia* for 60 days show superior zootechnical performance, including higher weight, total length, weight gain, and specific growth rate, than those fed commercial flakes (Moreira et al., 2011). Therefore adding live food supplements such as *Artemia*, *Daphnia*, worms, mosquito larva, or similar to the goldfish diet is recommended if the research applications require a rapid fish growth. In

TABLE 15.2 Composition of a typical commercial custom diet for goldfish (Blanco et al., 2016).

Protein (%)	**32.3**	**Minerals**	
Arginine (%)	1.66	Ash (%)	4.9
Histidine (%)	0.81	Calcium (%)	1.11
Isoleucine (%)	1.73	Phosphorus (%)	1.21
Leucine (%)	2.75	Phosphorus (available) (%)	1.09
Lysine (%)	2.44	Potassium (%)	0.11
Methionine (%)	0.90	Magnesium (%)	0.03
Cystine (%)	0.17	Sulfur (%)	0.21
Phenylalanine (%)	1.49	Sodium (%)	0.48
Tyrosine (%)	1.28	Chloride (%)	0.20
Threonine (%)	1.34	Fluorine (ppm)	18.0
Tryptophan (%)	0.33	Iron (ppm)	169
Valine (%)	2.16	Zinc (ppm)	181
Alanine (%)	1.26	Manganese (ppm)	86
Aspartic acid (%)	2.33	Copper (ppm)	13
Glutamic acid (%)	5.09	Cobalt (ppm)	0.27
Glycine (%)	1.90	Iodine (ppm)	1.52
Proline (%)	2.12	Chromium (ppm)	0.48
Serine (%)	1.43	Molybdenum (ppm)	n/s
Taurine (%)	0.04	Selenium (ppm)	0.29
Fat (ether extract) (%)	**6.0**	**Vitamins**	
Fat (acid hydrolysis) (%)	**6.2**	Carotene (ppm)	0.0
Cholesterol (ppm)	405	Vitamin A (IU/g)	9
Linoleic acid (%)	0.05	Vitamin D3 (added) (IU/g)	2.4
Linolenic acid (%)	0.07	Vitamin E (IU/kg)	105
Arachidonic acid (%)	0.08	Vitamin K (as menadione) (ppm)	4.7
Omega-3 fatty acids (%)	1.34	Thiamine hydrochloride (ppm)	36
Total saturated fatty acids (%)	2.18	Riboflavin (ppm)	34.2
Total monounsaturated fatty acids (%)	1.43	Niacin (ppm)	204

Continued

TABLE 15.2 Composition of a typical commercial custom diet for goldfish (Blanco et al., 2016).—cont'd

Polyunsaturated fatty acids (%)	1.06	Pantothenic acid (ppm)	99
Fiber (max) (%)	**6.8**	Folic acid (ppm)	6.6
Neutral detergent fiber[1] (%)	6.1	Pyridoxine (ppm)	28.24
Acid detergent fiber[2] (%)	5.4	Biotin (ppm)	0.4
Nitrogen-free extract (by difference) (%)	**40.1**	Vitamin B12 (mcg/kg)	72
Starch (%)	24.25	Choline chloride (ppm)	1228
Glucose (%)	0.00	Ascorbic acid (ppm)	200.2
Fructose (%)	0.00	**Energy (kcal/g)**	**3.44**
Sucrose (%)	0.00	From protein	1.290 kcal 37.6%
Lactose (%)	0.02	From fat (ether extract)	0.541 kcal 15.8%
Total digestible nutrients (%)	62.4	From carbohydrates	1.604 kcal 46.7%

[1]NDF, approximately cellulose, hemicellulose and lignin.
[2]ADF, approximately cellulose and lignin.

addition, algae-based feed, in particular fresh algal biomass of *Nostoc ellipsosporum* and *Navicula minima*, has also been reported as beneficial for body weight gain and specific growth rate in goldfish (Khatoon et al., 2010). For larvae, an exclusive feeding with *Artemia* during the first 2 weeks, followed by feeding a mixed diet of 50% *Artemia* and 50% dry feed during the consecutive weeks, was observed to lead to optimal growth results (Abi-Ayad and Kestemont, 1994). In an aquatic facility, goldfish are typically fed once daily. However, frequency of feeding can be increased up to six times/day to maximize utilization of the food ration and to improve the growth rate, as demonstrated by Priestley et al. (2006b).

Maintenance/cleaning and disinfection

A large part of keeping fish healthy is ensuring that their habitat and environment remain clean, healthy, and safe. Thus, following a regular maintenance routine is essential in every aquatic facility. Specific points to consider during maintenance vary, and include procedures to be carried out daily, weekly, or just periodically (monthly, bimonthly, yearly). Keeping records of each procedure on an aquarium journal or log is highly recommended. It will keep the maintenance organized when multiple people are working in the facility and, more importantly, it could help diagnose problems with the aquaria and any health issue that may appear.

Daily maintenance procedures include:

- Visual equipment check to ensure the filters and any other equipment (e.g., air pumps, lights, etc., if present) are running properly.
- Visual examination of the water quality (transparency/turbidity).
- Check water levels in tanks and, for RAS, also in the sump tank. If levels are low, they should be topped off with treated or aged water.
- Temperature check to ensure it is in the proper range.
- Fish counting and visual check of their physical appearance. Any unusual behavior, such as low activity, should also be monitored.
- Removal of uneaten food.

Weekly maintenance procedures include:

- Cleaning outside surfaces of tanks with a damp cloth and scrapping inside surfaces with a clean sponge or using a magnetic glass cleaner. Cleaning agents and chemicals should be avoided.
- Filter cleaning. As with the tanks, soap and chemical agents are not to be used; instead, each filter should be removed and rinsed with water.
- Partial water changes. If using independent tanks, a partial volume of water from each tank ($\approx$10%–15%) should be replaced by treated and, if possible, aged water. In RAS, fresh water should be incorporated to replace a large volume of water from the sump tank.
- Testing of water quality parameters (pH, dissolved oxygen, levels of ammonia, nitrite and nitrate, etc.) using strips, test kits, and/or electronic probes.

Periodic maintenance procedures include:

- Change of filters and/or filter media.
- Calibration of pH meter, O_2 probes, etc.

The procedures explained so far describe general protocols for routine maintenance. However, on occasion, tanks and aquatic facilities need to be sanitized because of an infection or disease. One of the most common disinfectants used for fish aquaria is bleach. While the protocol to sanitize a fish tank using bleach is variable, a common practice includes rinsing all infected equipment and tanks in a 10% bleach solution. Bleach is very dangerous to fish, so fish must be placed in a separate tank while sanitizing the dirty tanks. For the same reason, tanks and all equipment must be rinsed two or three times to remove any remaining bleach residue before reuse with fish again.

Breeding

Goldfish are not a common model fish used for breeding in the scientific field, as breeding this species is a long and quite difficult process. Nevertheless, goldfish can be bred in captivity if all appropriate conditions are met (Smartt and Bundell, 1996). The goldfish will become sexually mature in 1 year but will reach its prime in its third year. The first step in the breeding program is to create the right tank environment, which includes preparing a 75-L (at a minimum) tank with spawning mops or similar at the bottom. Then, goldfish must be sexed and distributed into the tank. Sexing goldfish is not simple, but there are some distinctive features between males and females during the breeding season that can be used to differentiate them. Thus females are rounder than males, they have round and short pectoral fins and swollen abdomens, and their vent protrudes when ready to spawn. On the other hand, males have pointier and longer pectoral fins, breeding tubercles or white stars covering their gills and pectoral fins, and their vent becomes more open and longer when ready to spawn. Ideally, more males than females should be introduced in the tank (e.g., three males and two females), although a single pair is also appropriate. Because goldfish breeding in the wild is triggered by the change in temperature from the cold winter to the warmer spring, simulating such temperature change is essential for inducing breeding in captivity. Thus water temperature needs to be dropped first to 10−12°C, and then, when breeding needs to be induced, it must be raised gradually by 2°C per day, until it is 20−23°C. Under these conditions, spawning behavior and natural spawning will likely occur. If natural spawning proves unsuccessful, a gentle rub can be applied to the vent of a male and a female to release the sperm and eggs, respectively, which will then be combined in swirled water. Goldfish in captivity have a tendency to eat the eggs. This makes it necessary to separate the parents from their eggs almost immediately to ensure a full batch of eggs is hatched successfully. Fertilized eggs should hatch within 4−7 days, depending on the temperature of the water (Smartt, 2001; Smartt and Bundell, 1996).

Diseases/pathogens/treatment

Healthy goldfish have a bright consistent color, with shiny, even, and clean scales that reflect the light well, clear eyes, and upright fins without rips or tears. They swim gracefully, eagerly gobble food at regular times and respond to changes in vibration and light. If these signs are not met, it is likely that goldfish are ill. Most of the time, health problems are accompanied by visible signs of illness that help diagnose the disease, and thus apply the appropriate treatment. The following is a list with the most common goldfish diseases, including their symptoms and most effective ways of treatment (Bassleer, 2009; Jepson, 2016). Most of the fish diseases are the result of poor water quality; therefore maintaining water in healthy conditions is the best way to prevent any disease.

Parasitic infections

Ich/white spot disease

Ichthyophthirius multifiliis, often termed "Ich," is a parasitic ciliate that represents the most common goldfish disease in freshwater aquaria (Sharma et al., 2014; Thilakaratne et al., 2003). Ich is usually caused by poor water quality, overcrowded aquaria, or when a sudden temperature change occurs. This disease also occurs when fish are stressed or when a new and infected fish is added to the community.

Symptoms: Clamped fins, small white spots on body, fins, and gills, irritation and inflammation that will make the goldfish itch against the side of the tank.

Treatment: Gradually raise the water temperature up to 24–27°C to speed up the ich's life cycle, and carry out a 0.3% salt treatment during 10 days.

Anchor worm

Anchor worms (*Lernaea* spp.) are external copepod parasites that attach to fish under the scales (Thilakaratne et al., 2003). This disease is caused by newly added fish already carrying the parasite or newly added plants that can have larvae on them.

Symptoms: Itching and flashing (rubbing against objects or walls in the aquarium), then visible parasite attached to the body.

Treatment: Remove any visible worms with tweezers and clean the wounds with hydrogen peroxide to help prevent infection. Then, bath all fish in the tank in water containing potassium permanganate (10 mg/L water) for 20–30 min. Transfer fish to clean water.

Flukes

Flukes are monogenean parasites. Freshwater fish, including goldfish, are commonly affected by monogeneans of the genera *Dactylogyrus* and *Gyrodactylus*. *Dactylogyrus* predominantly affects the gills, whereas *Gyrodactylus* is more commonly found on the skin (Mayer and Donnelly, 2013a). Flukes are usually introduced into the tank when new fish or plants that can carry eggs or the parasite itself are added.

Symptoms: Flashing, tiny red spots or yellow dusting, excessive production of mucus, shedding of the slime coat, clumped fins. When branchial infestations are present, respiratory signs (increased opercular rate, piping, and respiratory distress) may be seen.

Treatment: A number of treatments are available for flukes, but those containing praziquantel are the most effective. Remove carbon from the filter and add 1 g of praziquantel per 378 L of water. Repeat treatment twice more, spaced 4 or 5 days apart.

Fish lice

Fish lice (*Argulus* spp.) are branchiuran crustaceans that parasitize both marine and freshwater fishes (Wafer et al., 2015). As with anchor worms and flukes, usual causes for fish lice are new unquarantined fish or plants added to the aquarium.

Symptoms: Itching, visible parasites (seen as little green specks) moving around the fish in protected areas such as behind the fins, near the eyes or gills.

Treatment: Remove any visible lice with tweezers and disinfect the wounds with hydrogen peroxide. Then, treat the water in the tank. The most effective treatment against fish lice is organophosphates, which usually are given as two or three doses at 1-week intervals to kill emerging larvae and juveniles. For adults, diflubenzuron (Dimilin, Chemtura) has been proved an effective treatment, but it is a restricted-use pesticide. Other compounds with a similar mechanism of action to that of diflubenzuron, such as lufenuron (0.13 mg/L), have also been used successfully (Wafer et al., 2015).

Velvet disease

This is also known as "Rust Dust" or "Gold Dust" disease. Velvet is caused by tiny parasites of the genera *Oodinium* that give the fish a dusty, slimy look. Introduced to tanks by unquarantined infected new goldfish, this disease is highly contagious and can be easily spread by contaminated tanks, fish, or tools.

Symptoms: Fine rusty yellowish film on skin, slime shedding, clamped fins, and scratching.

Treatment: Gradually raise water temperature up to 24–27°C, dim lights (*Oodinium* is dependent on light), and treat water with 0.3% salt and 0.25 mg/L copper sulfate for 10 days.

Trichodiniasis

This disease is caused by opportunistic ciliophoran parasites (*Trichodina* spp.), usually found on the gills, skin, or fins, which infect a wide range of fish hosts. These parasites do not feed on the goldfish, but instead they use fish as home and transportation. However, in severe conditions and in large numbers, *Trichodina* can compromise the goldfish's immune system, promoting other complications, such as ulcers or other secondary bacterial infections (Sharma et al., 2014).

Symptoms: Flashing, irritation, ulcers, loss of appetite.

Treatment: Treat water with 0.3% salt. Medicated treatments, like water containing potassium permanganate (10 mg/L water), have also proven effective.

Costiasis

Protozoan flagellates of the genus *Ichthyobodo*, formally known as Costia, are known to attach to the skin and gill membrane of goldfish causing costiasis (Sharma et al., 2014). This disease can be harmless if protozoans are present in small numbers, but if a population explosion occurs (usually when the goldfish's immune system is weakened) it can kill fish quickly.

Symptoms: Slimy patches around the head and gills, small hemorrhages under chin, awkward swimming, gasping at the surface of the water.

Treatment: Gradually raise water temperature and treat water with 0.3% salt. If problem persists, formalin (50 ppm) or praziquantel (2 mg/L) can be added to the water.

Chilodonellasis

This disease is caused by *Chilodonella* spp., a prevalent ciliated protozoan. Like Costia, this microorganism can lie dormant for long periods but can be harmful if an outbreak in population occurs when a goldfish has a weak immune system (Bastos Gomes et al., 2017).

Symptoms: Web-like hemorrhages on skin, excessive slime production, clamped fins, gasping at the surface.

Treatment: Aquarium salt (0.3%) is the most effective *Chilodonella* treatment. Formalin and potassium permanganate are alternative options.

Bacterial infections

Fin rot

Fin rot is one of the most common diseases in aquarium fish, but it is also one of the most preventable and easy to cure if detected at an early stage. Although it can also be caused by certain types of fungus, it is usually caused by bacteria, typically *Aeromonas*, *Pseudomonas*, *Flovobacterium*, or *Vibrio*. The most common cause of fin rot is poor water quality, but overcrowding, rough handling, fighting, or very low temperatures may also lead to this disease in goldfish (Sharma et al., 2014).

Symptoms: Split and/or frayed fins, often with red streaks on them and a white edge.

Treatment: The most common treatment for fin rot is the addition of Melafix to the water (5 mL/40 L water). Melafix (a proprietary name) is an all-natural aquarium antibiotic containing an essential oil from a tree in the *Melaleuca* genus. Other compounds, such as malachite green or methylene blue, are also used to treat this disease.

Mouth rot

This disease, caused by bacteria from the genus *Flavobacterium*, appears in overcrowded tanks with poor quality water.

Symptoms: Red mouth at the beginning, then lips begin to separate from the mouth, and eventually the mouth might collapse in on itself leaving a gaping hole.

Treatment: Swab the infected area with hydrogen peroxide using a cotton ball, and treat water with 0.3% salt or Melafix.

Dropsy

Dropsy refers to a condition in which fish have an excessive amount of fluid inside the body cavity or tissues. As a symptom rather than a disease, it can indicate a number of underlying diseases, including bacterial or parasitic infections, or liver, kidney, or heart failure. However, systemic infections by bacteria of the genus *Aeromonas* are the most typical cause of this condition (Conroy, 1961).

Symptoms: Swollen body, raised scales that give the fish a pine cone-shaped appearance.

Treatment: Dropsy is not easily cured and most cases are fatal. However, if the infection is detected early, it is possible to save affected fish. Move sick fish to a separate tank, add salt (0.3%), and treat with antibiotics. Test water parameters to ensure quality is optimal.

Pop eye

Pop eye, medically known as exophthalmia, is a condition where one or both eyes of the fish are swollen and protrude abnormally from the socket. Like dropsy, pop eye is not technically a disease but a symptom of another underlying problem, typically a bacterial infection by *Aeromonas* or *Pseudomonas* spp.

Symptoms: Visible swelling of either one or both eyes, sometimes cloudy. In extreme cases, rupture of the eye is also observed.

Treatment: Move sick fish to a separate tank, add salt (0.3%), and treat with antibiotics. Ensure water quality parameters are in their optimal range.

Fungal infections

Cotton wool

Cotton wool is the general name for all fungal infections that are present in aquaria. Several fungus species can cause infections to fish, but *Saprolegnia* and *Achyla* spp. are the most common. Healthy fish do not have fungus as they protect themselves from fungal infections by producing mucous layers. Instead, fungus appears on fish weakened by stress, illness, or injury.

Symptoms: White, cottony growth on the body and fins.

Treatment: Treat sick fish in a separate tank using a salt bath (0.3%) or antifungal agents, such as malachite green or methylene blue.

Viral infections

Carp pox

Carp pox, also called fish pox, is caused by a viral strain of herpes called HPV-1 that affects the skin and scales (Waltzek et al., 2005). Many goldfish have the herpes gene at birth, and this will go through periods of dormancy and activity throughout the fish lifetime. However, during an outbreak it can be spread to fish not carrying the gene, although how the virus passes from fish to fish remains unknown.

Symptoms: Smooth, white, or pinkish warts on the edge of fins or on the body.

Treatment: Unfortunately, there is no known cure for fish pox and there is no way to fully eradicate the virus from affected fish. In most cases the condition will go away on its own. Raising the temperature of the water has been shown to accelerate recovery.

Lymphocystis

Caused by virus of the genus *Lymphocystivirus*, lymphocystis is a common viral disease of both freshwater and saltwater fish. This virus infects fish and transforms fibroblast of the skin and gills and internal connective tissue, resulting in remarkable hypertrophy of affected cells (named lymphocysts) (MacLachlan and Dubovi, 2017).

Symptoms: White, pearl-like lesions on the edge of the fins or on the body.

Treatment: Similar to carp pox, no particular treatment is known for this disease, which will likely disappear on its own in weeks or months. A weekly bacterial treatment can be helpful.

Other infections and diseases

Swim bladder disorder

Swim bladder disorder or disease is one of the most common goldfish diseases. It can be caused by several factors, including high nitrate levels in the water, sudden water temperature changes, a bacterial infection of the swim bladder, etc. However, the most common causes for this disorder are overfeeding and/or feeding a poor diet lacking in fiber, which sometimes cause gas in the gastrointestinal tract and can lead to severe constipation that would cause the abdomen to swell, preventing the swim bladder from functioning properly (Mayer and Donnelly, 2013b).

Symptoms: Swimming problems, including fish swimming on one side, head up or head down, floating on the surface (sometimes upside down), or resting on the bottom, or struggling to rise.

Treatment: Stop feeding for 3 days, then gradually start refeeding fish with a small amount of live food or cooked vegetables, peas, or beans.

Black spot disease

Black spot or black smudge disease has many potential causes, among which the most common are very high ammonia levels in the water or generally poor water conditions, or the development of parasites (mainly aquatic snails) in the tank.

Symptoms: Black spots on the body and/or fins.

Treatment: Test for water quality parameters, especially high ammonia levels, and treat the water with an antiparasitic agent.

Cloudy eye

Also called white eye, this condition can be caused by numerous factors, including poor water quality (especially very low pH), a diet lacking vitamins and minerals, eye damage, bacterial infections, and parasites.

Symptoms: Opaque, foggy-looking eye/eyes.

Treatment: Improve water quality in the aquarium (pH levels near 7 and zero ammonia and nitrates), add salt (0.3%) and feed fish a diet high in vitamins, minerals, and fiber.

Hole-in-the-head disease

This disease is one of the less common goldfish diseases. It is usually caused by prolonged periods of exposure to poor water quality or stress (e.g., that caused by overcrowding). Other possible causes include the intestinal parasite *Hemaxita* sp.

Symptoms: Cloudy appearance to the body, abdomen looking hollow, sores and lesions at the base of the fins, small holes on the head, and above the eyes.

Treatment: Place sick goldfish in a separate tank, add salt (0.3%), and feed a balanced and vitamin-enriched diet. The use of metronidazole (400–600 mg/100 L water for 3 days) was found effective (Bassleer, 2009).

Goldfish in biomedical research

Regulation and policies

Animal care and use in every country is regulated by ethics guidelines set by national agencies. For example, the Canadian Council of Animal Care (CCAC, 1984; https://www.ccac.ca/en/standards/guidelines/) oversees animal (including fish) use in research and teaching at Canadian institutions. Similarly, other countries have their own set guidelines. Since there are differences in these regulations among countries, the intention here is not to delve deeper into the guidelines of the global research community. Funding agencies and scientific journals or publishers always require ethical considerations and approval from appropriate regulatory entities for animal research. Irrespective of the agencies that regulate fish research, there must be approved standard operating protocols for all aspects, including fish housing, physical parameters, nutrition, stocking density, daily monitoring, disease prevention and treatment, veterinary care, analgesia, anesthesia, and specific study methods, including surgical and injection procedures. For goldfish, a large set of evidence is already available for ideal housing conditions and experimental protocols. Standard practices should be employed without fail for fish care, health, and experimental success.

Anesthesia, analgesia, and euthanasia

A large number of anesthetics are available for use in goldfish. A detailed review of anesthesia and analgesia in fish is provided in Sneddon (2012), and in the detailed guidelines of the American Veterinary Medical Association and the CCAC (cited in the reference list). Water-soluble anesthetics are commonly used for immersion anesthesia. An example of such an anesthetic is tricaine methanesulfonate-222. Since both analgesics and anesthetics are bound to affect fish physiology, attention must be given to the study goals and the general effects of anesthetics and analgesics. If anesthetics and analgesics are required, they should be chosen in consultation with the facility veterinarian. In most cases, a prescription from the veterinarian is required to order both anesthetics and analgesics. While several approaches of euthanasia exist, deep anesthesia, followed by spinal transection or pithing, is a widely accepted practice in goldfish research (American Veterinary Medical Association, 2020; CCAC, 1984).

Injections and blood collection

Intraperitoneal and intracerebroventricular injections under anesthesia are widely used in goldfish. The availability of a stereotaxic atlas of goldfish (Peter and Gill, 1975) helps with the accurate placement of a needle into the brain ventricles or specific brain regions, and directly injecting into the brain. Intraperitoneal injections can be carried out with ease. This is achieved by gently placing the needle attached to the syringe containing the solution at a 45-degree angle, in between two scales in the exterior of the peritoneum toward the anal pore, then slowly applying pressure to press the needle into the peritoneal cavity. Once the needle is inside the cavity, the solution could be released in a very slow manner. The injection volume depends on the size of the fish. Blood collection for obtaining serum or plasma is routinely employed in goldfish. The ideal approach is to collect blood by caudal vein puncture of anesthetized fish. Other options include collecting blood from the aorta or directly from the heart chamber. If spinal transection is employed, blood discharged from the point of dissection could be collected before clotting. Appropriate precautions must be taken in choosing anticoagulants (for plasma), peptide cleavage enzyme inhibitors, and storage conditions (especially temperature). It is possible to obtain approximately 1 mL of blood from a 25–30 g goldfish. A detailed compilation of standard operating protocols for blood collection is prepared by the Department of Fisheries and Oceans of Canada, and is made available through the CCAC website cited earlier.

Biomedical research applications

A recent (October 2020) search using the term "goldfish" yielded over 7100 articles in PubMed. A majority of these articles detail research that used goldfish as a model organism. A comprehensive review of all of these research articles is beyond the scope of this review. Next, we will focus only on brief summaries of some of the major areas of research that has employed goldfish as a model.

Neuroscience

Goldfish is routinely used in the study of brain and diseases affecting the central nervous system. Xing et al. (2017) found that both aromatase and radial glial cells influence neuronal degeneration and regeneration in goldfish. Pollard and colleagues (1993) reviewed the use of goldfish as a suitable model for Parkinson's disease and other neurodegenerative diseases. A more recent review on the same topic (Manasa et al., 2020) summarized the use of teleosts, especially cyprinids, in studying compounds that induce and remediate neuronal defects seen in Parkinson's diseases. A single dose of neurotoxin 1-methyl-4-phenyl-1,2,3,6 tetrahydropyridine induces Parkinson's disease in goldfish (Weinreb and Youdim, 2007). It has been shown that the advanced metabolomics approach could be used in goldfish models of Parkinson's disease (Lu et al., 2014). The large size of the goldfish brain, the ability to distinguish various regions within the brain, and the vasculature of the brain help with its use in studying brain diseases. Another area of research that employs goldfish as a model organism is the study of retina. Retinal neurons were studied extensively to understand their electrical properties (Country et al, 2019, 2020). The effects of factors such as steroid hormones on visual processing (Yue et al., 2018), the role of factor XIIIA on the retina and optic nerve after an optic nerve lesion (Sugitani et al., 2012), and Pax2 expression in the optic nerve head in retinal regeneration (Parrilla et al., 2012) are some examples of other studies conducted using goldfish. Together, the foregoing examples provide strong support for the use of goldfish as a model organism for the study of human health and diseases, especially those pertaining to neurology and vision.

Endocrinology, reproduction, and associated behaviors

Goldfish has been extensively used in endocrine research, and it appears that the endocrine system and/or processes involved are the most widely studied using goldfish as a model. We reviewed the merits and demerits of goldfish as a model in endocrinology in Blanco et al. (2017b), and readers are referred to this article for detailed information on this aspect. The studies in goldfish either focused on identifying the central or peripheral actions of hormones and neuropeptides on food intake and/or metabolic processes or the synthesis and secretion of hormones that regulate physiological processes (Volkoff, 2013; Volkoff et al., 1999; Blanco et al., 2017a, 2017b, 2018, 2020b; Kah, 2020). Goldfish pituitary and pituitary cells (Chang et al., 1990) were widely studied for the regulation of growth hormone and gonadotropin synthesis and secretion (Marchant et al., 1989; Peng et al., 1993). Several pheromones were originally identified from and characterized in goldfish (Sorensen et al., 2018; Levesque et al., 2011). In addition, goldfish cells were studied to identify intracellular signaling pathways that mediate the effects of peptidyl hormones in somatotrophs and gonadotrophs (Yunker et al., 2003). Goldfish gonadal processes, including gametogenesis and steroidogenesis, are widely characterized (Habibi et al., 2012; Rajeswari et al., 2020; Ma et al., 2020; Qi et al., 2013; Pati and Habibi, 2002). Sexual behavior is another aspect that is widely characterized in goldfish (Shahjahan et al., 2011; Levesque et al., 2011; Sachuriga et al., 2019; Kawaguchi et al., 2014). Physical activity and anxiolytic behavior in response in hormones were also studied using goldfish (Matsuda et al., 2020; Araishi et al., 2019; Saoshiro et al., 2013). The highly conserved structure and functions of many hormones, receptors, and bioactive peptides enable the use of goldfish as a model organism in hormone research. The knowledge obtained from this organism could be applied in many other vertebrates.

Toxicology and pharmacology

Goldfish has been used to understand the adverse pharmacological and toxic effects of compounds. There is a large set of data available on this topic. The focus here will be to offer some information on recent studies that focused on various goldfish tissues. Zhang et al. (2020) reported that benzophenone-3 adversely affects goldfish tissues, including the liver and gut, especially the gut microbiota. Moyson and colleagues (2016) found that copper exposure increases renal activity in goldfish. Meanwhile, Wang et al. (2019b) reported that bisphenol-A has negative effects on goldfish reproductive axis. Wang et al. (2019a) found that bisphenol-A also caused neurotoxicity in goldfish. Similarly, fluoxetine exposure leads to reproductive defects in goldfish (Mennigen et al., 2017). Xing et al. (2016) published that dehydroabietic acid is toxic to the radial glial cells in goldfish. Overall, goldfish continues to be a desired model in toxicology research because several of the physiological systems in this organism are widely characterized.

Other use of goldfish in research

In addition to the foregoing main areas, goldfish has been used in studying other organs or biological processes, including hydromineral balance (Kelly and Peter, 2006), gills (Chasiotis and Kelly, 2012), heart (Leo et al., 2019), hypoxia (Farhat et al., 2019), ionoregulation and kidney function (Fehsenfeld and Wood, 2018), immunology (Xie and Belosevic, 2018; Grayfer and Belosevic, 2009), circadian clock genes (Sánchez-Bretaño et al., 2017), and gut motility (Mensah et al., 2018). This list is definitively more extensive, considering some old literature and studies or reports that are not cataloged. However, an extensive discussion of all reports available is beyond the scope of this review. Readers are encouraged to consult available primary literature for a detailed learning of topics interesting to them.

Conclusions and perspectives

Goldfish continues to be a preferred model organism for research in several fields across the globe. While the public perception of goldfish as an aquarium fish or ornamental fish provides this species a coveted position in the public mind, its journey as a research model is not devoid of limitations. A major concern is genome duplication and resulting presence of duplicate genes. This results in more than one copy of genes and encoded proteins, making it difficult to conduct genetic manipulation. In addition, the presence of multiple forms of the same peptide complicates interpretations of results from physiological studies. The possibility of sexual dimorphism exists in this species. In addition, goldfish is a seasonal breeder, which introduces variations in its physiology during sexually mature versus gonadal regressed or recrudescent states. Despite these challenges, goldfish is still a useful model. Its relatively large size, possibility of collecting larger volumes of blood, and already characterized or well-studied body systems allow easy employment of this species for research. The handling and care of goldfish must be strictly adhered to, and this approach will assist in avoiding additional limitations in its use. The more recent availability of goldfish genome (Chen et al., 2019) is expected to help in the better use and care of goldfish in research.

References

Ababouch, L., 2014. Food safety assurance systems: good practices in Fisheries and aquaculture. In: Encyclopedia of Food Safety. Elsevier, pp. 159–167. https://doi.org/10.1016/B978-0-12-378612-8.00343-7.

Abi-Ayad, A., Kestemont, P., 1994. Comparison of the nutritional status of goldfish (*Carassius auratus*) larvae fed with live, mixed or dry diet. Aquaculture 128, 163–176.

Al-Dubakel, A., Najim, S.M., Abed, J.M., 1999. Nutritional evaluation of goldfish *Carassius auratus* feeding a commercial fish food. Basrah J. Vet. Res. 22, 57–62.

American Veterinary Medical Association, 2020. AVMA Guidelines for the Euthanasia of Animals, 2020 Edition.

Araishi, K., Watanabe, K., Yamazaki, T., Nakamachi, T., Matsuda, K., 2019. Intracerebroventricular administration of arginine vasotocin (AVT) induces anorexigenesis and anxiety-like behavior in goldfish. Peptides 119, 170118. https://doi.org/10.1016/j.peptides.2019.170118.

Badiola, M., Basurko, O.C., Piedrahita, R., Hundley, P., Mendiola, D., 2018. Energy use in recirculating aquaculture systems (RAS): a review. Aquacult. Eng. 81, 57–70. https://doi.org/10.1016/j.aquaeng.2018.03.003.

Baker, B.I., 1981. Biological role of the pars intermedia in lower vertebrates. Ciba Found. Symp. 81, 166–179. https://doi.org/10.1002/9780470720646.ch10.

Bassleer, G., 2009. The New Illustrated Guide to Fish Diseases in Ornamental Tropical and Pond Fish. Bassleer Biofish, Belgium.

Bastos Gomes, G., Jerry, D.R., Miller, T.L., Hutson, K.S., 2017. Current status of parasitic ciliates *Chilodonella* spp. (Phyllopharyngea: Chilodonellidae) in freshwater fish aquaculture. J. Fish Dis. 40, 703–715. https://doi.org/10.1111/jfd.12523.

Blanco, A.M., Bertucci, J.I., Delgado, M.J., Valenciano, A.I., Unniappan, S., 2016. Tissue-specific expression of ghrelinergic and NUCB2/nesfatin-1 systems in goldfish (*Carassius auratus*) is modulated by macronutrient composition of diets. Comp. Biochem. Physiol. A. Mol. Integr. Physiol. 195, 1–9. https://doi.org/10.1016/j.cbpa.2016.01.016.

Blanco, A.M., Bertucci, J.I., Ramesh, N., Delgado, M.J., Valenciano, A.I., Unniappan, S., 2017a. Ghrelin facilitates GLUT2-, SGLT1- and SGLT2-mediated intestinal glucose transport in goldfish (*Carassius auratus*). Sci. Rep. 7, 45024. https://doi.org/10.1038/srep45024.

Blanco, A.M., Bertucci, J.I., Unniappan, S., 2020a. Goldfish adipocytes are pancreatic beta cell-like, glucose-responsive insulin-producing cells. J. Cell. Physiol. 235, 6875–6886. https://doi.org/10.1002/jcp.29581.

Blanco, A.M., Bertucci, J.I., Velasco, C., Unniappan, S., 2020b. Growth differentiation factor 15 (GDF-15) is a novel orexigen in fish. Mol. Cell. Endocrinol. 505, 110720. https://doi.org/10.1016/j.mce.2020.110720.

Blanco, A.M., Gómez-Boronat, M., Madera, D., Valenciano, A.I., Alonso-Gómez, A.L., Delgado, M.J., 2018. First evidence of nocturnin in fish: two isoforms in goldfish differentially regulated by feeding. Am. J. Physiol. Regul. Integr. Comp. Physiol. 314, R304–R312. https://doi.org/10.1152/ajpregu.00241.2017.

Blanco, A.M., Sundarrajan, L., Bertucci, J.I., Unniappan, S., 2017b. Why goldfish? Merits and challenges in employing goldfish as a model organism in comparative endocrinology research. Gen. Comp. Endocrinol. https://doi.org/10.1016/j.ygcen.2017.02.001.

Brown, C., Wolfenden, D., Sneddon, L., 2018. Goldfish (*Carassius auratus*). In: Companion Animal Care and Welfare: The UFAW Companion Animal Handbook. Wiley-Blackwell, New Jersey, United States.

Canadian Council on Animal Care, 1984. Guide to the Care and Use of Experimental Animals, vol. 2.

Canesi, L., Fabbri, E., 2015. Environmental effects of BPA: focus on aquatic species. Dose-Response Publ. Int. Hormesis Soc. 13 https://doi.org/10.1177/1559325815598304.

Chang, J.P., Cook, H., Freedman, G.L., Wiggs, A.J., Somoza, G.M., de Leeuw, R., Peter, R.E., 1990. Use of a pituitary cell dispersion method and primary culture system for the studies of gonadotropin-releasing hormone action in the goldfish, *Carassius auratus*. I. Initial morphological, static, and cell column perifusion studies. Gen. Comp. Endocrinol. 77, 256–273. https://doi.org/10.1016/0016-6480(90)90310-i.

Chasiotis, H., Kelly, S.P., 2012. Effects of elevated circulating cortisol levels on hydromineral status and gill tight junction protein abundance in the stenohaline goldfish. Gen. Comp. Endocrinol. 175, 277–283. https://doi.org/10.1016/j.ygcen.2011.11.024.

Chavin, W., 1956. Thyroid distribution and function in the goldfish, *Carassius auratus* L. J. Exp. Zool. 133, 259–279.

Chavin, W., Bouwman, B.N., 1965. Metabolism iodine and thyroid hormone synthesis in the goldfish, *Carassius auratus* L. Gen. Comp. Endocrinol. 5, 493–503.

Chen, S.C., 1956. A history of the domestication and the factors of the varietal formation of the common goldfish, *Carassius auratus*. Sci. Sin. 6, 89–116.

Chen, Z., Omori, Y., Koren, S., Shirokiya, T., Kuroda, T., Miyamoto, A., Wada, H., Fujiyama, A., Toyoda, A., Zhang, S., Wolfsberg, T.G., Kawakami, K., Phillippy, A.M., NISC Comparative Sequencing Program, Mullikin, J.C., Burgess, S.M., 2019. De novo assembly of the goldfish (*Carassius auratus*) genome and the evolution of genes after whole-genome duplication. Sci. Adv. 5, eaav0547. https://doi.org/10.1126/sciadv.aav0547.

Conroy, D.A., 1961. A study of the bacterium associated with an outbreak of oedema amongst goldfish (*Carassius carassius* var. auratus). Microbiol. Esp. 14, 73–105.

Country, M.W., Campbell, B.F.N., Jonz, M.G., 2019. Spontaneous action potentials in retinal horizontal cells of goldfish (*Carassius auratus*) are dependent upon L-type Ca^{2+} channels and ryanodine receptors. J. Neurophysiol. 122, 2284–2293. https://doi.org/10.1152/jn.00240.2019.

Country, M.W., Htite, E.D., Samson, I.A., Jonz, M.G., 2020. Retinal horizontal cells of goldfish (*Carassius auratus*) display subtype-specific differences in spontaneous action potentials in situ. J. Comp. Neurol. https://doi.org/10.1002/cne.25054.

Delahunty, G., de Vlaming, V.L., 1980. Seasonal relationships of ovary weight, liver weight and fat stores with body weight in the goldfish, *Carassius auratus* (L.). J. Fish Biol. 16, 5–13. https://doi.org/10.1111/j.1095-8649.1980.tb03683.x.

Dror, M., Sinyakov, M.S., Okun, E., Dym, M., Sredni, B., Avtalion, R.R., 2006. Experimental handling stress as infection-facilitating factor for the goldfish ulcerative disease. Vet. Immunol. Immunopathol. 109, 279–287. https://doi.org/10.1016/j.vetimm.2005.08.022.

Fänge, R., Nilsson, S., 1985. The fish spleen: structure and function. Experientia 41, 152–158. https://doi.org/10.1007/bf02002607.

Farhat, E., Turenne, E.D., Choi, K., Weber, J.-M., 2019. Hypoxia-induced remodelling of goldfish membranes. Comp. Biochem. Physiol. B Biochem. Mol. Biol. 237, 110326. https://doi.org/10.1016/j.cbpb.2019.110326.

Fehsenfeld, S., Wood, C.M., 2018. Section-specific expression of acid-base and ammonia transporters in the kidney tubules of the goldfish *Carassius auratus* and their responses to feeding. Am. J. Physiol. Renal Physiol. 315, F1565–F1582. https://doi.org/10.1152/ajprenal.00510.2017.

Ford, T., Beitinger, T.L., 2005. Temperature tolerance in the goldfish, *Carassius auratus*. J. Therm. Biol. 30, 147–152.

Garofalo, F., Imbrogno, S., Tota, B., Amelio, D., 2012. Morpho-functional characterization of the goldfish (*Carassius auratus* L.) heart. Comp. Biochem. Physiol. A. Mol. Integr. Physiol. 163, 215–222. https://doi.org/10.1016/j.cbpa.2012.05.206.

Glasauer, S.M.K., Neuhauss, S.C.F., 2014. Whole-genome duplication in teleost fishes and its evolutionary consequences. Mol. Genet. Genomics 289, 1045–1060. https://doi.org/10.1007/s00438-014-0889-2.

Grayfer, L., Belosevic, M., 2009. Molecular characterization, expression and functional analysis of goldfish (*Carassius aurutus* L.) interferon gamma. Dev. Comp. Immunol. 33, 235–246. https://doi.org/10.1016/j.dci.2008.09.001.

Habibi, H.R., Nelson, E.R., Allan, E.R.O., 2012. New insights into thyroid hormone function and modulation of reproduction in goldfish. Gen. Comp. Endocrinol. 175, 19–26. https://doi.org/10.1016/j.ygcen.2011.11.003.

Helfman, G.S., Collette, B.B., Facey, D.E., Bowen, B.W., 2009. The diversity of fishes. In: Biology, Evolution, and Ecology, Second ed. Wiley-Blackwell, United Kingdom.

Howdeshell, K.L., Peterman, P.H., Judy, B.M., Taylor, J.A., Orazio, C.E., Ruhlen, R.L., Vom Saal, F.S., Welshons, W.V., 2003. Bisphenol A is released from used polycarbonate animal cages into water at room temperature. Environ. Health Perspect. 111, 1180–1187. https://doi.org/10.1289/ehp.5993.

Imanpoor, M.R., Enayat, G.T., Zolfaghari, M., 2011. Effect of light and music on growth performance and survival rate of goldfish (*Carassius auratus*). Iran. J. Fish. Sci. 10, 641–653.

Jepson, L., 2016. Goldfish and koi. In: Exotic Animal Medicine: A Quick Reference Guide, second ed. Elsevier, United Stated.

Kah, O., 2020. A 45-years journey within the reproductive brain of fish. Gen. Comp. Endocrinol. 288, 113370. https://doi.org/10.1016/j.ygcen.2019.113370.

Kaul, S., Vollrath, L., 1974. The goldfish pituitary: I. Cytology. Cell Tissue Res. 154, 211–230. https://doi.org/10.1007/BF00223165.

Kawaguchi, Y., Nagaoka, A., Kitami, A., Mitsuhashi, T., Hayakawa, Y., Kobayashi, M., 2014. Gender-typical olfactory regulation of sexual behavior in goldfish. Front. Neurosci. 8, 91. https://doi.org/10.3389/fnins.2014.00091.

Kelly, S.P., Peter, R.E., 2006. Prolactin-releasing peptide, food intake, and hydromineral balance in goldfish. Am. J. Physiol. Regul. Integr. Comp. Physiol. 291, R1474–R1481. https://doi.org/10.1152/ajpregu.00129.2006.

Khatoon, N., Sengupta, P., Homechaudhuri, S., Pal, R., 2010. Evaluation of algae based feed in goldfish (*Carassius auratus*) nutrition. Proc. Zool. Soc. 63, 109–114.

Kleeman, C.R., Cutler, R.E., 1963. The neurohypophysis. Annu. Rev. Physiol. 25, 385–432. https://doi.org/10.1146/annurev.ph.25.030163.002125.

Kottelat, M., Freyhof, J., 2007. Handbook of European Freshwater Fishes. Publications Kottelat, Berlin, Germany.

Leem, J.-B., Jeon, W., Yun, C.-Y., Lee, S.-H., 2012. Quantitative analysis of fish schooling behavior with different numbers of medaka (*Oryzias latipes*) and goldfish (*Carassius auratus*). Ocean Sci. J. 47, 445–451.

Leo, S., Gattuso, A., Mazza, R., Filice, M., Cerra, M.C., Imbrogno, S., 2019. Cardiac influence of the β3-adrenoceptor in the goldfish (*Carassius auratus*): a protective role under hypoxia? J. Exp. Biol. 222 https://doi.org/10.1242/jeb.211334.

Levesque, H.M., Scaffidi, D., Polkinghorne, C.N., Sorensen, P.W., 2011. A multi-component species identifying pheromone in the goldfish. J. Chem. Ecol. 37, 219–227. https://doi.org/10.1007/s10886-011-9907-6.

Lim, L.C., Dhert, P., Sorgeloos, P., 2007. Preparation for transport: fish packaging technology. In: International Transport of Live Fish in the Ornamental Aquatic Industry. Ornamental Fish International OFI Educational Publication, OFI, Maarssen, The Netherlands, pp. 25–45.

Lim, L.C., Dhert, P., Sorgeloos, P., 2003. Recent developments and improvements in ornamental fish packaging systems for air transport. Aquacult. Res. 34, 923–935.

Linnaeus, C., 1758. Systema naturae per regna tria naturae :secundum classes, ordines, genera, species, cum characteribus, differentiis, synonymis, locis. Holmiae, Impensis Direct, tenth ed. Laurentii Salvii (Salvius publ.), Stockholm.

López-Olmeda, J.F., Madrid, J.A., Sánchez-Vázquez, F.J., 2006. Melatonin effects on food intake and activity rhythms in two fish species with different activity patterns: diurnal (goldfish) and nocturnal (tench). Comp. Biochem. Physiol. A. Mol. Integr. Physiol. 144, 180–187. https://doi.org/10.1016/j.cbpa.2006.02.031.

Lu, Z., Wang, J., Li, M., Liu, Q., Wei, D., Yang, M., Kong, L., 2014. 1H NMR-based metabolomics study on a goldfish model of Parkinson's disease induced by 1-methyl-4-phenyl-1,2,3,6-tetrahydropyridine (MPTP). Chem. Biol. Interact. 223, 18–26. https://doi.org/10.1016/j.cbi.2014.09.006.

Luz, R.K., Martínez-Álvarez, R.M., De Pedro, N., Delgado, M.J., 2008. Growth, food intake regulation and metabolic adaptations in goldfish (*Carassius auratus*) exposed to different salinities. Aquaculture 276, 171–178. https://doi.org/10.1016/j.aquaculture.2008.01.042.

Lv, Y., Kawasaki, K., Li, J., Li, Y., Bian, C., Huang, Y., You, X., Shi, Q., 2017. A genomic survey of SCPP family genes in fishes provides novel insights into the evolution of fish scales. Int. J. Mol. Sci. 18 https://doi.org/10.3390/ijms18112432.

Ma, Y., Ladisa, C., Chang, J.P., Habibi, H.R., 2020. Multifactorial control of reproductive and growth axis in male goldfish: influences of GnRH, GnIH and thyroid hormone. Mol. Cell. Endocrinol. 500, 110629. https://doi.org/10.1016/j.mce.2019.110629.

MacLachlan, J., Dubovi, E.J., 2017. Asfarviridae and Iridoviridae. In: Fenner's Veterinary Virology, fifth ed., pp. 175–188.

Magurran, A.E., Pitcher, T.J., 1983. Foraging, timidity and shoal size in minnows and goldfish. Behav. Ecol. Sociobiol. 12, 147–152.

Manasa, K., Chitra, V., Tamilanban, T., 2020. Teleost model as an alternative in Parkinson's disease. Neurol. India 68, 979–984. https://doi.org/10.4103/0028-3886.294542.

Marchant, T.A., Chang, J.P., Nahorniak, C.S., Peter, R.E., 1989. Evidence that gonadotropin-releasing hormone also functions as a growth hormone-releasing factor in the goldfish. Endocrinology 124, 2509–2518. https://doi.org/10.1210/endo-124-5-2509.

Matsuda, K., Yoshida, D., Sachuriga, Watanabe, K., Yokobori, E., Konno, N., Nakamachi, T., 2020. Effect of intracerebroventricular administration of two molecular forms of sulfated CCK octapeptide on anxiety-like behavior in the zebrafish danio rerio. Peptides 130, 170330. https://doi.org/10.1016/j.peptides.2020.170330.

Mayer, J., Donnelly, T.M., 2013a. Flukes (monogenean parasites). In: Clinical Veterinary Advisor: Birds and Exotic Pets. Saunders, United States, pp. 22–24.

Mayer, J., Donnelly, T.M., 2013b. Swim bladder disease/buoyancy disorders. In: Clinical Veterinary Advisor: Birds and Exotic Pets. Saunders, United States, pp. 36–37.

M'balaka, M., Kassam, D., Rusuwa, B., 2012. The effect of stocking density on the growth and survival of improved and unimproved strains of *Oreochromis shiranus*. Egypt. J. Aquat. Res. 38, 205–211. https://doi.org/10.1016/j.ejar.2012.12.013.

Mcvay, J.A., Kaan, H.W., 1940. The digestive tract of *Carassius auratus*. Biol. Bull. 78, 53–67.

Mennigen, J.A., Zamora, J.M., Chang, J.P., Trudeau, V.L., 2017. Endocrine disrupting effects of waterborne fluoxetine exposure on the reproductive axis of female goldfish, *Carassius auratus*. Comp. Biochem. Physiol. C Toxicol. Pharmacol. 202, 70–78. https://doi.org/10.1016/j.cbpc.2017.08.003.

Mensah, E.T., Blanco, A.M., Donini, A., Unniappan, S., 2018. Galanin decreases spontaneous resting contractions and potentiates acetyl choline-induced contractions of goldfish gut. Neuropeptides 69, 92–97. https://doi.org/10.1016/j.npep.2018.04.010.

Moreira, R.L., Da Costa, J.M., Teixeira, E.G., Moreira, A.G.L., De Moura, P.S., Rocha, R.S., Vieira, R.H.S.F., 2011. Performance of *Carassius auratus* with different food strategies in water recirculation system. Arch. Zootec. 60.

Moyson, S., Liew, H.J., Fazio, A., Van Dooren, N., Delcroix, A., Faggio, C., Blust, R., De Boeck, G., 2016. Kidney activity increases in copper exposed goldfish (*Carassius auratus*

auratus). Comp. Biochem. Physiol. C Toxicol. Pharmacol. 190, 32–37. https://doi.org/10.1016/j.cbpc.2016.08.003.

Muus, B.J., Dahlström, P., 1981. Los peces de agua dulce de España y Europa: pesca, biología, importancia económica. Omega, Barcelona, Spain.

Nagahama, Y., 1983. The functional morphology of teleost gonads. In: Fish Physiology. Reproduction, Part A: Endocrine Tissues and Hormones, vol. 9. Academic Press, pp. 223–275.

Navarrete Salgado, N.A., Yuritzi, T., Hernández, C., Citlalli, M., Gallardo, L., Contreras Rivero, G., 2009. Alimentación de carpas (Pisces, Cyprinidae) en el Embalse La Goleta, Estado de México. Rev. Zool. 20, 7–16.

Nemato, C.M., 1957. Experiments with methods for air transport of live ¢sh. Progress. Fish Cult. 19, 147–157.

Ogawa, M., 1962. Comparative study on the internal structure of the teleostean kidney. Sci. Rep. Saitama Univ. B4 107–129.

Ota, K.G., Abe, G., 2016. Goldfish morphology as a model for evolutionary developmental biology. Wiley Interdiscip. Rev. Dev. Biol. 5, 272–295. https://doi.org/10.1002/wdev.224.

Parrilla, M., Lillo, C., Herrero-Turrión, M.J., Arévalo, R., Aijón, J., Lara, J.M., Velasco, A., 2012. Characterization of Pax2 expression in the goldfish optic nerve head during retina regeneration. PLoS One 7, e32348. https://doi.org/10.1371/journal.pone.0032348.

Pati, D., Habibi, H.R., 2002. Involvement of protein kinase C and arachidonic acid pathways in the gonadotropin-releasing hormone regulation of oocyte meiosis and follicular steroidogenesis in the goldfish ovary. Biol. Reprod. 66, 813–822. https://doi.org/10.1095/biolreprod66.3.813.

Peng, C., Chang, J.P., Yu, K.L., Wong, A.O., Van Goor, F., Peter, R.E., Rivier, J.E., 1993. Neuropeptide-Y stimulates growth hormone and gonadotropin-II secretion in the goldfish pituitary: involvement of both presynaptic and pituitary cell actions. Endocrinology 132, 1820–1829. https://doi.org/10.1210/endo.132.4.8462479.

Peter, R.E., Gill, V.E., 1975. A stereotaxic atlas and technique for forebrain nuclei of the goldfish, *Carassius auratus*. J. Comp. Neurol. 159, 69–101. https://doi.org/10.1002/cne.901590106.

Pitcher, T.J., Magurran, A.E., 1983. Shoal size, patch profitability and information exchange in foraging goldfish. Anim. Behav. 31, 546–555.

Pollard, H.B., Adeyemo, M., Dhariwal, K., Levine, M., Caohuy, H., Markey, S., Markey, C.J., Youdim, M.B., 1993. The goldfish as a drug discovery vehicle for Parkinson's disease and other neurodegenerative disorders. Ann. N. Y. Acad. Sci. 679, 317–320. https://doi.org/10.1111/j.1749-6632.1993.tb18314.x.

Priestley, S.M., Stevenson, A.E., Alexander, L.G., 2006a. Growth rate and body condition in relation to group size in black widow tetras (*Gymnocorymbus ternetzi*) and common goldfish (*Carassius auratus*). J. Nutr. 136, 2078S–2080S. https://doi.org/10.1093/jn/136.7.2078S.

Priestley, S.M., Stevenson, A.E., Alexander, L.G., 2006b. The influence of feeding frequency on growth and body condition of the common goldfish (*Carassius auratus*). J. Nutr. 136, 1979S–1981S.

Puzdrowski, R.L., 1989. Peripheral distribution and central projections of the lateral-line nerves in goldfish, *Carassius auratus*. Brain Behav. Evol. 34, 110–131.

Qi, X., Zhou, W., Lu, D., Wang, Q., Zhang, H., Li, S., Liu, X., Zhang, Y., Lin, H., 2013. Sexual dimorphism of steroidogenesis regulated by GnIH in the goldfish, *Carassius auratus*. Biol. Reprod. 88, 89. https://doi.org/10.1095/biolreprod.112.105114.

Rajeswari, J.J., Hatef, A., Unniappan, S., 2020. Nesfatin-1-like peptide suppresses hypothalamo-pituitary-gonadal mRNAs, gonadal steroidogenesis, and oocyte maturation in fish†. Biol. Reprod. 103, 802–816. https://doi.org/10.1093/biolre/ioaa106.

Reimschuessel, R., 2001. A fish model of renal regeneration and development. ILAR J. 42, 285–291.

Robison, H.W., Buchanan, T.M., 2020. Fishes of Arkansas, second ed. The University of Arkansas Press, Fayetteville, United States.

Ruchin, A.B., 2004. Influence of colored light on growth rate of juveniles of fish. Fish Physiol. Biochem. 30, 175–178.

Rupp, B., Wullimann, M.F., Reichert, H., 1996. The zebrafish brain: a neuroanatomical comparison with the goldfish. Anat. Embryol. 194, 187–203.

Sachuriga, Iinuma, N., Shibata, H., Yoshida, D., Konno, N., Nakamachi, T., Matsuda, K., 2019. Intracerebroventricular administration of sulphated cholecystokinin octapeptide induces anxiety-like behaviour in goldfish. J. Neuroendocrinol. 31, e12667. https://doi.org/10.1111/jne.12667.

Sampour, M., 2008. The study of adrenal chromaffin of fish, *Carassius auratus* (Toleostei). Pak. J. Biol. Sci. 11, 1032–1036.

Sánchez-Bretaño, A., Blanco, A.M., Alonso-Gómez, Á.L., Delgado, M.J., Kah, O., Isorna, E., 2017. Ghrelin induces clock gene expression in the liver of goldfish in vitro via protein kinase C and protein kinase A pathways. J. Exp. Biol. 220, 1295–1306. https://doi.org/10.1242/jeb.144253.

Saoshiro, S., Kawaguchi, Y., Hayakawa, Y., Kobayashi, M., 2013. Sexual bipotentiality of behavior in male and female goldfish. Gen. Comp. Endocrinol. 181, 265–270. https://doi.org/10.1016/j.ygcen.2012.08.022.

Sarbahi, D.S., 1951. Studies of the digestive tracts and the digestive enzymes of the goldfish, *Carassius auratus* (Linnaeus) and the largemouth black bass, *Micropterus salmoides* (Lacépède). Biol. Bull. 100, 244–257.

Sarkar, A., Upadhyay, B., 2011. Role of photoperiod in enhancement of reproduction in goldfish (*Carassius auratus*). Asian J. Exp. Biol. Sci. 2, 544–547.

Schwartz, F.J., 1964. Natural salinity tolerances of some freshwater fishes. Underw. Nat. 2, 13–15.

Shahjahan, M., Ikegami, T., Osugi, T., Ukena, K., Doi, H., Hattori, A., Tsutsui, K., Ando, H., 2011. Synchronised expressions of LPXRFamide peptide and its receptor genes: seasonal, diurnal and circadian changes during spawning period in grass puffer. J. Neuroendocrinol. 23, 39–51. https://doi.org/10.1111/j.1365-2826.2010.02081.x.

Sharma, M., Shrivastav, A.B., Sahni, Y.P., Pandey, G., 2014. Overviews of the treatment and control of common fish diseases. Int. Res. J. Pharm. 3.

Shelton, E., Smith, M., 1970. The ultrastructure of carp (*Cyprinus carpio*) immunoglobulin: a tetrameric macroglobulin. J. Mol. Biol. 54, 615–617. https://doi.org/10.1016/0022-2836(70)90133-6.

Shete, A.P., Verma, A.K., Kohli, M.P.S., Dash, A., Tandel, R., 2013. Optimum stocking density for growth of goldfish, *Carassius auratus* (Linnaeus, 1758), in an aquaponic system. Isr. J. Aquacult. Bamidgeh 65. IJA_65.2013.910.

Smartt, J., 2001. Goldfish Varieties and Genetics: Handbook for Breeders. Wiley-Blackwell, Cornwall, United Kingdom.

Smartt, J., Bundell, J.H., 1996. Goldfish Breeding and Genetics. TFH Publications.

Smith, A.C., Gray, H., 2011. Goldfish in a tank: the effect of substrate on foraging behaviour in aquarium. Anim. Welfare 20, 311–331.

Sneddon, L.U., 2012. Clinical anesthesia and analgesia in fish. J. Exot. Pet Med. 21, 32–43. https://doi.org/10.1053/j.jepm.2011.11.009.

Sorensen, P.W., Appelt, C., Stacey, N.E., Goetz, F.W., Brash, A.R., 2018. High levels of circulating prostaglandin F2α associated with ovulation stimulate female sexual receptivity and spawning behavior in the goldfish (*Carassius auratus*). Gen. Comp. Endocrinol. 267, 128–136. https://doi.org/10.1016/j.ygcen.2018.06.014.

Spieler, R.E., 1992. Feeding-entrained circadian rhythms in fishes. In: Rhythms in Fishes, NATO ASI Series (Series A: Life Sciences). Springer, Boston, MA, United States, pp. 137–147.

Spieler, R.E., Noeske, T.A., 1984. Effects of photoperiod and feeding schedule on diel variations of locomotor activity, cortisol, and thyroxine in goldfish. Trans. Am. Fish. Soc. 113, 528–539. https://doi.org/10.1577/1548-8659(1984)113<528:EOPAFS>2.0.CO;2.

Spotila, J.R., Terpin, K.M., Koons, R.R., Bonati, R.L., 1979. Temperature requirements of fishes from eastern Lake Erie and the upper Niagara River: a review of the literature. Environ. Biol. Fish. 4, 281–307.

Stout, C.C., Tan, M., Lemmon, A.R., Lemmon, E.M., Armbruster, J.W., 2016. Resolving Cypriniformes relationships using an anchored enrichment approach. BMC Evol. Biol. 16, 244. https://doi.org/10.1186/s12862-016-0819-5.

Sugitani, K., Oogai, K., Nakashima, H., Kato, S., 2012. Factor XIIIA induction in the retina and optic nerve after optic nerve lesion in goldfish. Adv. Exp. Med. Biol. 723, 443–448. https://doi.org/10.1007/978-1-4614-0631-0_56.

Takahashi, H., Takano, K., 1972. Morphogenesis of accessory reproductive organs in male goldfish, *Carassius auratus*. Bull. Fac. Fish. Hokkaido Univ. 23, 53–64.

Thilakaratne, I.D.S.I.P., Rajapaksha, G., Hewakopara, A., Rajapakse, R.P.V.J., Faizal, A.C.M., 2003. Parasitic infections in freshwater ornamental fish in Sri Lanka. Dis. Aquat. Org. 54, 157–162. https://doi.org/10.3354/dao054157.

Underwood, W.L., 1901. Goldfish as destroyers of mosquito larvae. Science 14, 1017–1018. https://doi.org/10.1126/science.14.365.1017.

van Muiswinkel, W.B., 1995. The piscine immune system: innate and acquired immunity. In: Fish Diseases and Disorders I: Protozoan and Metazoan Infections. CAB International, Wallingford, UK, pp. 729–750.

Volkoff, H., 2013. The effects of amphetamine injections on feeding behavior and the brain expression of orexin, CART, tyrosine hydroxylase (TH) and thyrotropin releasing hormone (TRH) in goldfish (*Carassius auratus*). Fish Physiol. Biochem. 39, 979–991. https://doi.org/10.1007/s10695-012-9756-4.

Volkoff, H., Bjorklund, J.M., Peter, R.E., 1999. Stimulation of feeding behavior and food consumption in the goldfish, *Carassius auratus*, by orexin-A and orexin-B. Brain Res. 846, 204–209. https://doi.org/10.1016/s0006-8993(99)02052-1.

Wafer, L.N., Whitney, J.C., Jensen, V.B., 2015. Fish lice (*Argulus japonicus*) in goldfish (*Carassius auratus*). Comp. Med. 65, 93–95.

Walker, R.M., Johansen, P.H., 1977. Anaerobic metabolism in goldfish, *Carassius auratus*. Can. J. Zool. 55, 1304–1311.

Wallen, I.E., 1951. The direct effect of turbidity on fishes. Bull. Okla. Agric. Mech. Coll. 48, 1–27.

Waltzek, T.B., Kelley, G.O., Stone, D.M., Way, K., Hanson, L., Fukuda, H., Hirono, I., Aoki, T., Davison, A.J., Hedrick, R.P., 2005. Koi herpesvirus represents a third cyprinid herpesvirus (CyHV-3) in the family Herpesviridae. J. Gen. Virol. 86, 1659–1667. https://doi.org/10.1099/vir.0.80982-0.

Wang, J.-T., Li, J.-T., Zhang, X.-F., Sun, X.-W., 2012. Transcriptome analysis reveals the time of the fourth round of genome duplication in common carp (*Cyprinus carpio*). BMC Genomics 13, 96. https://doi.org/10.1186/1471-2164-13-96.

Wang, Q., Lin, F., He, Q., Liu, X., Xiao, S., Zheng, L., Yang, H., Zhao, H., 2019a. Assessment of the effects of bisphenol A on dopamine synthesis and blood vessels in the goldfish brain. Int. J. Mol. Sci. 20 https://doi.org/10.3390/ijms20246206.

Wang, Q., Yang, H., Yang, M., Yu, Y., Yan, M., Zhou, L., Liu, X., Xiao, S., Yang, Y., Wang, Y., Zheng, L., Zhao, H., Li, Y., 2019b. Toxic effects of bisphenol A on goldfish gonad development and the possible pathway of BPA disturbance in female and male fish reproduction. Chemosphere 221, 235–245. https://doi.org/10.1016/j.chemosphere.2019.01.033.

Warr, G.W., DeLuca, D., Marchalonis, J.J., 1976. Phylogenetic origins of immune recognition: lymphocyte surface immunoglobulins in the goldfish, *Carassius auratus*. Proc. Natl. Acad. Sci. U.S.A. 73, 2476–2480. https://doi.org/10.1073/pnas.73.7.2476.

Wei, H., Cai, W.-J., Liu, H.-K., Han, D., Zhu, X.-M., Yang, Y.-X., Jin, J.-Y., Xie, S.-Q., 2019. Effects of photoperiod on growth, lipid metabolism and oxidative stress of juvenile gibel carp (*Carassius auratus*). J. Photochem. Photobiol., B 198, 111552. https://doi.org/10.1016/j.jphotobiol.2019.111552.

Weil, C., Lefèvre, F., Bugeon, J., 2013. Characteristics and metabolism of different adipose tissues in fish. Rev. Fish Biol. Fish. 23, 157–173. https://doi.org/10.1007/s11160-012-9288-0.

Weinreb, O., Youdim, M.B.H., 2007. A model of MPTP-induced Parkinson's disease in the goldfish. Nat. Protoc. 2, 3016–3021. https://doi.org/10.1038/nprot.2007.393.

Wendelaar Bonga, S.E., 1997. The stress response in fish. Physiol. Rev. 77, 591–625.

Xie, J., Belosevic, M., 2018. Characterization and functional assessment of the NLRC3-like molecule of the goldfish (*Carassius auratus* L.). Dev. Comp. Immunol. 79, 1–10. https://doi.org/10.1016/j.dci.2017.09.021.

Xing, L., Gutierrez-Villagomez, J.M., Da Fonte, D.F., Venables, M.J., Trudeau, V.L., 2016. Dehydroabietic acid cytotoxicity in goldfish radial glial cells in vitro. Aquat. Toxicol. 180, 78–83. https://doi.org/10.1016/j.aquatox.2016.09.009.

Xing, L., Venables, M.J., Trudeau, V.L., 2017. Role of aromatase and radial glial cells in neurotoxin-induced dopamine neuron degeneration and regeneration. Gen. Comp. Endocrinol. 241, 69–79. https://doi.org/10.1016/j.ygcen.2016.02.011.

Yamamoto, T., 1965. Some observations on the fine structure of the intrahepatic biliary passages in goldfish (*Carassius auratus*). Z. Zellforsch. Mikrosk. Anat. 65, 319–330.

Yue, S., Wadia, V., Sekula, N., Dickinson, P.S., Thompson, R.R., 2018. Acute effects of sex steroids on visual processing in male goldfish. J. Comp. Physiol. A Neuroethol. Sens. Neural. Behav. Physiol. 204, 17–29. https://doi.org/10.1007/s00359-017-1220-6.

Yun, S.G., Kim, N.N., Shin, H.S., Choi, Y.J., Choi, J.Y., Song, J.A., Choi, C.Y., 2015. The effects of different wavelengths of light-emitting diodes on the expression of reproduction-related genes in goldfish *Carassius auratus*. Fish. Aquat. Sci. 18, 211–220.

Yunker, W.K., Smith, S., Graves, C., Davis, P.J., Unniappan, S., Rivier, J.E., Peter, R.E., Chang, J.P., 2003. Endogenous hypothalamic somatostatins differentially regulate growth hormone secretion from goldfish pituitary somatotropes in vitro. Endocrinology 144, 4031–4041. https://doi.org/10.1210/en.2003-0439.

Zhang, P., Lu, G., Liu, J., Yan, Z., Wang, Y., 2020. Toxicological responses of *Carassius auratus* induced by benzophenone-3 exposure and the association with alteration of gut microbiota. Sci. Total Environ. 747, 141255. https://doi.org/10.1016/j.scitotenv.2020.141255.

Chapter 16

Danionella translucida, a tankful of new opportunities

Gokul Rajan[1,2], Karine Duroure[1,2], Filippo Del Bene[1,2]
[1]*Sorbonne University; INSERM, CNRS, Institut de la Vision, Paris, France;* [2]*Institut Curie, PSL Research University; INSERM, CNRS, Paris, France*

Biology and ecology of *Danionella translucida*

Danionella translucida (DT) are cyprinid fish belonging to the Danioninae subfamily, the same group of fish to which the popular model organism zebrafish (ZF) belongs. The adults of this species measure only 10−12 mm in length as seen in Fig. 16.1. DT was first reported by Roberts (1986) from the roots of floating aquatic plants in Myanmar. They were found in shallow (maximum 1 m deep), slow flowing streams in the Pegu Division of Myanmar. Some other fishes collected from this site included other cyprinids of the genera *Danio* and *Microrasbora*, and other fishes like *Hara* and *Oryzias*.

Both DT and ZF are found in the fresh waters of the tropical regions in South-East Asia. A comparison of recent reported sightings of ZF and DT based on Parichy (2015) and Roberts (1986), respectively, is shown in Fig. 16.2. DT share the unique features of danionins, which include a large indentation in the medial margin of mandibles (Howes, 1979), longer anal fin, and a higher number of caudal vertebrae when compared to abdominal vertebrae (Roberts, 1986). Their reduced adult size makes them an interesting candidate model organism. This reduction in size is a result of miniaturization, a developmental phenomenon observed in at least 36 species of Cyprinidae (Kottelat et al., 2006). It is suggested that this phenomenon might have evolved independently several times in the cyprinid family (Ruber et al., 2007). While in some cases, body miniaturization maintains the general morphology of nonminiature sister species, some of these miniature fish show developmental truncation and DT is one of them. Among these, DT along with *Danionella mirifa* and *Paedocypris* are some of the most developmentally truncated fish (Britz and Conway, 2009). DT have a two-chambered swim bladder with the anterior and posterior chambers widely

Laboratory Fish in Biomedical Research. https://doi.org/10.1016/B978-0-12-821099-4.00017-1
Copyright © 2022 Elsevier Inc. All rights reserved.

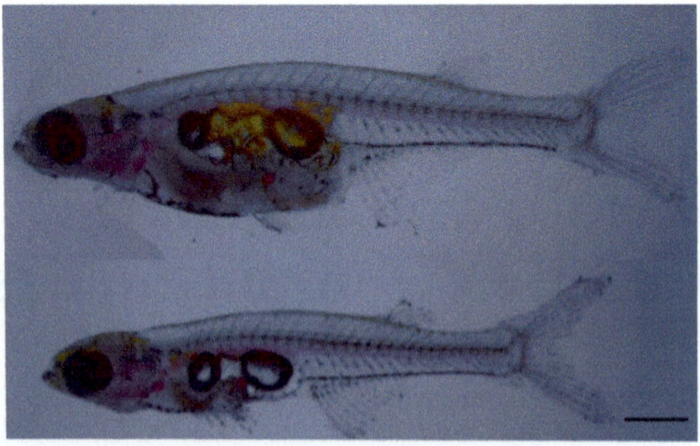

FIGURE 16.1 Eleven-month-old adult female (top) and male (bottom) *Danionella translucida*. Scale bar is 2 mm.

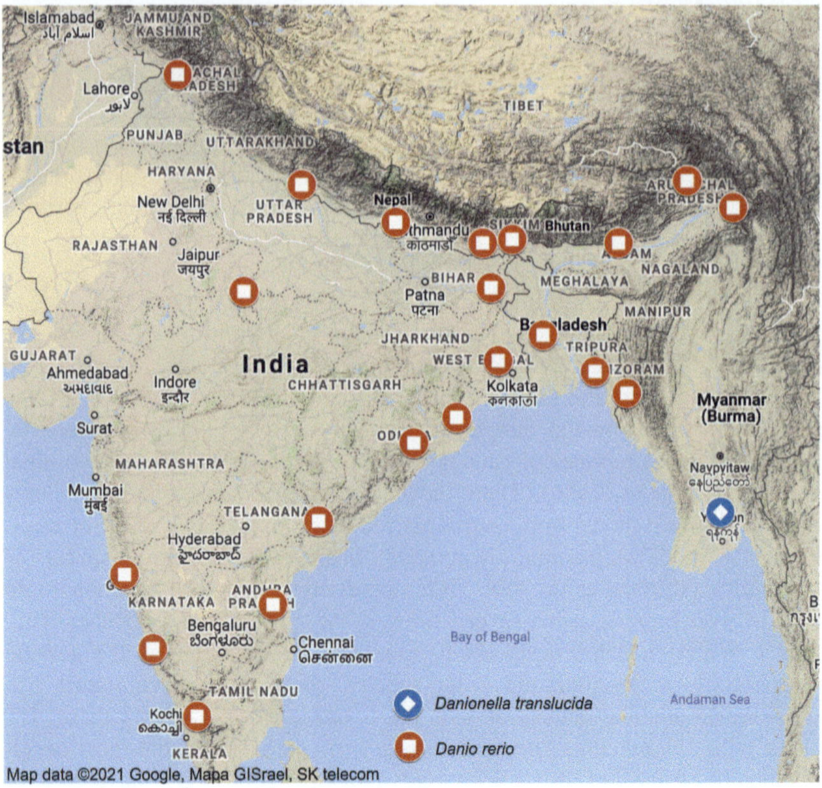

FIGURE 16.2 Distribution of *Danio rerio* and *Danionella translucida* in south and South-East Asia based on Parichy (2015) and Roberts (1986).

separated. Wild-type DT has very little pigmentation. More melanophores are present on the ventral side compared to the dorsal side where the few melanophores are mostly limited to the head.

Neuroscience

Larval ZF has become a very popular model organism in systems neuroscience. The popularity can be ascribed to their small size, transparency, and development of tools over the years to perform genetic manipulations. One of the major advantages is the ability to image whole brain activity at single cell resolution using genetically encoded calcium indicators (Ahrens et al., 2013). By virtue of the close evolutionary relationship between ZF and DT, a lot of these genetic tools are easily transferable from the former to the latter. To create a mutant lacking any pigmentation, including in the eye, Schulze et al. (2018) successfully used the CRISPR-Cas9 genome editing technique to target the *tyr* gene in DT. They have also been successful in generating a stable *Tg(NeuroD:GCaMP6f)* transgenic line using the Tol2-mediated transgenesis technique. Moreover, Kadobianskyi et al. (2019) published an assembled and annotated genome sequence of DT. This will aid all the future work on targeted genetic manipulations in DT.

In ZF, most studies have focused on simple behavioral questions pertaining to perception, locomotion, or sensorimotor transformations of a reflexive nature. This is primarily because most complex behaviors appear in the later stages of larval development. ZF larvae start to show reliable learning at 3 weeks (Valente et al., 2012). Similarly, social preference starts to appear in 3-week-old larvae (Dreosti et al., 2015). This is where the advantages of DT become more obvious. While the adults of DT are small in size and transparent, they also show a rich repertoire of behaviors. They perform visually mediated shoaling and schooling. The males also exhibit vocalization, which is likely related to male—male aggression (Schulze et al., 2018). Penalva et al. (2018) also showed socially reinforced learning in adult DT. As shown in Fig. 16.3, imaging of adult fish can be carried out noninvasively in moderately sedated fish embedded in agar where the gills are still free to move. This rich behavioral repertoire in adult DT combined with its small size, lack of pigmentation, and ease of transfer of genetic tools from ZF to DT make adult DT a very favorable system to understand neuronal underpinnings of complex behaviors.

In addition to the aforementioned neurobehavioral applications in adult DT, there can be at least two other interesting applications of DT in neuroscience: (1) To study the ontogeny of behaviors. In a vertebrate organism, it can often be challenging to closely follow the development of a behavior and its underlying neuronal circuit over various developmental stages in life. DT fares well here due to its small size and transparency throughout the life cycle. We can track the changes in neuronal networks as they mature. (2) To identify

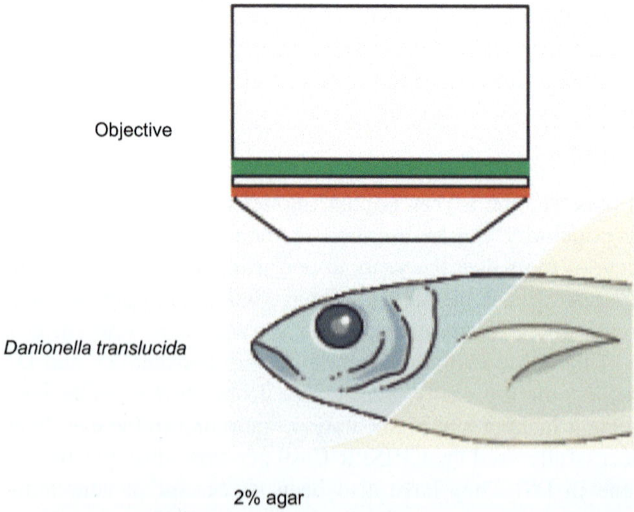

FIGURE 16.3 Mounting of adult *Danionella translucida* for noninvasive *in vivo* brain imaging.

general principles and evolution of neuronal circuits. Given the high degree of evolutionary closeness between DT and ZF, they can be used to understand functional divergence of neuronal circuit if divergent behaviors can be identified between the two species, which may or may not be a result of miniaturization in DT. As studies on ossification in *Danionella dracula* and *Paedocypris* show that most bones affected by truncation are those that appear late in the development of ZF, it suggests that developmental truncation occurs later in the ontogeny of this fish species. Hence, developmentally the two species are very much comparable in the early stages. This should come as an advantage to comparative studies of DT and ZF, which look at the larval and juvenile stages of the two fish species. Comparatives studies following this methodology can be seen in invertebrates like Nudipleura (Newcomb et al., 2012), but such studies at the level of neuronal circuits are not common in vertebrates due to the inherent difficulty of working with organisms of higher biological complexity.

Development and evolution

The advantages of a small, nonpigmented, adult fish are not limited to neuroscience. The transparency of the fish throughout its life does offer a great opportunity to developmental biologists to track the growth of various organs during its maturation. It also offers the advantage of studying developmental defects and other pathologies in a matured system.

On the other hand, miniaturization, which is more common among fishes, amphibians, and reptiles, can lead to size reduction and structural simplification in body plan (Hanken and Wake, 1993). Previous work has suggested that miniaturization can introduce morphological novelties that can increase variability (Hanken and Wake, 1993). For instance, a number of sexually dimorphic and nondimorphic morphological novelties have been described in DT (Britz and Conway 2009). From an organismal and evolutionary perspective, this is interesting and calls for more attention. At the genomic level, Kadobianskyi et al. (2019) reported that the genome size of DT is around half of that reported for ZF genome and this reduction can be a result of extensive intron truncation that is observed in DT. The origin and role of introns in eukaryotes have long been a source of much speculation (Rogozin et al., 2012) and the DT genome can be a possible tool to investigate this.

Animal husbandry: early larval stages

It is critical to pay attention to the early-stage larval fish as this is usually the period with the most mortality. We particularly pay more attention to the fish from 0 to 16 days postfertilization (dpf). From 0 to 4 dpf, we keep the larvae in incubators at 28°C. The fish is kept in E3 medium (without methylene blue). The density is maintained at a maximum of four clutches (~48 eggs) per 90 cm Petri plate and the E3 medium in the plates is changed every day. The incubator follows a light/dark cycle of approximately 14/10 h.

During 4–16 dpf, we use a coculture of rotifers and algae to grow the larvae as they start to autonomously feed on rotifers at 4–5 dpf. The rotifers are prepared in a 2 L instant ocean (Aquastore, catalog #342142) solution of 15 ppt concentration with 1.5 mL algae added twice a day. The rotifer system is continuously aerated using a pump. Algal stock used is RGcomplete (Planktovie, catalog #B001). To create the coculture, we prepare an 800 mL solution (in a 1 L beaker) of 3.75 ppt instant ocean solution (1/4th of the salt concentration used in the rotifer culture) along with 200 μL algae. Rotifers collected from the culture are sequentially filtered using filters of 250, 150, and 50 μm pore size (VWR, catalog #510-0507, 510-0520, 510-0525). The filtered rotifers collected on the 50 μm filter are mixed in a 3.75 ppt solution and added to the foregoing coculture beaker to make a final rotifer density of ~60 rotifers/mL in a volume of 800 mL. The larvae (at most 30) at 4 dpf are added to this beaker of rotifer–algae coculture and maintained until 16 dpf. The resulting culture is maintained at room temperature (~24°C) until day 16. Fig. 16.4 shows DT larvae at 10 dpf. The coculture is checked daily, and rotifers and algae are replenished when needed based on their density. Rotifer density is maintained at ~60 rotifers/mL. Dead algae is periodically removed and the beaker is changed occasionally when necessary.

Animal husbandry: early stages to adulthood

At around 16 dpf, the larvae are transferred to the system water in the main fish facility. The fish are transferred to nursery tanks with filters. Maximum

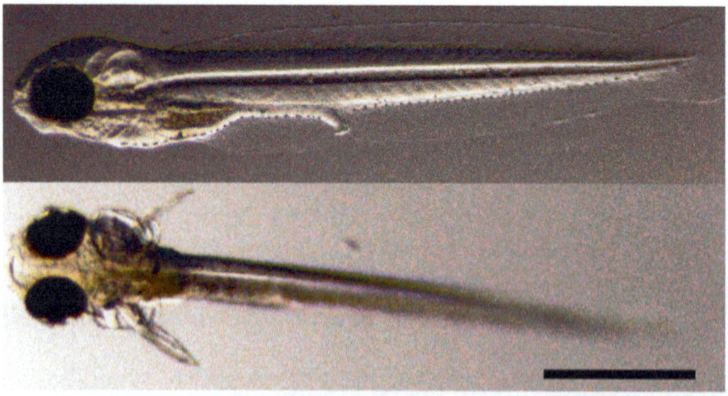

FIGURE 16.4 Larval *Danionella translucida* at 10 days postfertilization. Scale bar is 1 mm.

density at this stage is kept at 60 larvae/tank. Until 28 dpf, they remain in this tank and the water inflow is kept to a minimum (slow dripping). At 28 dpf, they are moved to adult tanks with a filter size of 800 μm and a constant stream of water inflow. We use adult tanks of 8 L volume with a density of at most 11.25 fish/L. Although it is still less than 12 fish/L as recommended for ZF (Castranova et al., 2011; Aleström et al., 2019), considering that the size of adult DT is less than 1/3 of ZF adult, it should be possible to accommodate more fish. We will now discuss more about the physicochemical parameters and the food regime in use after moving the fish into the system water of the animal house.

As most of the danionin species are found in the fresh waters of tropical and subtropical regions, they show a considerable degree of similarity in terms of tolerance to physicochemical parameters and food source. ZF, for instance, can tolerate a wide range of temperatures (10–40°C), pH (6–10), and conductivities (10–271 μS/cm) in nature (Aleström et al., 2019) but a narrower range of these parameters is used in the animal housing's system water. Similarly, we use a narrow range of these parameters for DT, which are quite near the values used for ZF husbandry. Hence, the housing systems can be shared by the two species if needed. For DT, as shown in Table 16.1, the room

TABLE 16.1 Physicochemical parameters used for *Danionella translucida* in the animal housing.

Parameter	Range
Room temperature	20–24°C
Water temperature	25–28°C
pH	6.3–8.3
Conductivity	250–450 μS

temperature of the facility is maintained from 21 to 23°C. The temperature of the water is in the range of 25–28°C. The pH is balanced at 7.3 ± 1. The conductivity of the system water can vary from 250 to 450 μS/cm. The system water contains 20 g of instant ocean (aquarium systems) per 1 L and 10 g $CaCl_2$ per 1 L. The pH is maintained with $NaHCO_3$. Water renewal is fixed at 10% in 24 h. The light/dark cycle is maintained at 14/10 h in our facility for both ZF and DT.

Regarding food, McClure et al. (2006) analyzed the gut content of 327 individual fish representing 17 populations of eight danionin species in the wild to conclude that insects were the primary source of food in these species. For DT in the laboratory setting, larval fish starting from 16 dpf are fed with a regular diet of Gemma Micro 75 (Skretting, USA) (twice a day) with occasional rotifers (three times a week). The water inlet at this point is maintained at a very slow rate with constant dripping. Once the fish are at 28 dpf, they are transferred to a feeding regime with regular Gemma Micro 150 (Skretting, USA) (twice a day) and live *Artemia* (once a day). The water inlet at this point is increased to a constant flow. Spawning, schooling, and vocalization—signs of social behavior—can be seen as a proxy for a good health status of the fish.

Our fish husbandry system and therefore the components thereof, including the tanks, come from Tecniplast, France.

Breeding

DT spawn eggs in clutches. Sexual maturity is reached in our facility at 6–8 weeks of age. They are also known to spawn in crevices. Hence, we add two to four silicone tubes (~5 cm long) in the adult tanks to aid spawning (Schulze et al., 2018). Fig. 16.5 shows an adult tank with such tubes. Supplementary Video S1.1 shows a video of the same. We also observe that spawning is often initiated immediately after the first feeding cycle of *Artemia* in the morning indicating that food may stimulate mating.

DT do not require special breeding tanks as they do not feed on their eggs like ZF. They lay eggs inside or around the silicone tubes in their home tanks. So, the eggs can be collected at a later time of the day. We perform breeding of the fish in the community in their home tanks. Mating in smaller groups has not been tried sufficiently as with the limited number of mutants and transgenics at the moment it was not necessary. However, we have had anecdotal success in breeding with as little as three fish (both two males/one female and one male/two females).

We have now carried out at most three generations of inbreeding of DT in our animal house and we have not observed any signs of inbreeding depression such as reduced fecundity or emergence of sick progenies. In fact, the 2–3-year-old fish stocks in our facility still appear mostly healthy. However, egg production can decrease with age as we note later.

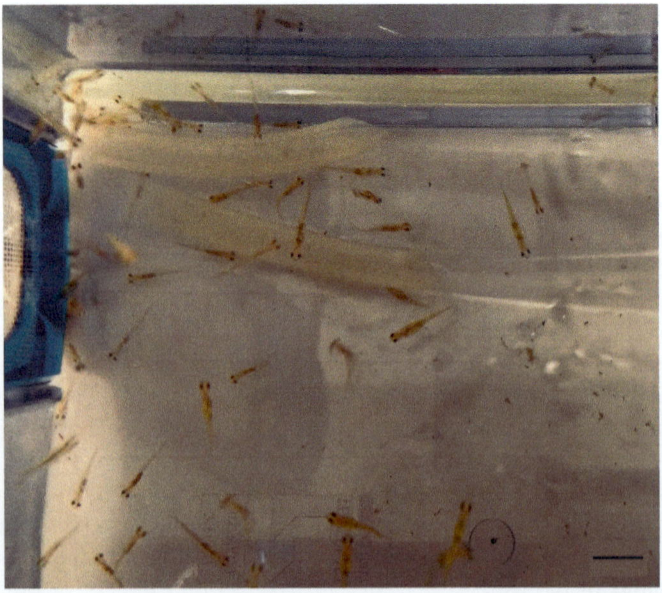

FIGURE 16.5 Adult *Danionella translucida* in their home tanks with two silicone tubes to aid spawning in the fish. Scale bar is 10 mm.

In other fish species, it has been observed that a reduction in body size is in general linked with a reduction in fecundity and an increase in size of the egg (Hanken and Wake, 1993; Duarte and Alcaraz, 1989). We carried out a small observational study to quantify the egg production in DT. All the housing parameters in the study are maintained at the aforementioned range and breeding methodology is also kept the same (number of silicone tubes = four per tank). We observed that 1-year-old adult DT produce on an average 8.26 clutches/50 females/day or 99.12 eggs/50 females/day (at an average of 12 eggs/clutch). Fig. 16.6 shows two clutches of DT eggs. We also observed that 1-year-old DT females produced more clutches than the 2-year-old fish that produce 1.57 clutches/50 females/day. A one female:two males sex ratio was maintained in both 1-year- and 2-year-old groups. With regard to the sex ratio, we had tanks with two different sex ratios in the 2-year-old group. The tanks with the one female:two males ratio produced 2.62 times more clutches per day than the tank with the one female:one male ratio.

The number of eggs produced per female in DT is very low if one compares it to ZF, but this had been mostly sufficient for our purposes. The best breeding outcome of 8.26 clutches/50 females/day was obtained in 1-year-old DT fish maintained at a one female:two males ratio. However, while breeding three fish (one female/two males) in a tank, we have had an almost constant production of one clutch/female/day. This might indicate that community

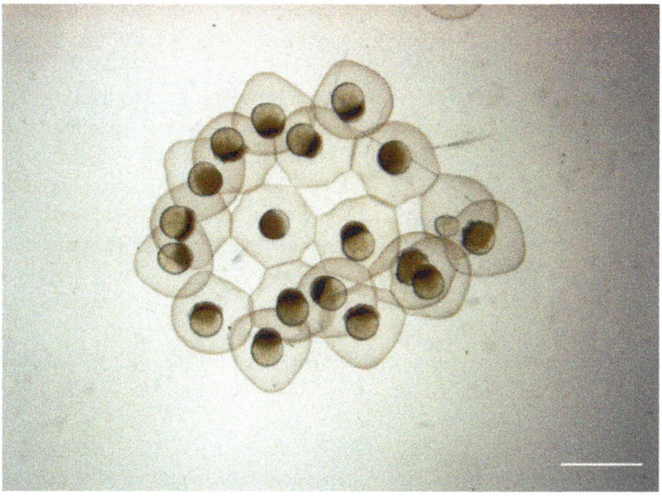

FIGURE 16.6 Clutches of eggs of *Danionella translucida*. Scale bar is 2 mm.

breeding comes at a cost of high intracommunity competition to mate. So, we believe that egg production can be eventually increased further with improvement in breeding techniques. Breeding in many smaller communities compared to one large community might be just one of the many changes that one can incorporate.

Supplementary data

Supplementary data related to this article can be found online at https://doi.org/10.1016/B978-0-12-821099-4.00017-1.

Acknowledgments

We thank the fish facility at Institut Curie for their support in animal husbandry. We also extend our thanks to the Judkewitz lab in Berlin for the initial stock of DT and advice on growing and breeding the fish.

References

Ahrens, M.B., Orger, M.B., Robson, D.N., Li, J.M., Keller, P.J., 2013. Whole-brain functional imaging at cellular resolution using light-sheet microscopy. Nat. Methods 10 (5), 413–420. https://doi.org/10.1038/nmeth.2434.

Aleström, P., D'Angelo, L., Midtlyng, P.J., Schorderet, D.F., Schulte-Merker, S., Sohm, F., Warner, S., 2019. Zebrafish: housing and husbandry recommendations. Lab. Anim. https://doi.org/10.1177/0023677219869037.

Britz, R., Conway, K.W., 2009. Osteology of *paedocypris*, a miniature and highly developmentally truncated fish (Teleostei: Ostariophysi: Cyprinidae). J. Morphol. 270 (4), 389–412. https://doi.org/10.1002/jmor.10698.

Castranova, D., Lawton, A., Lawrence, C., Baumann, D.P., Best, J., Coscolla, J., et al., 2011. The effect of stocking densities on reproductive performance in laboratory zebrafish (*Danio rerio*). Zebrafish 8 (3), 141–146. https://doi.org/10.1089/zeb.2011.0688.

Dreosti, E., Lopes, G., Kampff, A.R., Wilson, S.W., 2015. Development of social behavior in young zebrafish. Front. Neural Circ. 1, 39. https://doi.org/10.3389/fncir.2015.00039.

Duarte, C.M., Alcaraz, M., 1989. To produce many small or few large eggs : a size-independent reproductive tactic of fish. Oecologia 80 (3), 401–404.

Hanken, J., Wake, D.B., 1993. Miniaturization of body size: organismal consequences and evolutionary significance. Annu. Rev. Ecol. Syst. 24, 501–519.

Howes, G.J., 1979. Notes on the anatomy of *Macrochirichthys macrochirus* (Valenciennes), 1844, with comments on the Cultrinae (Pisces, Cyprinidae). Bull. Brit. Mus. Nat. Hist., ser. zool. 36, 147–200.

Kadobianskyi, M., Schulze, L., Schuelke, M., Judkewitz, B., 2019. Hybrid genome assembly and annotation of *Danionella translucida*, a transparent fish with the smallest known vertebrate brain. Sci. Data 6 (1), 1–7. https://doi.org/10.1038/s41597-019-0161-z, 31451709.

Kottelat, M., Britz, R., Tan, H.H., Witte, K.E., 2006. Paedocypris, a new genus of southeast Asian cyprinid fish with a remarkable sexual dimorphism, comprises the world's smallest vertebrate. Proc. R. Soc. B 273, 895–899. https://doi.org/10.1098/rspb.2005.3419.

McClure, M.M., McIntyre, P.B., McCune, A.R., 2006. Notes on the natural diet and habitat of eight danionin fishes, including the zebrafish *Danio rerio*. J. Fish Biol. 69 (2), 553–570. https://doi.org/10.1111/j.1095-8649.2006.01125.x.

Newcomb, J.M., Sakurai, A., Lillvis, J.L., Gunaratne, C.A., Katz, P.S., 2012. Homology and homoplasy of swimming behaviors and neural circuits in the Nudipleura (Mollusca, Gastropoda, Opisthobranchia). Proc. Natl. Acad. Sci. U. S. A. https://doi.org/10.1073/pnas.1201877109.

Parichy, D.M., 2015. Advancing biology through a deeper understanding of zebrafish ecology and evolution. eLife. https://doi.org/10.7554/eLife.05635.001.

Penalva, A., Bedke, J., Cook, E.S.B., Barrios, J.P., Bertram, E.P.L., Douglass, A.D., 2018. Establishment of the miniature fish species *Danionella translucida* as a genetically and optically tractable neuroscience model. BioRxiv 444026. https://doi.org/10.1101/444026.

Roberts, T.R., 1986. *Danionella translucida*, a new genus and species of cyprinid fish from Burma, one of the smallest living vertebrates. Environ. Biol. Fish. 16 (4), 231–241. https://doi.org/10.1007/BF00842977.

Rogozin, I.B., Carmel, L., Csuros, M., Koonin, E.V., 2012. Origin and evolution of spliceosomal introns. Biol. Direct. https://doi.org/10.1186/1745-6150-7-11. BioMed Central.

Rüber, L., Kottelat, M., Tan, H.H., Ng, P.K.L., Britz, R., 2007. Evolution of miniaturization and the phylogenetic position of Paedocypris, comprising the world's smallest vertebrate. BMC Evol. Biol. 7 (1), 38. https://doi.org/10.1186/1471-2148-7-38.

Schulze, L., Henninger, J., Kadobianskyi, M., Chaigne, T., Faustino, A.I., Hakiy, N., et al., 2018. Transparent *Danionella translucida* as a genetically tractable vertebrate brain model. Nat. Methods 15 (11), 977–983. https://doi.org/10.1038/s41592-018-0144-6.

Valente, A., Huang, K.H., Portugues, R., Engert, F., 2012. Ontogeny of classical and operant learning behaviors in zebrafish. Learn. Mem. 19 (4), 170–177. https://doi.org/10.1101/lm.025668.112.

Section 3

Databases

Chapter 17

Fish inventory databases

Jana Oltova[1,2]
[1]*Department of Cell Differentiation, Institute of Molecular Genetics of the Czech Academy of Sciences, Prague, Czech Republic;* [2]*CZ-OPENSCREEN, Institute of Molecular Genetics of the Czech Academy of Sciences, Prague, Czech Republic*

Introduction

Thanks to numerous experimental advantages, rapidly evolving new techniques, as well as the principle of replacement, an integral part of the 3R concept (Kirk, 2018; Russell and Burch, 1959), small fish species are constantly gaining popularity in the field of biomedicine, toxicology, and many other research areas. Moreover, in areas where there are experimental limitations that restrict the utility of rodents, fish can often present a relevant alternative, e.g., large-scale studies and screens (reviewed in Lieschke and Currie, 2007; Ablain and Zon, 2013; Rasighaemi et al., 2015). With more than 1000 laboratories using zebrafish as a model organism (zfin.org) and many more using other small fish species, the question of proper tracking and reporting of animals used in research is becoming more critical than before. The fish facility manager may face the challenge of tracking tens of thousands of animals from the moment of fertilization until their death, which can be a laborious task.

There are certain aspects of fish tracking that make it more challenging and complex compared to rodents. Currently, there is no cost-effective system to distinguish fish individuals in a tank. This means that groups of animals must be tracked as opposed to individuals, which are typically the basic trackable unit in rodents. Although emerging technologies like subdermal chipping may come into play in the future, they are not yet sufficiently miniaturized to be routinely used for small fish species (Cousin et al., 2012). It is therefore highly recommended to employ a dedicated and scalable database solution that can track groups of animals and their rearrangements, and is capable of storing the breeding history as well as creating animal usage reports, tailored specifically to the needs of a fish facility.

For a long time, proper animal tracking has been a neglected topic in the fish community, but gradually this subject is getting more attention. The use of nonmammalian vertebrates is a subject of legal regulation by EU Directive

Laboratory Fish in Biomedical Research. https://doi.org/10.1016/B978-0-12-821099-4.00005-5
Copyright © 2022 Elsevier Inc. All rights reserved.

2010/63 on the protection of animals for scientific purposes, and recommendations have been issued by the Federation for Laboratory Animal Science Associations and EuFishBioMed (Aleström, 2020), which emphasize the need for high standards of identification, tracking, and reporting of aquatic animals. Also, in the last few years, the focus on reproducibility, standardization, and proper identification of genetic modifications and backgrounds has become much more prominent in the community.

In this chapter, you will find an overview of the features currently available in purpose-designed fish databases in the market, as well as practical recommendations for animal tracking.

Basic operation principles

Before we delve deeper into the topic, let me briefly mention general considerations to focus on when choosing the database solution for a fish facility. Because the fish inventory database will be used by facility staff and researchers on a daily basis, careful evaluation of the basic operation principles of available options before finally choosing the most suitable one will certainly pay off in the long term.

First, it is a good idea to consider **platform** and **data accessibility**. There are solutions dedicated only to a single platform (e.g., iPad or desktop) but there are also cross-platform, web-based options, which offer one notable advantage: the data can be accessed from anywhere. This can be helpful because browsing stocks from a desktop sitting in an office can be as equally important as entering fish-related actions from a mobile phone while walking around a facility. Another relevant point, especially for web-based solutions, is **data security**. Secure sockets layer encryption for communication between users and an application, as well as storage of sensitive data at highly secure servers, should always be guaranteed.

Next, excellent **usability** via an intuitive user interface, with attributes like consistent color coding, advanced data visualization, drag-and-drop support, and mobile device optimization, is a plus. The rationale is that the more often users have to write notes by hand and then enter this into the software, or are even forced to find workarounds due to poor design of the database, the more mistakes will be generated in the long term. This will not only slow down the whole process, but potentially also compromise the overall accuracy of the data. Finally, a wide variety of **exporting, reporting, and backup** options will make a facility manager's life easier. Importantly, **sustainability** of a solution of choice should always be guaranteed by the provider, as transferring data from one solution to another is a long and burdensome task.

Finding a fish husbandry database compliant with all these requirements may indeed be a challenge, but as the inventory will be one of the cornerstones of smooth facility operation, all the aforementioned aspects deserve attention. Avoiding solutions with poor sustainability, limited exporting options, or a single platform commitment is a reasonable strategy, because all these aspects can slow down the workflow significantly or even compromise the data.

Fish characterization and grouping

Comprehensive spatial and temporal tracking, a foundation of any husbandry database, is based on systematic characterization and grouping of the animals. Therefore the crucial step in the process of animal tracking is the introduction of proper naming rules.

The common naming standard for genetically modified zebrafish lines, recommended by the Zebrafish Nomenclature Committee (see zfin.org), should be used consistently throughout reports, manuscripts, and the tracking system itself to provide sufficient details about the animals. Without a well-designed database system, naming consistency may be difficult to achieve in larger facilities, as users often unintentionally introduce different spelling variants of the same genomic feature. This may become unavoidable without proper consistency controls in the database system.

For routine daily work, in some cases it may be practical to use aliases or abbreviations, especially for lines that possess multiple genomic features, or that have a well-known and abundant historical "nickname." A common example might be the Casper zebrafish, which carries two mutations in the genome causing transparency, but there are many other examples where a trivial name is commonly used (for a list of examples, see Table 17.1). However, in these cases it is necessary to store both the alias and the full name of the line within the database to prevent confusion. When identified, the genomic features and alleles should always be defined in the database for each line.

In zebrafish, inbred lines as we know them from rodents have not been established, due to high sensitivity of the animals to inbreeding depression

TABLE 17.1 Examples of commonly used trivial names for zebrafish mutant lines.

Trivial name	Affected gene	Phenotypic defect
Casper	mitfa, mpv17	Pigmentation
Panther	csf1ra	Pigmentation
Sparse	kita	Pigmentation
Moonshine	trim33	Erythroid development
Cloche	npas4l	Hematopoietic and endothelial development
Vlad Tepes	gata1	Erythroid development
Spadetail	tbx16	Mesodermal and hematopoietic development

(Monson and Sadler, 2010). So-called backgrounds are based on the origins of the founder population and its characteristics, and they do not necessarily ensure genetic homogeneity.

Structure of a line name

To ensure naming consistency, it is recommended to structure the names of the genetically modified lines into multiple fields and avoid unstructured text whenever possible. Here is an example of a name structure for genetic modification. Please note that various database solutions may slightly differ in the designation of individual elements.

Type of modification: transgenic (Tg), mutant (Mut), other custom options, e.g., enhancer trap (Et)
Fields for transgenic modifications: driver, triggered gene/genetic element, allele/mutation (optional)
Fields for mutant modifications: affected gene, allele/mutation

In the software, the name of a genetically modified line can occur in several forms: **full official name**, e.g., Tg(*gata1*:DsRed), **short name**, e.g., *gata1*:DsRed, or an **alias**, e.g., *gata1*. On the other hand, for wild-type line names, simply designating the background, e.g., WT (AB), may be sufficient in most cases, without the need for abbreviation.

In case multiple fish species are kept within the same facility, it is recommended to use a prefix, suffix, or a dedicated field to distinguish between species in the database.

Animal grouping

Based on the genotype and background, animals of the same species should be divided into lines/strains, which are further subdivided into stocks, i.e., groups of fish sharing the same line, as well as parents and date of birth (DOB). Within the database, each line and stock should have a unique identifier (ID) that can be used for a quick reference in the physical facility but also for fast searching in the database.

In reality, the number of fish originating from a single mating event can, and often does, exceed the capacity of a single tank. Therefore the database should also track rearrangements of the tanks, like splitting and merging. For instance, in one of the currently available solutions, Zebrabase, this is solved by diving the virtual stock units further into so-called substocks, defined as physical subgroups of a stock sharing the same container. Substocks serve as the basic operation units within the software and can be further rearranged.

For each line/strain, it is also practical to store permission documents for experimental lines, information about the source and the date the line was imported or created, the user responsible for the proper breeding strategy of that particular line/strain, as well as attachments like publications, genotyping protocols, or even visual content (e.g., observed phenotype, or expression pattern of a transgene). For specific stocks or their subgroups, tracking of the owner, IDs of parents, and permission document IDs for experimental tanks are recommended.

Record browsing and spatial tracking

The most essential feature of a husbandry database is efficient browsing of the data that allows quick filtering and searching for specific subset of stocks based on various parameters, e.g., owner, line/strain, age, status, or location. Although for very small facilities, it may be sufficient to use a list view of all the tanks (Fig. 17.1A), to ensure comprehensive tracking in larger facilities, it is necessary to use a dedicated visual representation of the facility within the software (Fig. 17.1B). Details of the stock records then contain additional details, like action history or IDs of parents and siblings, that do not fit into the more compact views (Fig. 17.2A and B).

For efficient spatial tracking, it is essential to have a clear system in designating individual tank positions by unique identifiers. Typically, the facility comprises multiple rooms and racks. An example of a unique identifier of a specific tank position could be **F01:C11**, where the ID of the room (F) and rack (01) are part of the position name, together with the specific tank position within the rack (C11). Ideally, every change in position should be tracked within the software. However, in certain circumstances this may not be necessary or practical, and information about a room or rack may be sufficient. Some database solutions also allow advanced spatial operations, like reserving or blocking positions, splitting the tanks into multiple subtanks using a divider, or sharing positions by multiple independent stocks (Fig. 17.1B).

Tank labeling

It is recommended to employ a consistent tank labeling system, including the name of the line, DOB, and a unique identifier of the (sub)stock within the database (Fig. 17.2C). Most currently available databases provide the option of printing preconfigured labels directly from the software. QR codes providing a link to a specific tank record in the database can improve the efficiency of browsing the record details as well as entering actions (Fig. 17.2A).

Performing actions and monitoring parameters

Counting of the animals

Central monitoring of the animal numbers is vital for any facility. Initial counting of any tank should be performed at the developmental stage, which is required by law (e.g., 120 h postfertilization corresponding to independently

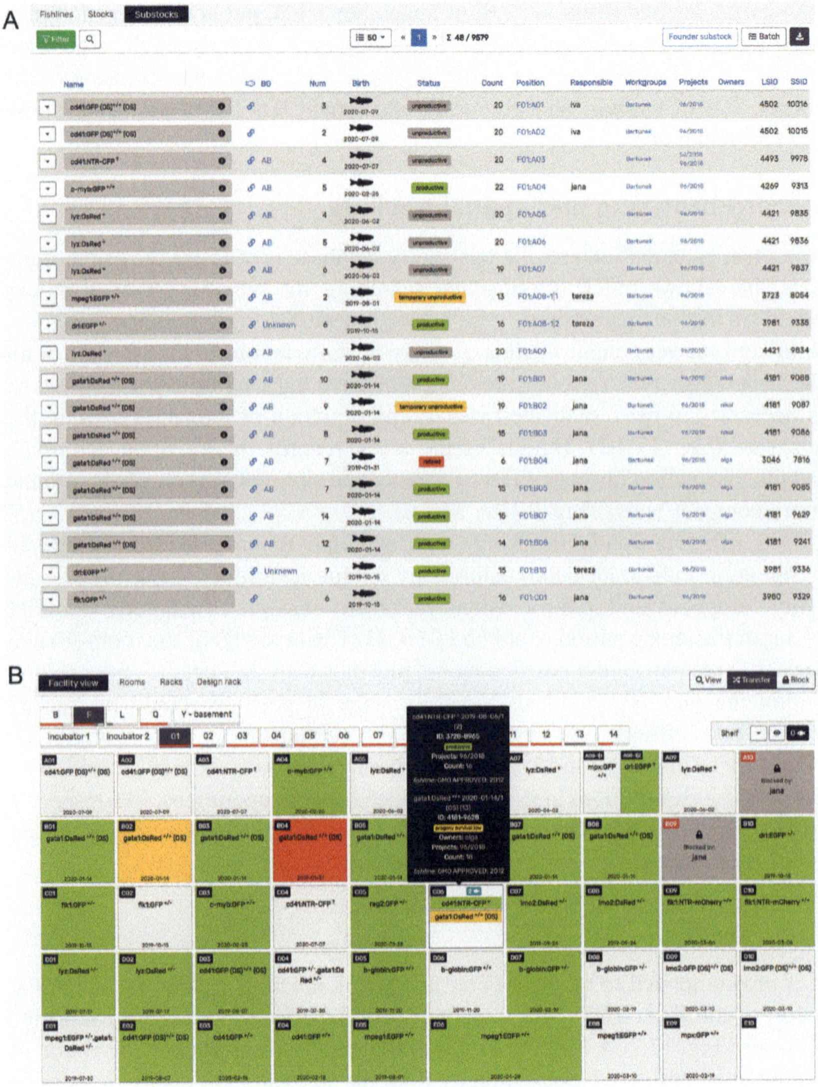

FIGURE 17.1 **Basic views of the data.** (A) **List view of fish stocks**. List view can be browsed, filtered, and searched to find a specific (sub)stock record. It displays a set of basic relevant information like number of fish in the tank, the owner, and location within the facility. When clicked, the record detail opens. The interface also performs actions via the action menu button on the left-hand side of each row. (B) **Visual representation of a facility**. Tanks are organized according to their position in rooms, e.g., F, and racks, e.g., 01. Tanks can be transferred by drag and drop from one position to another and additional information is shown in the tooltip on mouse-over. When clicked, the record detail opens. *All images acquired in Zebrabase 3.0 Oltova, J., et al., 2018. Zebrabase: an intuitive tracking solution for aquatic model organisms. Zebrafish 15 (6), 642–647; for more information, visit www.zebrabase.org.*

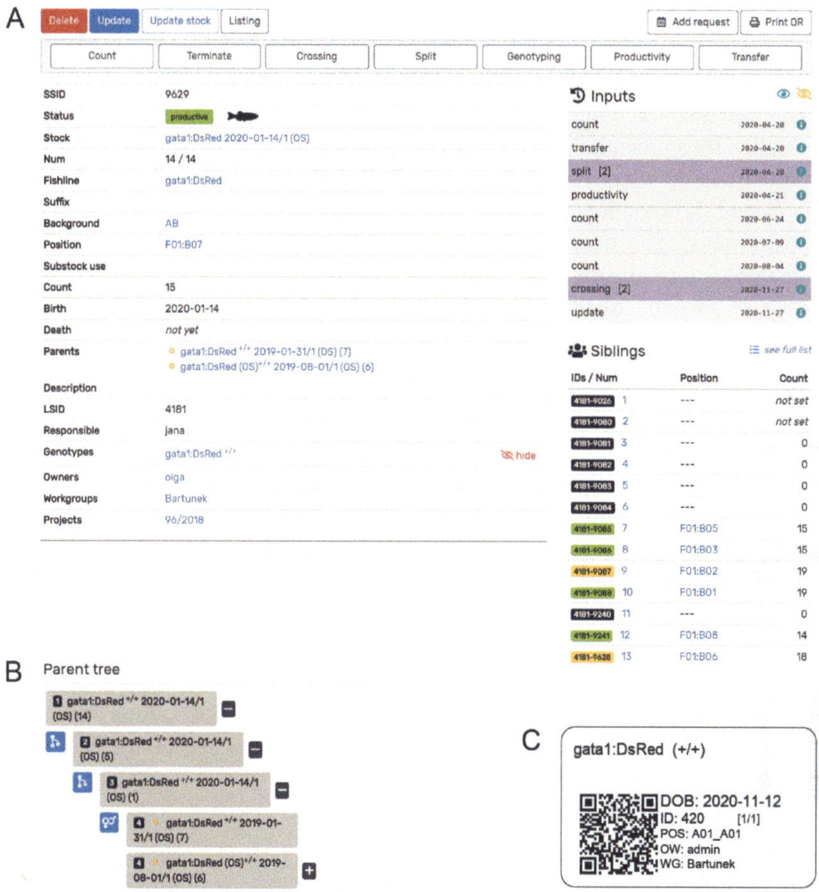

FIGURE 17.2 **Tank record detail.** (A) **Detailed stock record**, including the number of animals, date of birth, owner, location in the facility, and status. This view allows the most common actions to be performed like counting, crossing, splitting, or genotyping. (B) **Pedigree** shows past generations and actions from which the current substock originated. All parts of the pedigree serve as links to the respective ancestor records. (C) **An example of a tank label including a QR code.** The code serves as a direct link to the stock detail in the database. *All images acquired in Zebrabase 3.0 Oltova, J., et al., 2018. Zebrabase: an intuitive tracking solution for aquatic model organisms. Zebrafish 15 (6), 642–647; for more information, visit www.zebrabase.org.*

feeding larvae in zebrafish). Then, every change in the number of fish should be entered to the database, including details about the cause (animal found dead, animal euthanized, animal used for experiment) and the name of the user reporting the change. In case an animal is used for an experiment, additional details like procedure and severity assessment should be entered as well.

Proper fish densities in a tank are essential for optimal development and size of the animals. For zebrafish, it has been reported that the density can influence the sex ratio within the group (Ribas et al., 2017). Usually, it is recommended to keep 4−10 adult fish per liter (Lawrence and Manson, 2012; Castranova et al., 2011) and it is generally advisable to keep zebrafish as mixed sex groups to avoid oviduct inflammation in females (Aleström, 2020). Therefore also the monitoring of the numbers of animals of each sex in a tank is highly recommended. This information gives users a good idea about how many breeding tanks can be set up from a specific housing tank.

In certain situations, it might be necessary to isolate a single fish in a housing tank. The time a fish spends in solitude should be limited, as it is detrimental to its health, but sometimes the isolated fish can be mixed with a group of visually distinguishable animals of other stocks (e.g., transparent background). Such "positional sharing" between two independent stocks should also be indicated in the database, in case there is an option to do so.

Animal age, productivity, and status

Monitoring and visualization of animal age, health, and productivity is another essential feature of any husbandry database. The age can be visualized using icons as embryos, juvenile fish, and adult fish in list view (Fig. 17.1A). Moreover, consistent color coding can be used throughout the application to implicate tank status in different views, e.g., gray for unproductive fish, green for productive fish, red for retired fish, and yellow for user-defined alerts (Fig. 17.1A and B). The database should provide enough versatility to define custom statuses and rules when these are applied, e.g., sick animal, low productivity, low progeny survival, etc. That way, each user can immediately see potential issues associated with a particular stock.

Breeding management

High fecundity is one of the most attractive characteristics of zebrafish that make it a great research model and thus breeding is one of the most common activities in a facility and requires a dedicated strategy. Mating and its frequency should be tracked to ensure that a necessary regeneration period for the females is provided between mating (typically 1 week). Another aim of breeding management is the prevention of inbreeding, which can cause inbreeding depression as already mentioned. Inbreeding depression results in accumulation of deleterious effects (e.g., low survival, fecundity, and developmental defects) in the progeny of sibling parents. New generations should therefore arise preferentially from an outcross. Close relational breeding should be avoided whenever possible (Varga, 2016; Martins et al., 2016; Trevarrow and Robison, 2004).

In the database, a purpose-designed function should be used for incrossing and outcrossing of the animals. The internal structure of the database should automatically combine the genotypes of the parents when outcrossing to a different line, and when the genotype combination is unique, a new line should also be created automatically in the software. Moreover, the zygosity (e.g., +/+, +/−) of the stocks should always be indicated, both in the software and on the tank labels, and it is recommended to track changes of zygosity in the database using a dedicated genotyping function.

Furthermore, it is advisable to store breeding frequency, the date of last breeding, as well as breeding performance of each individual stock. Breeding history as well as interactive pedigrees should be displayed, showing the dates of crossings and stock rearrangements (Fig. 17.2B). Breeding history enables animals that shared the same container in the past to be tracked, which can be convenient when a disease outbreak or unintentional animal mix-up occurs and it is necessary to trace back the mistake to the source tank.

Monitoring of the conditions

In addition to storing the standard operating procedures for the reference of all facility users and especially the caretakers, it may be useful to monitor the following parameters in detail and keep the information in the database.

Temperature

The development of poikilothermic animals, including zebrafish, is temperature dependent, and for consistency when referring to developmental forms, it is recommended to keep zebrafish embryos at a defined reference temperature of 28.5°C (Kimmel et al.). Although adults generally tolerate a wider temperature range (24−29°C), it is important to avoid sudden changes in temperature (Aleström, 2020). In systems without internal monitoring, temperature should be checked regularly, and values should be stored in the database system.

Light cycle

A static dark/light cycle is used in most fish facilities, commonly using the recommended 10 h dark, 14 h light cycle for zebrafish. However, alternative or shifted light cycles can be introduced for various practical and experimental reasons. These should be tracked within the database to provide fast and easy identification of light cycle variants within the facility.

Diet

A combination of dry and live feed is used in most facilities to provide an optimal mixture supporting growth, productivity, and also to provide enrichment for the animals (Varga, 2016; Lawrence et al., 2012). As fish of different developmental stages require different feed compositions and feeding frequencies, it also may be useful to store the information independently for each tank, rack, or room for the caretaker's reference, especially in case multiple feeding regimes for the adults are employed within a single facility. These data can also be correlated with productivity or size of the animals, which can provide useful information when testing different diets.

Water quality

In housing systems, basic parameters of the water like pH and conductivity should be inspected daily. Even when the unit has internal monitoring of water quality, it might be a good idea to import the information in the husbandry database, and in the case of health or productivity issues, look for correlation of the problem with changes in water quality parameters. Filtering of the outgoing water and a constant water exchange in recirculating systems ensure proper control of ammonia, nitrite, and nitrate levels but as these parameters are critical for animal health, they should be monitored and documented at least once per week (Aleström, 2020).

Other actions

Also for other actions, like tank rearrangements, updates of the basic stock entry, or batch actions used for changing owner, permission ID, or printing labels, it is recommended to keep a full log, including the date, time, and user. All performed actions should be fully searchable in the action history.

Statistics and reporting

Based on the 3Rs (refinement, replacement, and reduction), part of the EU Directive 2010/63, the onset of protected life stage for all nonhuman vertebrates is the independently feeding free-living form, which has been set at 120 h postfertilization at 28.5°C for zebrafish (Strahle et al., 2012; Hernandez et al., 2018). From that time point, all experimental procedures as well as fish numbers should be recorded and summarized in periodical reports. The database should store and filter the fish census and export usage reports for the whole facility or a specific user, line/strain, permission license ID, location, and date range. The statistics should display an overview of the animals that were born, found dead, euthanized, and used for experiments.

According to EU Directive 2010/63, the experimental procedures should also include severity assessment, where each procedure in an experiment should be classified as "nonrecovery," "mild," "moderate," or "severe." This classification is based on several parameters like the type of procedure, frequency, intensity, and duration, effectiveness of refinements, and cumulative severity, and can differ for different species. For more information about severity assessment in fish, please refer to Hawkins et al. (2011).

Advanced features

User management

For efficient organization of larger facilities, it is necessary to set up several permission groups within the software, e.g., administrator, caretaker, standard user, or guest. The user rights of various permission groups can range from view-only to most advanced facility configuration rights. In the case when multiple research groups are sharing the same facility, some solutions assign the users to working groups ensuring effective data separation.

Communication, requesting, and experiment planning

Communication between users, which is one of the building blocks of smooth facility operation, can be streamlined by an in-app messaging system. Moreover, it is useful to configure in-built notifications in a way that each change related to a tank is reported to the owner. For planning of the animal-related actions, this is practical when the database supports different types of request to the facility manager/caretakers (e.g., tank termination, setting up fish in breeding tanks, or genotyping), and if users can use a built-in calendar system to plan their own fish-related activities, like genotyping/sorting and various experimental procedures.

Data management

Periodical backups (recommended minimum once per day) and exports of the data should be performed by the facility manager based on the facility's needs, unless they are performed automatically by the software.

Examples of current solutions

Commercial solutions

PyRAT Aquatics (Scionics)

https://www.scionics.com/pyrat_aquatic.html

Scionics has been well known for its PyRAT animal facility software, which became a widespread standard for rodent tracking. The dedicated aquatic version of its facility inventory system is web based and provides a wide range of features using the original PyRAT user interface. The product is ISO certified. Free demo installation is available upon request.

Hybrid solutions

Zebrabase (Oltova et al., 2018)

https://zebrabase.org/

Zebrabase is a nonprofit web-based application developed at the Institute of Molecular Genetics in Prague, provided for free to small facilities. Larger facilities are charged based on their size to ensure technical reinforcement, future development, and sustainability of the application. Zebrabase has been designed specifically for small fish species but can also be adjusted for use with other aquatic animals housed in tanks. The hosted character of Zebrabase allows automatic backups and instant version upgrades as part of the service. Free demo can be accessed directly from the webpage.

Open-source solutions

ZeBase (Hensley et al., 2012)

http://zebase.bio.purdue.edu/

ZeBase is an open-source relational database designed to store fishline-related genetic, breeding, and survival information. The application includes barcode generation and scanning, and provides an option for both single- and multilaboratory environments. The application can be accessed through web browsers in the network and allows customization by simple scripting. Demo can be accessed directly from the webpage.

Facile Fish (Yakulov and Walz, 2015)

https://zebrafish.jimdofree.com/

In addition to basic fish-tracking functionalities, Facile Fish plans experiments, monitors and displays water parameters, and prints labels directly from the database. Moreover, it provides an integrated calendar and allows customization. Demo can be accessed directly from the webpage.

Future of fish databases

Although keeping fish records organized seems like a straightforward assignment, without a proper tracking system it can become very laborious and time demanding. When it comes to features, the currently available solutions certainly offer significant improvements over those available several years ago, including more attractive operation principles (cross-platform character, mobile-optimized user interface, etc.). However, the development of fish inventory databases will unquestionably move forward in the coming years. One possible future direction is advanced management features for large facilities, like detailed enterprise resource planning and invoicing, which would streamline the charging of users for different services. Another improvement could be the generation of customized exports based on user-defined templates that could be used for reporting purposes or monitoring of various parameters and their correlation in the facility.

There is also room for future development in networking. Currently, the solutions do not allow any interactions between facilities; however, their concept could be easily expanded to develop a common repository of shared fish stocks, where users could upload their stock information directly from their own database. Contact information and physical location of the stock and its status could be synchronized with the original database and contained within a stock record in a "virtual stock center." Such a tool would help to streamline the process of animal importing and reinforce collaboration in the already very collaborative fish community, which would certainly be highly appreciated among researchers and facility managers.

Acknowledgment

I wish to express my gratitude to my supervisor Petr Bartunek and to all the members of our database development team, especially to Jindrich Jindrich, Martin Hykl, and Ctibor Skuta for their valuable input and numerous discussions on the topic. The work in this chapter was supported by the Ministry of Education, Youth and Sports of the Czech Republic (RVO 68378050-KAV-NPUI and LM2018130).

References

Ablain, J., Zon, L.I., 2013. Of fish and men: using zebrafish to fight human diseases. Trends Cell Biol. 23 (12), 584–586.

Aleström, P., et al., 2020. Zebrafish: housing and husbandry recommendations. Lab. Anim. 54 (3), 213–224. https://doi.org/10.1177/0023677219869037.

Castranova, D., et al., 2011. The effect of stocking densities on reproductive performance in laboratory zebrafish (*Danio rerio*). Zebrafish 8 (3), 141–146.

Cousin, X., et al., 2012. Electronic individual identification of zebrafish using radio frequency identification (RFID) microtags. J. Exp. Biol. 215 (Pt 16), 2729–2734.

Hawkins, P., et al., 2011. Guidance on the severity classification of scientific procedures involving fish: report of a Working Group appointed by the Norwegian Consensus-Platform for the Replacement, Reduction and Refinement of animal experiments (Norecopa). Lab. Anim. 45 (4), 219–224.

Hensley, M.R., et al., 2012. ZeBase: an open-source relational database for zebrafish laboratories. Zebrafish 9 (1), 44–49.

Hernandez, R.E., et al., 2018. Delay of initial feeding of zebrafish larvae until 8 days post-fertilization has no impact on survival or growth through the juvenile stage. Zebrafish 15 (5), 515–518.

Kimmel, C.B., et al., Stages of Embryonic Development of the Zebrafish.

Kirk, R.G.W., 2018. Recovering the principles of humane experimental technique: the 3Rs and the human essence of animal research. Sci. Technol. Hum. Val. 43 (4), 622–648.

Lawrence, C., Manson, T., 2012. Zebrafish housing systems: a review of basic operating principles and considerations for design and functionality. ILAR J. (2), 179–191.

Lawrence, C., et al., 2012. The effects of feeding frequency on growth and reproduction in zebrafish (*Danio rerio*). Aquaculture 368, 103–108.

Lieschke, G.J., Currie, P.D., 2007. Animal models of human disease: zebrafish swim into view. Nat. Rev. Genet. 8 (5), 353–367.

Martins, S., et al., 2016. Toward an integrated zebrafish health management program supporting cancer and neuroscience research. Zebrafish 13 (Suppl. 1), S47–S55.

Monson, C.A., Sadler, K.C., 2010. Inbreeding depression and outbreeding depression are evident in wild-type zebrafish lines. Zebrafish 7 (2), 189–197.

Oltova, J., et al., 2018. Zebrabase: an intuitive tracking solution for aquatic model organisms. Zebrafish 15 (6), 642–647.

Rasighaemi, P., et al., 2015. Zebrafish as a model for leukemia and other hematopoietic disorders. J. Hematol. Oncol. 8, 29.

Ribas, L., et al., 2017. Appropriate rearing density in domesticated zebrafish to avoid masculinization: links with the stress response. J. Exp. Biol. 220 (Pt 6), 1056–1064.

Russell, W.M.S., Burch, R.L., 1959. The Principles of Humane Experimental Technique.

Strahle, U., et al., 2012. Zebrafish embryos as an alternative to animal experiments–a commentary on the definition of the onset of protected life stages in animal welfare regulations. Reprod. Toxicol. 33 (2), 128–132.

Trevarrow, B., Robison, B., 2004. Genetic backgrounds, standard lines, and husbandry of zebrafish. Methods Cell Biol. 77, 599–616.

Varga, Z.M., 2016. Aquaculture, husbandry, and shipping at the Zebrafish International Resource Center. Methods Cell Biol. 135, 509–534.

Yakulov, T.A., Walz, G., 2015. Zebrafish database: customizable, free, and open-source solution for facility management. Zebrafish 12 (6), 462–469.

Index

'*Note*: Page numbers followed by "f" indicate figures and "t" indicate tables.'

B

G

H

I

N

O

FSC
www.fsc.org
MIX
Papier | Fördert
gute Waldnutzung
FSC® C083411

Kolibri 360 GmbH
Ferdinand-Jühlke-Straße 7
99095 Erfurt, Deutschland
produktsicherheit@kolibri360.de